Advanced Structured Materials

Volume 137

Common engineering materials reach in many applications their limits and new developments are required to fulfil increasing demands on engineering materials. The performance of materials can be increased by combining different materials to achieve better properties than a single constituent or by shaping the material or constituents in a specific structure. The interaction between material and structure may arise on different length scales, such as micro-, meso- or macroscale, and offers possible applications in quite diverse fields.

This book series addresses the fundamental relationship between materials and their structure on the overall properties (e.g. mechanical, thermal, chemical or magnetic etc.) and applications.

The topics of *Advanced Structured Materials* include but are not limited to

- classical fibre-reinforced composites (e.g. glass, carbon or Aramid reinforced plastics)
- metal matrix composites (MMCs)
- micro porous composites
- micro channel materials
- multilayered materials
- cellular materials (e.g., metallic or polymer foams, sponges, hollow sphere structures)
- porous materials
- truss structures
- nanocomposite materials
- biomaterials
- nanoporous metals
- concrete
- coated materials
- smart materials

Advanced Structured Materials is indexed in Google Scholar and Scopus.

More information about this series at http://www.springer.com/series/8611

Francesco dell'Isola · Leonid Igumnov
Editors

Dynamics, Strength of Materials and Durability in Multiscale Mechanics

Springer

Editors
Francesco dell'Isola
Dipartimento di Ingegneria Strutturale
Università degli Studi dell'Aquila
MEMOCS
L'Aquila, Italy

Università di Roma "La Sapienza"
Rome, Italy

Leonid Igumnov
Research Institute for Mechanics
National Research Lobachevsky State
University of Nizhny Novgorod
Nizhny Novgorod, Russia

ISSN 1869-8433 ISSN 1869-8441 (electronic)
Advanced Structured Materials
ISBN 978-3-030-53754-8 ISBN 978-3-030-53755-5 (eBook)
https://doi.org/10.1007/978-3-030-53755-5

This Springer imprint is published by the registered company Springer Nature Switzerland AG
The registered company address is: Gewerbestrasse 11, 6330 Cham, Switzerland

Contents

Chapter 1
Modeling Fatigue Life of Structural Alloys Under Block Asymmetric Loading

Ivan A. Volkov, Leonid Igumnov, Ivan S. Tarasov, Denis N. Shishulin, and Denis V. Kapitanov

Abstract The processes of plastic deformation and damage accumulation of polycrystalline structural alloys under block-type non-stationary asymmetric cyclic loading are considered. A mathematical model describing the processes of thermoplastic deformation and fatigue damage accumulation under low-cycle loading has been developed, based on the viewpoint of mechanics of damaged media (MDM). The MDM model consists of three interrelated parts: governing equations defining the cyclic thermoplastic behavior of the material, taking into account its dependence on the failure process; equations describing the kinetics of damage accumulation; a strength criterion of the damaged material. A version of the constitutive equations of elastoplasticity is based on the concept of the yield surface and the gradient principle of the plastic strain rate vector to the yield surface in the loading point. This version of equations of state reflects the main effects of the cyclic thermoplastic material deformation process for arbitrary complex deformation trajectories. A version of the kinetic equations of damage accumulation based on the introduction of a scalar damage parameter has been proposed. Based on the energy principles, it accounts for the main effects of nucleation, growth and merging of microdefects under random complex regimes of low-cycle loading. The condition for achieving the critical damage value is used as the strength criterion of a damaged material. To assess the reliability and determine the scope of applicability of the constitutive equations of MDM, the processes of plastic deformation and damage accumulation in a number of structural steels in low-cycle tests have been numerically analyzed, and the obtained numerical results have been compared with the data of full-scale experiments. It is shown that the proposed model of damaged media qualitatively and quantitatively, with the accuracy required for practical calculations, describes the main effects of plastic deformation processes and fatigue damage accumulation in structural alloys under block non-stationary asymmetric low-cycle loading.

I. A. Volkov (✉) · L. Igumnov · I. S. Tarasov · D. N. Shishulin · D. V. Kapitanov
Research Institute for Mechanics, National Research Lobachevsky State University of Niznhy Novgorod, Gagarin ave., 23, 603950 Nizhny Novgorod, Russian Federation
e-mail: pmptmvgavt@yandex.ru

F. dell'Isola and L. Igumnov (eds.), *Dynamics, Strength of Materials and Durability in Multiscale Mechanics*, Advanced Structured Materials 137,
https://doi.org/10.1007/978-3-030-53755-5_1

Keywords Soil · Plate · Paimushin V.nN model · Harmonic wave · Oscillation frequency · Vibration absorption · Vibrational acceleration

1.1 Introduction

Cyclic properties of structural materials are extremely important for the reliable assessment of strength and service life of structural elements and supporting units subjected to alternating combined thermomechanical effects. Evaluation of life of structural elements using finite element analysis of inelastic strains in hazardous zones of structural elements requires formulating the constitutive relations of thermoplasticity, which account for real cyclic properties of materials (Mitenkov et al. 2007).

Special attention is currently paid to experimental study of laws of cyclic deformation processes. It has been found that stationary cyclic deformation (if it takes place) is preceded by a transition stage determined by cyclic hardening, softening or relaxation of the memory of the material about the preceding cyclic deformation history. Parameters of a stabilized plastic hysteresis loop do not depend on the place of its stabilization. In asymmetric cyclic deformation, the material can show one-sided accumulation of plastic deformation. During hard cyclic loading with initial anisotropy of the stress amplitude at half-cycles of tension and compression, average cycle stresses are observed to relax up to zero in a finite number of loading cycles. When mechanical loads and temperature act simultaneously but do not change in phase, processes of cyclic change of stresses, total and plastic strains are multi-axial and non-proportional, leading to additional effects of cyclic behavior of materials. The results of experimental studies of these processes show that the behavior of structural materials under cyclic proportional loading differs significantly from that under monotone deformation processes (the laws of cyclic hardening substantially differ from those of monotone deformation). In their turn, multi-axial non-proportional cyclic processes substantially differ from the proportional cyclic ones (Lamba 1978; Macdowell 1985; Ohasi et al. 1985; Tanaka et al. 1985a, b; Hassan et al. 2008; Huang et al. 2014; Jiang and Zhang 2008; Taleb et al. 2014).

Equations of state constructed on the basis of monotone loading processes and not accounting for specific features of cyclic deformation under proportional and non-proportional loading may lead to big errors in determining the main parameters of the stressed–strained state, which are then used for evaluating service life characteristics of materials. Formulation of the reliable constitutive equations of thermoplasticity for the above processes requires, in the first place, experimental studies of the effects of cyclic behavior of structural materials under proportional and non-proportional loading (Volkov and Korotkikh 2008; Mitenkov et al. 2015; Bondar and Danshin, 2008; Chaboche 1989; Bodner and Lindholm 1976; Lemaitre 1985).

Classical methods for predicting service life of materials using semiempirical formulas (rules), based on a stable analysis of the deformation process and connecting the parameters of plastic hysteresis loops with a number of cycles prior to failure, require a large amount of experimental data and are valid only for a narrow range of loading conditions within the available basic experimental data (Kollinz 1984).

Some new on damage modeling can be found in (Placidi 2015; Placidi et al. 2018, 2019). There are also works on damage and plasticity for granular materials (Misra and Poorsolhjouy 2015; Misra and Singh 2015; Zhao et al. 2018). Information about generalized continuum theories can be found in (Alibert et al. 2003; dell'Isola et al. 2012; Auffray et al. 2013; dell'Isola et al. 2015; Abali et al. 2017; dell'Isola et al. 2017). The most fascinating application of theories is in designing new artificial microstructured metamaterials (Del Vescovo and Giorgio, 2014; Placidi et al. 2016; Barchiesi et al. 2018; dell'Isola et al. 2019).

In recent years, a new scientific direction of mechanics of damaged media (MDM) for solving such problems has been successfully developed (Murakami 1983; Volkov and Igumnov, 2017). The current practice of using MDM equations for different mechanisms of worked out service life suggests that this approach is efficient enough for practical applications and quite accurate in evaluating the process of working out of service life of structural elements and components of load-bearing structures (Mitenkov et al. 2007; Volkov and Korotkikh 2008; Volkov and Igumnov 2017).

In the present paper, a mathematical model of mechanics of damaged media (Volkov and Korotkikh 2008; Volkov and Igumnov 2017) is developed, which describes the processes of plastic deformation and fatigue accumulation in structural steels (SS304, 12X18H9) under block-type non-stationary asymmetric low-cycle loading. The obtained numerical results are compared with the data of full-scale experiments and with numerical results, obtained by other researchers (Guozheng et al. 2002).

1.2 Constitutive Equations of MDM

The model of damaged media developed in (Volkov and Korotkikh 2008; Volkov and Igumnov 2017) consists of three interrelated parts as follows:

- equations describing thermoplastic behavior of materials, taking into account its dependence on the failure process;
- evolutionary equations describing the kinetics of damage accumulation;
- strength criterion of the damaged material.

1.2.1 Constitutive Equations in Plasticity

Constitutive equations in plasticity are based on the following main assumptions (Mitenkov et al. 2007; Volkov and Korotkikh 2008; Volkov and Igumnov 2017):

- components of strain tensors e_{ij} and of strain rates $\dot{e}_{ij}$ include elastic e^e_{ij}, $\dot{e}^e_{ij}$ and plastic strains e^p_{ij}, $\dot{e}^p_{ij}$, i.e., reversible and irreversible components;
- the initial yield surface for different temperatures is described by a Mises-type surface. The evolution of the yield surface is described by the change of its radius C_p and by the displacement of its center ρ_{ij};
- the body volume changes elastically;
- initially isotropic media are considered;
- only anisotropy due to the processes of plasticity is accounted for (anisotropy due to the processes of damaged material is not accounted for);
- processes characterized by small deformations are considered.

In the elastic region, the relation between spherical and deviatoric components of stress and strain tensors and their rates is described by Hooke's law:

$$\sigma = 3K[e - \alpha(T - T_0)],\ \sigma_{ij} = 2Ge'^{e}_{ij},$$
$$\dot{\sigma} = 3K(\dot{e} - \dot{\alpha}T - \alpha\dot{T}) + \frac{\dot{K}}{K}\sigma,\ \dot{\sigma}'_{ij} = 2G\dot{e}'^{e}_{ij} + \frac{\dot{G}}{G}\sigma'_{ij}, \tag{1.1}$$

where T is temperature, T_0 is initial temperature, $K(T)$ is volumetric compression modulus, $G(T)$ is shear modulus, and $\alpha(T)$ is linear thermal expansion coefficient of the material.

To describe the effects of monotone and cyclic deformation, a yield surface is introduced:

$$F_s = S_{ij}S_{ij} - C_p^2 = 0,\ S_{ij} = \sigma'_{ij} - \rho_{ij}. \tag{1.2}$$

To describe complex cyclic deformation regimes, a cyclic 'memory' surface is introduced in the stress space. The equation of the 'memory' surface is as follows:

$$F_\rho = \rho_{ij}\rho_{ij} - \rho^2_{\max} = 0, \tag{1.3}$$

where $\rho_{\max}$ is maximal modulus ρ_{ij} during the entire loading history.

In the range of temperatures T, at which annealing effects can be neglected, it is assumed that isotropic hardening (evolution of C_p) may be of three types: monotone, cyclic and that connected with the change in temperature T. The concretization of the evolutionary equation for the yield surface radius has the form (Volkov and Korotkikh 2008; Mitenkov et al. 2015):

$$\dot{C}_p = [q_\chi H(F_\rho) + a(Q_s - C_p)\Gamma(F_\rho)]\dot{\chi} + q_3\dot{T}, \tag{1.4}$$

$$C_p = C_p^0 + \int_0^t \dot{C}_p \mathrm{d}t,\ \dot{\chi} = \left(2\dot{e}_{ij}^p \dot{e}_{ij}^p/3\right)^{1/2},$$
$$\chi_m = \int_0^t \dot{\chi} H(F_\rho)\mathrm{d}t,\ \chi = \int_0^t \dot{\chi}\mathrm{d}t. \tag{1.5}$$

$$q_\chi = \frac{q_2 A\psi_1 + (1-A)q_1}{A\psi_1 + (1-A)},\ Q_s = \frac{Q_2 A\psi_2 + (1-A)Q_1}{A\psi_2 + (1-A)},$$
$$a = \frac{a_2 A\psi_3 + (1-A)a_1}{A\psi_3 + (1-A)},\ 0 \le \psi_i \le 1,\ i = 1, 2, 3,$$
$$A = 1 - \cos^2\theta,\ \cos\theta = n_{ij}^e n_{ij}^e,\ n_{ij}^e = \frac{\dot{e}'_{ij}}{(\dot{e}'_{ij}\dot{e}'_{ij})^{1/2}},\ n_{ij}^s = \frac{S_{ij}}{(S_{ij}S_{ij})^{1/2}},$$
$$H(F_\rho) = \begin{Bmatrix} 1,\ F_\rho = 0 \wedge \rho_{ij}\dot{\rho}_{ij} > 0 \\ 0,\ F_\rho < 0 \vee \rho_{ij}\dot{\rho}_{ij} \le 0 \end{Bmatrix},\ \Gamma(F_\rho) = 1 - H(F_\rho), \tag{1.6}$$

where q_1, q_2 and q_3 are moduli of monotone isotropic hardening, Q_1 and Q_2 are cyclic isotropic hardening moduli, a is a constant defining the rate of the process of stationing of the hysteresis loop of cyclic deformation of the material, Q_s is stationary value of the yield surface radius for the given $\rho_{\max}$ and T and C_p^0 is initial value of the yield surface radius.

It is postulated that the evolution of internal variable ρ_{ij} has the form:

$$\dot{\rho}_{ij} = f(\chi_m)\left[g_1\dot{e}_{ij}^p - g_2\rho_{ij}\dot{\chi}\right] + g_T\rho_{ij}\langle\dot{T}\rangle + \dot{\rho}_{ij}^*,\ \rho_{ij} = \int_0^t \dot{\rho}_{ij}dt \tag{1.7}$$

$$f(\chi_m) = 1 + k_1\left(1 - e^{-k_2\chi_m}\right),$$
$$\dot{\rho}_{ij}^* = g_3\dot{e}_{ij}^p H(F_\rho) - g_4\rho_{ij}\dot{\chi}\Gamma(F_\rho)\langle\cos\beta\rangle, \tag{1.8}$$

$$\langle\cos\beta\rangle = \dot{\rho}_{ij}\rho_{ij}\Big/\left(\left(\dot{\rho}_{ij}\dot{\rho}_{ij}\right)^{1/2}\left(\rho_{ij}\rho_{ij}\right)^{1/2}\right) \tag{1.9}$$

where $g_1, g_2, g_3, g_4, g_T, k_1$ and k_2 are experimentally found material parameters (moduli of anisotropic hardening).

For asymmetric hard and soft cyclic loading, member $\dot{\rho}_{ij}^*$, Eq. (1.7) describes the processes of setting and ratcheting of the cyclic plastic hysteresis loop. At $g_T = g_3 = g_4 = k_1 = 0$, we derive from (1.7) a special case of Eq. (1.7)—the Armstrong–Frederic–Kadashevich equation.

$$\dot{\rho}_{ij} = g_1\dot{e}_{ij}^p - g_2\rho_{ij}\dot{\chi}. \tag{1.10}$$

To describe the evolution of the 'memory' surface, an equation for $\rho_{\max}$ is formulated:

$$\dot{\rho}_{\max} = \frac{(\rho_{ij}\dot{\rho}_{ij})H(F_\rho)}{(\rho_{mn}\rho_{mn})^{1/2}} - g_2\rho_{\max}\dot{\chi} - g_T\rho_{\max}\langle\dot{T}\rangle. \tag{1.11}$$

The components of the plastic strain rate tensor obey the gradientality principle of the plastic strain rate vector to the yield surface in the loading point:

$$\dot{e}_{ij}^p = \lambda S_{ij}, \tag{1.12}$$

where λ is proportionality coefficient determined from the condition that a new yield surface passes through the end of the stress deviator vector at the end of the loading stage.

At the stage of the growth of defects scattered over the volume, the effect of damage on the physical–mechanical properties of the material is observed. This effect may be accounted for by introducing the effective stresses (Volkov and Igumnov 2017):

$$\tilde{\sigma}'_{ij} = F_1(\omega)\sigma'_{ij} = \frac{G}{\tilde{G}}\sigma'_{ij} = \frac{\sigma'_{ij}}{(1-\omega)[1-(6K+12G)\omega/(9K+8G)]}, \tag{1.13}$$

$$\tilde{\sigma} = F_2(\omega)\sigma = \frac{K}{\tilde{K}}\sigma = \frac{\sigma}{4G(1-\omega)/(4G+3K\omega)}, \tag{1.14}$$

where $\tilde{G}$ and $\tilde{K}$ are effective elasticity moduli, defined by McKenzie formulas (Mackenzie 1950).

The effective variable $\tilde{\rho}_{ij}$ is determined by analogy:

$$\tilde{\rho}_{ij} = F_1(\omega)\rho_{ij}. \tag{1.15}$$

1.2.2 *Evolutionary Equations Describing Fatigue Damage Accumulation*

We postulate that the damage accumulation rate during low-cycle fatigue (LCF) is determined by the evolution equation of the form (Volkov and Korotkikh 2008; Lemaitre 1985; Volkov and Igumnov 2017):

$$\dot{\omega} = f_1(\beta)f_2(\omega)f_3(W)f_4(\dot{W}) \tag{1.16}$$

where functions $f, i = 1 \ldots 4$ account for: volumetric stress state ($f_1(\beta)$), accumulated level of damage degree ($f_2(\omega)$), accumulated relative energy, spent on defect formation ($f_3(W)$) and the rate of change of damage energy ($f_4(\dot{W})$).

In Eq. (1.16):

$$f_1(\beta) = \exp(\beta),$$
$$f_2(\omega) = \begin{cases} 0, & W \leq W_a \\ \omega^{1/3}(1-\omega)^{2/3} \wedge W > W_a \wedge \omega \leq 1/3 \\ \sqrt[3]{16}\omega^{-1/3}(1-\omega)^{-2/3}/9 \wedge W > W_a \wedge \omega > 1/3 \end{cases} \tag{1.17}$$

$$\dot{W} = \rho_{ij}\dot{e}^p_{ij},\ W = \int \rho_{ij} de^p_{ij},\ f_3(W) = \frac{W - W_a}{W_f},\ f_4(\dot{W}) = \dot{W}/W_f, \tag{1.18}$$

where β is volumetric stress state parameter ($\beta = \sigma/\sigma_u$), W_a is value of the energy at the end of scattered damage nucleation under LCF condition, and W_f is value of the energy corresponding to microcrack formation.

The duration of microdefect nucleation phase will be associated with the value W_a.

When microdefects become comparable with the average distance between them, the merging process begins (breakage of the remaining continuous areas between the defects). In the present work, a detailed model of cavity merging was not constructed; however, to account for this process, the kinetic equation (at the expense of the member $f_2(\omega)$) was formulated so that upon reaching $\omega = 1/3$, the dependence $\dot{\omega} = f_1(\omega)$ takes into account the 'avalanche-like' increase in damage value.

1.2.3 *Strength Criterion of Damaged Material*

The condition at which damage degree ω reaches its critical value is taken as a criterion of the phase end of the development of scattered microdefects:

$$\omega = \omega_f \leq 1. \tag{1.19}$$

1.3 Numerical Results

The reliability of the proposed version of the thermoplasticity model was assessed using the results of the experimental studies of laboratory specimens made of stainless steels of austenitic type (SS304) (Guozheng et al. 2002) under conditions of soft (controlled stress) block-type asymmetric low-cycle loading.

1.3.1 Block-Type Asymmetric Soft Cyclic Loading

The first example presents the results of numerical modeling of deformation processes in specimens of stainless steel SS304 (Guozheng et al. 2002) under block-type asymmetric soft cyclic loading:

- in the first block, 50 cycles of asymmetric soft loading with the stress amplitude of $\sigma_{11} = 248$ MPa and average cycle stress of $\sigma_{11}^{aver} = 78$ MPa are realized;
- the second block consists of 50 cycles of asymmetric soft loading with the amplitude of $\sigma_{11} = 248$ MPa and average stress of $\sigma_{11}^{aver} = 117$ MPa;
- the third block includes 20 loading cycles with the stress amplitude of $\sigma_{11} = 248$ MPa and average stress of $\sigma_{11}^{aver} = 78$ MPa.

Tables 1.1, 1.2 and 1.3 present the main physical–mechanical characteristics and material parameters of thermoplasticity model for SS304 steel, used in the computations.

Figure 1.1 presents the results of comparing the calculated and experimental curves of the average strain as a function of number of loading cycles, and Fig. 1.2 depicts the curves of cyclic deformation constructed on the basis of the numerical results.

Table 1.1 Main physical–mechanical characteristics and material parameters of the model of plastic deformation for SS304 steel

G, MPa	K, MPa	c_p^o	g_1, MPa	g_2	g_3, MPa	g_4	k_1, MPa
65,384	141,700	228–249	22,000	300	1000	10	0,68

k_2, MPa	a_1	a_2	Q_2, MPa
50	15	50	800

Table 1.2 Monotone isotropic hardening modulus q_1 as a function of plastic deformation path length χ for SS304 steel ($q_2 = 0$)

q_1, MPa	−13.462	−8508	−8738	−6027	−2690	−1173	−118
χ	0	0.001	0.002	0.003	0.005	0.007	0.01

q_1, MPa	−489	0	0
χ	0.015	0.02	0.03

Table 1.3 Value of cyclic hardening modulus Q_1 as a function of maximum value of displacement of the yield surface center ρ_{max} for SS304 steel ($Q_2 = 0$)

Q_1, MPa	228–241	220–241	220–241
ρ_{max}, MPa	0	10	80

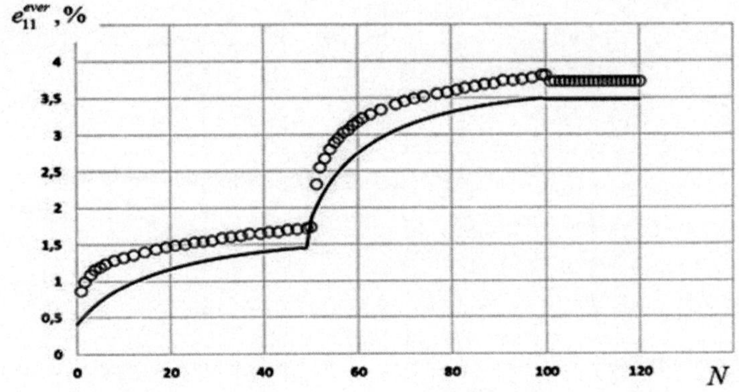

Fig. 1.1 Computed and experimental dependencies of changes in the average deformations depending on the load number

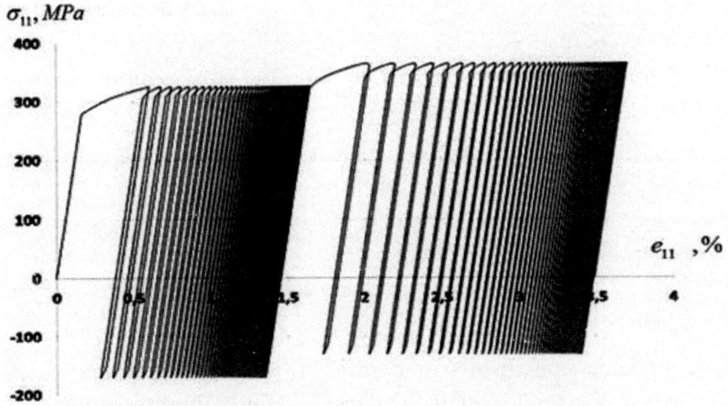

Fig. 1.2 Computed curves of cyclic deformation

1.3.2 *Multi-axial Proportional and Non-proportional Regimes of Soft Block-Type Cyclic Loading*

The second example presents the results of numerical modeling of the deformation processes in the specimens of SS304 steel under multi-axial proportional (Fig. 1.3a) and non-proportional (Fig. 1.3b) regimes of soft block-type cyclic loading.

For the loading history, presented in Fig. 1.3a, the stress amplitude σ_{11} in blocks at constant stress $\sqrt{3}\sigma_{12} = \text{const}$ was changed as follows:

- the first block includes 50 loading cycles with the amplitude of $\sigma_{11} = 248\,\text{MPa}$ and average stress of $\sigma_{11}^{\text{aver}} = 78\,\text{MPa}$;
- the second block consists of 50 loading cycles with the amplitude of $\sigma_{11} = 248\,\text{MPa}$ and average stress of $\sigma_{11}^{\text{aver}} = 117\,\text{MPa}$;

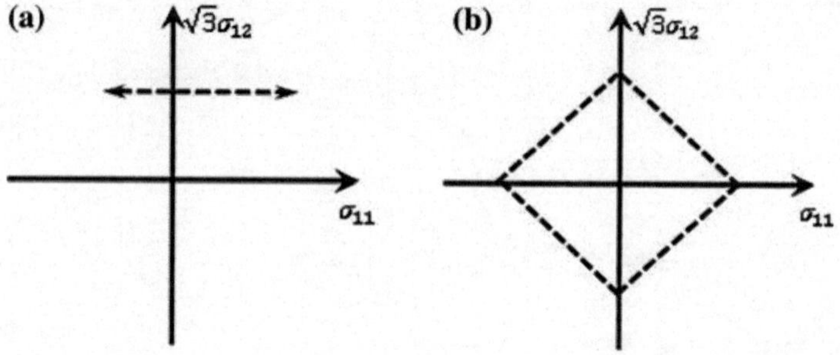

Fig. 1.3 The history of the loading at block proportional and non-proportional deformation

- the third block includes 50 loading cycles with the amplitude of $\sigma_{11} = 248$ MPa and average stress $\sigma_{11}^{\text{aver}} = 78$ MPa.

Figure 1.4 depicts the deformation curves, constructed in the strain path on the basis of the numerical results.

For the loading history, presented in Fig. 1.3b, the stress amplitude σ_{11} in blocks at constant stress σ_{12} with the amplitude $\sqrt{3}\sigma_{12} = \text{const}$ was changed as follows:

- the first block consists of 50 loading cycles with the amplitude of $\sigma_{11} = 248$ MPa and average cycle stress $\sigma_{11}^{\text{aver}} = 78$ MPa;
- the second block includes 50 loading cycles with the amplitude of $\sigma_{11} = 248$ MPa and average stress $\sigma_{11}^{\text{aver}} = 117$ MPa;
- the third block consists of 50 loading cycles with the amplitude of $\sigma_{11} = 248$ MPa and average stress $\sigma_{11}^{\text{aver}} = 78$ MPa.

Figure 1.5 presents the results of comparing the calculated and experimental curves of the average strain (axial and shear) as a function of the number of loading

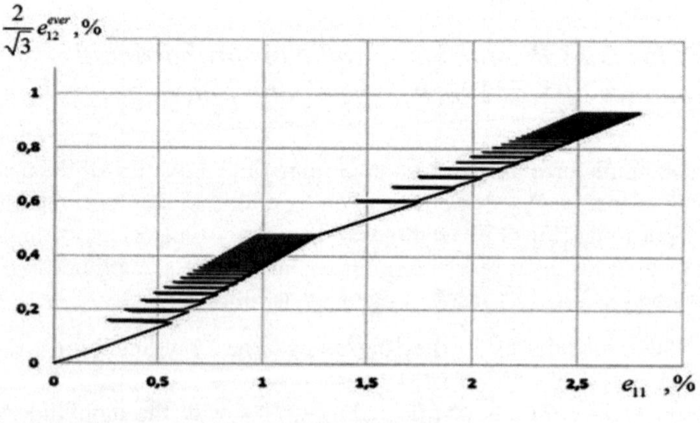

Fig. 1.4 Computed deformation trajectories in the strain space

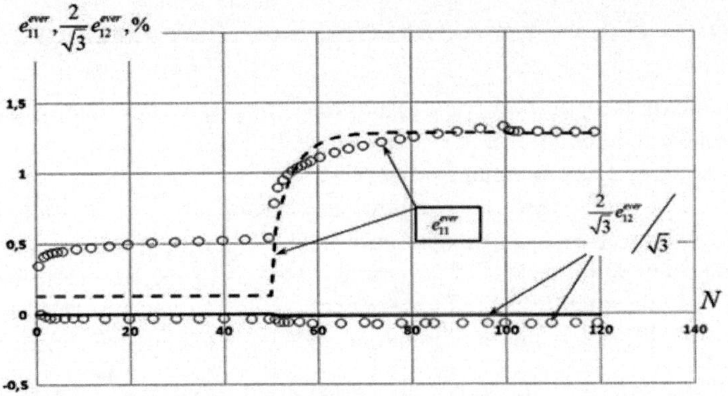

Fig. 1.5 Computed and experimental dependencies of changes in the average deformations depending on the load number

cycles, and Fig. 1.6 depicts the deformation curves, constructed in the strain path on the basis of the numerical results.

It is evident that the experimental and modeled representations for describing the effect of ratcheting of the plastic hysteresis loop under uniaxial and multi-axial, proportional and non-proportional regimes of soft cycling loading agree reasonably well both qualitatively and, reliably enough for engineering purposes, quantitatively, which testifies to the adequacy of the modeling the process of non-stationary asymmetric cyclic loading and determining the material parameters of the constitutive equations of plasticity.

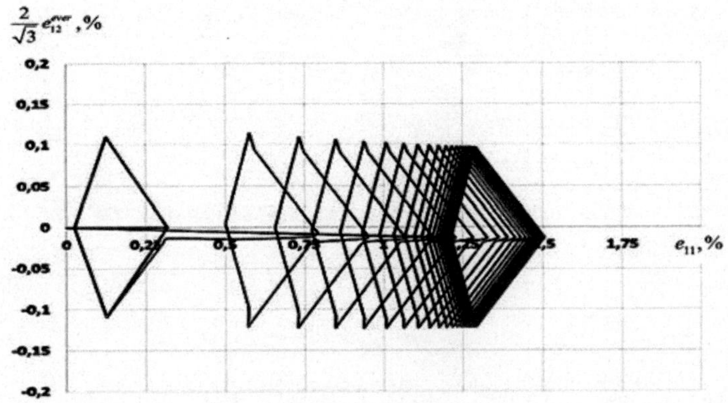

Fig. 1.6 Computed deformation trajectories in the deformation space

1.3.3 Hard Block-Type Asymmetric Low-Cycle Loading

Further studies present the numerical results of cyclic plastic deformation process and fatigue damage accumulation during hard (controlled strains) block-type asymmetric low-cycle loading. Experimental studies have been performed on the samples made of stainless 12X18H9 steel under conditions of hard (controlled strains) uniaxial tension–compression at room temperature in the laboratory for testing the physical and mechanical properties of structural materials of the Research Institute for Mechanics of Lobachevsky State University of Nizhny Novgorod. The test program consists of two blocks, including monotone and cyclic loadings (Mackenzie 1950):

- in the first block, the sample is compressed up to strain of $e_{11} = 0.01$, and then stretched up to strain of $e_{11} = 0.05$;
- in the second block, the sample is subjected to asymmetric hard cyclic loading with the strain amplitude of $\Delta e_{11} = e_{11}^{(+)} - e_{11}^{(-)} = 0.01$ up to failure ($N_f = 850$). Here, the setting of the plastic hysteresis loop takes place (Fig. 1.5) and after the 500th loading cycle the loop becomes practically symmetrical.

Tables 1.4, 1.5 and 1.6 present the main physical–mechanical characteristics of 12X18H9 steel and the material parameters of the MDM model, determined from the results of the basic experiment (Volkov and Korotkikh 2008; Mitenkov et al. 2015), used in calculations.

Figure 1.7 depicts the deformation process of 12X18H9 steel in the second loading block (the 500th cycle). The experimental and numerical data agree both qualitatively and quantitatively.

Table 1.4 Physical–mechanical characteristics and material parameters of the MDM model of 12X18H9 steel

K	G	C_p^o, MPa	g_1, MPa	g_2	g_3, MPa	g_4	k_1, MPa	k_2	a_1	a_2	W_a	W_f
165.277	76.282	190	24.090	286	800	2	10.000	0.2	5	0	0	800

Table 1.5 Cyclic hardening modulus $Q_1(\rho_{max})$(MPa) for 12X18H9 steel ($Q_2 = 0$)

Q_1, MPa	190	205	210	215	220	225	225
ρ_{max}, MPa	0	20	40	60	80	100	120

Table 1.6 Monotone hardening modulus q_1(MPa) for 12X18H9 steel ($q_2 = 0$)

q_1, MPa	−5000	−4471	−4188	−3859	−2460	−182	888
χ	0	0.002	0.004	0.006	0.008	0.01	0.015

q_1, MPa	1531	1274	913	913	913
χ	0.02	0.03	0.04	0.05	0.06

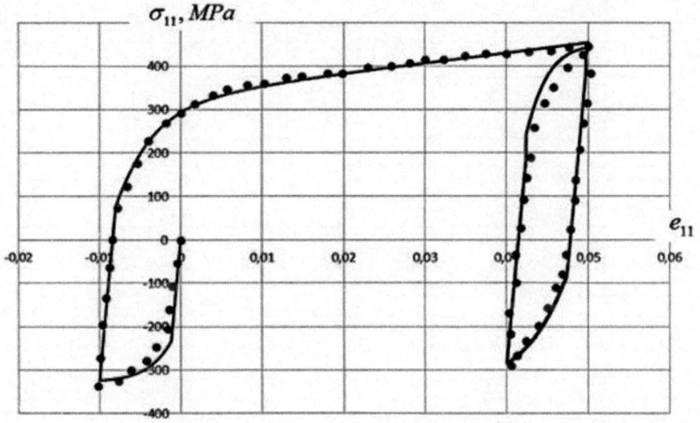

Fig. 1.7 Computed and experimental deformation trajectories in the second loading block

Figure 1.8 shows the history of average stress in the cycle $\sigma_{11}^{(m)}$ in the process of cyclic loading in the second block.

It can be seen that the used thermoplasticity model qualitatively and quantitatively describes the process after setting.

Figure 1.9 presents fatigue curve of stainless 12X18H9 steel under hard asymmetric cyclic loading. Here the solid curve corresponds to the experimental curve, while markers show the computational results using the constitutive equations of MDM (Volkov and Korotkikh 2008; Volkov and Igumnov 2017; Volkov et al. 2016; Korotkikh et al. 2016, 2015).

The comparison of computational and experimental results proves that the developed MDM model describes both qualitatively and, reliably enough for engineering purposes, quantitatively, the process of fatigue life of polycrystalline structural alloys under block-type asymmetric low-cycle loading.

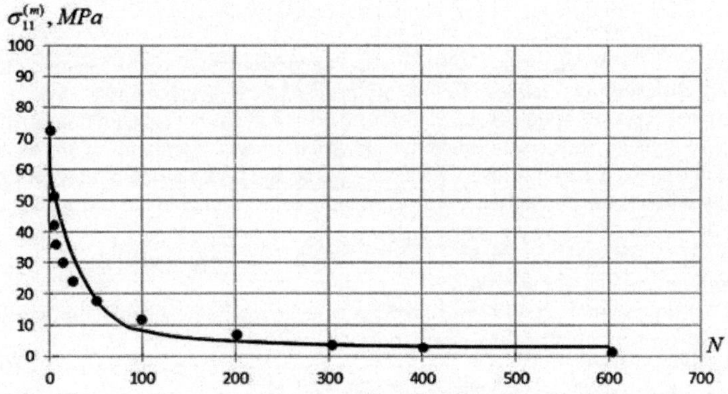

Fig. 1.8 History of changes in the average voltage in the second loading block

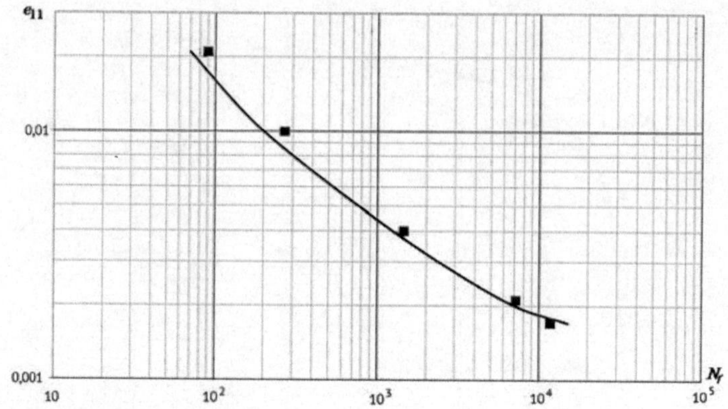

Fig. 1.9 Fatigue curve for stainless steel

1.4 Conclusion

To describe the fatigue life of polycrystalline structural alloys under block-type asymmetric low-cycle loading, the assessment of the reliability of the constitutive equations of MDM has been carried out by comparing the results of numerical experiments with experimental data on plastic deformation and damage accumulation in stainless steels (SS304, 12X18H9) that confirmed the accuracy of modeling and determining material parameters.

Acknowledgements The reported study was funded by RFBR, project numbers 18-08-00881, 20-08-00450.

References

Abali, B. E., Muller, W., & dell'Isola, F. (2017). Theory and computation of higher gradient elasticity theories based on action principles. *Archive of Applied Mechanics, 87,* 1495–1510.
Alibert, J. J., Seppecher, P., & dell'Isola, F. (2003). Truss modular beams with deformation energy depending on higher displacement gradients. *Mathematics and Mechanics of Solids, 8,* 51–73.
Auffray, N., dell'Isola, F., Eremeyev, V., Madeo, A., & Rosi, G. (2013). Analytical continuum mechanics à la Hamilton-Piola: Least action principle for second gradient continua and capillary fluids. *Mathematics and Mechanics of Solids.*
Barchiesi, E., Spagnuolo, M., & Placidi L. (2018). Mechanical metamaterials: A state of the art. *Mathematics and Mechanics of Solids.*
Bodner S. R., & Lindholm, U. S. (1976). Kriterii prirashcheniia povrezhdeniia dlia zavisiashchego ot vremeni razrusheniia materialov. *Trudy Amer. ob-va inzh.-mekh. Ser. D. Teoret. osnovy inzh. Raschetov, 100*(2), 51–58.
Bondar, V. S., & Danshin V. V. (2008) Plastichnost. Proportsionalnye i neproportsionalnye nagruzheniya. (in Rus.) M.: Fizmatlit. – 176 s.

Chaboche, J. L. (1989). Constitutive equation for cyclic plasticity and cyclic viscoplasticity. *International Journal of Plasticity, 5*(3), 247–302.

Del Vescovo, D., & Giorgio, I. (2014). Dynamic problems for metamaterials: Review of existing models and ideas for further research. *International Journal of Engineering Science, 80,* 153–172.

dell'Isola, F., Della, C. A., & Giorgio, I. (2017). Higher-gradient continua: The legacy of Piola, Mindlin, Sedov and Toupin and some future research perspectives. *Mathematics and Mechanics of Solids, 22,* 852–872.

dell'Isola, F., Seppecher, P., Alibert, J. J., Lekszycki, T., Grygoruk, R., Pawlikowski, M., et al. (2019). Pantographic metamaterials: An example of mathematically driven design and of its technological challenges. *Continuum Mechanics and Thermodynamics, 31*(4), 851–884.

dell'Isola, F., Seppecher, P., & Madeo, A. (2012). How contact interactions may depend on the shape of Cauchy cuts in Nth gradient continua: Approach "à la D'Alembert." *Zeitschrift für Angewandte Mathematik und Physik, 63,* 1119–1141.

dell'Isola, F., Andreaus, U., & Placidi, L. (2015). At the origins and in the vanguard of peridynamics, non-local and higher gradient continuum mechanics. An underestimated and still topical contribution of Gabrio Piola. *Mathematics and Mechanics of Solids, 20,* 887–928.

Guozheng, K., Qing, G., Lixun, C., & Yafang, S. (2002). Experimental study on uniaxial and nonproportionally multiaxial ratcheting of SS304 stainless steel at room and high temperatures. *Nuclear Engineering and Design, 216,* 13–26.

Hassan, T., Taleb, L., & Krishna, S. (2008). Influence of non-proportional loading on ratcheting responses and simulations by two recent cyclic plasticity models. *International Journal of Plasticity, 24,* 1863–1889.

Huang, Z. Y., Chaboche, J. L., Wang, Q. Y., Wagner, D., & Bathias C. (2014). Effect of dynamic strain aging on isotropic hardening in low cycle fatigue for carbon manganese steel. *Materials Science and Engineering, A589,* 34–40.

Jiang, Y., & Zhang, J. (2008). Benchmark experiments and characteristic cyclic plasticity deformation. *International Journal of Plasticity, 24,* 1481–1515.

Kollinz, G. (1984). Povrezhdeniye materialov v konstructsiyakh. Analiz. Predskazaniye. Predotvrashcheniye. M.: Mir.

Korotkikh, Yu. G., Volkov, I. A., Igumnov, L. A., Shishulin, D. N., & Guseva, M. A. (2016). Modelirovanie slozhnogo plasticheskogo deformirovaniya i nakopleniya ustalostnyh povrezhdenij v zharoprochnyh splavah pri kombinirovannom termomekhanicheskom nagruzhenii (in Rus). Vol. 78. No. 1. P. 45–59.

Korotkikh, Yu. G., Volkov, I. A., Igumnov, L. A., Shishulin, D. N., & Tarasov, I. S. (2015). Modelirovanie processov neuprugogo deformirovaniya i razrusheniya zharoprochnyh splavov pri ciklicheskom termomekhanicheskom nagruzhenii (in Rus). Vol. 77. No.4. P. 329–343.

Lamba, S. (1978). Plastichnost pri tsiklicheskom nagruzhenii po neproportsionalnym traektoriyam (in Rus.). *Teoreticheskiye osnovy inzhenernyh raschetov, 100*(1), 108–126.

Lemaitre, G. (1985) Kontinualnaya model povrezhdeniya, ispolzuemaya dlya rascheta razrusheniya plastichnykh materialov. *Trudy Amer. ob-va inzh.-mekh. Ser. D. Teoret. osnovy inzh. Raschetov, 107*(1), 90–98.

Macdowell. (1985). Eksperimentalnoye izucheniye struktury opredelyayushih uravneniy dlya neproportsionalnoy tsiklicheskoy plastichnosti (in Rus.). *Teoreticheskiye osnovy inzhenernyh raschetov* (4), 98–111.

Mackenzie, J. K. (1950). The elastic constants of a solid containing spherical holes. *Proceedings of the Physical Society. Section B, 63*(2).

Misra, A., & Poorsolhjouy, P. (2015). Granular micromechanics model for damage and plasticity of cementitious materials based upon thermomechanics. *Mathematics and Mechanics of Solids.*

Misra, A., & Singh, V. (2015). Thermomechanics-based nonlinear rate-dependent coupled damage-plasticity granular micromechanics model. *Continuum Mechanics and Thermodynamics, 27,* 787–817.

Mitenkov, F. M., Kaydalov, V. B., & Korotkikh, Yu. G. i dr. (2007). Metody obosnovaniya resursa YaEU.(in Rus.) – M.: Mashinostroyeniye, 2007. – 445s.

Mitenkov, F. M., Volkov, I. A., Igumnov, L. A., & Korotkikh, Yu. G. i dr. (2015). Prikladnaya teoriya plastichnosti. (in Rus.) – M.: Fizmatlit, 2015. – 284 s.

Murakami, E. (1983). Sushchnost mehaniki povrezhdennoi sredy I eye prilozheniye k teorii anizotropnykh povrezhdeniy pri polzuchesti. *TOIR, 2,* 44–50.

Ohasi, Kavan, Kaito. (1985). Neuprugoye povedeniye stali 316 pri mnogoosnyh neproportsionalnyh tsiklicheskih nagruzheniyah pri povyshennoy temperature (in Rus.). *Teoreticheskiye osnovy inzhenernyh raschetov. 107*(2), 6–15.

Placidi, L. (2015). A variational approach for a nonlinear one-dimensional damage-elasto-plastic second-gradient continuum model. *Continuum Mechanics and Thermodynamics, 28,* 119–137.

Placidi, L., Barchiesi, E., & Misra, A. (2018). A strain gradient variational approach to damage: A comparison with damage gradient models and numerical results. *Mathematics and Mechanics of Complex Systems, 6,* 77–100.

Placidi, L., Barchiesi, E., Turco, E., & Rizzi, N. L. (2016). A review on 2D models for the description of pantographic fabrics. *Zeitschrift für angewandte Mathematik und Physik, 67*.

Placidi, L., Misra, A., & Barchiesi, E. (2019). Simulation results for damage with evolving microstructure and growing strain gradient moduli. *Continuum Mechanics and Thermodynamics, 31,* 1143–1163.

Taleb, L., Cailletaud, G., & Sa'i, K. (2014). Experimental and numerical analysis about the cyclic behavior of the 304L and 316L stainless steels at 350 °C. *International Journal of Plasticity, 61,* 32–48.

Tanaka, E., Murakami, S., & Ooka, M. (1985a). Effects of strain path shapes on nonproportional cyclic plasticity. *Journal of the Mechanics and Physics of Solids, 33*(6), 559–575.

Tanaka, E., Murakami, S., & Ooka, M. (1985b). Effects of plastic strain amplitudes on non-proportional cyclic plasticity. *ActaMechanica, 57*, 167–182.

Volkov, I. A., & Igumnov, L. A. (2017). Vvedeniye v kontinualnyyu mekhaniku povrezhdennoy sredy (in Rus). M.: Fizmatlit. P. 304.

Volkov, I. A., Igumnov, L. A., Korotkih, Yu. G., Kazakov, D. A., Emel'yanov, A. A., Tarasov, I. S., et al. (2016). Programmnaya realizatsiya protsessov vyazkoplasticheskogo deformirovaniya i nakopleniya povrezhdeniy v konstrukcionnyh splavah pri termomekhanicheskom nagruzhenii (in Rus). Vol. 78. No. 2. P. 188–207.

Volkov I.A., Korotkikh Yu.G., 2008, Uravneniya sostoyaniya vyazkouprugoplasticheskih sred s povrezhdeniyami. (in Rus). M.: Fizmatlit. P. 424.

Zhao, C.-F., Yin, Z.-Y., Misra, A., & Hicher, P.-Y. (2018). Thermomechanical formulation for micromechanical elasto-plasticity in granular materials. *International Journal of Solids and Structures, 138,* 64–75.

Chapter 2
Excitation of the Waves with a Focused Source, Moving Along the Border of Gradient-Elastic Half-Space

Artem M. Antonov, Vladimir I. Erofeev, Aleksey O. Malkhanov, and Nadezhda A. Novoseltseva

Abstract In the frames of the mathematical model of the gradient-elastic continuum, i.e., the medium with the stress–strain state described by the strain tensor, the second gradients of the displacement vector, the asymmetric stress tensor, and the moment stress tensor, we consider the problem of generating disturbances by a moving source. It is assumed that the source moves at a constant speed along the border of the half-space. The problem is considered in a two-dimensional formulation, when all processes are homogeneous along the horizontal transverse coordinate axis. The displacement vector contains two components: longitudinal and vertical transverse. As a result of analytical studies, it has been shown that a moving source will generate waves propagating along the border of a half-space and decreasing exponentially into its depth. Such a wave, in contrast to the classical surface Rayleigh wave, has a dispersion, since its phase velocity is not constant, but depends on the frequency. The displacement amplitudes vary depending on the magnitude of the load of the moving source and its speed.

Keywords Gradient-elastic half-space · Moving source · Surface wave

2.1 Introduction

In 1885, the English scientist Lord Rayleigh theoretically showed that along the flat border of a solid elastic half-space with a vacuum or a sufficiently diluted medium (for example, with air) waves can propagate, whose amplitude rapidly decreases with depth (Lord Rayleigh 1885). These waves, later called Rayleigh surface waves, are the main type of waves observed during earthquakes. Therefore, they are studied in detail in seismology (Aki and Richards 2009).

A. M. Antonov · V. I. Erofeev (✉) · A. O. Malkhanov
Mechanical Engineering Research Institute of RAS, Nizhny Novgorod, Russia
e-mail: erof.vi@yandex.ru

V. I. Erofeev · N. A. Novoseltseva
Research Institute for Mechanics, National Research Lobachevsky State University of Nizhny Novgorod, 23, Gagarin Av., Nizhny Novgorod 603950, Russia

F. dell'Isola and L. Igumnov (eds.), *Dynamics, Strength of Materials and Durability in Multiscale Mechanics*, Advanced Structured Materials 137,
https://doi.org/10.1007/978-3-030-53755-5_2

The main laws of propagation of Rayleigh waves are as follows: the absence of dispersion, i.e., wave speed does not depend on its frequency and is constant for each material; this speed reaches 0.87–0.96 from the velocity of a volume shear wave; the displacement vector has longitudinal and transverse components, while the transverse component always exceeds the longitudinal one.

In recent years, Rayleigh waves of ultrasonic range have found wide application. With their help, it is possible to monitor the state of the surface layer of the sample (detection of surface and near-surface defects in samples of metal, glass, plastic, and other materials—ultrasonic surface flaw detection). The influence of the properties of the surface layer of the sample on the velocity and attenuation of Rayleigh waves allows the latter to be used to determine the residual stresses of the surface metal layer and the thermal and mechanical properties of the surface layer of the sample (Klyuev 2004; Uglov et al. 2009).

Along with the model of the classical continuum, the models of generalized continua are widely used in the mechanics of a deformable solid body (Maugin and Metrikine 2010; Altenbach and Eremeyev 2013; Altenbach et al. 2011, 2013, 2016, 2018a, b, 2019; Bagdoev et al. 2016; Maugin 2017; dell'Isola et al. 2012, 2015, 2016a, b, c, 2018; Abali et al. 2017, 2019; Neff et al. 2014; Alibert et al. 2003; Auffray et al. 2013; Sciarra et al. 2007; Rahali et al. 2015).

The appearence of the model of the generalized continua that belong, in particular, to the gradient-elastic medium, dates back to the beginning of the twentieth century and is associated with the names of Le Roux (Le Roux 1911, 1913) and Jaramillo (Jaramillo 1929).

The famous Cosserat continuum model (Cosserat et al. 1909), when the dependence of the rotation vector on the displacement rotor (constrained rotation) (Erofeyev 2003) is rigidly fixed, also reduces to the gradient-elastic medium model.

One of the great applications of generalized continuum theories is designing of new artificial microstructured metamaterials (Barchiesi et al. 2019; Del Vescovo and Giorgio 2014). An example of mechanical metamaterial is pantographic structure (dell'Isola et al. 2016, 2019a, b; Placidi et al. 2016, 2017).

The Rayleigh surface waves in the framework of the gradient-elastic model have not been practically studied. An exception is the work (Sabodash and Filippov 1971), in which, on the basis of the studies performed, it is stated that the velocity of a surface wave in a gradient-elastic medium can exceed the velocity of a bulk shear wave.

The paper studies the main laws of propagation of the Rayleigh waves along the boundary of the gradient-elastic half-space, in particular, it verifies the assertion contained in (Sabodash and Filippov 1971). The problem of the generation of a surface wave by a source moving at superspeeds along the border of a gradient-elastic half-space is considered. The dependence of the amplitude of the wave, the Mach cone on the source load, and its velocity is determined.

2.2 The Basic Equations of Gradient Theory of Elasticity

The deformed state of the gradient-elastic medium is described by the strain tensor and the second gradients of the displacement vector

$$\varepsilon_{kl} = \frac{1}{2}\left[\frac{\partial u_k}{\partial x_l} + \frac{\partial u_l}{\partial x_k}\right], \quad \chi_{klm} = -\frac{\partial^2 u_k}{\partial x_l \partial x_m}. \tag{2.1}$$

When considering the adiabatic processes of elastic deformation, it is necessary to postulate the dependence of the internal energy U on invariant of deformation measure (2.1).

Let us expand function U in the vicinity of the natural state ($\varepsilon_{kl} = 0$, $\chi_{klm} = 0$) in the Taylor series, neglecting the values of the third order. For an isotropic homogeneous and centrally symmetric body, we obtain the decomposition of the following form (Erofeyev 2003):

$$U = \frac{\lambda}{2}\varepsilon_{kk}^2 + \mu\varepsilon_{ik}^2 + 2\mu L^2\left(\chi_{klm}^2 + \tilde{\nu}\chi_{klm}\chi_{lkm}\right), \tag{2.2}$$

where λ and μ—Lame elastic constants, L^2—the ratio of the curvature modulus to the shear modulus μ, which has dimension of length square, $\tilde{\nu}$—dimensionless constant, ρ—density of the medium.

In displacements, the vector equation of the dynamics of a gradient-elastic medium has the form:

$$\rho\frac{\partial^2\vec{u}}{\partial t} - (\lambda + \mu)\text{grad}\,\text{div}\vec{u} - \mu\Delta\vec{u} + 4\mu L^2\Delta(\Delta\vec{u} + \tilde{\nu}\,\text{grad}\,\text{div}\vec{u}) = 0, \tag{2.3}$$

It can be easily seen that this equation contains the fourth order of derivatives with respect to coordinates, in contrast to the classical Lame equation, which describes the dynamics of a deformable solid body containing second derivatives with respect to coordinates.

2.3 The Statement and Solution to the General Problem of Waves Propagation in Gradient-Elastic Medium

Let us consider an elastic isotropic half-space $y \geq 0$ (we study the two-dimensional case when all processes are homogeneous along the z axis). Suppose that the surface wave propagates in the direction of the x axis.

Equations of dynamics, equivalent to the vector equation (2.3), in the two-dimensional case can be written as (Sabodash and Filippov 1971):

$$\frac{\partial \sigma_{xx}}{\partial x} + \frac{\partial \sigma_{yx}}{\partial y} = \rho \frac{\partial^2 u}{\partial t^2}, \quad \frac{\partial \sigma_{xy}}{\partial x} + \frac{\partial \sigma_{yy}}{\partial y} = \rho \frac{\partial^2 v}{\partial t^2}. \tag{2.4}$$

Here, u—longitudinal and v—transversal components of displacements vector.

We assume that at the boundary $y = 0$ there are no stresses and moment stresses, i.e., boundary conditions for the system of equations (2.4) have the form (Sabodash and Filippov 1971):

$$\mu_y = 0, \sigma_{yy_{y=0}} = 0, \sigma_{yx_{y=0}} = 0, \sigma_{yz_{y=0}} = 0, \tag{2.5}$$

Moreover, the third condition is identically satisfied by virtue of the assumption that deformations are independent of the variable z.

The components of the stress tensor included in (2.4), (2.5) are related to displacements u, the following relations:

$$\sigma_{xx} = \lambda\left(\frac{\partial u}{\partial x} + \frac{\partial v}{\partial y}\right) + 2\mu\frac{\partial u}{\partial x}, \sigma_{yy} = \lambda\left(\frac{\partial u}{\partial x} + \frac{\partial v}{\partial y}\right) + 2\mu\frac{\partial v}{\partial y},$$
$$\sigma_{xy} = \mu\left[\frac{\partial v}{\partial x} + \frac{\partial u}{\partial y} - L^2\left(\frac{\partial^2}{\partial x^2} + \frac{\partial^2}{\partial y^2}\right)\left(\frac{\partial v}{\partial x} - \frac{\partial u}{\partial y}\right)\right],$$
$$\sigma_{yx} = \mu\left[\frac{\partial v}{\partial x} + \frac{\partial u}{\partial y} + L^2\left(\frac{\partial^2}{\partial x^2} + \frac{\partial^2}{\partial y^2}\right)\left(\frac{\partial v}{\partial x} - \frac{\partial u}{\partial y}\right)\right]. \tag{2.6}$$

It can be easily seen that $\sigma_{xy} \neq \sigma_{yx}$.

Moment stresses μ_x and μ_y can be expressed through u and v:

$$\mu_x = 2\mu L^2 \frac{\partial}{\partial x}\left(\frac{\partial v}{\partial x} - \frac{\partial u}{\partial y}\right), \quad \mu_y = 2\mu L^2 \frac{\partial}{\partial y}\left(\frac{\partial v}{\partial x} - \frac{\partial u}{\partial y}\right). \tag{2.7}$$

We introduce scalar φ and vector ψ potentials so that the displacement vector u can be written as (Erofeyev 2003):

$$u = \nabla\varphi + \nabla * \psi, \tag{2.8}$$

Since the displacement does not depend on the coordinate z, the only nonzero component of vector potential is the component along the z axis, and we denote this component by ψ.

With the help of (2.8), the system (2.4) is reduced to the equations

$$\Delta\varphi - \frac{1}{c_1^2}\frac{\partial^2 \varphi}{\partial t^2} = 0, = 0, \tag{2.9}$$

$$\Delta(1 - L^2\Delta)\psi - \frac{1}{c_2^2}\frac{\partial^2 \psi}{\partial t^2} = 0, \tag{2.10}$$

where $c_1^2 = \frac{(\lambda+2\mu)}{\rho}$—the square of longitudinal wave velocity, $c_2^2 = \frac{\mu}{\rho}$—the square of shear wave velocity.

Their solution will be sought in the form of waves that are harmonic in time and propagate in the direction of the x axis:

$$\varphi = Ae^{\zeta y+i(\omega t-kx)}, \quad \psi = Be^{\eta y+i(\omega t-kx)}. \tag{2.11}$$

The amplitudes of these waves depend on the y coordinate. Substituting formulas (2.11) into (2.9), (2.10), we obtain the equations for the definition of ζ and η:

$$\zeta^2 + \left(\frac{\omega}{c_1}\right)^2 - k^2 = 0, \quad L^2\eta^4 - (1+2L^2k^2)\eta^2 + \left[k^2 + L^2k^4 - \left(\frac{\omega}{c_2}\right)^2\right] = 0. \tag{2.12}$$

In order for the perturbations to decrease from the boundary inside the medium and correspond to the surface wave, it is necessary to determine such roots of equations (2.12) so that ζ and η are positive. As a result, we get:

$$\zeta = \sqrt{k^2 - \left(\frac{\omega}{c_1}\right)^2}, k^2 > \left(\frac{\omega}{c_1}\right)^2,$$

$$\eta_{1,2} = \sqrt{\frac{(1+2L^2k^2) \pm \sqrt{(1+2L^2k^2)^2 - 4L^2\left[k^2 + L^2k^4 - \left(\frac{\omega}{c_2}\right)^2\right]}}{2L^2}}. \tag{2.13}$$

Then, relations (2.11) will have the following form:

$$\varphi = Ae^{\zeta y+i(\omega t-kx)}, \quad \psi = B_1e^{\eta_1 y+i(\omega t-kx)} + B_2e^{\eta_2 y+i(\omega t-kx)}. \tag{2.14}$$

Displacements u and v, stresses σ_{yx}, σ_{yy} and moment stress μ_y can be expressed through potentials φ and ψ:

$$u = \frac{\partial\varphi}{\partial x} + \frac{\partial\psi}{\partial y}, v = \frac{\partial\varphi}{\partial y} - \frac{\partial\psi}{\partial x}, \quad \sigma_{yy} = \lambda\left(\frac{\partial^2\varphi}{\partial x^2} + \frac{\partial^2\varphi}{\partial y^2}\right) + 2\mu\left(\frac{\partial^2\varphi}{\partial y^2} - \frac{\partial^2\psi}{\partial x\partial y}\right),$$

$$\sigma_{yx} = \mu\left(2\frac{\partial^2\varphi}{\partial x\partial y} - \frac{\partial^2\psi}{\partial x^2} + \frac{\partial^2\psi}{\partial y^2} - L^2\left(\frac{\partial^4\psi}{\partial x^4} + 2\frac{\partial^4\psi}{\partial x^2\partial y^2} + \frac{\partial^4\psi}{\partial y^4}\right)\right),$$

$$\mu_y = -2\mu L^2\left(\frac{\partial^3\psi}{\partial x^2\partial y} + \frac{\partial^3\psi}{\partial y^3}\right). \tag{2.15}$$

Substituting into (2.15) the expressions (2.14) and using the boundary conditions (2.5), we obtain the following system of equations:

$$\begin{aligned}
\sigma_{yy} &= Ae^{\zeta y+i(\omega t-kx)}\left(\lambda\zeta^2-\lambda k^2+2\mu\zeta^2\right)+\\
&\quad+2B_1\mu ik\eta_1 e^{\eta_1 y+i(\omega t-kx)}+2B_2\mu ik\eta_2 e^{\eta_2 y+i(\omega t-kx)}=0,\\
\sigma_{yx} &= -2A\mu\zeta kie^{\zeta y+i(\omega t-kx)}+B_1\mu e^{\eta_1 y+i(\omega t-kx)}\left(k^2+\eta_1^2+2\eta_1^2k^2L^2-\eta_1^4L^2-k^4L^2\right)\\
&\quad+B_2\mu e^{\eta_2 y+i(\omega t-kx)}\left(k^2+\eta_2^2+2\eta_2^2k^2L^2-\eta_2^4L^2-k^4L^2\right)=0,\\
&\quad B_2\mu e^{\eta_2 y+i(\omega t-kx)}\left(k^2+\eta_2^2+2\eta_2^2k^2L^2-\eta_2^4L^2-k^4L^2\right)=0,\\
\mu_y &= 2B_1L^2\eta_1\mu e^{\eta_1 y+i(\omega t-kx)}\left(k^2-\eta_1^2\right)+2B_2L^2\eta_2\mu e^{\eta_2 y+i(\omega t-kx)}\left(k^2-\eta_2^2\right)=0.
\end{aligned}\tag{2.16}$$

representing a homogeneous system of algebraic equations for the definition of A, B_1 and B_2. This system has nonzero solutions if its determinant equals zero.

$$\begin{aligned}
&\eta_2\left(\lambda\zeta^2-\lambda k^2+2\mu\zeta^2\right)\left(k^2+\eta_1^2+2\eta_1^2k^2L^2-\eta_1^4L^2-k^4L^2\right)\left(k^2-\eta_2^2\right)\\
&\quad-\eta_1\left(\lambda\zeta^2-\lambda k^2+2\mu\zeta^2\right)\left(k^2+\eta_2^2+2\eta_2^2k^2L^2-\eta_2^4L^2-k^4L^2\right)\left(k^2-\eta_1^2\right)\\
&\quad+2k^2\zeta\eta_1\eta_2\left(\eta_2^2-\eta_1^2\right)=0,
\end{aligned}\tag{2.17}$$

For further research, let us introduce the following notions $\varsigma = C_R^2 = \frac{\omega^2}{k^2c_2^2}$, $\alpha = L^2k^2$, $\beta = \frac{1-2\nu}{2-2\nu}$, where ν—Poisson's coefficient, C_R—the velocity of surface wave. When $L = 0$ or $\alpha = 0$ from (2.17) we can get the equation for determining the velocity of a surface wave in the classical case (Erofeyev 2003).

$$16(1-\beta\varsigma)(1-\varsigma) = (2-\varsigma)^2(3+(1-\varsigma)^2-2\varsigma) \tag{2.18}$$

Note that the wave number α is included in Eq. (2.18), therefore the surface wave has a dispersion, in contrast to the classical case in which the Rayleigh surface wave does not have dispersion.

Figure 2.1 shows the dependence of the square of the velocity of the surface wave C_R on wave number α.

The curves are represented in dimensionless form: the square of the velocity of the surface wave is related to the square of the velocity of the shear wave c_2^2. The curves are calculated for two values of Poisson's ratio: $\nu = 0.2$ (solid line) and $\nu = 0.5$ (dashed line). From Fig. 2.1, it is clear that with the increase of wave number α the square of velocity of the surface wave growths and when $\alpha \to \infty$, $\varsigma \to 2$ or $C_R \to \sqrt{2}$.

Let us obtain the expression for the phase velocity of a plane shear wave $v_{ph}(\alpha)$ and compare it with the speed of the surface wave $C_R(\alpha)$.

The equation for a plane shear wave, taking into account (2.10), takes the form:

$$\frac{\partial^2\psi}{\partial x^2} - L^2\frac{\partial^4\psi}{\partial x^4} = \frac{1}{c_2^2}\frac{\partial^2\psi}{\partial t^2}. \tag{2.19}$$

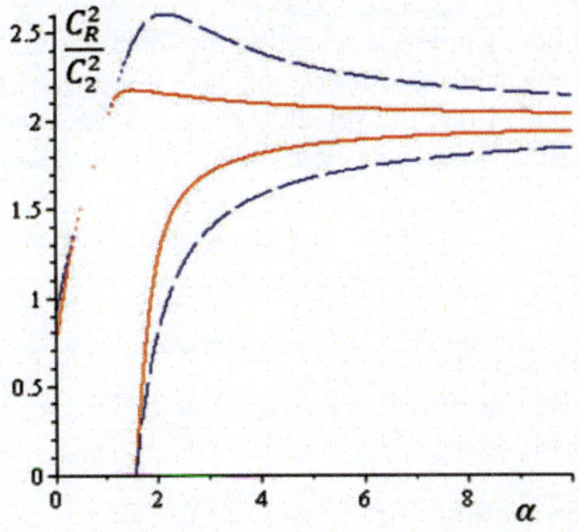

Fig. 2.1 Dependence of the square of the velocity of the surface wave from the wave number

We will look for the solution of this equation in the form: $\psi = Be^{i(\omega t - kx)}$. We get:

$$\omega^2 = c_2^2 k^2 \left(1 + L^2 k^2\right), \quad v_{\text{ph}}^2 = \frac{\omega^2}{k^2} = c_2^2 \left(1 + L^2 k^2\right), \quad \frac{v_{\text{ph}}^2}{c_2^2} = (1 + \alpha). \tag{2.20}$$

From (2.20), it follows that the shear wave in a gradient-elastic medium has dispersion, its phase velocity does not coincide with c_2 and exceeds it for any nonzero value of the wave number.

Figure 2.2 presents two dependencies: the square of the surface wave velocity C_R^2 and the square of the shear wave phase velocity v_{ph}^2. The curves are presented in

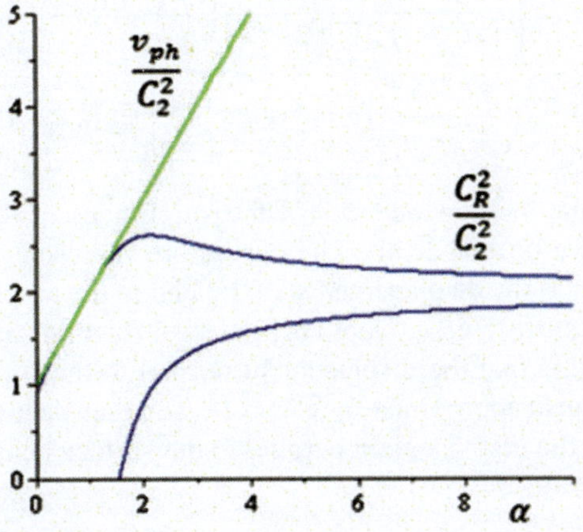

Fig. 2.2 Dependence of the square of the velocity of the surface wave and the square of the phase velocity of the shear wave from the wave number

dimensionless form: both values are related to the square of the shear wave velocity c_2^2. Surface wave speed calculated for Poisson's coefficient $\nu = 0.5$. From the above graph, it can be concluded that the velocity of the surface wave cannot exceed the phase velocity of the shear wave, but can reach it at certain values of the wave number α.

From (2.14) and (2.15), it follows that displacements u and v can be written in the form:

$$\begin{aligned} u &= -iAke^{\zeta y+i(\omega t-kx)} + B_1\eta_1 e^{\eta_1 y+i(\omega t-kx)} + B_2\eta_2 e^{\eta_2 y+i(\omega t-kx)} \\ &= \left(-iAke^{\zeta y} + B_1\eta_1 e^{\eta_1 y} + B_2\eta_2 e^{\eta_2 y}\right)e^{i(\omega t-kx)}, \\ v &= A\zeta e^{\zeta y+i(\omega t-kx)} + iB_1ke^{\eta_1 y+i(\omega t-kx)} + iB_2ke^{\eta_2 y+i(\omega t-kx)} \\ &= \left(A\zeta e^{\zeta y} + iB_1ke^{\eta_1 y} + iB_2ke^{\eta_2 y}\right)e^{i(\omega t-kx)}. \end{aligned} \tag{2.21}$$

The system of equations (2.16) allows us to express the constant B_1 and A through B_2:

$$B_1 = -B_2\frac{\eta_2}{\eta_1}\left(\frac{k^2-\eta_2^2}{k^2-\eta_1^2}\right), \quad A = \frac{2iB_2\eta_2k\mu}{(\lambda\zeta^2-\lambda k^2+2\mu\zeta^2)}\left(\frac{k^2-\eta_2^2}{k^2-\eta_1^2}-1\right). \tag{2.22}$$

Relations (2.22), after taking the real parts (2.21), allow us to write displacements in the form:

$$\begin{aligned} u &= B_2\eta_2\left[\frac{2k^2\mu}{(\lambda\zeta^2-\lambda k^2+2\mu\zeta^2)}\left(\frac{f^2-\eta_2^2}{f^2-\eta_1^2}-1\right)e^{-\zeta y}\right. \\ &\quad \left. -\frac{f^2-\eta_2^2}{f^2-\eta_1^2}e^{-\eta_1 y} + e^{-\eta_2 y}\right]\cos(kx-\omega t), \\ v &= -B_2\left[\frac{2k^2\mu\eta_2\zeta}{(\lambda\zeta^2-\lambda k^2+2\mu\zeta^2)}\left(\frac{f^2-\eta_2^2}{f^2-\eta_1^2}-1\right)e^{-\zeta y}\right. \\ &\quad \left. -k\frac{\eta_2}{\eta_1}\left(\frac{f^2-\eta_2^2}{f^2-\eta_1^2}\right)e^{-\eta_1 y} + ke^{-\eta_2 y}\right]\sin(kx-\omega t). \end{aligned} \tag{2.23}$$

Figure 2.3 presents the dependences of the amplitudes of displacements u and v in the surface wave on the depth. The curves are represented in dimensionless form: the amplitudes of the displacements are related to the amplitude of the normal displacement on the surface $v_{y=0}$. Depth is plotted in fractions of wavelength. It can be seen from the graph that the displacement normal to the surface first increases, reaching its maximum at approximately $y = 0.1\lambda$, and then decreases monotonically with depth, whereas the displacement parallel to the surface changes sign at a depth of approximately $y = 0.15\lambda$.

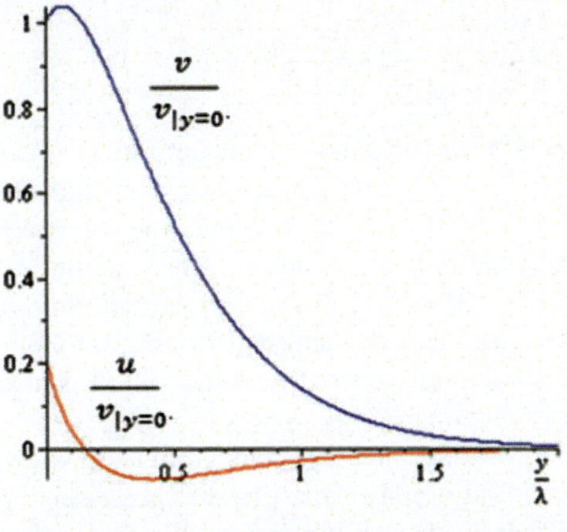

Fig. 2.3 Dependence of the amplitudes of displacements in the surface wave from the depth

Stresses can be represented through the potentials φ and ψ, and similarly to the calculation of displacements, expressing the constants B_1 and A through B_2, after taking the real parts, we get:

$$\sigma_{yy} = B_2\left[-2\eta_2 f\mu\left(\frac{f^2-\eta_2^2}{f^2-\eta_1^2}-1\right)e^{-\zeta y} + 2\eta_2 f\mu\left(\frac{f^2-\eta_2^2}{f^2-\eta_1^2}\right)e^{-\eta_1 y}\right.$$
$$\left.-2\eta_2 f\mu e^{-\eta_2 y}\right]\sin(kx-\omega t),$$
$$\sigma_{xx} = B_2\left[2\eta_2 f\mu\frac{(\lambda\zeta^2-\lambda k^2+2\mu k^2)}{(\lambda\zeta^2-\lambda k^2+2\mu\zeta^2)}\left(\frac{f^2-\eta_2^2}{f^2-\eta_1^2}-1\right)e^{-\zeta y}\right.$$
$$\left.+2\eta_2 f\mu\left(\frac{f^2-\eta_2^2}{f^2-\eta_1^2}\right)e^{-\eta_1 y} - 2\eta_2 f\mu e^{-\eta_2 y}\right]\cos(kx-\omega t),$$
$$\sigma_{yx} = B_2\left[\frac{4\eta_2 f^2\mu^2}{(\lambda\zeta^2-\lambda k^2+2\mu\zeta^2)}\left(\frac{f^2-\eta_2^2}{f^2-\eta_1^2}-1\right)e^{-\zeta y}\right.$$
$$-\mu\frac{\eta_2}{\eta_1}\left(\frac{f^2-\eta_2^2}{f^2-\eta_1^2}\right)\left(k^2+\eta_1^2+2\eta_1^2k^2L^2-\eta_1^4L^2-k^4L^2\right)e^{-\eta_1 y}$$
$$\left.+\mu\left(k^2+\eta_2^2+2\eta_2^2k^2L^2-\eta_2^4L^2-k^4L^2\right)e^{-\eta_2 y}\right]\cos(kx-\omega t),$$
$$\sigma_{xy} = B_2\left[\frac{4\eta_2 f^2\mu^2}{(\lambda\zeta^2-\lambda k^2+2\mu\zeta^2)}\left(\frac{f^2-\eta_2^2}{f^2-\eta_1^2}-1\right)e^{-\zeta y}\right.$$
$$-\mu\frac{\eta_2}{\eta_1}\left(\frac{f^2-\eta_2^2}{f^2-\eta_1^2}\right)\left(k^2+\eta_1^2-2\eta_1^2k^2L^2+\eta_1^4L^2+k^4L^2\right)$$
$$\left.+\mu\left(k^2+\eta_2^2-2\eta_2^2k^2L^2+\eta_2^4L^2+k^4L^2\right)e^{-\eta_2 y}\right]\cos(kx-\omega t),$$
$$\mu_y = 2B_2L^2\mu\eta_2\left(f^2-\eta_2^2\right)\left(e^{-\eta_2 y}-e^{-\eta_1 y}\right)\cos(kx-\omega t),$$

$$\mu_x = 2B_2L^2\mu f\left(f^2 - \eta_2^2\right)\left(\frac{\eta_2}{\eta_1}e^{-\eta_2 y} + e^{-\eta_1 y}\right)\sin(kx - \omega t). \tag{2.24}$$

Figure 2.4 represents the dependencies of magnitudes of stresses σ_{xx}, σ_{xy}, σ_{yx}, σ_{yy} and moment stresses μ_x, μ_y in the surface wave on the depth. The curves are presented in dimensionless form: the amplitudes of the voltages and the moment voltages are related to the amplitude of the normal stress on the surface $\sigma_{xx|y=0}$, $\mu_{x|y=0}$, respectively. Depth is plotted in fractions of wavelength. From the graphs, it is clear that σ_{xx} changes its sign, when σ_{yy} and σ_{yx} reach a maximum at approximately $y = 0.2\lambda$ and then exponentially decrease with depth. It is also clear from Fig. 2.4 that the stress tensor is asymmetric ($\sigma_{yx} \neq \sigma_{xy}$).

Figure 2.5 represents the dependencies of magnitudes of stresses σ_{xx}, σ_{xy}, σ_{yx}, σ_{yy} in the surface wave on the depth in the classic case (when $L = 0$). Curves are presented in dimensionless form. From the graphs, it can be seen that the stress tensor has become symmetrical, since the stress amplitudes coincide σ_{xy}, σ_{yx} and reach a maximum at approximately $y = 0.2\lambda$.

The set of curves shown in Figs. 2.3, 2.4, and 2.5 illustrates that the surface wave is localized in a thin surface layer.

Based on the above considerations, it can be concluded that the velocity of a surface wave propagating along the free border of a gradient-elastic half-space can exceed the velocity of a bulk shear wave, calculated as the radical of the ratio of shear modulus to material density. However, in the medium under consideration, the shear wave also has dispersion and the value of the indicated velocity is only the lower limit of its phase velocity. Thus, in a gradient-elastic medium, the phase velocity of a surface wave cannot exceed the phase velocity of a bulk shear wave, but, at certain values of the wave number, it can reach it.

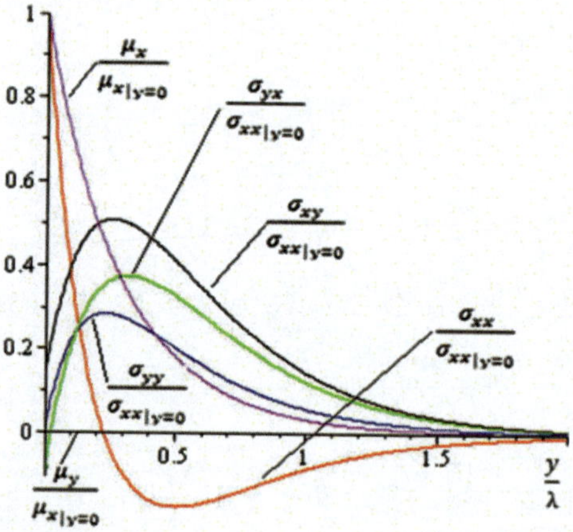

Fig. 2.4 Dependence of the amplitudes of the stresses in the surface wave from the depth

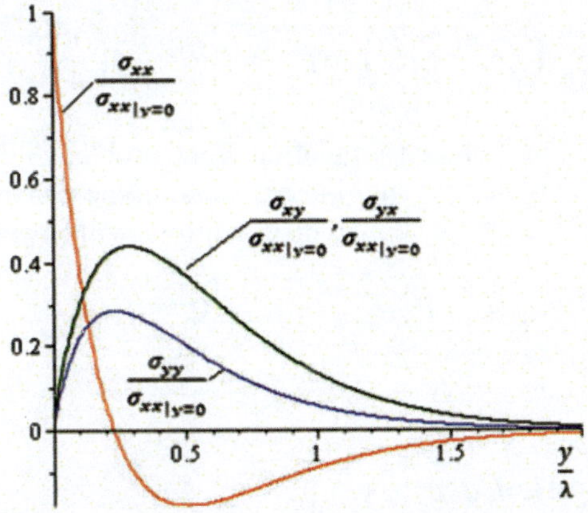

Fig. 2.5 Dependence of the amplitudes of the stresses in the surface wave on the depth in the classical case (when $L = 0$)

2.4 The Statement and Solution to the Problem of a Gradient-Elastic Medium with a Moving Source Generating Surface Waves

Let us denote the fixed coordinate system as x', y'.

Along with the above problem, there is an interest in the question of the generation of surface waves by a source moving along the border of a gradient-elastic half-space with a constant velocity D which is greater the velocity of shear $c_2 = \sqrt{\frac{\mu}{\rho}}$ and longitudinal $c_1 = \sqrt{\frac{\lambda+2\mu}{\rho}}$ waves.

The equations of dynamics equivalent to the vector equation in the two-dimensional case are written as:

$$\frac{\partial \sigma_{x'x'}}{\partial x'} + \frac{\partial \tau_{y'x'}}{\partial y'} = \rho \frac{\partial^2 u}{\partial t^2}, \quad \frac{\partial \tau_{x'y'}}{\partial x'} + \frac{\partial \sigma_{y'y'}}{\partial y'} = \rho \frac{\partial^2 v}{\partial t^2}. \tag{2.25}$$

Components of stress tensor and moment stresses $\mu_{x'}$ и $\mu_{y'}$ expressed through displacements u and v:

$$\sigma_{x'x'} = \lambda\left(\frac{\partial u}{\partial x'} + \frac{\partial v}{\partial y'}\right) + 2\mu\frac{\partial u}{\partial x'}, \quad \sigma_{y'y'} = \lambda\left(\frac{\partial u}{\partial x'} + \frac{\partial v}{\partial y'}\right) + 2\mu\frac{\partial v}{\partial y'},$$

$$\sigma_{x'y'} = \mu\left[\frac{\partial v}{\partial x'} + \frac{\partial u}{\partial y'} - l^2\left(\frac{\partial^2}{\partial x'^2} + \frac{\partial^2}{\partial y'^2}\right)\left(\frac{\partial v}{\partial x'} - \frac{\partial u}{\partial y'}\right)\right],$$

$$\sigma_{y'x'} = \mu\left[\frac{\partial v}{\partial x'} + \frac{\partial u}{\partial y'} + l^2\left(\frac{\partial^2}{\partial x'^2} + \frac{\partial^2}{\partial y'^2}\right)\left(\frac{\partial v}{\partial x'} - \frac{\partial u}{\partial y'}\right)\right],$$

$$\mu_{x'} = 2\mu l^2 \frac{\partial}{\partial x'}\left(\frac{\partial v}{\partial x'} - \frac{\partial u}{\partial y'}\right), \quad \mu_{y'} = 2\mu l^2 \frac{\partial}{\partial y'}\left(\frac{\partial v}{\partial x'} - \frac{\partial u}{\partial y'}\right). \tag{2.26}$$

The principal difference in the formulation of the problem will be observed in the process of setting the boundary conditions for solving the dynamic equation. In the problem under consideration, the boundary conditions will have the following form:

$$\sigma_{y'y'} = -P\delta(x), \quad \sigma_{y'x'} = 0, \quad \mu_{y'} = 0. \tag{2.27}$$

2.4.1 The Subsonic Case

Before proceeding to the solution of the problem with a moving source at supersonic speeds ($c_1 > c_2 > D$), let us consider the case of the motion of a source of disturbances with a velocity less than the velocities of the longitudinal and shear waves—the subsonic case.

To do this, we introduce a moving coordinate system (x, y), in which the source of perturbations resides and which is connected with the fixed coordinate system by the famous Galileo transformation:

$$x = x' - Dt, \quad y = y'. \tag{2.28}$$

Relations (2.26), (2.28) allow to write down equations of dynamics in displacements:

$$\begin{gathered}
(\lambda + 2\mu - \rho D^2)\frac{\partial^2 u}{\partial x^2} + \lambda \frac{\partial^2 v}{\partial x \partial y} + \mu\left[\frac{\partial^2 v}{\partial x \partial y} + \frac{\partial^2 v}{\partial y^2} + l^2 \Delta\left(\frac{\partial^2 v}{\partial x \partial y} - \frac{\partial^2 u}{\partial y^2}\right)\right] = 0,\\
(\lambda + 2\mu)\frac{\partial^2 v}{\partial y^2} + \lambda \frac{\partial^2 u}{\partial x \partial y} + \mu\left[\left(1 - \frac{\rho D^2}{\mu}\right)\frac{\partial^2 v}{\partial x^2} + \frac{\partial^2 u}{\partial x \partial y} - l^2 \Delta\left(\frac{\partial^2 v}{\partial x^2} - \frac{\partial^2 u}{\partial x \partial y}\right)\right] = 0.
\end{gathered} \tag{2.29}$$

The solution to Eqs. (2.29) will look in the form:

$$u = Ae^{kqy}\sin(kx), \quad v = Be^{kqy}\cos(kx). \tag{2.30}$$

After substitution (2.30) into (2.29) we get the system of algebraic equations with respect to A and B

$$\begin{aligned}
&\left[-\mu l^2 k^2 q^4 + \mu q^2(1 + l^2 k^2) - (\lambda + 2\mu - \rho D^2)\right]\\
&\quad + \left[\mu l^2 k^2 q^3 + q(\lambda + \mu - \mu l^2 k^2)\right]B = 0,\\
&\left[-\mu l^2 k^2 q^3 - q(\lambda + \mu - \mu l^2 k^2)\right]A
\end{aligned}$$

$$+\left[(\lambda+2\mu-\rho D^2)q^2-(\mu+\mu l^2k^2-\rho D^2)\right]B=0, \tag{2.31}$$

which has nonzero solutions when:

$$\begin{aligned}&(l^2k^2-\gamma+\beta_1^2)l^2k^2q^6\\&+\left(l^2k^2(1+l^2k^2-\beta_1^2)+\gamma(1+l^2k^2)(1-\gamma\beta_1^2)+2l^2k^2(\gamma-l^2k^2)\right)q^4\\&+\left((1+l^2k^2)(1+l^2k^2-\beta_1^2)-(\gamma-\beta_1^2)^2+(\gamma-l^2k^2)^2\right)q^2\\&+(\gamma-\beta_1^2)(1+l^2k^2-\beta_1^2)=0.\end{aligned} \tag{2.32}$$

Here, $\beta_1^2=\rho D^2/\mu$—the ratio of the square of the velocity of the source to the square of the velocity of the shear wave; $\gamma=\lambda+2\mu/\mu$—the ratio of the square of the velocity of the longitudinal wave to the square of the velocity of the shear wave.

The roots of equations (2.32) q_1,q_2,q_3 should be sought when $Re(q)>0$. Longitudinal and transverse displacements will be expressed as sums:

$$u=\sum_{i=1}^{3}\int_0^\infty A_ie^{kq_iy}\sin kx\mathrm{d}k,\quad v=\sum_{i=1}^{3}\int_0^\infty \alpha_iA_ie^{kq_iy}\cos kx\mathrm{d}k. \tag{2.33}$$

Here,

$$\alpha_i=\frac{\mu l^2k^2q_i^4-\mu q_i^2(1+l^2k^2)+(\lambda+2\mu-\rho D^2)}{\mu l^2k^2q_i^3+(\lambda+\mu-\mu l^2k^2)q_i},\quad B_i=\alpha_iA_i. \tag{2.34}$$

Let us substitute (2.33) into boundary conditions (2.27), we will get the system for A_1, A_2, A_3 identification.

$$\begin{aligned}&\sum_{i=1}^{3}[\lambda+\alpha_iq_i(\lambda+2\mu)]A_i=-\frac{P}{\pi k},\\&\sum_{i=1}^{3}[-\alpha_i(1-l^2k^2)+q_i(1-lk^2\alpha_iq_i)+lk^2q_i(1-q_i^2)]A_i=0,\\&\sum_{i=1}^{3}q_i(\alpha+q_i)A_i=0.\end{aligned} \tag{2.35}$$

It is known that the effect of moment stresses is especially pronounced at short waves (Erofeyev 2003). Therefore, we introduce a dimensionless small parameter $\varepsilon=1/lk$.

Accurate to values of order ε^2 from (2.32) we get:

$$q_1 = a_{01}\varepsilon, \quad q_2 = \frac{(a_{02} + a_{12}\varepsilon)}{\sqrt{\varepsilon}}, \quad q_3 = \frac{(a_{02} - a_{12}\varepsilon)}{\sqrt{\varepsilon}}, \tag{2.36}$$

where

$$a_{01} = -\frac{\gamma\sqrt{1-\beta_0^2}}{\sqrt{2}}, \quad a_{02} = -\frac{1+i}{\sqrt[4]{128}},$$
$$a_{12} = \frac{i(1+i)(3-\beta_0^2)\sqrt{2}}{2\sqrt[4]{8}}, \quad \beta_0^2 = \frac{\rho D^2}{\lambda + 2\mu}. \tag{2.37}$$

From (2.34), (2.35), and (2.36), we get the following relations:

$$\alpha_1 = \alpha_{01} + \alpha_{11}\varepsilon, \alpha_2 = q_2, \alpha_3 = q_3, \quad A_1 = -\frac{P}{\pi\lambda k}, \quad A_2 = A_3 = 0,$$
$$\alpha_{01} = \frac{(1+\beta_0^2)\gamma^2 - 2\beta_1^2}{\gamma\sqrt{1-\beta_0^2}}, \quad \alpha_{11} = \gamma\sqrt{1-\beta_0^2}\left[\frac{(1+\beta_0^2)\gamma^2}{2} - \beta_1^2\right] \tag{2.38}$$

which allow to write down displacements (2.33) in the following form:

$$u = -\frac{P}{\pi\lambda}e^{-\beta_2 y}\int\limits_{k_0}^{\infty}\frac{\sin kx}{k}\mathrm{d}k, \tag{2.39}$$

$$v = -\frac{P\alpha_{01}}{\pi\lambda}e^{-\beta_2 y}\int\limits_{k_0}^{\infty}\frac{\cos kx}{k}\mathrm{d}k - \frac{P\alpha_{11}k_0}{\pi\lambda}e^{-\beta_2 y}\int\limits_{k_0}^{\infty}\frac{\cos kx}{k^2}\mathrm{d}k, \tag{2.40}$$

where $k_0 = 1/l$, $\beta_2 = a_{01}k_0$.

From the expressions (2.39) and (2.40), it can be seen that displacement amplitudes will vary depending on the magnitude of the load of the moving source P and its velocity D.

$$V_1 = -\frac{v\pi\lambda e^{\beta_2 y}}{P} = \alpha_{01}\int\limits_{k_0}^{\infty}\frac{\cos kx}{k}\mathrm{d}k + \alpha_{11}k_0\int\limits_{k_0}^{\infty}\frac{\cos kx}{k^2}\mathrm{d}k, \tag{2.41}$$

Figure 2.6 shows the dependence of the normalized amplitude of the transverse displacement on the square of the dimensionless velocity of a moving source of disturbances. It is clear from the graphic image that as the speed of a moving source reaches the shear wave speed, V_1 increases unlimitedly.

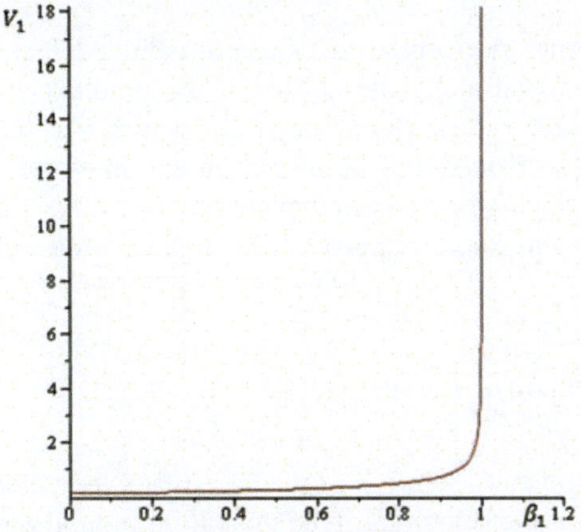

Fig. 2.6 Dependence of the amplitude of the transverse displacement of the speed of a moving source

Figure 2.7 presents the dependences of the longitudinal U and transversal V displacements on depth. The curves are given in dimensionless form: the displacement amplitudes are related to the normal displacement amplitude on the surface $V_{y=0}$. Depth is plotted in fractions of wavelength. It is clear from the graphic image that the displacement components decrease exponentially with depth, the transverse component of the displacement vector always exceeds the longitudinal one, and the particles rotate as the surface wave propagates in the x-axis direction along an elliptical trajectory.

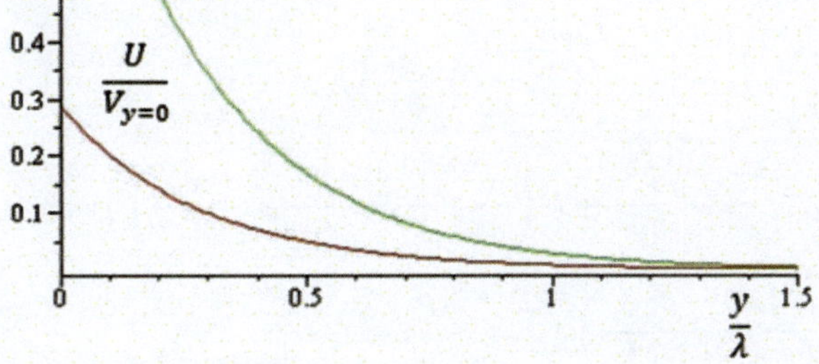

Fig. 2.7 Dependence of the amplitudes of displacements in the surface wave on the depth

Based on the studies performed, it can be concluded that a source moving at a constant subsonic speed along the border of the gradient-elastic half-space will generate surface elastic waves. Such waves, in contrast to the classical Rayleigh surface waves, have a dispersion. The displacement amplitudes change depending on the magnitude of the load of the moving source, as well as its speed, and increase without limit as the source speed approaches the shear wave velocity.

2.4.2 The Supersonic Case

We next proceed to the consideration of the case when the velocity of the source of disturbances D exceeds the velocities of the shear longitudinal waves—the supersonic case. For that we introduce potentials (φ, ψ), which satisfy the following relations:

$$u = \frac{\partial \varphi}{\partial x'} - \frac{\partial \psi}{\partial y'},\ v = \frac{\partial \varphi}{\partial y'} + \frac{\partial \psi}{\partial x'}, \tag{2.42}$$

and converting the system of equations (2.25) with (2.26) into the system of equations

$$\left(\frac{\partial^2}{\partial x'^2} + \frac{\partial^2}{\partial y'^2} - \frac{1}{c_1^2}\frac{\partial^2}{\partial t^2}\right)\varphi = 0, \tag{2.43}$$

$$\left(\frac{\partial^2}{\partial x'^2} + \frac{\partial^2}{\partial y'^2} - l^2\left(\frac{\partial^4}{\partial x'^4} + 2\frac{\partial^4}{\partial x'^2 \partial y'^2} + \frac{\partial^4}{\partial y'^4}\right) - \frac{1}{c_2^2}\frac{\partial^2}{\partial t^2}\right)\psi = 0 \tag{2.44}$$

Similar to the subsonic case, we introduce a moving coordinate system (x, y) in which the source of disturbances rests and which is connected with a fixed coordinate system by the Galilean transformation:

$$x = x' - Dt,\ y = y'. \tag{2.45}$$

As a result the boundary problem (2.25), (2.27) will take the following form:

$$\left(\lambda_1^2\frac{\partial^2}{\partial x} - \frac{\partial^2}{\partial y}\right)\varphi = 0, \tag{2.46}$$

$$\left(\lambda_2^2\frac{\partial^2}{\partial x^2} - \frac{\partial^2}{\partial y^2} + l^2\left(\frac{\partial^4}{\partial x^4} + 2\frac{\partial^4}{\partial x^2 \partial y^2} + \frac{\partial^4}{\partial y^4}\right)\right)\psi = 0, \tag{2.47}$$

$$\left((\eta_2^2 - 2)\frac{\partial^2 \varphi}{\partial x^2} + 2\frac{\partial^2 \psi}{\partial x \partial y}\right)_{y=0} = -\frac{P}{\mu}\delta(x),$$
$$\left(2\frac{\partial^2 \varphi}{\partial x \partial y} + \eta_2^2\frac{\partial^2 \psi}{\partial x^2} - 2\frac{\partial^2 \psi}{\partial y^2}\right)_{y=0} = 0,\quad \left(\frac{\partial^3 \psi}{\partial x^2 \partial y} + \frac{\partial^3 \psi}{\partial y^3}\right)_{y=0} = 0, \tag{2.48}$$

where $\eta_\alpha = \frac{D}{c_\alpha}, \lambda_1^2 = \sqrt{(\eta_\alpha^2 - 1)}, \alpha = 1, 2.$

Note that in the third equation of the system (2.48) there is only the potential of shear waves. Let us integrate this expression with respect to y:

$$\frac{\partial^2\psi}{\partial x \partial y} + \frac{\partial^2\psi}{\partial y^2} + \text{const} = 0. \tag{2.49}$$

Assuming the integration constant to be zero, expression (2.49) is substituted into the second equation of system (2.48). We obtain the following system of equations:

$$\left(\left(\eta_2^2 - 2\right)\frac{\partial^2\varphi}{\partial x^2} + 2\frac{\partial^2\psi}{\partial x \partial y}\right)_{y=0} = -\frac{P}{\mu}\delta(x), \quad \left(2\frac{\partial^2\varphi}{\partial x \partial y} + \left(\eta_2^2 + 2\right)\frac{\partial^2\psi}{\partial x^2}\right)_{y=0} = 0. \tag{2.50}$$

The solution to the differential equations (2.25), (2.27) will be integrals

$$\varphi(x, y) = \frac{1}{\sqrt{2\pi}}\int_{-\infty}^{\infty} Ae^{-ia(x-\lambda_1 y)}\mathrm{d}a, \quad \psi(x, y) = \frac{1}{\sqrt{2\pi}}\int_{-\infty}^{\infty} Be^{-ia(x-\lambda_2 y)}\mathrm{d}a \tag{2.51}$$

found by applying the integral Fourier transform.

It is easy to verify that these functions satisfy the conditions of radiation at infinity.

Applying the Fourier transform to the boundary conditions (2.50) and given that the Fourier transform for the function $\delta(x)$ has the form:

$$F[\delta(x)] = \frac{1}{\sqrt{2\pi}}\int_{-\infty}^{\infty} \delta(x)Be^{-iax}\mathrm{d}x = \frac{1}{\sqrt{2\pi}}$$

we get the following system of equations:

$$-a\left(\eta_2^2 - 2\right)A + 2a^2\lambda_2 B = -\frac{P}{\mu\sqrt{2\pi}}, \quad 2a^2\lambda_1 A - a^2\left(\eta_2^2 + 2\right)B = 0. \tag{2.52}$$

We define the values of A and B. For this, from the second equation of system (2.52), we express B through A:

$$B = \frac{2\lambda_1}{\left(\eta_2^2 + 2\right)}A,$$

then we find

$$A = \frac{(\eta_2^2 - 2)P}{\mu a^2\sqrt{2\pi}\left[(\eta_2^2 + 2)(\eta_2^2 - 2) + 4\lambda_1\lambda_2\right]},$$

$$B = \frac{2\lambda_1 P}{\mu a^2\sqrt{2\pi}\left[(\eta_2^2 + 2)(\eta_2^2 - 2) + 4\lambda_1\lambda_2\right]}. \tag{2.53}$$

We introduce the following variables: $\gamma_1 = \eta_2^2+2$, $\gamma_2 = \eta_2^2-2$, $\Delta = \gamma_1\gamma_2+4\lambda_1\lambda_2$, then expressions (2.53) take the following form:

$$A = \frac{\gamma_2}{\mu a^2\sqrt{2\pi}\,\Delta},\ B = \frac{2\lambda_1}{\mu a^2\sqrt{2\pi}\,\Delta}. \tag{2.54}$$

In accordance with (2.51):

$$u = -\frac{iP}{2\pi\mu\Delta}\int\limits_{-\infty}^{\infty}\left(\gamma_2 e^{i\lambda_1 ay} + 2\lambda_1\lambda_2 e^{i\lambda_2 ay}\right)\frac{e^{-iax}}{a}da,$$

$$v = \frac{iP}{2\pi\mu\Delta}\int\limits_{-\infty}^{\infty}\left(\gamma_2\lambda_1 e^{i\lambda_1 ay} - 2\lambda_1 e^{i\lambda_2 ay}\right)\frac{e^{-iax}}{a}da. \tag{2.55}$$

Since displacements are real functions, it follows from (2.55) that

$$u = -\frac{P}{2\pi\mu\Delta}\left[\gamma_2\int\limits_{-\infty}^{\infty}\frac{\sin a(x-\lambda_1 y)}{a}da - 2\lambda_1\lambda_2\int\limits_{-\infty}^{\infty}\frac{\sin a(x-\lambda_2 y)}{a}da\right],$$

$$v = \frac{P}{2\pi\mu\Delta}\left[\gamma_2\lambda_1\int\limits_{-\infty}^{\infty}\frac{\sin a(x-\lambda_1 y)}{a}da - 2\lambda_1\int\limits_{-\infty}^{\infty}\frac{\sin a(x-\lambda_2 y)}{a}da\right]. \tag{2.56}$$

We define the step function equal to zero for negative values of z and one for positive values. So that the domain of definition of the function contains all points of the real axis, at zero the function is defined by the number ½. This function, as is known, is called the Heaviside function:

$$H(z) = \begin{Bmatrix} 0 \text{ when } z < 0 \\ \frac{1}{2} \text{ when } z = 0 \\ 1 \text{ when } z > 0 \end{Bmatrix} = \frac{1}{2} + \frac{1}{\pi}\int\limits_0^{\infty}\frac{\sin az}{a}da \tag{2.57}$$

From (2.56) and (2.57), we get the expression for displacements:

$$u = -\frac{P}{\mu\Delta}\left[\gamma H(x-\lambda_1 y) - 2\lambda_1\lambda_2 H(x-\lambda_2 y)\right] + \text{const},$$

$$v = -\frac{P}{\mu\Delta}\left[\gamma\lambda_1 H(x - \lambda_1 y) + 2\lambda_1 H(x - \lambda_2 y)\right] + \text{const}'. \tag{2.58}$$

Here const—constant value which does not affect stresses distribution.

Figure 2.8 shows the dependence of the displacement amplitudes U (a) and V (b) in the Rayleigh wave on the depth. The curves are given in dimensionless form: the displacement amplitudes are related to the normal displacement amplitude on the surface $V_{y=0}$. Depth is plotted in fractions of wavelength. It is seen that when the longitudinal component decreases, the transverse component begins to increase. In the process of distance from the surface layer, an increase in the amplitude of the displacement of particles in the wave and only at a depth *2L-3L* the attenuation of disturbances occurs.

Similarly, we determine the stresses:

$$\begin{aligned}
\sigma_{xx} &= -\frac{P}{\Delta}\left[\gamma\left(2 + \frac{\lambda}{\mu}\eta_1^2\right)\delta(x - \lambda_1 y) - 4\lambda_1\lambda_2\delta(x - \lambda_2 y)\right],\\
\sigma_{yy} &= -\frac{P}{\Delta}\left[\gamma\left(2\lambda_1^2 + \frac{\lambda}{\mu}\eta_1^2\right)\delta(x - \lambda_1 y) + 4\lambda_1\lambda_2\delta(x - \lambda_2 y)\right],\\
\sigma_{yx} &= -\frac{2P}{\Delta}\lambda_1\gamma[\delta(x - \lambda_1 y) - \delta(x - \lambda_2 y)],\\
\sigma_{xy} &= -\frac{2P}{\Delta}\lambda_1\gamma[\delta(x - \lambda_1 y) + 2\delta(x - \lambda_2 y)],\\
\mu_x &= -\frac{2aL^2P}{\Delta}(\lambda_1 + \lambda_1\lambda_2^2)\delta(x - \lambda_2 y),\\
\mu_y &= -\frac{2aL^2P}{\Delta}(\lambda_1\lambda_2 + \lambda_2^3)\delta(x - \lambda_2 y).
\end{aligned} \tag{2.59}$$

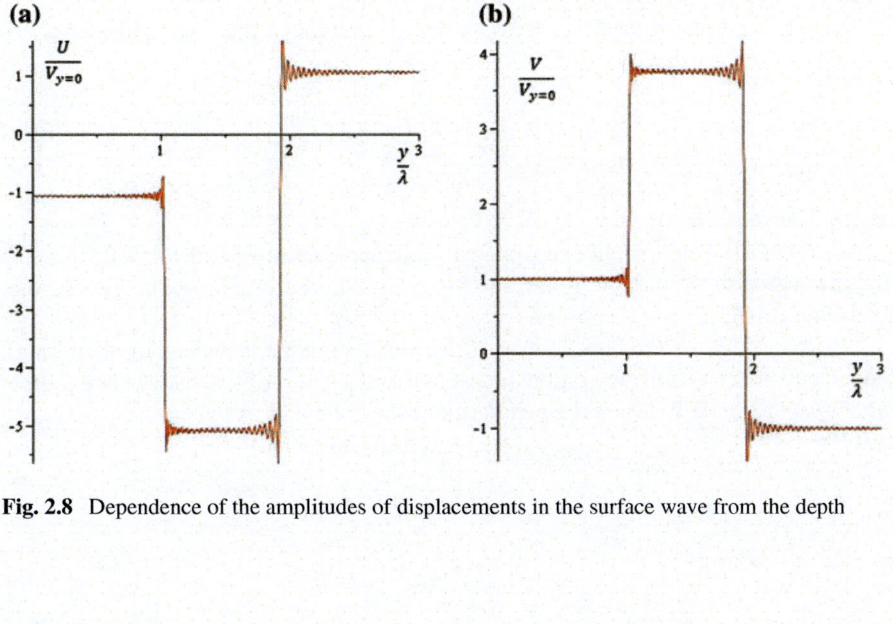

Fig. 2.8 Dependence of the amplitudes of displacements in the surface wave from the depth

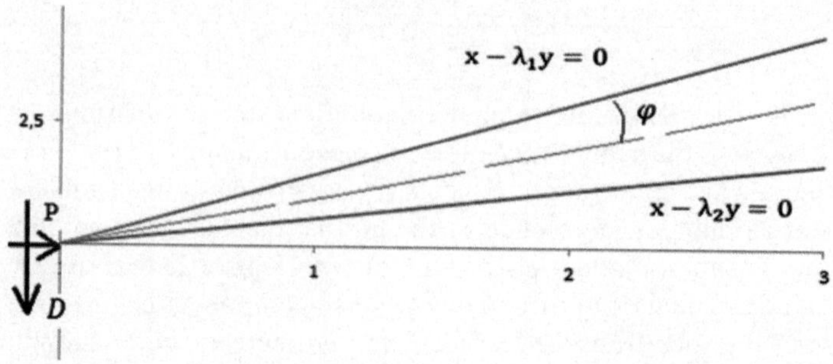

Fig. 2.9 Mach cone

Here, $\frac{dH(z)}{dz} = \delta(z)$, where $\delta(z)$—Dirac function. From (2.58) and (2.59), it is obvious that disturbance is caused by a normal load, characterized by two Mach waves:

$$x - \lambda_1 y = 0,$$

$$x - \lambda_2 y = 0,$$

where $\lambda_\alpha^2 = \sqrt{\left(\eta_\alpha^2 - 1\right)}$, $\eta_\alpha = \frac{D}{c_\alpha}$, $\alpha = 1, 2$.

When a normal load moves at supersonic speeds, a conical surface (Mach cone—Fig. 2.9) is formed after the disturbance source, bounding the region in which disturbances are concentrated from the unperturbed region of the elastic half-space. The surface of the Mach cone is the envelope of a system of waves generated by a normal load.

The angle φ (Mach angle)—between the generators of the cone and its axis is defined as

$$\sin\varphi = \frac{c_\alpha}{D} = \frac{1}{\eta_\alpha}, \alpha = 1, 2,$$

where η_α—Mach numbers.

As the speed of movement of the normal load increases, the Mach angle decreases, and the region of perturbation concentration shifts to the line of application of the load (Fig. 2.10).

Figure 2.11 shows the dependence of the Mach angle on the speed of the normal load. It can be seen from the graphic image that as the speed of the disturbance source increases, the angle between the generatrix of the cone and its axis decreases.

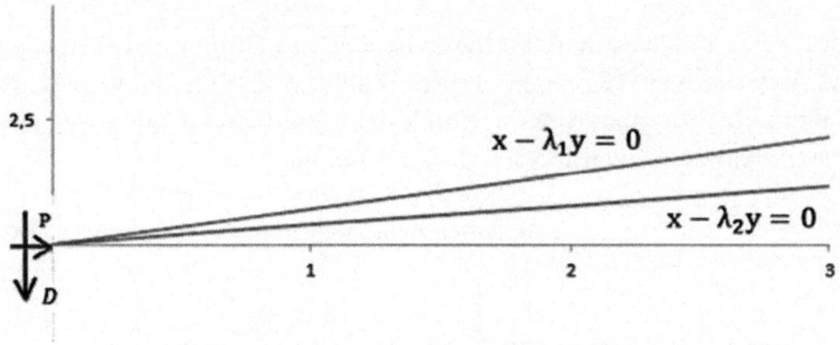

Fig. 2.10 Behavior of the Mach cone with increasing speed of movement of the load

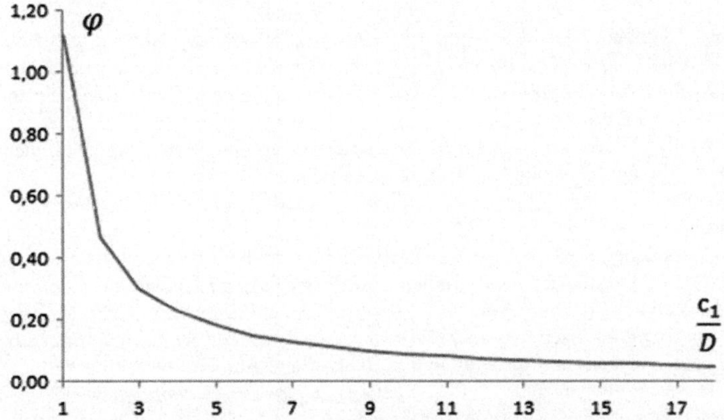

Fig. 2.11 Dependence of the Mach angle on the speed of the normal load

2.5 Conclusion

As a result of the study, it can be concluded that a source moving along the border of a gradient-elastic half-space with a constant velocity, which is greater in magnitude of the velocity of the shear elastic wave, will generate surface elastic waves. Moreover, the transverse component of the displacement vector exceeds the longitudinal component only in the near-surface layer of the half-space, which is a distinctive feature of the propagation of waves from a moving source at high speeds. It is also necessary to note that with an increase in the depth ($2\lambda - 3\lambda$) of the wave propagation, the disturbances attenuate and the wave process stabilizes.

When the source of disturbances moves at supersonic speeds, a Mach cone is formed, which contains the region of propagation of elastic waves. Moreover, the region of concentration of disturbances directly depends on the speed of movement of the source of disturbances.

In conclusion, we note that the excitation of the free edge of two-dimensional and three-dimensional elastic bodies, as well as edge contact of various materials, along with Rayleigh waves, may be the existence of bending waves of the Rayleigh type (Konenkov's waves) (Konenkov 1960; Zakharov 2002; Wilde et al. 2010). For structural elements, the materials of which obey the laws of the gradient theory of elasticity, such waves are waiting for their researchers.

Acknowledgements This work was supported by a grant from the Government of the Russian Federation (contract No. 14.Y26.31.0031).

References

Abali, B. E., Altenbach, H., dell'Isola, F., Eremeyev, V. A., & Ochsner, A. (Eds.). (2019). *New achievements in continuum mechanics and thermodynamics. A tribute to Wolfgang H. Muller. Advanced Structured Materials* (Vol. 108, 564 p.). Cham, Switzerland: Springer Nature Switzerland AG. Part of Springer.

Abali, B. E., Muller, W., & dell'Isola, F. (2017). Theory and computation of higher gradient elasticity theories based on action principles. *Archive of Applied Mechanics, 87*(9).

Aki, K., & Richards, P. (2009). *Quantitative siesmology* (2nd ed., p. 700). Mill Valley, CA: University Science Books.

Alibert, J. J., Seppecher, P., & dell'Isola, F. (2003). Truss modular beams with deformation energy de-pending on higher displacement gradients. *Mathematics and Mechanics of Solids, 8*(1).

Altenbach, H., Belyaev, A., Eremeyev, V. A.., Krivtsov, A., & Porubov, A. V. (Eds). (2019). *Dynamical processes in generalized continua and structures. Advanced Structured Materials* (Vol. 103, 526 p.). Cham, Switzerland: Springer Nature Switzerland AG. Part of Springer.

Altenbach, H., & Eremeyev V. A. (Eds.). (2013). *Generalized continua—from the theory to engineering applications* (388 p.). Wien: Springer.

Altenbach, H., & Forest, S. (Eds.). (2016). *Generalized continua as models for classical and advanced materials. Advanced Structured Materials* (Vol. 42, 458 p.). Switzerland: Springer-Verlag.

Altenbach, H., Forest, S., & Krivtsov, A. (Eds.). (2013). *Generalized continua as models with multi-scale effects or under multi-field actions. Advanced Structured Matherials* (Vol. 22, 332 p.). Berlin-Heidelberg: Springer-Verlag.

Altenbach, H., Maugin, G. A., Erofeev, V. (Eds). (2011). *Mechanics of generalized continua. Advanced Structured Materials* (Vol. 7, 350 p.). Berlin-Heidelberg: Springer-Verlag.

Altenbach, H., Pouget, J., Rousseau, M., Collet, B., & Michelitsch, T. (Eds.). (2018a). *Generalized models and non-classical approaches in complex materials 1. Advanced Structured Materials* (Vol. 90, 760 p). Springer International Publishing AG, part of Springer Nature.

Altenbach, H., Pouget, J., Rousseau, M., Collet, B., & Michelitsch T. (Eds.). (2018b). *Generalized models and non-classical approaches in complex materials 2. Advanced Structured Materials* (Vol. 90, 328 p.). Springer International Publishing AG, part of Springer Nature.

Auffray, N., dell'Isola, F., Eremeyev, V., Madeo, A., & Rossi, G. (2013). Analytical continuum mechanics à la Hamilton-Piola: least action principle for second gradient continua and capillary fluids. *Mathematics and Mechanics of Solids.*

Bagdoev, A. G., Erofeyev, V. I., & Shekoyan, A. V. (2016). Wave dynamics of generalized continua. Advanced Structured Materials (Vol. 24, 274 p.). Berlin-Heidelberg: Springer-Verlag.

Barchiesi, E., Spagnuolo, M., & Placidi, L. (2018). Mechanical metamaterials: a state of the art. *Mathematics and Mechanics of Solids.*

Cosserat, E., et al. (1909). *Theorie des Corp Deformables* (226 p.). Paris: Librairie Scientifique A. Hermann et Fils.

Del Vescovo, D., & Giorgio, I. (2014). Dynamic problems for metamaterials: Review of existing models and ideas for further research. *International Journal of Engineering Science, 80*, 153–172.

dell'Isola, F., Andreaus, U., & Placidi, L. (2015). At the origins and in the vanguard of peridynamics, non-local and higher-gradient continuum mechanics: An underestimated and still topical contribution of Gabrio Piola. *Mathematics and Mechanics of Solids, 20*(8).

dell'Isola, F., Della Corte, A., & Giorgio, I. (2016a). Higher-gradient continua: The legacy of Piola, Mindlin, Sedov and Toupin and some future research perspectives. *Mathematics and Mechanics of Solids*.

dell'Isola, F., Della Corte, A., Greco, L., Luongo, A. (2016b). Plane bias extension test for a continuum with two inextensible families of fibers: a variational treatment with Lagrange multipliers and a perturbation solution. *International Journal of Solids and Structures*.

dell'Isola, F., Giorgio, I., Pawlikowski, M., & Rizzi, N. (2016c). Large deformations of planar extensible beams and pantographic lattices: heuristic homogenization, experimental and numerical examples of equilibrium. *Proceedings of The Royal Society A, 472*(2185).

dell'Isola, F., Eremeyev, V., & Porubov, A. (Eds). (2018). *Advanced in mechanics of microstructured media and structures. Advanced Structured Materials* (Vol. 87, 370 p.). Cham, Switzerland: Springer.

dell'Isola, F., Seppecher, P., Alibert, J. J., Lekszycki, T., Grygoruk, R., Pawlikowski, M., et al. (2019a). Pantographic metamaterials: An example of mathematically driven design and of its technological challenges. *Continuum Mechanics and Thermodynamics, 31*(4), 851–884.

dell'Isola, F., Seppecher, P., & Alibert, J. J. (2019b). Pantographic metamaterials: An example of mathematically driven design and of its technological challenges. *Continuum Mechanics and Thermodynamics, 31*(4), 851–884.

dell'Isola, F., Seppecher, P., & Madeo, A. (2012). How contact interactions may depend on the shape of Cauchy cuts in Nth gradient continua: approach "à la D'Alembert". *Zeitschrift für angewandte Mathematik und Physik, 63*(6).

Erofeyev, V. I. (2003). *Wave processes in solids with microstructure* (p. 256). New Jersey, London, Singapore, Hong Kong: World Scientific.

Jaramillo, T. J. (1929). *A generalization of the energy function of elasticity theory*. Dissertation, Department of Mathematics, University of Chicago.

Klyuev, V. V. (Ed.). (2004). Non-destructive testing: A handbook in 7t. In I. N. Ermolov & Yu. V. Lange (Eds.), *Ultrasonic control* (Vol. 3, 864 p.). Moscow: Mashinostroenie. (in Russian).

Konenkov, Yu. K. (1960). On the flexural wave of the Rayleigh type. *Soviet physics. Acoustics, 6*(1), 122–124.

Le Roux, J. (1911). Etude geometrique de la flexion, dans les deformations infinitesimaleg d'nn milien continu. *Annales Scientifiques de l'École Normale Supérieure, 28*, 523–579.

Le Roux, J. (1913). Recherchesg sur la geometrie beg deformatios finies. *Annales Scientifiques de l'École Normale Supérieure, 30*, 193–245.

Lord Rayleigh. On waves propagated along the plane surface of an elastic solid. Proceedings of the London Mathematical Society. 1885. s1–17(1). 4–11.

Maugin G.A. Non-Classical Continuum Mechanics. Advanced Structured Matherials.Vol. 51. Springer, Singapore, 2017. 260 p.

Maugin, G. A., & Metrikine A. V. (Eds.). (2018). Mechanics of generalized continua: On hundred years after the Cosserats. Advances in Mathematics and Mechanics (Vol. 21, 338 p.). Berlin: Springer.

Neff, P., Ghiba, I.-D., Madeo, A., Placidi L., & Rosi G. (2014). A unifying perspective: the relaxed linear micromorphic continuum. *Continuum Mechanics and Thermodynamics, 26*(5).

Placidi, L., Andreaus, U., & Giorgio, I. (2017). Identification of two-dimensional pantographic structure via a linear D4 orthotropic second gradient elastic model. *Journal of Engineering Mathematics*.

Placidi, L., Barchiesi, E., Turco, E., Rizzi, N. L. (2016). A review on 2D models for the description of pantographic fabrics. *Zeitschrift für angewandte Mathematik und Physik, 67*(5).

Rahali, Y., Giorgio, I., Ganghoffer, J.-F., & dell'Isola, F. (2015). Homogenization à la Piola produces second gradient continuum models for linear pantographic lattices. *International Journal of Engineering Science, 97*.

Sabodash, P. F., & Filippov I. G. (1971). *On the effect of a moving load on an elastic half-space taking into account moment stresses. Durability and Plasticity* (pp. 317–321). Moscow: Nauka. (in Russian).

Sciarra, G., dell'Isola, F., & Coussy, O. (2007). Second gradient poromechanics. *International Journal of Solids and Structures, 44*(20).

Uglov, A. L., Erofeev, V. I., & Smirnov, A. N. (2009). *Acoustic control of equipment in the manufacture and operation* (280 p.). Moscow: Nauka. (in Russian).

Wilde, M. V., Kaplunov, Yu. D., Kossovich, L. Yu. (2010). *Edge and interface resonance phenomena in elastic bodies* (280 p.). Moscow: Fizmatlit. (in Russian).

Zakharov, D. D. (2002). Konenkov's waves in anisotropic layered plates. *Acoustical Physics, 48*(2), 171–175.

Chapter 3
On the Spectrum of Relaxation Times in Coupled Diffusion and Rheological Processes in Metal Alloys

Dmitry Dudin and Ilya Keller

Abstract The paper is concerned with studying the relaxation times of coupled diffusion and rheological processes in metals within the framework of a linear model, which takes into account elastic and viscous bulk and shear modulus. Here, we consider the spatial perturbations of a homogeneous stationary solution of field equations for a one-dimensional model problem, in which only the diffusion and rheological relations remain nontrivial. The study made allowed us to obtain the asymptotic expressions for the coefficients of interdiffusion at vanishingly small and infinitely large characteristic lengths of perturbations, as well as the dependences on the ratios of the elastic to viscous moduli. Consideration for the material elasticity leads to the appearance of an additional relaxation time, which is responsible for the diffusion mechanism. In this case, conventional thermal diffusion can be accompanied by fast diffusion or it can exist against the background of slow diffusion, depending on the ratio of the characteristic values of the elastic and thermal energies. A viscous shear flow in the region of small wavelengths of spatial perturbations plays a key role in the mechanism of interdiffusion in a binary alloy, in which vacancy diffusion is not taken into account. The relation between the diffusion and rheological processes is maintained by stresses. The method can be used for qualitative analysis of much more complex, coupled diffusion-rheological systems.

Keywords Metal alloys · Interdiffusion · Rheological processes · Stresses · Coupled processes · Asymptotics · Perturbation analysis

---

D. Dudin
Perm National Research Polytechnic University, Perm, Russia
e-mail: dmitryovj@yandex.ru

I. Keller (✉)
Institute of Continuous Media Mechanics of the UB RAS, Perm, Russia
e-mail: kie@icmm.ru

F. dell'Isola and L. Igumnov (eds.), *Dynamics, Strength of Materials and Durability in Multiscale Mechanics*, Advanced Structured Materials 137,
https://doi.org/10.1007/978-3-030-53755-5_3

3.1 Introduction

The study of coupled processes in multicomponent metal alloys including diffusive mass transfer, viscoelastic deformation, chemical reactions, and the evolution of microstructures is a necessary condition for successful solution of various problems in material mechanics. For example, such studies will allow researchers to optimize the technology of surface treatment of machine parts to increase their durability either by exposing them to the action of high-energy particle fluxes or subjecting them to mechanical, ultrasonic, or laser peening followed by a diffusion-chemical stage of formation of a gradient layer. The problems of predicting the resistance of loaded machine parts and structures to chemical corrosion also require coupled formulations.

Elements constituting an alloy might form a microstructure. And this is itself a very interesting subjects in modern continuum mechanics and engineering: the development of newly (scientifically) conceived materials ("metamaterials") with mechanical properties that cannot be found in nature (Barchiesi et al. 2018; Del Vescovo and Giorgio 2014). These (macroscopic) properties are mainly determined by the microstructure or nanostructure of the considered metamaterial rather than by the chemical and physical properties of the materials constituting it at the microscopic level. The reader is explicitly warned here: with the prefixes "macro" or "micro" and we do not necessarily refer to a specific absolute characteristic length. We, instead, refer generically to the length-scale where phenomena are observed and need to be controlled (macro) and to one or more than one length-scales used to build the architecture of the metamaterial (micro): in particular, the prefix "micro" does not refer to micrometers or to any other specific length.

An example of mechanical metamaterials is pantographic structures (dell'Isola et al. 2016a, b, c, 2017, 2019a, 2019b; Placidi et al. 2016, 2017).

In order to account for multiscale mechanical interactions, which taking place for metamaterials, higher order gradient continuum theories can be a choice (Alibert et al. 2003; Auffray et al. 2013; dell'Isola et al. 2012, 2015, 2016a, b, c; Rahali et al. 2015; Sciarra et al. 2007).

Rather complicated processes of transformations take place in solid and powder metal materials under intense plastic deformations (Straumal et al. 2004). In the literature, there are few studies on this subject (Knyazeva 2003, 2004, 2005), which, however, do not close the gaps in the problem of interpreting the qualitative behavior of coupled processes under the action of various external factors, which, in addition, intends to settle a number of methodological issues.

To do this, the study of wave processes can be effectively used. Stephenson (1988) and later (Brassart et al. 2018) used, as a framework for a qualitative investigation of rather slow (non-dynamic) coupled diffusion and rheological processes, a one-dimensional model problem, in which the relaxation of small spatial perturbations of a uniform stationary solution was studied. Within this problem, the perturbed behavior of the fields is described only by the diffusion and rheological relations, which makes it possible to obtain the relationships between the relaxation times of the perturbations and their wavelength in a closed form and investigate in great depth the relaxation mechanisms in the asymtotic cases. The method is equally applicable for both the linear and nonlinear models allowing linearization in the vicinity of a homogeneous stationary solution.

The focus of this study is the phenomenological model, proposed by Brassart et al (2018), describing the interdiffusion of the components of a binary metal alloy accompanied by the bulk and shear viscous flows, supplemented in this study by the bulk and shear elastic strains. Here, consideration is given to a geometrically and physically linear version of the formulation, which is justified by the purpose of the successive studies into relaxation of small perturbations of a uniform stationary state with zero stresses. The subject of research is the qualitative effects, which introduced by elastic stresses into the system of coupled processes of interdiffusion and viscous flow. The processes are considered isothermal and occur in the absence of bulk sources of the component substance and the kinetic moment.

3.2 The Brassart's Model Supplemented with Elastic Strains

Consider a binary metal alloy characterized by concentrations C_A, C_B of the atoms of type A and B per a unit volume of the material in the reference frame.

3.2.1 Deformation and Volumetric Expansion

Here, it is assumed that the processes of perturbation and relaxation of the equilibrium state of the medium are accompanied by small deformations of the material elements. Therefore, we consider the tensors of small strains $\boldsymbol{\varepsilon}$, consisting of the elastic $\boldsymbol{\varepsilon}^e$ and viscous $\boldsymbol{\varepsilon}^v$ parts

$$\boldsymbol{\varepsilon} = \boldsymbol{\varepsilon}^e + \boldsymbol{\varepsilon}^v. \tag{3.1}$$

Each of these tensors is represented by the sum of the spherical and deviatoric components

$$\boldsymbol{\varepsilon} = \tfrac{1}{3}\varepsilon_m \boldsymbol{I} + \boldsymbol{e}, \quad \boldsymbol{\varepsilon}^e = \tfrac{1}{3}\varepsilon_m^e \boldsymbol{I} + \boldsymbol{e}^e, \quad \boldsymbol{\varepsilon}^v = \tfrac{1}{3}\varepsilon_m^v \boldsymbol{I} + \boldsymbol{e}^v. \tag{3.2}$$

Let the expression

$$1+\varepsilon_m = \Omega\big(C_A, C_B, \varepsilon_m^e\big) \tag{3.3}$$

defines the ratio of volumetric expansion, which depends on the concentrations of the atoms of both types and the bulk elastic strain. It is assumed that this quantity is a homogeneous first-order function of concentrations

$$\Omega = V_A(\xi)C_A + V_B(\xi)C_B + \varepsilon_m^e, \tag{3.4}$$

where V_A, V_B are the partial volumes, $\xi = C_B/(C_A + C_B)$ is the variable of the alloy composition. Expressions (3.3), (3.4) interpret the volumetric expansion of the material caused by a change in the concentration of the alloy components and elastic strains.

Differentiating (3.3) and (3.4) with respect to time yields the following equalities

$$\dot{\Omega} = \frac{\partial \Omega}{\partial C_A}\dot{C}_A + \frac{\partial \Omega}{\partial C_B}\dot{C}_B + \dot{\varepsilon}_m^e, \tag{3.5}$$

$$\dot{\Omega} = \dot{V}_A C_A + \dot{V}_B C_B + V_A \dot{C}_A + V_B \dot{C}_B + \dot{\varepsilon}_m^e,$$

which, when fulfilled simultaneously, lead to the relations

$$V_A = \frac{\partial \Omega}{\partial C_A}, \quad V_B = \frac{\partial \Omega}{\partial C_B}, \tag{3.6}$$

$$\dot{V}_A C_A + \dot{V}_B C_B = 0,$$

consistent with the relations obtained in (Brassart et al. 2018) without considering elastic strains.

3.2.2 *Free Energy*

The Helmholtz free energy is assumed to be the following function

$$\psi = \psi\big(C_A, C_B, \boldsymbol{\varepsilon}^e\big), \tag{3.7}$$

which is a homogeneous function of the first order in concentrations

$$\psi = F_A(\xi)C_A + F_B(\xi)C_B + F_e(\boldsymbol{\varepsilon}^e), \tag{3.8}$$

where F_A, F_B are the specific mixing energies of the A-type and B-type atoms and F_e is the energy of elastic strains.

Differentiating (3.7) and (3.8) with respect to time yields the following relations

$$\dot{\psi} = \frac{\partial \psi}{\partial C_A}\dot{C}_A + \frac{\partial \psi}{\partial C_B}\dot{C}_B + \frac{\partial \psi}{\partial \boldsymbol{\varepsilon}^e} : \dot{\boldsymbol{\varepsilon}}^e, \tag{3.9}$$

$$\dot{\psi} = \dot{F}_A C_A + \dot{F}_B C_B + F_A \dot{C}_A + F_B \dot{C}_B + \frac{\partial F_e}{\partial \boldsymbol{\varepsilon}^e} : \dot{\boldsymbol{\varepsilon}}^e,$$

which, when fulfilled simultaneously, lead to the relations

$$F_A = \frac{\partial \psi}{\partial C_A}, \quad F_B = \frac{\partial \psi}{\partial C_B}, \tag{3.10}$$

$$\dot{F}_A C_A + \dot{F}_B C_B = 0,$$

consistent with relations obtained in (Brassart et al. 2018) without considering elasticity.

3.2.3 Thermodynamic Inequality

The second law of thermodynamics requires that the Helmholtz free energy of isolated system should never increase

$$\int_V \dot{\psi} \mathrm{d}V + \int_S (\mu_A \boldsymbol{J}_A + \mu_B \boldsymbol{J}_B) \cdot \boldsymbol{n} \mathrm{d}S - \int_V \boldsymbol{\sigma} : \dot{\boldsymbol{\varepsilon}} \Omega \mathrm{d}V \leq 0, \tag{3.11}$$

where μ_A, μ_B are the chemical potentials of the A-type and B-type atoms, $\boldsymbol{J}_A$, $\boldsymbol{J}_B$ are the diffusion fluxes of the A and B matters, $\boldsymbol{\sigma}$ is the Cauchy stress tensor, S and V are the surface and volume of the body in the reference configuration.

The mass balance equation for the material components is written as

$$\frac{\mathrm{d}C_A}{\mathrm{d}t} = -\nabla \cdot \boldsymbol{J}_A, \; \frac{\mathrm{d}C_B}{\mathrm{d}t} = -\nabla \cdot \boldsymbol{J}_B. \tag{3.12}$$

The application of the divergence theorem (3.11) gives in view of (3.12) the following expression

$$\int_V (\Omega \boldsymbol{\sigma} : \dot{\boldsymbol{\varepsilon}} + \mu_A \dot{C}_A + \mu_B \dot{C}_B - \boldsymbol{J}_A \cdot \nabla \mu_A - \boldsymbol{J}_B \cdot \nabla \mu_B - \dot{\psi}) \mathrm{d}V \geq 0.$$

Then its application to any arbitrary local volume of the material taking into account (3.9) and (3.10) leads to

$$\Omega \sigma : \dot{\boldsymbol{\varepsilon}} + (\mu_A - F_A)\dot{C}_A + (\mu_B - F_B)\dot{C}_B - \boldsymbol{J}_A \cdot \nabla \mu_A - \boldsymbol{J}_B \cdot \nabla \mu_B - \frac{\partial F_e}{\partial \boldsymbol{\varepsilon}^e} : \boldsymbol{\varepsilon}^e \geq 0. \tag{3.13}$$

Considering $\boldsymbol{\sigma} : \dot{\boldsymbol{\varepsilon}} = \sigma_m \dot{\varepsilon}_m + \boldsymbol{s} : \dot{\boldsymbol{e}}$, where the Cauchy stress tensor is decomposed into the spherical and deviatoric parts $\boldsymbol{\sigma} = \sigma_m \boldsymbol{I} + \boldsymbol{s}$, and using relations (3.5), (3.6), we can rewrite the thermodynamic inequality (3.13) as

$$\left(\frac{\mu_A - F_A}{V_A} + \sigma_m\right) V_A \dot{C}_A + \left(\frac{\mu_B - F_B}{V_B} + \sigma_m\right) V_B \dot{C}_B - \boldsymbol{J}_A \cdot \nabla \mu_A - \boldsymbol{J}_B \cdot \nabla \mu_B$$
$$+ \Omega \boldsymbol{s} : \dot{\boldsymbol{e}}^v + \Omega \boldsymbol{s} : \dot{\boldsymbol{e}}^e + \Omega \sigma_m \dot{\varepsilon}_m^e - \frac{\partial F_e}{\partial \boldsymbol{\varepsilon}^e} : \dot{\boldsymbol{\varepsilon}}^e \geq 0. \tag{3.14}$$

Then, following (Brassart et al. 2018), we introduce in terms of the current configuration the volume concentrations of atoms $c_A, c_B, C_A = c_A \Omega, C_B = c_B \Omega$, the fluxes and gradients $\boldsymbol{J}_A \cdot \nabla \mu_A = \Omega \boldsymbol{j}_A \cdot \widehat{\nabla} \mu_A$, $\boldsymbol{J}_B \cdot \nabla \mu_B = \Omega \boldsymbol{j}_B \cdot \widehat{\nabla} \mu_B$, and denote the rates of volumetric insertion as $i_A = V_A \dot{C}_A / \Omega$, $i_B = V_B \dot{C}_B / \Omega$. As the result, relation (3.14) can be finally written as

$$\left(\frac{\mu_A - F_A}{V_A} + \sigma_m\right) i_A + \left(\frac{\mu_B - F_B}{V_B} + \sigma_m\right) i_B + \boldsymbol{s} : \dot{\boldsymbol{e}}^v - \boldsymbol{j}_A \cdot \widehat{\nabla} \mu_A - \boldsymbol{j}_B \cdot \widehat{\nabla} \mu_B$$
$$+ \boldsymbol{s} : \dot{\boldsymbol{e}}^e + \sigma_m \dot{\varepsilon}_m^e - \frac{1}{\Omega} \frac{\partial F_e}{\partial \boldsymbol{\varepsilon}^e} : \dot{\boldsymbol{\varepsilon}}^e \geq 0. \tag{3.15}$$

3.2.4 Elastic Relations and Functions of State

In the absence of irreversible processes, (3.15) degenerates into the equality

$$\boldsymbol{s} : \dot{\boldsymbol{e}}^e + \sigma_m \dot{\varepsilon}_m^e - \frac{1}{\Omega} \frac{\partial F_e}{\partial \boldsymbol{\varepsilon}^e} : \dot{\boldsymbol{\varepsilon}}^e = 0. \tag{3.16}$$

Taking into account the smallness of strains, the expression for the energy of elastic strains can be written as

$$F_e = G \boldsymbol{e}^e : \boldsymbol{e}^e + \frac{1}{2} K \varepsilon_m^{e2}, \tag{3.17}$$

where G is the shear modulus, K is the bulk modulus of elasticity.

Substituting (3.17) into Eq. (3.16) leads to the relation

$$\left(\boldsymbol{s} - 2G\boldsymbol{e}^e\right) : \dot{\boldsymbol{e}}^e + \left(\sigma_m - K\varepsilon_m^e\right)\dot{\varepsilon}_m^e = 0. \tag{3.18}$$

The independence of the bulk and shear elastic strain rates suggests the validity of the generalized Hooke law

$$\boldsymbol{s} = 2G\boldsymbol{e}^e, \quad \sigma_m = K\varepsilon_m^e. \tag{3.19}$$

The dependence of the elastic shear moduli G and bulk compression moduli K on the component concentrations is neglected here.

The expressions for the specific energies of component mixing in the framework of the ideal mixing hypothesis take the following form (Brassart et al. 2016):

$$F_A = kT\left(\log\frac{1-\xi}{V_m} + \frac{\xi(V_B - V_A)}{V_m}\right),\quad F_B = kT\left(\log\frac{\xi}{V_m} + \frac{(1-\xi)(V_A - V_B)}{V_m}\right), \tag{3.20}$$

where $V_m = (1-\xi)V_A + \xi V_B$ is the average partial volume, $\xi = c_B/(c_A + c_B)$ is the composition variable,k is the Boltzmann constant, T is the absolute temperature.

3.2.5 Kinetic Equations

The kinetics of the shear flow of a crystal is described in terms of the rheology of a linear-viscous Newtonian fluid

$$\boldsymbol{s} = 2\eta\dot{\boldsymbol{e}}^v, \tag{3.21}$$

where η is the coefficient of shear viscosity. In the absence of other processes, the elastic (3.19) and viscous (3.21) elements constitute the rheological Maxwell model.

The diffusion kinetics without considering the cross terms is defined as

$$\boldsymbol{j}_A = -c_A M_A \widehat{\nabla}\mu_A,\quad \boldsymbol{j}_B = -c_B M_B \widehat{\nabla}\mu_B, \tag{3.22}$$

where M_A, M_B are the mobility factors.

The relation between the rheological and diffusion kinetics is governed by the laws

$$\frac{\mu_A - F_A}{V_A} + \sigma_m = \beta_A i_A,\quad \frac{\mu_B - F_B}{V_B} + \sigma_m = \beta_B i_B, \tag{3.23}$$

introduced in (Brassart et al. 2018), which also ignore the cross-terms. Here, β_A, β_B are the bulk viscosities.

Relations (3.21)–(3.23) are some solutions of thermodynamic inequality (3.15).

3.2.6 Balance Equations

Relations (3.19)–(3.23) are closed by the matter balance equations and the equilibrium equation

$$\frac{dc_A}{dt} + c_A \widehat{\nabla} \cdot \boldsymbol{v} = -\widehat{\nabla} \cdot \boldsymbol{j}_A, \quad \frac{dc_B}{dt} + c_B \widehat{\nabla} \cdot \boldsymbol{v} = -\widehat{\nabla} \cdot \boldsymbol{j}_B, \tag{3.24}$$

$$\widehat{\nabla} \cdot \boldsymbol{\sigma} = 0. \tag{3.25}$$

The system of Eqs. (3.19)–(3.25) is a nonlinear formulation of the coupled problem of diffusion and rheology, which takes into account the elastic properties.

3.3 Analysis of Relaxation of Spatial Perturbations

3.3.1 Model Problem

To carry out a qualitative analysis of the relaxation of perturbations described by the equations given above, we consider a model problem (Stephenson 1988; Brassart et al. 2018), in which the following hypotheses are accepted:

1. The atoms can diffuse only along the x-coordinate

$$c_A = c_A(x, t), \quad c_B = c_B(x, t). \tag{3.26}$$

2. The components of the total strain tensor are zero except for

$$\varepsilon_{xx} = \varepsilon(x, t) \neq 0. \tag{3.27}$$

3. The components of the stress tensor are zero except for

$$\sigma_{yy} = \sigma_{zz} = \sigma(x, t) \neq 0. \tag{3.28}$$

Hypotheses (3.26)–(3.28) as applied to the system (3.19)–(3.25) allow us to proceed to a one-dimensional problem of diffusion with accompanying rheological processes. Equation (3.25) is satisfied identically by virtue of (3.28). The non-zero components of the elastic strain deviator $e^e_{xx} = \frac{2}{3}(\varepsilon^e - \varepsilon_\perp)$, $e^e_{yy} = e^e_{zz} = -\frac{1}{3}(\varepsilon^e - \varepsilon_\perp)$ and viscous strain deviator $e^v_{xx} = \frac{2}{3}(\varepsilon^v + \varepsilon_\perp)$, $e^v_{yy} = e^v_{zz} = -\frac{1}{3}(\varepsilon^v + \varepsilon_\perp)$ can be expressed in terms of the decomposition component $\varepsilon = \varepsilon^e + \varepsilon^v$ in (3.27) and the component $\varepsilon_\perp = \varepsilon^e_{yy} = \varepsilon^e_{zz} = -\varepsilon^v_{yy} = -\varepsilon^v_{zz}$. Similarly, the non-zero components of the stress deviator are $s_{xx} = -\sigma_m$, $s_{yy} = s_{zz} = \frac{1}{2}\sigma_m$. Then, from relations (3.19) and (3.21), we can successively obtain $\varepsilon^e = (1/(3K) - 1/(3G))\sigma_m$, $\varepsilon_\perp = (1/(3K) + 1/(3G))\sigma_m$, $\dot{\varepsilon}^v = -(1/(3K) + 1/(3G))\dot{\sigma}_m - 3/(4\eta)\sigma_m$, $\dot{\varepsilon}^v_m = -(1/K + 3/(4G))\dot{\sigma}_m - 3/(4\eta)\sigma_m$,

$$\dot{\varepsilon} = -\frac{3\dot{\sigma}_m}{4G} - \frac{3}{4\eta}\sigma_m. \tag{3.29}$$

Then, the components of the elastic strain energy can be written as $G\boldsymbol{e}^e : \boldsymbol{e}^e = 3/(8G)\sigma_m^2$, $K\varepsilon_m^{e2} = 1/(2K)\sigma_m^2$, and the dissipation rate of the shear viscous flow is $\eta\dot{\boldsymbol{e}}^v : \dot{\boldsymbol{e}}^v = 3/(8\eta)\sigma_m^2$.

3.3.2 Field Equations

Equation (3.29) is written in terms of the velocity of the material point displacement $v(x, t)$ along the spatial coordinate x

$$\frac{\partial v}{\partial x} = -\frac{3}{4G}\frac{\partial \sigma_m}{\partial t} - \frac{3}{4\eta}\sigma_m, \tag{3.30}$$

where (and in the following) we set $\mathrm{d}/\mathrm{d}t \approx \partial/\partial t$ on the assumption of small v. The remaining field equations take the following form:

$$\frac{\partial v}{\partial x} = -V_A\frac{\partial j_A}{\partial x} - V_B\frac{\partial j_B}{\partial x}, \tag{3.31}$$

$$\frac{\partial c_B}{\partial t} + c_B\frac{\partial v}{\partial x} = -\frac{\partial j_B}{\partial x}, \tag{3.32}$$

$$j_A = -c_A M_A\frac{\partial \mu_A}{\partial x}, \quad j_B = -c_B M_B\frac{\partial \mu_B}{\partial x},$$

$$i_A = V_A\frac{\partial c_A}{\partial t} + c_A V_A\frac{\partial v}{\partial x}, \quad i_B = V_B\frac{\partial c_B}{\partial t} + c_B V_B\frac{\partial v}{\partial x}. \tag{3.33}$$

They are closed by the constitutive Eq. (3.23) with regard to (3.20).

3.3.3 Perturbed System and Its Analysis

The system of coupled Eqs. (3.20), (3.23), (3.30)–(3.33) has a homogeneous stationary solution

$$\sigma_m(x, t) \equiv \sigma_0, \quad c_A(x, t) \equiv c_{A0}, \quad c_B(x, t) \equiv c_{B0}, \tag{3.34}$$

corresponding to some equilibrium state.

The spectrum of the relaxation times of the system is determined using the perturbation method. To this end, this system is linearized in the vicinity of (3.34)

$$\frac{\partial \phi}{\partial t} - kT\,\Xi(\xi_0 M_A + (1-\xi_0)M_B)\frac{\partial^2 \phi}{\partial x^2} - \phi_0(1-\phi_0)(M_A V_A \beta_\mathrm{A} + M_B V_B \beta_\mathrm{B})\frac{\partial^3 \phi}{\partial t\,\partial x^2}$$

$$= \phi_0(1-\phi_0)\left(M_A V_A\left(1+\frac{3\beta_A(1-\phi_0)}{4\eta}\right) - M_B V_B\left(1+\frac{3\beta_B\phi_0}{4\eta}\right)\right)\frac{\partial^2\sigma_m}{\partial x^2}$$
$$+ \phi_0(1-\phi_0)(M_A V_A \beta_A(1-\phi_0) - M_B V_B \beta_B \phi_0)\frac{3}{4G}\frac{\partial^3\sigma_m}{\partial t \partial x^2}, \tag{3.35}$$

$$\frac{3}{4G}\frac{\partial\sigma_m}{\partial t} - (M_A V_A \beta_A (1-\phi_0)^2 + M_B V_B \beta_B \phi_0^2)\left(\frac{3}{4G}\frac{\partial^3\sigma_m}{\partial t\partial x^2} + \frac{3}{4\eta}\frac{\partial^2\sigma_m}{\partial x^2}\right) -$$
$$- (M_A V_A(1-\phi_0) + M_B V_B \phi_0)\frac{\partial^2\sigma_m}{\partial x^2} + \frac{3\sigma_m}{4\eta} = kT\Xi V_m\left(\frac{M_A}{V_B} - \frac{M_B}{V_A}\right)\frac{\partial^2\phi}{\partial x^2}$$
$$+ (M_A V_A \beta_A(1-\phi_0) - M_B V_B \beta_B \phi_0)\frac{\partial^3\phi}{\partial t\partial x^2}, \tag{3.36}$$

where $\phi = c_B V_B = 1 - c_A V_A$ is the volume fraction of the B component of the binary alloy, $\Xi = [V_A(1-\phi_0) + V_B\phi_0]/V_m$, $\phi_0 = c_{B0}V_B = 1 - c_{A0}V_A$, $\xi_0 = c_{B0}/(c_{A0} + c_{B0})$ are the volume fraction of the B component of the binary alloy and the value of the composition variable in the equilibrium state, $\sigma_0 = 0$ is the mean stress value in the equilibrium state.

The perturbations imposed on the equilibrium distribution of the volume fraction ϕ_0 and mean stress σ_0, corresponding to (3.33) are given as

$$\phi(x,t) = \phi_0 + \mathrm{Re}\left(\hat{\phi}\exp\left(-\frac{t}{\tau}\right)\exp\left(i\frac{2\pi x}{\lambda}\right)\right), \tag{3.37}$$

$$\sigma_m(x,t) = \sigma_0 + \mathrm{Re}\left(\hat{\sigma}\exp\left(-\frac{t}{\tau}\right)\exp\left(i\frac{2\pi x}{\lambda}\right)\right). \tag{3.38}$$

They have characteristic dimension λ, characteristic relaxation time τ, and satisfy the coupled system of linear parabolic Eqs. (3.35), (3.36). In (3.37), (3.38), $\hat{\phi}, \hat{\sigma} \in \mathbb{C}$, $|\hat{\phi}|, |\hat{\sigma}| \ll 1$.

Substituting relations (3.37), (3.38) into the system of differential Eqs. (3.35), (3.36) leads to an eigenvalue problem, which imposes the constraint $\tau(\lambda)$, at which $\hat{\phi}$ and $\hat{\sigma}$ cannot be simultaneously equal to zero. In this case, this dependence has two branches $\tau_\pm(\lambda)$, each of which corresponds to a certain characteristic straight line in the plane (Re $\hat{\phi}$, Re $\hat{\sigma}$). If its orientation does not coincide with the direction of any of the coordinate axes, the relaxation process corresponding to the branch of characteristic times under consideration turns out to be coupled. In this case, the diffusion is partially or completely controlled by stresses, which reflect the course of the rheological processes in the system. This method makes it possible to study the physics of relaxation processes in a coupled system taking into account various factors.

3.3.4 The Relaxation Time of Perturbations and Their Asymptotics

The functions $\tau_+(\lambda)$ and $\tau_-(\lambda)$, depending on the dimensionless factors M_B/M_A, V_B/V_A, $kT/(GV_m)$, β/η (it is assumed that $\beta_A = \beta_B = \beta$), have been obtained in a closed form and are not given here because of the cumbersome form of their representation. The resulting expressions are analyzed using the computer algebra tools. Figures 3.1 and 3.2 show the plots of $\tau_-(\lambda)$ and $\tau_+(\lambda)$ as a function of dimensionless bulk viscosity β/η.

Both dependences $\tau_-(\lambda)$, $\tau_+(\lambda)$ have asymptotics at $\lambda \to 0$ and $\lambda \to \infty$ (Figs. 3.1 and 3.2).

For $\tau_-(\lambda)$ at $\lambda \to \infty$, the stress gradient does not affect the process of interdiffusion, which occurs by the Darken mechanism (Darken 1948; Mehrer 2007; Paul et al. 2014) with the coefficient

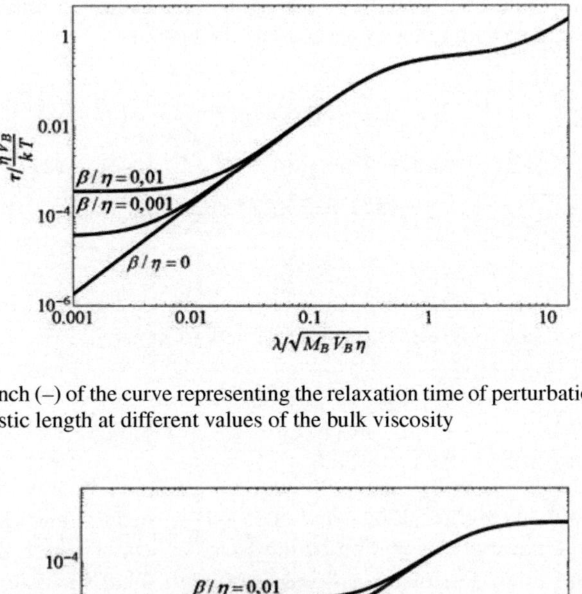

Fig. 3.1 First branch (–) of the curve representing the relaxation time of perturbations as a function of their characteristic length at different values of the bulk viscosity

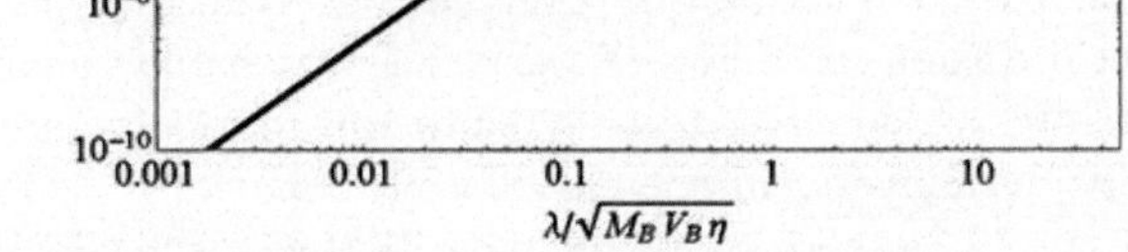

Fig. 3.2 Second branch (+) of the curve representing the relaxation time of perturbations as a function of their characteristic length at different values of the bulk viscosity

$$D_\infty^- = kT\,\Xi(M_A\xi_0 + M_B(1-\xi_0)), \tag{3.39}$$

corresponding to a series connection of the Fick bodies, at which the process is limited by the fast component. This result was obtained earlier in (Stephenson 1988; Brassart et al. 2018).

For $\tau_+(\lambda)$ at $\lambda \to \infty$, the perturbation relaxation obeys the equation

$$\frac{\partial \sigma_m}{\partial t} = -\frac{G}{\eta}\frac{\partial^2 \sigma_m}{\partial x^2}, \tag{3.40}$$

describing the nondiffusive viscous shear mechanism, which is insensitive to perturbations of the component concentrations.

At $\lambda \to 0$, both relaxation times are controlled by the ratio of the bulk to the shear viscosities β/η provided it is finite. If this ratio is vanishingly small and the characteristic insertion energy GV_m is commensurable with the thermal energy kT, the coefficients of interdiffusion for both branches $\tau(\lambda)$ show a complex dependence on the diffusion and rheological properties of the system

$$\begin{aligned} D_0^\pm = \frac{1}{6}&(4G(M_AV_A(1-\phi_0) + M_BV_B\phi_0) + 3kT\,\Xi(M_A\xi_0 + M_B(1-\xi_0)) \\ &\pm\big((4G(M_AV_A(1-\phi_0) + M_BV_B\phi_0) + 3kT\,\Xi(M_A\xi_0 + M_B(1-\xi_0)))^2 \\ &\qquad - 48kT\,\Xi G M_A M_B V_m)^{1/2}\big). \end{aligned} \tag{3.41}$$

At $\beta/\eta = 0$ and $kT/GV_m \ll 1$, the asymptotics of the first relaxation time at $\lambda \to 0$ corresponds to interdiffusion with a coefficient

$$D_0^- = kT\,\Xi \frac{M_A M_B V_m}{M_AV_A(1-\phi_0) + M_BV_B\phi_0}, \tag{3.42}$$

which is consistent with the coefficient of diffusion by the Nazarov–Gurov vacancy mechanism (Mehrer 2007; Paul et al. 2014; Nazarov and Gurov 1974) and corresponds to a parallel connection of the Fick bodies, at which the process rate is limited by the slowest diffusion process. Although shear viscosity does not enter into expression (3.42), the fluxes of components are moderated by the average stress gradients generated by a shear viscous flow. This conclusion was previously made in (Stephenson 1988; Brassart et al. 2018), so elasticity has no effect on this mechanism. In this case, instead of vacancy diffusion, the fast relaxation mechanism can be performed by the diffusion creep flow of vacancies, which plays an important role at elevated temperatures and in materials with microcrystalline structure. In the absence of shear viscosity, the above-mentioned asymptotic behavior is characterized by the diffusion coefficient (3.39) corresponding to the Darken mechanism.

At $\beta/\eta = 0$ и $kT/GV_m \ll 1$, the asymptotics of the second relaxation time at $\lambda \to 0$ corresponds to the mechanism of interdiffusion with a coefficient

$$D_0^+ = \frac{4G}{3}(M_A V_A(1-\phi_0) + M_B V_B \phi_0), \tag{3.43}$$

which in the framework of the asymptotics under consideration significantly exceeds the coefficient of conventional thermal diffusion: $D_0^+ \gg D_0^-$. In this case, (3.43) corresponds to a sequental connection of diffusing bodies. If there is a considerable difference in mobility between the components of a binary system, for example, $M_A \gg M_B$, interdiffusion is determined by diffusion of the fast component (*A*). As for the first branch, the diffusion mechanism due to the mean stress gradients is an important part of the relaxation process.

Similar asymptotics of the branches of the function $\tau(\lambda)$ at $\beta/\eta = 0$ and $\lambda \to 0$ for $GV_m/kT \ll 1$ can be interpreted as follows. The second branch corresponds to thermal diffusion with the interdiffusion coefficient

$$D_0^+ = kT\,\Xi(M_A\xi_0 + M_B(1-\xi_0)) \tag{3.44}$$

corresponding to the Darken mechanism. In this case, the first branch should be associated with the mechanism of slow diffusion with the interdiffusion coefficient

$$D_0^- = \frac{4G}{3}\frac{M_A M_B V_A V_B}{M_A V_A \phi_0 + M_B V_B(1-\phi_0)}, \tag{3.45}$$

with $D_0^- \ll D_0^+$ denoting a parallel connection of diffusing bodies. If there is an essential difference in the mobility between the components of a binary system, for example, $M_A \gg M_B$, interdiffusion is determined by diffusion of the slow component (*B*). As before, the barodiffusion process is an important part of the relaxation mechanism.

3.4 Conclusion

In this study, we have demonstrated the efficiency of the method for qualitative study of rather slow (non-dynamic)-coupled diffusion and rheological processes. The approach is based on the analysis of the relaxation times of a spatially perturbed homogeneous stationary solution of the field equations for a one-dimensional model problem, in which only the diffusion and rheological relations remain nontrivial. Mathematically, the method is reduced to an eigenvalue problem and can be easily studied using computer algebra systems. It has been used by Stephenson (1988) and Brassart et al. (2018) and proved to be an effective tool for qualitative study of the complex, coupled diffusion-rheological processes, accompanied by chemical reactions and changes in the microstructure of a deformable metal alloy. The asymptotics of the relaxation times can be associated with the structural schemes integrating the diffusion and rheological elements. Here, consideration is given to a linear model, although the method is suitable for studying geometrically and physically nonlinear equations with the parameters independent of spatial coordinates and time.

The nontrivial result of this study is the conclusion that the elastic strains depending on the ratio of the characteristic elastic and thermal energies can control the fast or slow diffusion in a binary metal alloy accompanying the ordinary interdiffusion by the thermal-fluctuation mechanism. The observed fast diffusion is controlled by a mean stress gradient and is not associated with the microstructure of metal crystal lattice defects (which is undoubtedly the most important factor responsible for the activation of the diffusion processes).

Acknowledgements This work was performed in the framework of the state assignment (reg. No. AAAA-A16-116121410009-8) and under partial financial support by the grant of the Russian Foundation for Basic Research (project No. 17-08-01085).

References

Alibert, J. J., Seppecher, P., & dell'Isola, F. (2003). Truss modular beams with deformation energy depending on higher displacement gradients. *Mathematics and Mechanics of Solids, 8*(1).

Auffray, N., dell'Isola, F., Eremeyev, V., Madeo, A., & Rossi, G. (2013). Analytical continuum mechanics à la Hamilton-Piola: Least action principle for second gradient continua and capillary fluids. *Mathematics and Mechanics of Solids.*

Barchiesi, E., Spagnuolo, M., & Placidi, L. (2018). Mechanical metamaterials: A state of the art. *Mathematics and Mechanics of Solids.*

Brassart, L., Liu, Q., & Suo, Z. (2016). Shear, dilation and swap: Mixing in the limit of fast diffusion. *Journal of the Mechanics and Physics of Solids, 96,* 48–64.

Brassart, L., Liu, Q., & Suo, Z. (2018). Mixing by shear, dilation, swap and diffusion. *Journal of the Mechanics and Physics of Solids, 112,* 253–272.

Darken, L. S. (1948). Diffusion, mobility and their interrelation through free energy in binary metallic systems. *Transactions of the American Institute of Mining, Metallurgical, and Petroleum Engineers, 175,* 184–201.

dell'Isola, F., Seppecher, P., & Madeo, A. (2012). How contact interactions may depend on the shape of Cauchy cuts in Nth gradient continua: Approach "à la D'Alembert". *Zeitschrift für angewandte Mathematik und Physik, 63*(6).

dell'Isola, F., Andreaus, U., & Placidi, L. (2015). At the origins and in the vanguard of peridynamics, non-local and higher-gradient continuum mechanics: An underestimated and still topical contribution of Gabrio Piola. *Mathematics and Mechanics of Solids, 20*(8).

dell'Isola, F., Della, C. A., & Giorgio, I. (2016a). *Higher-gradient continua: The legacy of Piola, Mindlin.* Mathematics and Mechanics of Solids: Sedov and Toupin and some future research perspectives.

dell'Isola, F., Giorgio, I., Pawlikowski, M., & Rizzi, N. (2016b). Large deformations of planar extensible beams and pantographic lattices: Heuristic homogenization, experimental and numerical examples of equilibrium. *Proceedings of The Royal Society A, 472*(2185).

dell'Isola, F., Della Corte, A., Greco, L., & Luongo, A. (2016c). Plane bias extension test for a continuum with two inextensible families of fibers: A variational treatment with Lagrange multipliers and a perturbation solution. *International Journal of Solids and Structures.*

dell'Isola, F., Cuomo, M., Greco, L., & Della Corte, A. (2017). Bias extension test for pantographic sheets: Numerical simulations based on second gradient shear energies. *Journal of Engineering Mathematics*.

dell'Isola, F., Seppecher, P., Alibert, J. J., Lekszycki, T., Grygoruk, R., Pawlikowski, M., et al. (2019a). Pantographic metamaterials: An example of mathematically driven design and of its technological challenges. *Continuum Mechanics and Thermodynamics, 31*(4), 851–884.

dell'Isola, F., Seppecher, P., Spagnuolo, M., Barchiesi, E., Hils, F., Lekszycki, T., et al. (2019b). Advances in pantographic structures: Design, manufacturing, models, experiments and image analyses. *Continuum Mechanics and Thermodynamics, 31*(4), 1231–1282.

Del Vescovo, D., & Giorgio, I. (2014). Dynamic problems for metamaterials: Review of existing models and ideas for further research. *International Journal of Engineering Science, 80,* 153–172.

Knyazeva, A. G. (2003). Cross effects in solid media with diffusion. *Journal of Applied Mechanics and Technical Physics, 44*(3), 373–384.

Knyazeva, A. G. (2004). Model of medium with diffusion and internal surfaces and some applied problems. *Materials Physics and Mechanics, 7*(1), 29–36.

Knyazeva, A. G. (2005). Diffusion by the vacancy mechanism in the materials with the large number of internal surfaces. *Chemistry for Sustainable Development, 2,* 233–242.

Mehrer, H. (2007). *Diffusion in solids*. Springer Series in Solid-State Sciences (Vol. 155, 637 p.). Heidelberg: Springer.

Nazarov, A. V., & Gurov, K. P. (1974). The kinetic theory of interdiffusion in binary system. Concentration of vacancies during mutual diffusion. *The Physics of Metals and Metallography, 37*, 496–503.

Paul, A., Laurila, T., Vuorinen, V., & Divinski, S. V. (2014). *Thermodynamics, diffusion and the Kirkendall effect in solids* (530 p.) Springer.

Placidi, L., Barchiesi, E., Turco, E., & Rizzi, N. L. (2016). A review on 2D models for the description of pantographic fabrics. *Zeitschrift für angewandte Mathematik und Physik, 67*(5).

Placidi, L., Andreaus, U., & Giorgio, I. (2017). Identification of two-dimensional pantographic structure via a linear D4 orthotropic second gradient elastic model. *Journal of Engineering Mathematics*.

Rahali, Y., Giorgio, I., Ganghoffer, J.-F., & dell'Isola, F. (2015). Homogenization à la Piola produces second gradient continuum models for linear pantographic lattices. *International Journal of Engineering Science, 97*.

Sciarra, G., dell'Isola, F., & Coussy, O. (2007). Second gradient poromechanics. *International Journal of Solids and Structures, 44*(20).

Stephenson, G. B. (1988). Deformation during interdiffusion. *Acta Metallurgica, 36,* 2663–2683.

Straumal, B. B., Baretzky, B., Mazilkin, A. A., Phillippa, F., Kogtenkova, O. A., Volkov, M. N., & Valiev, R. Z. (2004). Formation of nanograined structure and decomposition of supersaturated solid solution during high pressure torsion of Al–Zn and Al–Mg alloys. *Acta Materialia, 52*(15), 4469–4478.

Chapter 4
Finite Element Method Study of the Protection Damping Elements Dynamic Deformation

Anastasia V. Demareva, Aleksandr I. Kibets, Maria V. Skobeeva, Oleg G. Savichin, and Aleksandr F. Lyakhov

Abstract The problem of dynamic deformation of structures damping elements made of metal hollow spheres (MHS) filler is considered. MHS filler is a porous material obtained by joining homogeneous metal hollow spheres. The MHS filler is modeled by a continually homogeneous, orthotropic, physically nonlinear medium. The solution of the determining equation system is based on the finite element method moment scheme and the explicit finite-difference time integration "cross" type scheme. The problem of stability loss and supercritical behavior of a titanium spherical shell under compression loading between two non-deformable plates approaching with a constant velocity is considered. According to the problem numerical solution results, the dependence of the contact force on the plates movement was built, on the basis of which the deformation diagram and the parameters of the MHS filler mathematical model of the were determined. The problem of plate falling on a spherical shells' set located on a fixed base is solved using the obtained data. As it is shown by the calculation results analysis, the developed computational model allows to determine with acceptable accuracy the integral deformation parameters of the MHS filler (contact forces, displacements, displacement velocities) and to evaluate its damping properties.

Keywords Damper · Porous filler · Impact · Finite element method

4.1 Introduction

Porous metals are so-called functional materials, which are used because of their special properties, among which can be noted the low density and the ability to effectively absorb impact energy (Thornton and Magee 1975; Davies and Zhen 1983; Baumeister and Banhart 1998; Gibson and Ashby 1997; Banhart 2001; Ramchandra et al. 2003). This combination of characteristics is due to the fact that porous metals

A. V. Demareva · A. I. Kibets (✉) · M. V. Skobeeva · O. G. Savichin · A. F. Lyakhov
Research Institute for Mechanics, National Research Lobachevsky State University of Niznhy Novgorod, Gagarin ave., 23, 603950 Nizhny Novgorod, Russia
e-mail: kibec@mech.unn.ru

F. dell'Isola and L. Igumnov (eds.), *Dynamics, Strength of Materials and Durability in Multiscale Mechanics*, Advanced Structured Materials 137,
https://doi.org/10.1007/978-3-030-53755-5_4

consist of a metal matrix and pores in an amount of up to 80–90%. The technological expansion requires the expansion of the porous materials range and the production promising methods development. Preference is given to simple and economical technologies that allow to regulate the total porosity, pore diameter, and its configuration. Recently, the engineers' attention attracted metal hollow spheres (MHS) filler—porous material obtained by joining homogeneous metal hollow spheres (Ramchandra et al. 2003; Ruan et al. 2006; Caty et al. 2009; Liu et al. 2012). This multifunctional material is characterized by a nonlinear deformation diagram, and it significantly absorbs impact energy and has good prospects for use in high-impact structures, in particular, in containers for dangerous cargo transportation.

More information about porous materials one can find in (dell'Isola and Batra 1997; dell'Isola et al. 2000, 2009; Madeo et al. 2013; Sciarra et al. 2007). Generalized continua (Abali et al. 2017; Alibert et al. 2003; Auffray et al. 2013; dell'Isola et al. 2012, 2015, 2017) found its application in developing new micsrostructured metamaterials (Barchiesi et al., 2018; Del Vescovo and Giorgio 2014). An example of mechanical metamaterials is pantographic structure (dell'Isola et al. 2019; Placidi et al. 2016; Giorgio 2016).

In general, the choice of a porous material as a damper should be based on theoretical and experimental studies of its stress state under the corresponding loading. Currently, elasto-plastic buckling even of separated spherical shell is studied insufficiently. Therefore, direct calculations of the MHS filler dynamics with a detailed account of its structure are justified only in its compact fragments (representative volumes) deformation study. This determines the relevance of the development and justification of the mathematical model and the method of numerical stress–strain state study.

Below is a mathematical model and method for solving the deformation of structures with MHS filler problem. The results of its usage for the analysis of damping properties of individual spherical shells and for shells set are presented.

4.2 Constitutive System of Equations and Problem Solution Method

The motion of the structure is described from the standpoint of continuum mechanics using the current Lagrangian formulation (Bathe 1996). The equation of motion is derived from the balance of virtual power

$$\int\limits_{\Omega} \sigma_{ij}\delta\dot{\varepsilon}_{ij}\mathrm{d}V + \int\limits_{\Omega} \rho\ddot{U}_i\delta\dot{U}_i\mathrm{d}V = \int\limits_{\Gamma_p} P_i\delta\dot{U}_i\mathrm{d}\gamma + \int\limits_{\Gamma_q} P_i^q\delta\dot{U}_i\mathrm{d}\gamma, \qquad (4.1)$$

where $\dot{\varepsilon}_{ij}$ and σ_{ij} are strain rate and stress rate tensor components, U_i are displacements in a common coordinate system X, ρ is density, p_i^q is a contact pressure, p_i is a distributed load, Ω is the area occupied by the construction, Γ_q is a contact surface,

Γ_p is an area of external pressure action, and the dot above the symbol means a partial derivative of time t. Over repeated index is summation. Strain rates are defined in the current state metric:

$$\dot{\varepsilon}_{ij} = \left(\dot{U}_{i,j} + \dot{U}_{j,i}\right)/2, \quad \left(i, j = \overline{1,3}\right),$$

$$\dot{U}_{i,j} = \partial \dot{U}_i / \partial X_j, \quad X_j = X_j|_{t=0} + \int_0^t \dot{U}_j \mathrm{d}t. \tag{4.2}$$

Elastic–plastic deformation of metals and alloys is described by the relations of flow theory with kinematic and isotropic hardening (Volkov and Korotkikh 2008).

4.2.1 MHS Filler Modeling

MHS filler is modeled by a continually homogeneous, orthotropic, physically nonlinear medium (Hallquist 1998; Demareva et al. 2014). Strains and stresses are defined in the local coordinate system $x = [x_1, x_2, x_3]$. The relationship of elastic deformations and stresses is established on the basis of the generalized Hooke's law:

$$\dot{\sigma}_{ij} = C_{ijkl}\dot{\varepsilon}_{kl}, \quad i, j = \overline{1,3},$$

$$C_{ijkl} = \begin{bmatrix} C_{1111} & C_{1122} & C_{1133} & & & \\ C_{1122} & C_{2222} & C_{2233} & & 0 & \\ C_{1133} & C_{2233} & C_{3333} & & & \\ & & & C_{1212} & & \\ & 0 & & & C_{2323} & \\ & & & & & C_{3131} \end{bmatrix}. \tag{4.3}$$

For highly porous metals, it can be assumed that in the initial state, the Poisson's ratio is negligible. With this in mind, the ratio (4.3) takes the form:

$$\dot{\sigma}_{ij} = A_{ij}\dot{\varepsilon}_{ij}, \quad A_{ij} = C_{ijij}, \ i, j = \overline{1,3}. \tag{4.4}$$

As the porous material is compressed, its elastic characteristics are changed. The following ratio is accepted:

$$A_{ij} = A_{ij}^0 + \beta\left(A_{ij}^s - A_{ij}^0\right) \tag{4.5}$$

where indices "0" and "s" mark the elastic constants of the filler in the initial state and when the pores are completely closed. The parameter β is determined by the current value of the relative volume $\upsilon = V/V_0$

$$\beta = \max\left[\min\left(\frac{1 - \upsilon_{\min}}{1 - \upsilon_f}, 1\right), 0\right], \tag{4.6}$$

where υ_f is the limit value of the relative volume corresponding to the state of the material with completely closed pores. In the initial state, β is 0, and for fully compressed filler ($\nu \leq \upsilon_f$) $\beta = 1$.

If the stress component values obtained by integrating the generalized Hooke's law (4.5) over time violate the yield conditions:

$$\left|\sigma_{ij}\right| > \lambda\sigma_{ij}^T(\upsilon_{\min}) \tag{4.7}$$

correction is performed (Hallquist 1998).

$$\sigma_{ij} = \sigma_{ij}^T(\upsilon_{\min})\lambda\sigma_{ij}/\left|\sigma_{ij}\right|, \tag{4.8}$$

where $\sigma_{ij}^T(\upsilon_{\min})$ are limit filler stress components' values of the filler stress component, which depend on the degree of compression. The parameter λ in (4.8) describes the deformation rate influence to the analyzed process. The parameters of the homogeneous model used in (4.5)–(4.8) are determined from the numerical analysis of porous material typical blocks (representative volumes) deformation taking into account its structure or from the experimental dynamic deformation diagram.

After complete closure of the pores ($\beta = 1$): The relationship of stress and strain is described by the incremental plasticity theory equations (Volkov and Korotkikh 2008).

The deformable structural elements contact interaction is modeled by a one-way connection that allows temporary interruption and resumption of contact. The determining equation system (4.1)–(4.8) is supplemented by initial conditions and kinematic boundary conditions.

4.2.2 *Finite Element Analysis*

The determining equation system solution is based on the finite element method moment scheme and the explicit finite-difference "cross" type time integration scheme (Golovanov et al. 2006; Bazhenov et al. 2014, 2016a, b). The time integration step is determined from the condition of the Courant stability. Quadrature formulas are used for integration by spatial variables (Bathe 1996).

The algorithm (Bazhenov et al. 1995) is used to solve the deformable bodies on inconsistent finite element grids contact problem. The finite element method is implemented in the computer complex "Dinamika-3" (Russian Certificate of Conformity №ROSS RU.ME.20.H00338). To verify the MHS filler computational model, calculations of elastic–plastic buckling of individual spherical shells under quasi-static compression and dynamic shock loading was performed. The calculation results are in good agreement with the experimental data (Bazhenov et al. 2014, 2016a, b).

4.3 Results of Computational Experiments

The axisymmetric problem of stability loss and supercritical behavior of a spherical shell under compression loading between two non-deformable plates approaching with a constant velocity of 1 m/s is considered. The shell is made of titanium alloy (elastic modulus $E = 120\,\text{GPa}$, Poisson's ratio $\mu = 0.25$, density $\rho = 4.5\,\text{g/sm}^3$, yield limit $\sigma_T = 0.64\,\text{GPa}$, strain hardening modulus $g = 0.6\,\text{GPa}$).

Figure 4.1 shows graphs of the change in the force of the contact interaction F of the shell and plate depending on the plates convergence $\Delta = (D - D_0)/D_0$, where D_0 is the initial value of the inner diameter of the shell, D is the current distance between the upper and lower plates without two shell thicknesses. The solid line in Fig. 4.1 represents graphs of the contact force on the plate change in time, and the dashed line represents a graph of the contact interaction of the upper and lower half of the spherical shell on the inner surface.

Calculation results analysis showed the following. For the shell on the graph of $F(\Delta)$, it can be divided into three stages. In the first stage ($\Delta < 0.68$) as convergence of the plates at $\Delta \approx 0.03$, two dents are formed on the shell in the contact areas. The pole shell points depart from the plates, and contact areas take the ring form. The spherical shell bending is accompanied by the plastic deformations occurrence. The maximum values of plastic deformations (~60%) are achieved in the compression areas: on the outer surface in the contact areas and on the areas of the inner surface. In tensile areas, plastic deformation does not exceed 16%, which is acceptable for this material.

Since the shell buckling is accompanied by the plastic deformations occurrence, the growth of the contact force first slows down. After $\Delta \approx 0.4$, stabilization stage is occurred. When reaching the deflection of one-fourth of the diameter, the second stage of shell deformation begins. The upper and lower parts of the shell inner surface are closed, and a third contact zone is formed. This leads to an increase in the resistance of the shell to the plates convergence. Upon reaching $\Delta \approx 0.95$, the contact force increased approximately 2.5 times in relation to the value reached at

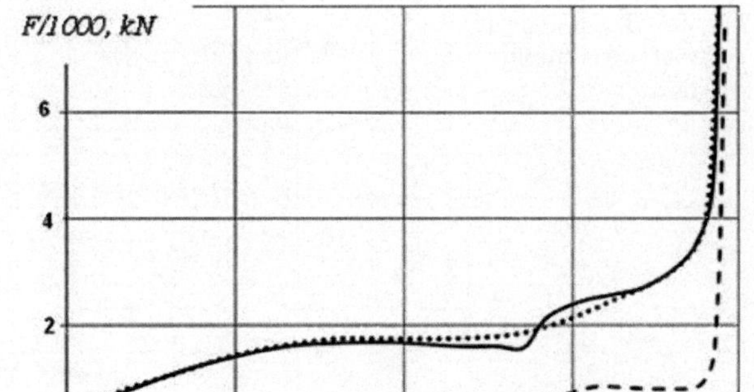

Fig. 4.1 Graphs of the change in the force of the contact interaction F of the shell and plate depending on the plates convergence

the stabilization stage. In this position, the shell inner surface is almost closed. The third stage of the shell deformation begins, and it is characterized by a sharp increase in the contact force.

As it is shown in Fig. 4.1, the behavior of spherical shells reaction force on the plates movement is nonlinear. Due to this, spherical shells have good energy-absorbing properties and can be used for shock damping in protective elements. The effective damper development is very relevant for the design of modern containers for the transport of radioactive materials and other potentially dangerous cargo. According to the IAEA regulations, in case of an emergency drop of a container on a rigid plate, the acceleration experienced by the cargo shall not exceed the limit value. In this regard, numerical studies of the damping properties of a titanium alloy spherical shell under shock loading were carried out. The shell is located on a fixed plate, and on top of it falls plate, simulating the transported cargo. The mass of the falling plate is 0.889 t, the initial speed of its fall is 13 m/s.

To verify the above computational model of MHS filler, calculations were carried out in which the shell was replaced by an equivalent in size, weight, and stiffness prototype (cylinder $R = 9$ cm, height $H = 18$ cm) of MHS filler. The problem was solved in three-dimensional formulation. Taking into account the symmetry of geometry, initial, and boundary conditions, one-fourth part of the shell prototype and plate was considered in the calculations.

The porous material stress–strain diagram (Fig. 4.2) was determined on the basis of a smoothed graph of the change in the contact force acting on the plate (dotted line in Fig. 4.1) at quasi-static compression of the shell. For comparison, we considered the problem of the fall plate on four shells, which are arranged vertically one above the other.

Figures 4.3, 4.4 and 4.5 shows time dependence graphs of the contact force on the plate, the displacement, and the displacement velocity of the plate. Solid lines in Figs. 4.3, 4.4, and 4.5 correspond to the solution of the problem with the finite elements shell discretization, and dashed lines correspond to the calculation with simulation of the MHS filler shell prototype. Numbers 1 and 2 are used to mark the

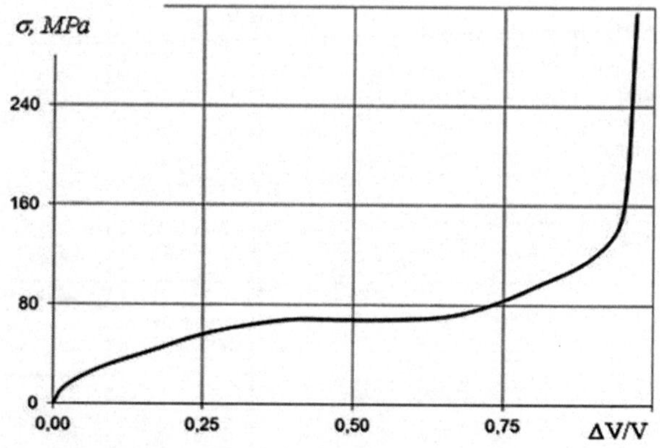

Fig. 4.2 The porous material stress-strain diagram

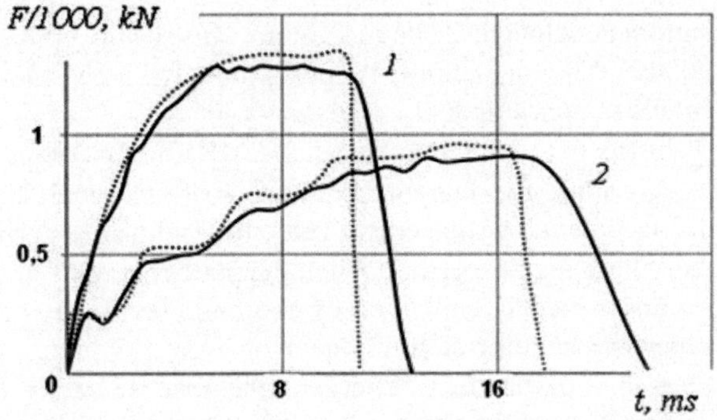

Fig. 4.3 Time dependence graph of the contact force on the plate

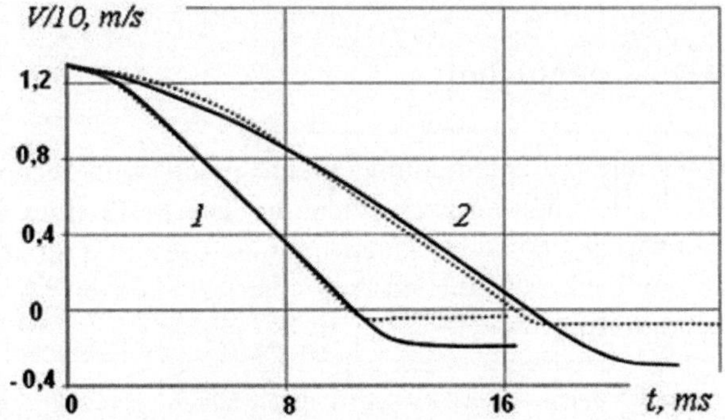

Fig. 4.4 Time dependence graph the displacement velocity of the plate

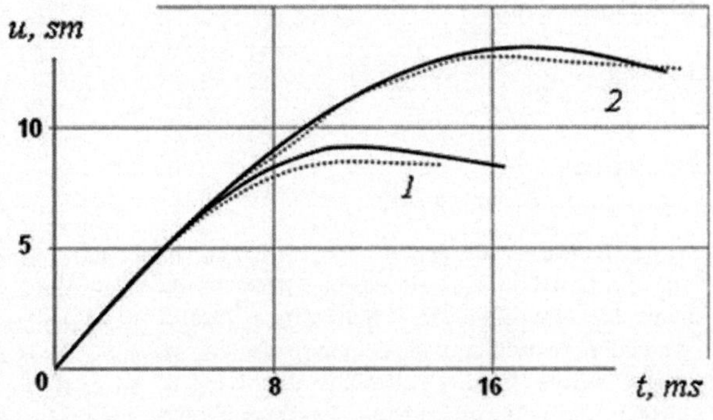

Fig. 4.5 Time dependence graph displacement of the plate

results of the problem solution for the first variant (plate falls on one shell) and the second variant (plate falls on four shells).

In the deformation process, shell closure is not occurred. Shell deformation does not exceed 26%. The maximum contact force F_{max} reaches 1200 kN, and the impact time is 12.5 ms. The average deceleration acceleration of the falling plate, calculated in Fig. 4.4, is equal to approximately 140 g ($g = 9.8$ m/s^2 is acceleration of gravity). In the second case, the damper of four shells does not experience significant changes. The maximum value of the contact force $F_{\max}$ does not exceed 900 kN. The impact time is approximately 22 ms, and average falling plate overload equals to approximately 80 g.

In the active loading stage, the MHS filler computational model, MHS, allows to determine the integral parameters of spherical shells buckling (contact forces, displacements, displacement velocities) with an accuracy of up to 5% and to evaluate their damping properties. When the plate rebounds, there is some discrepancy in the results, which is explained by the complex behavior of the spherical shell at the supercritical deformation stage. However, it should be taken into account that by this time, 90% of the kinetic energy of the plate is already absorbed by the elastic–plastic deformation of the shell.

4.4 Conclusion

Thus, the presented computational model with acceptable accuracy describes the MHS filler nonlinear deformation. The MHS filler computational model stress–strain diagram can be obtained on the basis of a numerical solution of the problem of nonlinear deformation and spherical shell buckling under appropriate loading conditions.

Acknowledgements The work is financially supported by the Federal Targeted Program for Research and Development in Priority Areas of Development of the Russian Scientific and Technological Complex for 2014–2020 under the contract No. 075-15-2019-1702 (unique identifier RFMEFI60519X0183).

References

Abali, B. E., Muller, W., & dell'Isola, F. (2017). Theory and computation of higher gradient elasticity theories based on action principles. *Archive of Applied Mechanics, 87*, 1495–1510.

Alibert, J. J., Seppecher, P., & dell'Isola, F. (2003). Truss modular beams with deformation energy depending on higher displacement gradients. *Mathematics and Mechanics of Solids, 8*, 51–73.

Auffray, N., dell'Isola, F., Eremeyev, V., Madeo, A., & Rosi, G. (2013). Analytical continuum mechanics à la Hamilton-Piola: Least action principle for second gradient continua and capillary fluids. *Mathematics and Mechanics of Solids.*

Banhart, J. (2001). Manufacture, characterization and application of cellular metals and metal foams. *Progress in Materials Science., 46*, 559–632.

Barchiesi, E., Spagnuolo, M., & Placidi, L. (2018) Mechanical metamaterials: A state of the art. *Mathematics and Mechanics of Solids.*

Bathe, K. Y. (1996). *Finite element procedures*. Upper Saddle River, NJ: «Prentice Hall» 1037p.

Baumeister, J., & Banhart, J. (1998). Deformation characteristics of metal foam. *Journal of Materials Science, 33*, 1431–1440.

Bazhenov, V. G., Kibets, A. I., & Tsvetkova, I. N. (1995). Numerical simulation of transient processes of shock interaction of deformable elements of constructions. *Problemy Mashinostroeniia I Nadezhnosti Mashin, 2*, 20–26.

Bazhenov, V. G., Gonik, E. G., Kibets, A. I., & Shoshin, D. V. (2014). Stability and limit states of elastoplastic spherical shells under static and dynamic loading. *Journal of Applied Mechanics and Technical Physics, 55*(1), 8–15.

Bazhenov, V. G., Demareva, A. V., Zhestkov, M. N., & Kibets, A. I. (2016a). Special features of numerical simulation for elastic-plastic buckling of hemispherical shells under loading with the rigid indenter. *PNRPU Mechanics Bulletin, 2,* 22–33.

Bazhenov, V. G., Demareva, A. V., Baranova, M. S., Kibets, A. I., Ryabov, A. A., & Romanov, V. I. (2016b). Finite-element modeling of large elastoplastic deformations of a spherical shell in a scaphander under an overload pulse. *Problems of Strenght and Plasticity, 78*(3), 322–332.

Caty, O., Maire, E., Douillard, T., Bertino, P., Dejaeger, R., & Bouchet, R. (2009). Experimental determination of the macroscopic fatigue properties of metal hollow sphere structures. *Materials Letters, 63,* 1131–1134.

Davies, G. J., & Zhen, S. (1983). Metallic foams-their production, properties and applications. *Journal of Materials Science, 18*(7), 1899–1911.

dell'Isola, F., Seppecher, P., & Madeo, A. (2012). How contact interactions may depend on the shape of Cauchy cuts in Nth gradient continua: Approach "à la D'Alembert." *Zeitschrift Für Angewandte Mathematik Und Physik, 63,* 1119–1141.

dell'Isola, F., Della, C. A., & Giorgio, I. (2017). Higher-gradient continua: The legacy of Pio-la, Mindlin, Sedov and Toupin and some future research perspectives. *Mathematics and Mechanics of Solids, 22,* 852–872.

dell'Isola, F., Seppecher, P., Alibert, J. J., Lekszycki, T., Grygoruk, R., Pawlikowski, M., et al. (2019). Pantographic metamaterials: An example of mathematically driven design and of its technological challenges. *Continuum Mechanics and Thermodynamics, 31*(4), 851–884.

dell'Isola, F., & Batra, R. (1997). Saint-Venant's problem for porous linear elastic materials. *Journal of Elasticity, 47*(1).

dell'Isola, F., Guarascio, M., & Hutter, K. (2000). A variational approach for the deformation of a saturated porous solid. A second-gradient theory extending Terzaghi's effective stress principle. *Archive of Applied Mechanics, 70*(5).

dell'Isola, F., Madeo, A., & Seppecher, P. (2009). Boundary conditions at fluid-permeable interfaces in porous media: A variational approach. *International Journal of Solids and Structures, 246*(17).

dell'Isola, F., Andreaus, U., & Placidi, L. (2015). At the origins and in the vanguard of peri-dynamics, non-local and higher gradient continuum mechanics. An underestimated and still topical contribution of Gabrio Piola. *Mathematics and Mechanics of Solids, 20,* 887–928.

Del Vescovo, D., & Giorgio, I. (2014). Dynamic problems for metamaterials: Review of existing models and ideas for further research. *International Journal of Engineering Science, 80,* 153–172.

Demareva, A. V., Ivanov, V. A., Zhestkov, M. N., Kibets, A. I., Kibets, Yu. I., & Shoshin, D. V. (2014). Numerically analyzing dynamic problems of deformation of multilayered shells with a porous filling. *Problems of Strenght and Plasticity, 76*(1), 46–54.

Gibson, L. J., & Ashby, M. F. (1997). *Cellular solids: Structure and properties* (p. 528). Cambridge: Cambridge University Press.

Giorgio, I. (2016). Numerical identification procedure between a micro-Cauchy model and a macro-second gradient model for planar pantographic structures. *Zeitschrift für an-gewandte Mathematik und Physik.*

Golovanov, A. I., Tiuleneva, O. N., & Shigabutdinov, A. F. (2006). *The finite element method in statics and dynamics of thin-walled structures* (p. 391). Moscow: Fizmatlit.

Hallquist, J. O. (1998). *LS-DYNA: Theoretical manual* (p. 498p). Livermore: Livermore Software Technology Corporation.

Liu YB, Wu HX, Wang B (2012) Gradient design of metal hollow sphere (MHS) foams with density gradients. *Composites Part B: Engineering 43,* 1346–1352.

Madeo, A., dell'Isola, F., & Darve, F. (2013). A continuum model for deformable, second gradient porous media partially saturated with compressible fluids. *Journal of the Mechanics and Physics of Solids, 61*(11).

Placidi, L., Barchiesi, E., Turco, E., & Rizzi, N.L. (2016). A review on 2D models for the description of pantographic fabrics. *Zeitschrift für angewandte Mathematik und Physik, 67*.

Ramchandra, S., Ramamurty, U., & Sudheer Kumar, P. (2003). Impact energy absorption in an Al foam at low velocities. *Scripta Materialia, 49*(8), 741–745.

Ruan, H. H., Gao, Z. Y., & Yu, T. X. (2006). Crushing of thin-walled spheres and sphere arrays. *International Journal of Mechanical Sciences., 48,* 117–133.

Sciarra, G., dell'Isola, D., & Coussy, O. (2007). Second gradient poromechanics. *International Journal of Solids and Structures, 44*(20).

Thornton, P. H., & Magee, C. L. (1975). The deformation of aluminium foams. *Metallurgical and Materials Transactions A, 6*(6), 1253–1263.

Volkov, I. A., & Korotkikh, Yu. G. (2008). *State equation for Viscoelastoplastic media with defects* (p. 424). Moscow: FIZMATLIT.

Chapter 5
Analyzing the Problem of a Spherical Cavity Expansion in a Medium with Mohr-Coulomb-Tresca's Plasticity Condition

Vasiliy L. Kotov, Elena Yu. Linnik, and Tatiana A. Sabaeva

Abstract An analytical solution of the one-dimensional problem of a spherical cavity expanding at a constant velocity from a point in a half-space occupied by a plastic medium has been obtained. Impact compressibility of the medium is described using linear Hugoniot's adiabat. Plastic deformation obeys the Mohr-Coulomb yield criterion with constraints on the value of maximum tangential stresses according to Tresca's criterion. In the assumption of rigid-plastic deformation (the elastic precursor being neglected), incompressibility behind the shockwave front and the equality of the propagation velocities of the fronts of the plastic wave and the plane shockwave defined by linear Hugoniot's adiabat, a boundary-value problem for a system of two first-order ordinary differential equations for the dimensionless velocity and stress depending on the self-similar variable is formulated. A closed-form solution of this problem has been obtained in the form of a stationary running wave—a plastic shockwave propagating in an unperturbed half-space. The solution is a generalization of the earlier obtained analytical solution for a medium with the Mohr-Coulomb plasticity condition. A formula for determining a critical pressure (a minimal pressure required for the nucleation of a cavity, accounting for internal friction in the framework of Mohr-Coulomb yield criterion), which is a generalization of the earlier solution for an ideal plastic medium with Tresca's criterion, has been derived. The resulting critical pressure was compared with a numerical solution in a full formulation at cavity propagation velocities close to zero in a wide range of the parameters of the Mohr-Coulomb yield criterion. The approximation inaccuracy of the introduced formula does not exceed 6% for the internal friction coefficient varying over the entire permissible range and the initial value of yield strength changing by three orders of magnitude. The effect of constraining the limiting value of maximal tangential stresses on the distribution of dimensionless stresses behind the shockwave front has been examined. Formulas for determining the range of cavity expansion velocities, within which a simple solution for a medium with Tresca's plasticity condition is applicable, have been derived. The obtained solution can be used for

---

V. L. Kotov (✉) · E. Yu. Linnik · T. A. Sabaeva
Research Institute for Mechanics, National Research Lobachevsky State University of Niznhy Novgorod, Gagarin ave. 23, 603950 Nizhny Novgorod, Russia
e-mail: vkotov@inbox.r

F. dell'Isola and L. Igumnov (eds.), *Dynamics, Strength of Materials and Durability in Multiscale Mechanics*, Advanced Structured Materials 137,
https://doi.org/10.1007/978-3-030-53755-5_5

evaluating resistance to high-velocity penetration of rigid strikers into low-strength soil media.

Keywords Expansion of a spherical cavity · Analytical solution · Elasticity · Plasticity · Mohr-Coulomb-Tresca's plasticity condition

5.1 Introduction

A special role in the high-velocity impact theory is given to determine parameters of contact interaction of bodies. Practice shows (Ben-Dor et al. 2005; Veldanov et al. 2011; Omidvar et al. 2012, 2014, 2015) that a good approximation of the pressure on the contact surface of a rigid striker with a resisting medium is a solution of the spherical or cylindrical cavity expansion problem. To determine stress and rate fields in the plastic deformation region adjoining the cavity, an effective algorithm of numerical analysis was developed (Forrestal and Longcope 1982; Forrestal and Luk 1992), that makes it possible to obtain a solution of the problem with an accuracy sufficient for its practical application. Numerical analyses of the spherical cavity expansion problem and their application to evaluating contact stresses and forces resisting penetration of rigid bodies into soil, concrete, and metal are given in (Forrestal and Longcope 1982; Forrestal and Luk 1992; Shi et al. 2014, 2015; Kong et al. 2017; Sun et al. 2017; Kotov 2008, 2019). Papers (Kotov et al. 2011; Linnik et al. 2012) present an analytical solution of the above problem, based on the assumption of incompressibility of the medium behind the shockwave front. Dynamic compressibility of a medium is characterized by a shock adiabat, and shear resistance is defined by the Mohr-Coulomb plasticity condition. Based on the proposed analytical solution, a methodology for evaluating forces resisting penetration of a rigid body into soft soil was developed (Kotov et al. 2013). Relations for the maximal value of the force resisting penetration of a rigid sphere and a conic-nosed striker into dry and water-saturated sands were derived (Bragov et al. 2018). Comparison of the analytical, numerical, and experimental results of determining resistance to penetration testifies to their good qualitative and quantitative agreement in the impact velocity range of 50–400 m/s.

On contact interactions via generalized continua framework, one can read in (dell'Isola et al. 2012, 2016). More information about analysis of saturated soil and porous media can be found in (dell'Isola and Hutter 1998, 1999; dell'Isola et al. 2000). Some of these works shares the same framework of generalized continua (Abali et al. 2017; Auffray et al. 2013; Alibert et al. 2003; dell'Isola et al. 2015, 2017; Rahali et al. 2015), nowadays, widely used on designing new artificial microstructured composites whose name is metamaterials (Barchiesi et al. 2018; Del Vescovo and Giorgio 2014; dell'Isola et al. 2019; Placidi et al. 2016, 2017).

Nevertheless, analytical solutions of this problem or approximations of the results of numerical solutions are still of interest to numerous applications of impact dynamics. The existing closed-form solutions were mainly obtained in the assumption of Tresca's yield condition. This is explained by the fact that the problematic part of the applications of the methodology based on analyzing the cavity expansion problem consists mainly of problems of impact deformation of metals and alloys, where the model of elastic-ideally-plastic media with Tresca's or Mises yield conditions is widely used.

The present paper introduces a formula derived for determining a critical pressure (a minimal pressure required for the nucleation of a cavity), accounting for internal friction in the framework of the Mohr-Coulomb yield criterion, which is a generalization of the earlier solution for an elastic-ideally-plastic medium with Tresca's criterion. The obtained solution can be used for analyzing spherical cavity expansion problems in porous materials and geo-materials.

Papers (Grigoryan 1960; Bazhenov and Kotov 2008; Bazhenov et al. 2009) present experimental and theoretical data testifying to the limited nature of yield strengths of soils under high pressures. As is demonstrated in (Bazhenov et al. 2009), the nonlinear dependence of yield strength on pressure, with the account of the scatter of the data and measuring inaccuracies, can be represented as a double-link broken line—a linear relation under low pressures, as it is assumed in the Mohr-Coulomb plasticity condition, and a limited maximal value of yield strength under high pressures according to Tresca's condition. In this connection, effective analyses of problems of penetration of rigid bodies into soft soils, using the cavity expansion methodology, require an analytical solution of the spherical cavity expansion problem in a medium with Mohr-Coulomb-Tresca's plasticity condition.

5.2 Formulation of an Initial Boundary-Value Problem for a System of Partial Differential Equations

A mathematical model of an elastoplastic medium is described by a system of differential equations expressing laws of continuity and change of kinetic momentum, which, in spherical Eulerian coordinates, will be written as:

$$\rho\left(\frac{\partial \upsilon}{\partial r}+2\frac{\upsilon}{r}\right)=-\left(\frac{\partial \rho}{\partial t}+\upsilon\frac{\partial \rho}{\partial r}\right),\quad \frac{\partial \sigma_r}{\partial r}+2\frac{(\sigma_r-\sigma_\theta)}{r}=-\rho\left(\frac{\partial \upsilon}{\partial t}+\upsilon\frac{\partial \upsilon}{\partial r}\right),\tag{5.1}$$

where ρ is density in a deformed state, υ is velocity, σ_r and σ_θ are radial and circumferential components of Cauchy stress tensor (which are assumed positive in compression), r is radial coordinate.

Equation system (5.1) is closed by the relation between pressure and volumetric strain, which is linear or close to linear for small strains

$$p=f(\theta)\equiv K\theta+O\left(\theta^2\right),\tag{5.2}$$

where K is elastic modulus of volumetric compression.

It is assumed that, in the region limited by radii $r_0 = Vt$ and $r = ct$, the medium deforms plastically, with a linear dependence of yield strength on Mohr-Coulomb's pressure

$$\sigma_r - \sigma_\theta = Y + kp, \tag{5.3}$$

where Y and k are initial value of yield strength and internal friction coefficient, respectively.

In the adjacent region of elastic deformation, which is limited by coordinate $r_e = c_e t$, stress tensor components are related to strains through Hooke's law with elastic moduli K and G, where $c_e = \sqrt{(K + 4G/3)/\rho_0}$ is propagation velocity of a plane elastic wave, G is shear modulus.

Yield condition $s_r - s_\theta = Y + kp$ gives equalities $s_r = 2(Y + kp)/3$, $s_\theta = -(Y + kp)/3$, wherefrom

$$\sigma_r - \sigma_\theta = Y + \frac{k}{1 + \frac{2}{3}k}\sigma_r = Y + \mu\sigma_r, \quad \mu = \frac{k}{1 + \frac{2}{3}k} \tag{5.4}$$

A one-dimensional problem of spherical cavity expansion in the plastic deformation region is then formulated. Partial differential equation system (5.1) for σ_r and υ, accounting for (5.4), will take the following form:

$$\begin{aligned} &\frac{1}{K_1}\left(\frac{\partial \sigma_r}{\partial t} + \upsilon\frac{\partial \sigma_r}{\partial r}\right) + (1-\theta)\left(\frac{\partial \upsilon}{\partial r} + \frac{2\upsilon}{r}\right) = 0, \\ &\frac{\partial \sigma_r}{\partial r} + \frac{2(Y + \mu\sigma_r)}{r} = -\frac{\rho_0}{1-\theta}\left(\frac{\partial \upsilon}{\partial t} + \upsilon\frac{\partial \upsilon}{\partial r}\right), \end{aligned} \tag{5.5}$$

where:

$$1 - \theta = 1 - f_1^{-1}(\sigma_r), \ \sigma_r = f_1(\theta) \equiv \frac{2}{3}Y + \left(1 + \frac{2}{3}k\right) f(\theta),$$
$$K_1 = \frac{\partial f_1(\theta)}{\partial \theta} \equiv \left(1 + \frac{2}{3}k\right)\frac{\partial f(\theta)}{\partial \theta}.$$

On the boundary of the expanding cavity of radius $r_0 = Vt$, velocity V is assigned; the outer surface of spherical layer r_∞ is free of stresses, at an initial time the velocity and stresses in the medium are equal to zero:

$$\upsilon(r_0, t) = V, \ \sigma_r(r_\infty, t) = 0, \ \upsilon(r, 0) = \sigma_r(r, 0) = 0 \tag{5.6}$$

5.3 Formulating a Boundary-Value Problem for a System of Two First-Order Ordinary Differential Equations in the Plastic Region

Consider a self-similar solution of the system for variable $\xi = r/ct$, introducing dimensionless variables $U = \frac{v}{c}$, $S = \frac{\sigma_r}{\rho_0 c^2}$ and designating $T = \frac{Y}{\rho_0 c^2}$, $\tilde{f}_1() = \frac{f_1()}{\rho_0 c^2}$, $\tilde{K}_1 = \frac{\partial \tilde{f}_1(\theta)}{\partial \theta}$

As a result of the substitution, partial differential equation system (5.5) is transformed into a system of ordinary differential equations:

$$U' + 2\frac{U}{\xi} = \frac{(\xi - U)}{(1-\theta)\tilde{K}_1} S', \quad S' + 2\frac{T + \mu S}{\xi} = \frac{(\xi - U)}{1 - \theta} U',$$

where $1 - \theta = 1 - \tilde{f}_1^{-1}(S)$, the stroke shows differentiation for

Finally, the boundary-value problem for a system of two ordinary differential equations (ODE), written in the normal form, will be as follows:

$$U' = \frac{2\left(U\tilde{K}_1 + \tilde{f}_2\phi\right)}{\xi\left(\phi^2 - \tilde{K}_1\right)}, \quad S' = \frac{2\tilde{K}_1\left(\tilde{f}_2 + U\phi\right)}{\xi\left(\phi^2 - \tilde{K}_1\right)}, \quad \varepsilon < \xi < 1, \tag{5.7}$$

$$U(\xi = \varepsilon) = \varepsilon, U(\xi = 1) = U^e, S(\xi = 1) = S^e \tag{5.8}$$

where $\phi = (\xi - U)/(1-\theta)$, $\varepsilon = V/c$.

Equation systems (5.7) and (5.8) include unknown parameter c—propagation velocity of the interface between the elastic and plastic regions, or the plastic shock-wave front. The unknown velocity is found by iterations, until boundary condition $|U(\xi = \varepsilon) - \varepsilon| < \delta$ is satisfied to assigned accuracy δ. At each iteration step, the fourth-order accuracy Runge–Cutta numerical method is used when self-similar variable ξ varies from the elastic–plastic interface boundary ($\xi = 1$) to the cavity boundary ($\xi = \varepsilon$).

The values of U^e, S^e in boundary conditions (5.8) are determined from the condition of continuity of the velocity and stress along the boundary with the region of elastic behavior of the material for $\xi = 1$.

5.4 Formulation and Solution of the Boundary-Value Problem for Second-Order ODE's in the Elastic Deformation Region

Consider the equation of motion in equation system (5.1), neglecting the convective components in the time derivative of velocity:

$$\frac{\partial\sigma_r}{\partial r} + 2\frac{(\sigma_r - \sigma_\theta)}{r} = -\rho\left(\frac{\partial \upsilon}{\partial t}\right), \upsilon = \frac{\partial u}{\partial t}$$

The radial and circumferential small strains in an elastic medium are related to displacements through Cauchy relations. As before, stresses in compression are assumed to be positive.

Applying Cauchy relations and Hooke's law, the dynamic equation of elastic media can be transformed into the following form

$$\frac{\partial^2 u}{\partial r^2} + \frac{2}{r}\frac{\partial u}{\partial r} - \frac{2u}{r^2} = \frac{1}{c_e^2}\frac{\partial^2 u}{\partial t^2},$$

where $c_e = \sqrt{(\lambda + 2G)/\rho_0} = \sqrt{(K + 4G/3)/\rho_0}$ is propagation velocity of the longitudinal wave front in an elastic medium.

Following (Forrestal and Luk 1988), it is assumed that $\xi = \frac{r}{ct}$, $\tilde{u} = \frac{u}{ct}$ are dimensionless coordinate and displacement, respectively, c is velocity of the plastic wave front (elastic–plastic interface); the derivatives will be transformed, taking into account the change of the variables:

$$\frac{u}{r} = \frac{\tilde{u}}{\xi}, \frac{\partial u}{\partial r} = \tilde{u}', \frac{\partial^2 u}{\partial r^2} = \frac{1}{ct}\tilde{u}'', \frac{\partial^2 u}{\partial t^2} = \frac{c}{t}\xi^2\tilde{u}''$$

where the stroke designates differentiation for.

Consider the conditions along the boundaries of the elastic deformation region. Along the boundary with the unperturbed region, displacement is equal to zero. Along the elastic–plastic interface for $\xi = 1$, plastic yield condition $(\sigma_r - \sigma_\theta)|_{\xi=1} = Y + kK\theta$ holds, as well as Hooke's law $\sigma_r - \sigma_\theta = 2G\left(\frac{u}{r} - \frac{\partial u}{\partial r}\right)$, whence, due to continuity of stresses, one has

$$\left(\frac{u}{r} - \frac{\partial u}{\partial r}\right) = \frac{Y + kK\theta}{2G}, \theta = -\left(\frac{\partial u}{\partial r} + 2\frac{u}{r}\right)$$

Transformation of the derivatives and substitution into Eq. (5.7) and boundary conditions (5.8) yields the following boundary-value problem for the second-order ODE for the dimensionless displacement

$$\tilde{u}(\xi = 1/\alpha) = 0, \left.\left(\frac{\tilde{u}}{\xi} - \frac{\partial \tilde{u}}{\partial \xi}\right)\right|_{\xi=1} = \frac{Y + kK\theta(\xi = 1)}{2G} \tag{5.9}$$

where $\alpha = c/c_e$.

To find a general solution of the differential equation, a number of transformations are carried out (Forrestal and Luk 1988) that yield an expression for the dimensionless displacement

$$\tilde{u} = A\left(1/3\alpha^2\xi^2 - 1\right) + B\alpha\xi, \tag{5.10}$$

where A, B are integration constants.

To determine constants A, B, boundary conditions (5.9) will be used. From the first boundary condition, it follows that $B = 2A/3$, thus,

$$\tilde{u} = A\left(\frac{1}{3\alpha^2\xi^2} + \frac{2}{3}\alpha\xi - 1\right) = A\frac{(1-\alpha\xi)^2(1+2\alpha\xi)}{3\alpha^2\xi^2}. \tag{5.11}$$

Then, expressions for determining the dimensionless velocity

$$U = \tilde{u} - \xi\frac{d\tilde{u}}{d\xi},\ U(\xi) = A\left(\frac{1-\alpha^2\xi^2}{\alpha^2\xi^2}\right), \tag{5.12}$$

the radial component of the stress tensor

$$\sigma_r(\xi) = 2A\frac{(1-\alpha\xi)}{\alpha^2\xi^3}\left(K\alpha^2\xi^2 + \frac{2G}{3}(1+\alpha\xi)\right) \tag{5.13}$$

and the volumetric strain

$$\theta = -\left(\frac{\partial\tilde{u}}{\partial\xi} + 2\frac{\tilde{u}}{\xi}\right) = 2A\left(\frac{1-\alpha\xi}{\xi}\right) \tag{5.14}$$

are derived.

Taking account of equality (5.14), equations for the dimensionless velocity (5.12) and stress (5.13) can be transformed into the following form

$$U(\xi) = \left(\frac{1+\alpha\xi}{2\alpha^2\xi}\right)\theta(\xi), \tag{5.15}$$

$$\sigma_r(\xi) = \left(K + \frac{2G}{3}\frac{(1+\alpha\xi)}{\alpha^2\xi^2}\right)\theta(\xi) \tag{5.16}$$

To determine integration constant A, the second boundary condition in boundary-value problem (5.9) will be considered. The difference of strains for $\xi = 1$ is defined as $\left(\tilde{u}/\xi - \partial\tilde{u}/\partial\xi\right)\big|_{\xi=1} = A\left(\alpha^{-2} - 1\right)$. Thus,

$$A = \frac{Y\alpha^2}{2(1-\alpha)\left(G(1+\alpha) - kK\alpha^2\right)}. \tag{5.17}$$

A solution of the problem in the assumption of Tresca's yield condition was earlier obtained in (Forrestal and Luk 1988). In the case of the Mohr-Coulomb plasticity condition for $\xi = 1$, one has

$$\theta(\xi = 1) = \frac{Y\alpha^2}{G(1+\alpha) - kK\alpha^2}$$

Expressions (5.15) and (5.16) for $\xi = 1$ define the boundary conditions as a function of the value of α:

$$U^e = U(\xi = 1) = \frac{(1+\alpha)}{2\alpha^2}\theta = \frac{Y(1+\alpha)}{2\left(G(1+\alpha) - kK\alpha^2\right)},$$

$$S^e = S(\xi = 1) = \left(\widetilde{K} + 2/3\widetilde{G}\frac{(1+\alpha)}{\alpha^2}\right)\theta = \frac{\left(K\alpha^2 + 2/3G(1+\alpha)\right)}{\left(G(1+\alpha) - kK\alpha^2\right)}T, \quad (5.18)$$

where $\tilde{K} = K/r_0c^2$, $\tilde{G} = G/r_0c^2$.

In the case of Mohr-Coulomb-Tresca's plasticity condition, function

$$(\sigma_r - \sigma_\theta)\big|_{\xi=1} \equiv \begin{cases} \tau_0 + kK\theta, & 0 < \theta \le \theta_\mathrm{M}, \\ \tau_M, & \theta \ge \theta_M \end{cases}, \quad \theta_M = \frac{\tau_M - \tau_0}{kK}$$

and, for $\xi = 1$, one has

$$\theta = \begin{cases} \tau_0/\left(G(\alpha) - kK\right), & G(\alpha) > G^*, \\ \tau_M/G(\alpha), & G(\alpha) \le G^*, \end{cases} \quad (5.19)$$

where $G(\alpha) = (1+\alpha)G/\alpha^2$, $G^* = \tau_M kK/(\tau_M - \tau_0) = \tau_M/\theta_M$.

Expressions (5.15) and (5.16), for $\xi = 1$ and taking into account (5.19), will define boundary conditions for the problem from p. 2 as a function of the value of α:

$$U^e = U(\xi = 1) = \frac{(1+\alpha)}{2\alpha^2}\theta,\ S^e = S(\xi = 1) = \left(\tilde{K} + 2/3\tilde{G}\right)\theta \quad (5.20)$$

where θ is defined by expression (5.19), $\widetilde{K} = K/r_0c^2$, $\widetilde{G} = G(\alpha)/r_0c^2$.

5.5 Determining the Critical Pressure

The issue of evaluating a minimal stress along the cavity boundary, necessary for the cavity to expand (critical pressure P_c) is now considered. Earlier (Rosenberg and Dekel 2008), an equation for a critical pressure in a linearly compressible elastic-ideally-plastic medium with Tresca's plasticity condition was derived

$$P_c = \frac{2}{3}Y\left(1 + \ln\frac{E}{3(1-\nu)Y}\right), \quad (5.21)$$

where E is Young's modulus, ν is Poisson's coefficient.

To evaluate a critical pressure in a medium with the Mohr-Coulomb plasticity criterion, the second equation in system (5.7) will be transformed into an equation for a dimensional stress and velocity. Taking into account the fact that elastic–plastic interface velocity $c = O(V)$, for $V \to 0$ one has a Cauchy problem

$$\frac{d\sigma_r}{d\xi} + 2\frac{Y + \mu\sigma_r}{\xi} = 0, \varepsilon \le \xi \le 1,$$

$$\sigma_r(\xi = 1) = 2Y/3, \tag{5.22}$$

where the boundary conditions are determined from Eq. (5.18) for $\alpha = 0$.

The solution of Cauchy problem (5.22) is function $\sigma_r(\xi) = \frac{2}{3}Y\xi^{-2\mu} + Y\left(\frac{\xi^{-2\mu}-1}{\mu}\right)$, which, for $\xi = \varepsilon$, determines the critical pressure

$$P_c = \frac{2}{3}Y\varepsilon^{-2\mu} + Y\left(\frac{\varepsilon^{-2\mu} - 1}{\mu}\right). \tag{5.23}$$

The value of the critical pressure for $\mu = 0$ can be found by passing to the limit $\mu \to 0$ in (5.23)

$$P_c = \frac{2}{3}Y\left(1 + \ln \varepsilon^{-3}\right). \tag{5.24}$$

In relations (5.23) and (5.24), the value of $\varepsilon = V/c$ remains undetermined. To determine it, an equation for the dimensionless velocity in equations system (5.7) will be considered. Neglecting the convective summand and assuming $1 - \theta \approx 1$, yields

$$U' + 2\frac{U}{\xi} = \xi\frac{\sigma_r'}{K_1},$$

where $\sigma_r' = -\frac{2Y}{3}(3 + 2\mu)\xi^{(-2\mu-1)}$, $K_1 = \left(1 + \frac{2}{3}k\right)K = K\left(1 - \frac{2}{3}\mu\right)^{-1}$.

Finally, the problem is obtained with the initial conditions of the form

$$U' + 2\frac{U}{\xi} = -2\frac{Y}{K}\left(1 - \frac{4}{9}\mu^2\right)\xi^{-2\mu}, \varepsilon \le \xi \le 1, \tag{5.25}$$

$$U(\xi = 1) = \frac{Y}{2G}$$

The solution of Cauchy problem (5.25) is the function

$$U(\xi) = \frac{Y}{18}\frac{(12G + 9K + 8G\mu)}{KG\xi^2} - \frac{2Y}{9}\frac{(3 + 2\mu)}{K\xi^2}\xi^{3-2\mu}.$$

To determine the value of ε, it is necessary to use the condition for $\xi = \varepsilon$ in equation set (5.8) $U(\xi = \varepsilon) = \varepsilon$:

$$\frac{Y}{18}\frac{(12G + 9K + 8G\mu)}{KG} - \frac{2Y}{9}\frac{(3 + 2\mu)}{K}\varepsilon^{3-2\mu} = \varepsilon^3. \tag{5.26}$$

Nonlinear equation (5.26) is solved numerically by iterations.

In a special case of $\mu = 0$:

$$\frac{Y}{18}\frac{(12G + 9K)}{KG} - \frac{2Y}{3K}\varepsilon^3 = \varepsilon^3,$$

The solution of Eq. (5.26) can be written in the following closed form

$$\varepsilon^3 = \frac{(4G + 3K)Y}{6KG\left(1 + \frac{2Y}{3K}\right)} = \frac{3(1 - \nu)Y}{E\left(1 + \frac{2Y}{3K}\right)} \approx \frac{3(1 - \nu)Y}{E},$$

using the relations for the elastic constants:

Young's modulus $E = \frac{9KG}{3K+G}$,

Poisson's coefficient $\nu = \frac{3K-2G}{6K+2G}$, $1 - \nu = \frac{3K+4G}{6K+2G}$.

This solution coincides with the earlier obtained solution (5.21) taking account of Eq. (5.24).

Consider the results of numerically analyzing the problem for the following values of elasticity moduli: $K = 220$ MPa, $G = 150$ MPa, the parameters of the plasticity condition can vary.

Figure 5.1 presents the diagrams of distribution of the cavity expansion velocity as a function of the elastic–plastic interface velocity, depending on the variation of

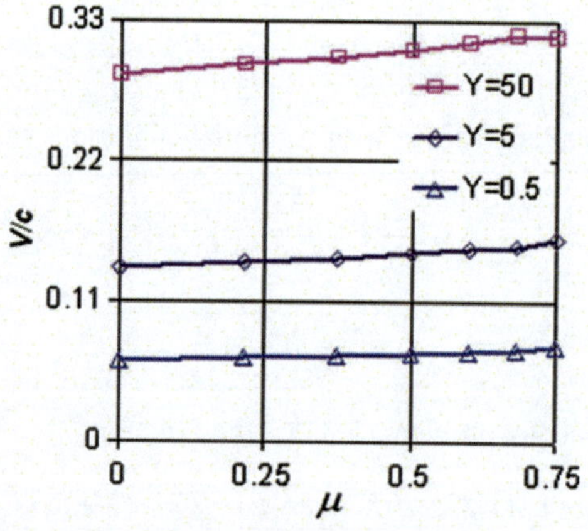

Fig. 5.1 Relation $\varepsilon = V/c$ as a function of the internal friction coefficient for the cavity expansion velocities close to 0

the internal friction coefficient, for the values of initial yield strength $Y = 50, 5$ and 0.5 kgs/cm^2 (the solid lines with a square, a rhomb, and a triangle, respectively).

It can be noted that, at cavity expansion velocities close to 0, the value of $\varepsilon = V/c$ depends on the initial value of yield strength. At the same time, a fairly weak (not more than 10%) change of ε with changing μ is observed in the range of $0 \leq \mu \leq 0.75$ for all the values of yield strength considered.

Considering that the value of ε weakly depends on the change of the internal friction coefficient, it is proposed to use Eqs. (5.21) and (5.24) to approximately determine ε, assuming that

$$\varepsilon^{-3} = \frac{E}{3(1-\nu)Y}, \quad \varepsilon = \sqrt[3]{\frac{3(1-\nu)Y}{E}}, \tag{5.27}$$

Thus, Eq. (5.23) and (5.27) will be a generalization of the known Eq. (5.21) for the case of the Mohr-Coulomb plasticity condition.

Figure 5.2 depicts the diagrams of distribution of critical pressure, relative to the initial value of yield strength, as a function of the value of the internal friction coefficient for the values of $Y = 50, 5$ and 0.5 kgs/cm^2 (the solid lines with a square, a rhomb, and a triangle, respectively). In contrast with ε, critical pressure substantially depends on the parameters of the Mohr-Coulomb plasticity condition.

Figure 5.3 presents the values of critical pressure acting on the cavity wall, as numerically determined for different levels of the initial value of yield strength in a medium, where a logarithmic scale is used for yield strength. The solution shown in Fig. 5.3 by a dotted line was obtained using formula (5.23), when evaluating ε as a result of numerically solving Eq. (5.26). Somewhat lower inaccuracy is observed only for the values of $Y < 0.1$ MPa, whereas for higher initial values of yield strength, solution (5.23) and (5.27) appears more preferable.

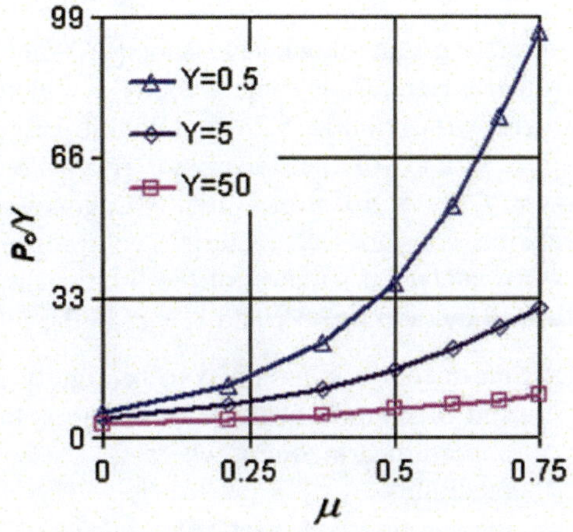

Fig. 5.2 Dimensionless critical pressure as a function of internal friction coefficient

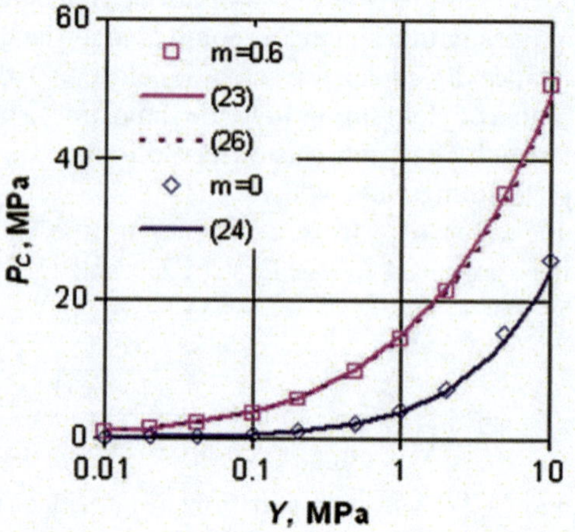

Fig. 5.3 Critical pressure as a function of the initial value of yield strength: numerical solution (marked with squares and rhombs, respectively) and solution using formulas (5.23), (5.24) and (5.26) (solid lines and the dotted line)

5.6 An Analytical Solution of the Cavity Expansion Problem in a Medium with a Linear Shock Adiabat

Experiments (Lagunov and Stepanov 1963; Dianov et al. 1977; Bragov and Grushevskii 1993; Arlery et al. 2010; Bragov et al. 2005; Balandin et al. 2015) show that dynamic compressibility of many media is characterized by Hugoniot's shock adiabat in the form of linear relation

$$U_s = C_0 + s u_p, \tag{5.28}$$

correlating plane shockwave velocity U_s and velocity of the particles behind the wave front u_p. Here, C_0 is sonic velocity in a medium under zero pressure, s is constant.

Applying relation (5.28) and Rankin-Hugoniot's conditions, a correlation between stress σ_r and volumetric strain θ is obtained, which will have the following form: $\sigma_r = f_1(\theta) \equiv \rho_0 C_0^2 \theta (1 - s\theta)^{-2}$ (Lagunov and Stepanov 1963). In this form of the relation, constant s characterizes compressive strength of the medium.

An analytical solution of the problem is constructed in the plastic yield region limited by radii $r = Vt$ and $r = ct$ in the following assumptions:

(a) the medium is assumed to be rigidly plastic (elastic deformation of the soft soil medium is neglected), i.e., the plastic yield region adjoins the region of unperturbed medium, where $\sigma_r = \upsilon = 0$, υ is velocity of the particles of the medium;

(b) the solution is a plastic shockwave propagating through an unperturbed half-space;

(c) density ρ_s behind the shockwave front is assumed constant, i.e., incompressibility is assumed behind the shockwave front, the value of ρ_s depends on the cavity expansion velocity;
(d) plastic wave front propagation velocity c is equal to the propagation velocity of the plane shockwave front, determined by Hugoniot's linear adiabat (5.28), i.e., it is assumed that $U_s \equiv c$.

Such assumptions appear to be justified for soft soils characterized by low cohesion and high porosity (Bragov et al. 2018).

Using self-similar substitution $\xi = r/ct$, the system of partial differential equations is transformed (Kotov 2019) into a system of ordinary differential equations (ODE).

Consider a boundary-value problem for a system of two first-order ODEs for dimensionless velocity $U = v/c$ and dimensionless stress $S = \sigma_r / \rho_s c^2$, which, in the above assumptions, will take the following form:

$$U' + 2\frac{U}{\xi} = 0, \quad \varepsilon < \xi < 1 \tag{5.29}$$

$$U(\xi = \varepsilon) = \varepsilon, \tag{5.30}$$

$$U(\xi = 1) = \theta_s \tag{5.31}$$

$$S' + 2\frac{\tilde{f}_2}{\xi} = (\xi - U)U', \varepsilon < \xi < 1 \tag{5.32}$$

$$S(\xi = 1) = \theta_s - \theta_s^2, \tag{5.33}$$

where $\varepsilon = V/c$ is value of the dimensionless coordinate, corresponding to the cavity boundary, $\tilde{f}_2() = f_2()/\rho_s c^2$ is dimensionless function in the plasticity condition, the volumetric strain along the shockwave front takes the value $\theta_s = (1 - C_0/c)/s$, the stroke shows differentiation for ξ.

Apart from dimensionless velocity U and stress S, equation system (5.29)–(5.33) includes unknown parameter c—propagation velocity of the plastic shockwave front.

Equation (5.29) is an equation with separable variables $\mathrm{d}U/U = -2\mathrm{d}\xi/\xi$, the solution of which yields $U = c_1/\xi^2$. The integration constant is determined from boundary condition (5.30) as $c_1 = \varepsilon^3$, and the dimensionless velocity has the form:

$$U = \varepsilon^3/\xi^2. \tag{5.34}$$

To find the unknown value of ε, accounting for boundary condition (5.30) $\varepsilon^3 = (1 - C_0/c)/s = 1 - (C_0 V/Vc)/s = (1 - \varepsilon/M)/s$, the following cubic equation is obtained:

$$\varepsilon^3 + \frac{\varepsilon}{sM} - \frac{1}{s} = 0, \tag{5.35}$$

where $M = V/C_0$.

The coefficients in Eq. (5.35) depend only on the parameters of the shock adiabat of the medium C_0, s and cavity expansion velocity V.

Consider a linear approximation of ε following from Eq. (5.35), using Taylor's series expansion $\varepsilon = \frac{1}{\sqrt[3]{s}}\left(1 + \left(-\frac{\varepsilon}{M}\right)\right)^{1/3} \approx \frac{1}{\sqrt[3]{s}}\left(1 - \frac{\varepsilon}{3M}\right)$

$$\varepsilon \approx 3M/\left(1 + 3M\sqrt[3]{s}\right). \tag{5.36}$$

Designating $\varepsilon = V/c$, the following expression is obtained:

$$c \approx \sqrt[3]{s}V + C_0/3. \tag{5.37}$$

A solution for a rigid-plastic medium was earlier obtained for the cylindrical cavity expansion problem (Forrestal and Longcope 1982).

It is also to be noted that the dimensionless velocity of a medium is determined only by Hugoniot's shock adiabat and does not depend on the plasticity condition of the medium.

5.7 Determining Stresses in a Medium with the Mohr-Coulomb Yield Condition

It is assumed, in what follows, that the plasticity criterion of the medium is described by Mohr-Coulomb law

$$f_2() \equiv \tau_0 + \kappa p = \tau_0 + \mu\sigma_r, \tag{5.38}$$

where τ_0 is cohesion, κ is internal friction coefficient, $p = \left(\sigma_r + 2\sigma_\phi\right)/3$ is pressure, $\mu = \kappa/1 + 2\kappa/3$.

With the account of solution (5.34) and yield criterion (5.38), ODE (5.32) and boundary condition (5.33) can be written as:

$$S' + 2\frac{T_0 + \mu S}{\xi} = -2\left(\xi - \frac{\varepsilon^3}{\xi^2}\right)\frac{\varepsilon^3}{\xi^3} = -2\frac{\varepsilon^3}{\xi^2} + 2\frac{\varepsilon^6}{\xi^5},\ \varepsilon < \xi < 1 \tag{5.39}$$

$$S(\xi = 1) = \varepsilon^3 - \varepsilon^6 \tag{5.40}$$

where $T_0 = \tau_0/\rho_s c^2$.

With the account of boundary conditions (5.40), the dimensionless stress will take the form

$$S(\xi) = -\frac{T_0}{\mu} - 2\frac{\varepsilon^3}{(2\mu-1)\xi} + \frac{\varepsilon^6}{(\mu-2)\xi^4} + \left(\frac{T_0}{\mu} + 2\frac{\varepsilon^3}{(2\mu-1)} - \frac{\varepsilon^6}{(\mu-2)} + \varepsilon^3 - \varepsilon^6\right)\xi^{-2\mu} \tag{5.41a}$$

Equation (5.41a) is not defined for $\mu = 0$ and 0.5. A solution in those cases can be obtained by passing to the limit in (5.41a) with μ tending to 0 and 0.5, respectively,

$$S(\xi) = -2T_0 \ln \xi - \varepsilon^3\left(1 - \frac{2}{\xi}\right) - \frac{\varepsilon^6}{2}\left(1 + \frac{1}{\xi^4}\right),\ \mu = 0, \tag{5.41b}$$

$$S(\xi) = -2T_0\left(1 - \frac{1}{\varepsilon}\right) - \varepsilon^3\left(2\frac{\ln \xi}{\xi} - \frac{1}{\xi}\right) - \frac{\varepsilon^6}{3}\left(\frac{1}{\xi} + \frac{2}{\xi^4}\right),\ \mu = 0.5 \tag{5.41c}$$

In a dimensional form, the relation for stress as a function of the self-similar variable will have the form:

$$\sigma_r(\xi) = \tau_0\left(\frac{\xi^{-2\mu} - 1}{\mu}\right) + \frac{\rho_0 V^2 \varepsilon}{1 - \varepsilon^3}\left(\frac{2}{(2\mu-1)}\left(\frac{\xi^{-2\mu+1} - 1}{\xi}\right) + \frac{\varepsilon^3}{(\mu-2)}\left(\frac{1 - \xi^{-2\mu+4}}{\xi^4}\right) + (1 - \varepsilon^3)\xi^{-2\mu}\right). \tag{5.42}$$

Expression (5.42) uses equality $\rho_s = \rho_0/(1 - \theta_s) = \rho_0/(1 - \varepsilon^3)$ that follows from boundary condition (5.31) with the account of solution (5.34).

In a dimensional form, the stress along the boundary of the cavity expanding at velocity V will be:

$$\sigma_r(\xi = \varepsilon) = \tau_0\left(\frac{\varepsilon^{-2\mu} - 1}{\mu}\right) + \frac{\rho_0 V^2}{1 - \varepsilon^3}\left(\frac{3}{(\mu-2)(2\mu-1)} + \frac{2\mu+1}{2\mu-1}\cdot\varepsilon^{1-2\mu} - \frac{\mu-1}{\mu-2}\cdot\varepsilon^{4-2\mu}\right), \tag{5.43a}$$

$$\sigma_r(\xi = \varepsilon) = -2\tau_0 \ln \varepsilon + \frac{\rho_0 V^2}{1 - \varepsilon^3}(3/2 - \varepsilon - \varepsilon^4/2).,\ \mu = 0, \tag{5.43b}$$

$$\sigma_r(\xi = \varepsilon) = 2\tau_0(\varepsilon^{-1} - 1) + \frac{\rho_0 V^2}{1 - \varepsilon^3}(1/3 - 2\ln \varepsilon - \varepsilon^3/3),\ \mu = 0.5 \tag{5.43c}$$

In Eqs. (5.42) and (5.43), the value of ε is determined based on Eq. (5.36). Thus, closed forms of relations have been obtained that make it possible to find stress in a medium with the Mohr-Coulomb plasticity condition.

5.8 Determining Stress in a Medium with Mohr-Coulomb-Tresca's Yield Condition

The function of plasticity for the Mohr-Coulomb condition, with the account of a constraint on the maximal value of Tresca's yield strength, in a dimensional and dimensionless form has the following form:

$$f_2 \equiv \begin{cases} \tau_0 + \mu\sigma_r, & 0 < \sigma_r \le \sigma_{\mathrm{M}}, \\ \tau_M, & \sigma_r > \sigma_M \end{cases}, \quad \tilde{f}_2 \equiv \begin{cases} T_0 + \mu S, & 0 < S \le S_{\mathrm{M}}, \\ T_M, & S > S_M, \end{cases}$$

where $\sigma_M = (\tau_M - \tau_0)/\mu$ and dimensionless quantities $T_M = \tau_M/\rho_s c^2$, $S_M = (T_M - T_0)/\mu$ are introduced.

Stress S monotonely decreases with dimensionless coordinate ξ changing from ε to 1, i.e., it has its minimal value for $\xi = 1$. The value of cavity expansion velocity V_M, at which $S(\xi = 1) = S_M$, will be now determined.

From the relations on the shockwave for $\xi = 1$, it follows that $\sigma_r = \rho_0 c \upsilon = \rho_0 c^2\theta$. The values of the shockwave velocity, volumetric strain, and ε, corresponding to $V = V_M$, will be designated as c_M, θ_M and ε_M, respectively. To define c_M, formula (5.37) will be used. Then, to

$$c_M = \sqrt[3]{s}V_M + C_0/3, \theta_M = \varepsilon_M^3 = \frac{V_M^3}{c_M^3}, \sigma_M = \rho_0 c_M^2\theta_M = \rho_0\frac{V_M^3}{c_M}$$

determine V_M, the following cubic equation is obtained:

$$\left(\frac{\tau_M - \tau_0}{\mu}\right)\left(\sqrt[3]{s}V_M + C_0/3\right) = \rho_0 V_M^3 \tag{5.44}$$

In a similar way, the stress has its maximal value for $\xi = \varepsilon$. Symbol V_0 will designate the value of cavity expansion velocity, for which equalities $S(\xi = \varepsilon) = S_M$ or $\sigma_r(\xi = \varepsilon) = \sigma_M$ hold. Using value $V = V_0$ in Eq. (5.43a): $\sigma_M = \sigma_r(\xi = \varepsilon)\big|_{V=V_0}$, the following nonlinear equation for determining V_0 is obtained

$$\frac{\tau_M - \tau_0}{\mu} = \frac{\tau_0}{\mu}\left(1 - \varepsilon_0^{-2\mu}\right) + \frac{\rho_0 V_0^2}{1 - \varepsilon_0^3}\left(\frac{3}{(\mu - 2)(2\mu - 1)} + \frac{2\mu + 1}{2\mu - 1}\cdot\varepsilon_0^{1-2\mu} - \frac{\mu - 1}{\mu - 2}\cdot\varepsilon_0^{4-2\mu}\right), \tag{5.45}$$

where the following designations are introduced $\varepsilon_0 = V_0/c_0$, $c_0 = \sqrt[3]{s}V_0 + C_0/3$.

For the cavity expansion velocity varying in the range of $V_0 < V < V_M$, it is necessary to determine the value of dimensionless coordinate $\xi = \xi_M$, for which the dimensionless stress has the value $S(\xi = \xi_M) = S_M$.

Finally, dimensional stress along the cavity boundary $\sigma_C \equiv S(\xi = \varepsilon)\rho_s c^2$

$$\sigma_r(V) = \begin{cases} \frac{\tau_0}{\mu}\left(\varepsilon^{-2\mu} - 1\right) + \frac{\rho_0 V^2}{1-\varepsilon^3}\left(\frac{3}{(\mu-2)(2\mu-1)} + \right. \\ \left. + \frac{2\mu+1}{2\mu-1} \cdot \varepsilon^{1-2\mu} - \frac{\mu-1}{\mu-2} \cdot \varepsilon^{4-2\mu}\right), 0 < V < V, \\ \sigma_M + \tau_M \ln\left(\frac{\xi_M}{\varepsilon}\right)^2 + \frac{\rho_0 V^2}{1-\varepsilon^3}\left(\frac{3}{2} - \frac{2\varepsilon}{\xi_M} + \frac{\varepsilon^4}{2\xi_M^4}\right), V_0 \le V \le V_M, \\ -2\tau_M \ln\varepsilon + \frac{\rho_0 V^2}{1-\varepsilon^3}\left(\frac{3}{2} - \varepsilon - \frac{\varepsilon^4}{2}\right), V > V_M. \end{cases} \quad (5.46)$$

In Eq. (5.46), the value of ε is determined based on Eq. (5.36). Thus, closed forms of relations have been obtained, that make it possible to determine stress in a medium with the Mohr-Coulomb plasticity condition and Tresca's condition.

5.9 Comparative Analysis of the Results of Analyzing the Cavity Problem in Media with Tresca's, Mohr-Coulomb, and Mohr-Coulomb-Tresca's Plasticity Conditions

The table below summarizes the values of the parameters of the equation of state for dry and water-saturated sand, earlier obtained based on the results of a series of impact experiments and numerical computations (Bragov et al. 2018).

The first and the second lines in Table 5.1 correspond to dry and water-saturated sand followed by the results of computations for the values of cavity expansion 400 and 250 m/s, respectively, for dry and water-saturated sand.

Figure 5.4 presents the stresses along the cavity boundary as a function of its expansion velocity in dry (a) and water-saturated (b) sand: the solid line, the dashed line with a triangle, the dashed-dotted lines, and the dashed line with a square correspond to the results obtained using Eqs. (5.46), (5.43a), (5.43b) and (5.43b) for $\tau_0 = \tau_M$.

It can be noted that stresses along the cavity boundary in a medium with Mohr-Coulomb-Tresca's plasticity condition can be determined accurately enough for practical purposes without using Eqs. (5.45) and (5.46), resorting to convex interpolation with Hermit or Bezier's cubic polynomials (Bazhenov et al. 2001) in the range of cavity expansion velocities of $V_0 \le V \le V_M$.

Table 5.1 Parameters of the equation of state for dry and water-saturated sand

№	ρ_0, kg/m^3	C_0, m/s	s	τ_0, MPa	μ	τ_M, MPa	σ_M, MPa
1	1730	460	2.3	0.042	0.6	180	300
2	2080	1700	3.4	0.021	0.25	25	1000

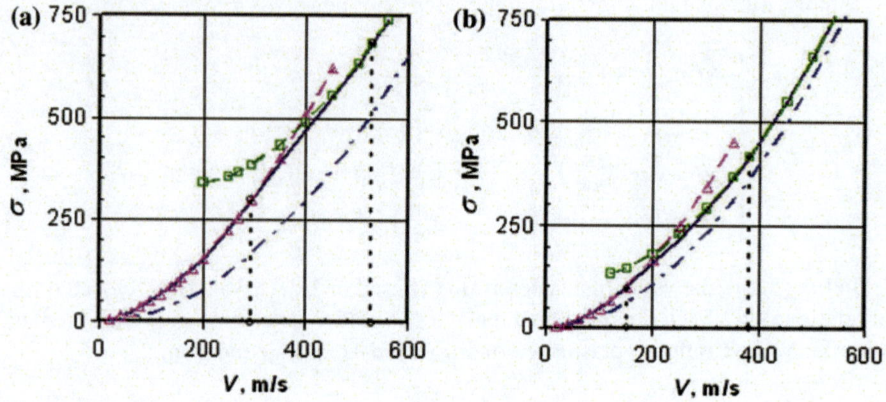

Fig. 5.4 Approximations of the stress along the cavity boundary as a function of its expansion velocity in dry (**a**) and water-saturated (**b**) sands

5.10 Conclusion

A formula has been derived for determining a minimal pressure leading to the nucleation of a cavity (a critical pressure), with the account of internal friction in the framework of the Mohr-Coulomb yield criterion, which is a generalization of the well-known solution for elastic ideally plastic media with Tresca's criterion. The approximation inaccuracy does not exceed 6% and the initial value of yield strength by three orders of magnitude. It is to be noted that Eqs. (5.23) and (5.27) yield good results for Y changing in the range of three orders of magnitude and for the internal friction coefficient changing over the entire permissible range, and that the critical pressure grows with internal friction.

An analytical solution of the spherical cavity expansion problem in a rigid-plastic medium with Mohr-Coulomb-Tresca's plasticity condition has been obtained in the assumption of incompressibility behind the shockwave front. The effect of constraining the value of maximal tangential stresses on the distribution of dimensionless stresses behind the shockwave front has been analyzed. Formulas for determining a maximum cavity expansion velocity have been derived that define the upper limits of possible values of the cavity expansion velocities, for which the solution with Mohr-Coulomb plasticity condition and a minimal velocity holds. For velocities exceeding this minimal velocity, Tresca's plasticity condition holds. The obtained generalization can be used for evaluating forces resisting high-velocity penetration of rigid strikers into low-strength soil media.

Acknowledgements The work was done under financial support from the Grant of Russian Fund of Fundamental Researches (Project No. 19-08-00430).

References

Abali, B. E., Muller, W., & dell'Isola, F. (2017). Theory and computation of higher gradient elasticity theories based on action principles. *Archive of Applied Mechanics, 87,* 1495–1510.

Alibert, J. J., Seppecher, P., & dell'Isola, F. (2003). Truss modular beams with deformation energy depending on higher displacement gradients. *Mathematics and Mechanics of Solids, 8,* 51–73.

Arlery, M., Gardou, M., Fleureau, J. M., & Mariotti, C. (2010). Dynamic behaviour of dry and water-saturated sand under planar shock conditions. *The International Journal of Impact Engineering, 37,* 1–10. https://doi.org/10.1016/j.ijimpeng.2009.07.009.

Auffray, N., dell'Isola, F., Eremeyev, V., Madeo, A., & Rosi, G. (2013). Analytical continuum mechanics à la Hamilton-Piola: Least action principle for second gradient continua and capillary fluids. *Mathematics and Mechanics of Solids.*

Balandin, V. V., Bragov, A. M., Igumnov, L. A., Konstantinov, A. Yu., Kotov, V. L., & Lomunov, A. K. (2015). Dynamic deformation of soft soil media: Experimental studies and mathematical modeling. *Mechanics of Solids, 50*(3), 286–293. https://doi.org/10.3103/S002565441503005X.

Barchiesi, E., Spagnuolo, M., & Placidi, L. (2018). Mechanical metamaterials: A state of the art. *Mathematics and Mechanics of Solids.*

Bazhenov, V. G., & Kotov, V. L. (2008). Method for identifying elastoplastic properties of ground media by penetration of impactors. *Mechanics of Solids, 4*(43), 678–686. https://doi.org/10.3103/S002565440804016X.

Bazhenov, V. G., Kotov, V. L., Kochetkov, A. V., Krylov, S. V., & Feldgun, V. R. (2001). On wave processes in soil subjected to a surface charge explosion. *Mechanics of Solids, 2*(36), 62–68.

Bazhenov, V. G., Bragov, A. M., & Kotov, V. L. (2009). Experimental-theoretical study of the penetration of rigid projectiles and identification of soil properties. *Journal of Applied Mechanics and Technical Physics, 6*(50), 1011–1019. https://doi.org/10.1007/s10808-009-0135-6.

Ben-Dor, G., Dubinsky, A., & Elperin, T. (2005). Ballistic impact: Recent advances in analytical modeling of plate penetration dynamics–A review. *Applied Mechanics Reviews, 58,* 355–371.

Bragov, A. M., & Grushevskii, G. M. (1993). Influence of the moisture content and granulometric composition on the shock compressibility of sand. *Technical Physics Letters, 19,* 385–386.

Bragov, A. M., Lomunov, A. K., Sergeichev, I. V., Proud, W., Tsembelis, K., & Church, P. (2005). A method for determining the main mechanical properties of soft soils at high strain rates (103–105 s–1) and load amplitudes up to several gigapascals. *Technical Physics Letters, 31*(6), 530–531. https://doi.org/10.1134/1.1969791.

Bragov, A. M., Balandin, V. V., Igumnov, L. A., Kotov, V. L., Krushka, L., & Lomunov, A. K. (2018). Impact and penetration of cylindrical bodies into dry and water-saturated sand. *International Journal of Impact Engineering, 122,* 197–208.

dell'Isola, F., & Hutter, K. (1998). A qualitative analysis of the dynamics of a sheared and pressurized layer of saturated soil. *Proceedings of the Royal Society of London A: Mathematical, Physical and Engineering Sciences, 454*(1980), 3105–3120.

dell'Isola, F., & Hutter, K. (1999). Variations of porosity in a sheared pressurized layer of saturated soil induced by vertical drainage of water. *Proceedings of the Royal Society of London A: Mathematical, Physical and Engineering Sciences, 455*(1988), 2841–2860.

dell'Isola, F., Guarascio, M., & Hutter, K. (2000). A variational approach for the deformation of a saturated porous solid. A second-gradient theory extending Terzaghi's effective stress principle. *Archive of Applied Mechanics, 70*(5), 323–337.

dell'Isola, F., Seppecher, P., & Madeo, A. (2012). How contact interactions may depend on the shape of Cauchy cuts in Nth gradient continua: Approach "à la D'Alembert". *Zeitschrift für Angewandte Mathematik und Physik (ZAMP), 63*(6), 1119–1141.

dell'Isola, F., Andreaus, U., & Placidi, L. (2015). At the origins and in the vanguard of peridynamics, non-local and higher gradient continuum mechanics. An underestimated and still topical contribution of Gabrio Piola. *Mathematics and Mechanics of Solids, 20,* 887–928.

dell'Isola, F., Madeo, A., & Seppecher, P. (2016). Cauchy tetrahedron argument applied to higher contact interactions. *Archive for Rational Mechanics and Analysis, 219*(3).

dell'Isola, F., Della Corte, A., & Giorgio, I. (2017). Higher-gradient continua: The legacy of Pio-la, Mindlin, Sedov and Toupin and some future research perspectives. *Mathematics and Mechanics of Solids, 22,* 852–872.

dell'Isola, F., Seppecher, P., Alibert, J. J., Lekszycki, T., Grygoruk, R., Pawlikowski, M., et al. (2019). Pantographic metamaterials: An example of mathematically driven design and of its technological challenges. *Continuum Mechanics and Thermodynamics, 31*(4), 851–884.

Del Vescovo, D., & Giorgio, I. (2014). Dynamic problems for metamaterials: Review of existing models and ideas for further research. *International Journal of Engineering Science, 80,* 153–172.

Dianov, M. D., Zlatin, N. A., Mochalov, S. M., et al. (1977). Shock compressibility of dry and water-saturated sand. *Soviet Technical Physics Letters, 2,* 207–208.

Forrestal, M. J., & Longcope, D. B. (1982). Closed-form solution for forces on conical-nosed penetrators into geological targets with constant shear strength. *Mechanics of Materials, 1,* 285–295.

Forrestal, M. J., & Luk, V. K. (1988). Dynamic spherical cavity-expansion in a compressible elastic-plastic solid. *The Journal of Applied Mechanics, 55*(2), 275–279 https://doi.org/10.1115/1.3173672.

Forrestal, M. J., & Luk, V. K. (1992). Penetration into soil targets. *The International Journal of Impact Engineering, 12*(3), 427–444. https://doi.org/10.1016/0734-743X(92)90167-R.

Grigoryan, S. S. (1960). On basic concepts of soil dynamics. *Journal of Applied Mathematics and Mechanics, 24*(6), 1604–1627. https://doi.org/10.1016/0021-8928(60)90013-7.

Kong, X. Z., Wu, H., Fang, Q., & Peng, Y. (2017). Rigid and eroding projectile penetration into concrete targets based on an extended dynamic cavity expansion model. *The International Journal of Impact Engineering, 100,* 13–22.

Kotov, V. L. (2008). Studying the applicability of the auto-modeling analysis of the problem of expansion of a spherical cavity in a compressible medium for determining the pressure on the surface of the striker-soil contact. *Problems of Strength and Plasticity, 70,* 123–131. (in Russian).

Kotov, V. L. (2019). Approximation of stresses in the vicinity of the cavity, expanding at a constant speed in a medium with the Mohr-Coulomb yield criterion. *Problems of Strength and Plasticity, 81*(2), 1–10. (in Russian).

Kotov, V. L., Linnik, E. Yu., Tarasova, A. A., & Makarova, A. A. (2011). The analysis of approximate solutions of a spherical cavity expansion in a soil medium problem. *Problems of Strength and Plasticity, 73,* 58–63. (in Russian).

Kotov, V. L., Balandin, V. V., Bragov, A. M., Linnik, E. Yu., & Balandin, V. V. (2013). Using a local-interaction model to determine the resistance to penetration of projectiles into sandy soil. *Journal of Applied Mechanics and Technical Physics, 54*(4), 612–621 https://doi.org/10.1134/S0021894413040123.

Lagunov, V. A., & Stepanov, V. A. (1963). Measurements of the dynamic compressibility of sand under high pressures. *Zh. Prikl. Mekh. Tekhn. Fiz., 1,* 88–96. [J. Appl. Mech. Tech. Phys. (Engl. Transl.)].

Linnik, E. Yu., Kotov, V. L., Tarasova, A. A., & Gonik, E. G. (2012). The solution of the problem of the expansion of a spherical cavity in a soil medium assuming incompressibility beyond the shock front. *Problems of Strength and Plasticity., 74,* 49–58. (In Russian).

Omidvar, M., Iskander, M., & Bless, S. (2012). Stress-strain behavior of sand at high strain rates. *International Journal of Impact Engineering, 49,* 192–213. https://doi.org/10.1016/j.ijimpeng.2012.03.004.

Omidvar, M., Iskander, M., & Bless, S. (2014). Response of granular media to rapid penetration. *International Journal of Impact Engineering, 66,* 60–82. https://doi.org/10.1016/j.ijimpeng.2013.12.004.

Omidvar, M., Malioche, Jeanne D., Bless, S., & Iskander, M. (2015). Phenomenology of rapid projectile penetration into granular soils. *International Journal of Impact Engineering, 85,* 146–160. https://doi.org/10.1016/j.ijimpeng.2015.06.002.

Placidi, L., Barchiesi, E., Turco, E., & Rizzi, N. L. (2016). A review on 2D models for the description of pantographic fabrics. *Zeitschrift für angewandte Mathematik und Physik, 67*(5).

Placidi, L., Andreaus, U., & Giorgio, I. (2017). Identification of two-dimensional pantographic structure via a linear D4 orthotropic second gradient elastic model. *Journal of Engineering Mathematics*.

Rahali, Y., Giorgio, I., Ganghoffer, J.-F., & dell'Isola, F. (2015). Homogenization à la Piola produces second gradient continuum models for linear pantographic lattices. *International Journal of Engineering Science, 97*.

Rosenberg, Z., & Dekel, E. (2008). A numerical study of the cavity expansion process and its application to long-rod penetration mechanics. *International Journal of Impact Engineering, 35*(3), 147–154.

Shi, C., Wang, M., Li, J., & Li, M. (2014). A model of depth calculation for projectile penetration into dry sand and comparison with experiments. *International Journal of Impact Engineering, 73,* 112–122. https://doi.org/10.1016/j.ijimpeng.2014.06.010.

Shi, C., Wang, M., Zhang, K., Cheng, Y., & Zhang, X. (2015). Semi-analytical model for rigid and erosive long rods penetration into sand with consideration of compressibility. *International Journal of Impact Engineering, 83,* 1–10.

Sun, Q., Sun, Y., Liu, Y., Li, R., & Zhao, Y. (2017). Numerical analysis of the trajectory stability and penetration ability of different lateral-abnormal projectiles for non-normal penetration into soil based on modified integrated force law method. *International Journal of Impact Engineering, 103,* 159–168.

Veldanov, V. A., Markov, V. A., Pusev, V. I., Ruchko, A. M., Sotskii, M. Yu., & Fedorov, S. V. (2011). Computation of non-deformable striker penetration into low strength obstacles using piezoelectric accelerometry data. *Technical Physics, 56*(7), 992–1002. https://doi.org/10.1134/S1063784211070231.

Chapter 6
Construction of the Solutions of Non-stationary Dynamic Problems for Linear Viscoelastic Bodies with a Constant Poisson's Ratio

Leonid Igumnov, Ekaterina A. Korovaytseva, and Sergey G. Pshenichnov

Abstract The problems of propagation of non-stationary waves in linear viscoelastic bodies on condition that the Poisson's ratio of the material does not change through the time are considered. The issues of finding of the solutions of such problems by the method of Laplace transform in time are discussed. Some properties of the solution in Laplace transforms, which relate to their singular points and simplify finding the originals, have also been considered. The case when a hereditary kernel is an exponential two-parametrical one is considered. We have demonstrated that in such case the singular points of the Laplace transforms are connected by a simple relation with the singular points of the Laplace transforms for the corresponding elastic body. The conditions under which the poles of the solution in transforms have the first order have been established. As an example, the solution of the problem of one-dimensional non-stationary wave propagation in a linear viscoelastic cylinder is presented.

Keywords Viscoelastic bodies · Non-stationary waves · Laplace transform

6.1 Introduction

In studies of transitional wave processes in viscoelastic bodies, an important role is played by the analytical methods of building the solutions of non-stationary dynamic problems of linear viscoelasticity. One of such methods developed by Ilyasov involves a special type of convolution of solution of a dynamic problem of the elasticity theory with the solution of a one-dimensional problem where hereditary kernels of a viscoelastic body take part (Ilyasov 2011). There is a numerical

---

L. Igumnov
Research Institute for Mechanics, National Research Lobachevsky State University of Niznhy Novgorod, Niznhy Novgorod, Russia

E. A. Korovaytseva (✉) · S. G. Pshenichnov
Institute of Mechanics, Lomonosov Moscow State University, Moscow, Russia
e-mail: katrell@mail.ru

F. dell'Isola and L. Igumnov (eds.), *Dynamics, Strength of Materials and Durability in Multiscale Mechanics*, Advanced Structured Materials 137,
https://doi.org/10.1007/978-3-030-53755-5_6

analytical method of modal expansion developed by Zheltkov (Jeltkov et al. 1993). A method being developed by Lychev is based on representing the solutions of the problems under consideration in the form of a spectral expansion with biorthogonal systems of the eigenfunctions of mutually conjugated pencils of differential operators (Lycheva and Lychev 2016). Laplace time transformation followed by inversion is the most common procedure of constructing the solutions of the considered problems (Christensen 1982; Filippov and Cheban 1988; Egorychev and Poddayeva 2006; Colombaro et al. 2017).

In addition to the works mentioned above, one can refer to a lot of others (for instance, Lokshin 1982; Amendola et al. 2012; Kurbanov and Nasibzada 2015), however, let us remark here that the mathematical complexity significantly limits the class of the studied problems. So, even for one-dimensional non-stationary wave processes the most results were received either in a limited time range or with low viscosity or they were represented in a hardly analyzable form.

The purpose of this work is to consider the issues related to constructing the solutions of problems of the above-noted types with a time-independent Poisson's ratio. We note that the process of constructing of the solution of non-stationary dynamic problems for linear viscoelastic bodies with a time-independent Poisson's ratio using the Laplace transform in time was considered earlier in the work (Christensen 1982) and some others. In this paper, the attention is paid to the conditions under which the poles of the solution in transforms are simple, and there are no branch points.

6.2 Mathematical Statement of Problem

Let us consider a non-stationary dynamic problem for a linear viscoelastic homogeneous isotropic body which occupies an area Ω with boundary Σ for a case when Poisson's ratio is time-independent: $\nu \equiv \nu_0$ (const). Let us set volumetric and shear relaxations of the material as $T(t)$, which in this case are identical. Mathematically, such problem will be stated with the equations:

$$(1 - \hat{T})\hat{L}\mathbf{u}(\mathbf{x}, t) + \mathbf{f}(\mathbf{x}, t) = \rho\ddot{\mathbf{u}}(\mathbf{x}, t), \tag{6.1}$$

$$\tilde{\sigma}(\mathbf{x}, t) = (1 - \hat{T})\,\hat{l}\mathbf{u}(\mathbf{x}, t), \mathbf{x}(x_1, x_2, x_3) \in \Omega, \tag{6.2}$$

boundary conditions ($\Sigma = \Sigma_1 \cup \Sigma_2$)

$$\tilde{\sigma}(\mathbf{x}, t)\mathbf{n} = \mathbf{p}^{(1)}(\mathbf{x}, t), \mathbf{x} \in \Sigma_1;\ \mathbf{u}(\mathbf{x}, t) = \mathbf{p}^{(2)}(\mathbf{x}, t), \mathbf{x} \in \Sigma_2, t > 0, \tag{6.3}$$

and initial conditions

$$\mathbf{u}(\mathbf{x}, 0) = \mathbf{b}^{(1)}(\mathbf{x}), \quad \dot{\mathbf{u}}(\mathbf{x}, 0) = \mathbf{b}^{(2)}(\mathbf{x}), \quad \mathbf{x} \in \Omega. \tag{6.4}$$

Here, $\hat{L}$ and $\hat{l}$ are differential operators, $\hat{T}$ is integral operator:

$$\hat{L}\mathbf{u}(\mathbf{x}, t) = (\lambda_0 + \mu_0)\text{grad div }\mathbf{u}(\mathbf{x}, t) + \mu_0 \Delta \mathbf{u}(\mathbf{x}, t), \tag{6.5}$$

$$\hat{l}\mathbf{u}(\mathbf{x}, t) = 2\mu_0 \text{def }\mathbf{u}(\mathbf{x}, t) + \lambda_0 \text{ div }\mathbf{u}(\mathbf{x}, t)\tilde{\mathbf{I}},\ \hat{T}\xi(t) = \int_0^t T(t-\tau)\xi(\tau)\,\mathrm{d}\tau, \tag{6.6}$$

$\tilde{\sigma}$ is stress tensor; $\mathbf{u}$, $\mathbf{p}^{(1)}$, $\mathbf{p}^{(2)}$, $\mathbf{f}$, $\mathbf{b}^{(1)}$, $\mathbf{b}^{(2)}$ are vectors of displacements, boundary actions, volumetric forces, initial displacements and velocities; $\mathbf{n}$ is outer unit normal; ρ is density; Δ is Laplace operator; $\tilde{\mathbf{I}}$ is unit tensor; λ_0, μ_0 are Lame elastic constants; the dot denotes a time derivative t.

Here and further, let us assume that the area of disturbance is limited, the displacement of the body as a rigid whole is excluded, and the creep of the material is limited.

6.3 Representation of the Problem in Transform Domain

Let us apply the Laplace integral transform in time to the Eqs. (6.1), (6.2) and boundary conditions (6.3) taking account of the initial conditions (6.4). In the transform domain, the following equations will be obtained

$$[1 - \Theta(s)]\hat{L}\mathbf{U}(\mathbf{x}, s) - \rho s^2 \mathbf{U}(\mathbf{x}, s) + \rho[s\mathbf{b}^{(1)}(\mathbf{x}) + \mathbf{b}^{(2)}(\mathbf{x})] + \mathbf{F}(\mathbf{x}, s) = \mathbf{0}, \tag{6.7}$$

$$\tilde{\mathbf{S}}(\mathbf{x}, s) = [1 - \Theta(s)]\hat{l}\mathbf{U}(\mathbf{x}, s), \mathbf{x} \in \Omega, s \in C \tag{6.8}$$

and boundary conditions

$$\tilde{\mathbf{S}}(\mathbf{x}, s)\mathbf{n} = \mathbf{P}^{(1)}(\mathbf{x}, s), \mathbf{x} \in \Sigma_1;\ \mathbf{U}(\mathbf{x}, s) = \mathbf{P}^{(2)}(\mathbf{x}, s), \mathbf{x} \in \Sigma_2 \tag{6.9}$$

where

$$\mathbf{U}(\mathbf{x}, s), \tilde{\mathbf{S}}(\mathbf{x}, s), \mathbf{F}(\mathbf{x}, s), \mathbf{P}^{(1)}(\mathbf{x}, s), \mathbf{P}^{(1)}(\mathbf{x}, s), \Theta(s)$$

are the transforms of the values

$$\mathbf{u}(\mathbf{x}, t), \tilde{\sigma}(\mathbf{x}, t), \mathbf{f}(\mathbf{x}, t), \mathbf{p}^{(1)}(\mathbf{x}, t), \mathbf{p}^{(2)}(\mathbf{x}, t), T(t)$$

Let us write the solution $\mathbf{U}(\mathbf{x}, s)$ of the problem (6.7)–(6.9) in the following form:

$$\mathbf{U}(\mathbf{x}, s) = \mathbf{U}^{(0)}(\mathbf{x}, s) + \mathbf{V}(\mathbf{x}, s), \tag{6.10}$$

where $\mathbf{U}^{(0)}(\mathbf{x}, s)$ is the solution of a problem involving equations

$$\hat{L}\mathbf{U}^{(0)}(\mathbf{x}, s) + \frac{1}{[1-\Theta(s)]}\{\rho[s\mathbf{b}^{(1)}(\mathbf{x}) + \mathbf{b}^{(2)}(\mathbf{x})] + \mathbf{F}(\mathbf{x}, s)\} = \mathbf{0}, \tag{6.11}$$

$$\tilde{\mathbf{S}}^{(0)}(\mathbf{x}, s) = \hat{l}\mathbf{U}^{(0)}(\mathbf{x}, s), \mathbf{x} \in \Omega, s \in C \tag{6.12}$$

and boundary conditions

$$\tilde{\mathbf{S}}^{(0)}(\mathbf{x}, s)\mathbf{n} = \frac{\mathbf{P}^{(1)}(\mathbf{x}, s)}{1-\Theta(s)}, \mathbf{x} \in \Sigma_1; \mathbf{U}^{(0)}(\mathbf{x}, s) = \mathbf{P}^{(2)}(\mathbf{x}, s), \mathbf{x} \in \Sigma_2 \tag{6.13}$$

Let us notice that (6.11)–(6.13) can be interpreted as a statistic problem of the elasticity theory with a complex parameter s as well as with complex volumetric forces and boundary actions.

Substituting the expression (6.10) in (6.7)–(6.9), we will obtain the following equation for $\mathbf{V}(\mathbf{x}, s)$:

$$\hat{L}\mathbf{V}(\mathbf{x}, s) - \rho\beta^2\mathbf{V}(\mathbf{x}, s) = \rho\beta^2\mathbf{U}^{(0)}(\mathbf{x}, s), \mathbf{x} \in \Omega, \tag{6.14}$$

$$\beta^2 = s^2/[1-\Theta(s)] \tag{6.15}$$

with homogeneous boundary conditions

$$[\hat{l}\mathbf{V}(\mathbf{x}, s)]\mathbf{n} = \mathbf{0}, \mathbf{x} \in \Sigma_1; \mathbf{V}(\mathbf{x}, s) = \mathbf{0}, \mathbf{x} \in \Sigma_2 \tag{6.16}$$

Alongside with the Eq. (6.14), let us consider the homogeneous equation

$$\hat{L}\mathbf{V}(\mathbf{x}, s) - \rho\beta^2\mathbf{V}(\mathbf{x}, s) = \mathbf{0}, \mathbf{x} \in \Omega. \tag{6.17}$$

With the boundary conditions (6.16), this equation describes the process of free oscillations of a viscoelastic body for a large period of time from an initial moment when the character of the oscillations does not depend on the way of their excitation. A complex eigenvalue s determines the frequency and coefficient of damping of oscillations, while $\mathbf{V}(\mathbf{x}, s)$ determines their form. The function $\Theta(s)$ is included only into the expression β^2 in the Eq. (6.17) but not into the boundary conditions (6.16). Therefore, the eigenfunctions of the problem (6.16), (6.17) for a viscoelastic body will be the same as in the case of linear elasticity.

Let us assume that $s_k^{(e)} = i\chi_k$, and $\mathbf{V}^{(k)}(\mathbf{x}) = \mathbf{V}(\mathbf{x}, s_k^{(e)})$ are countable sets of eigenvalues and eigenfunctions of the problem (6.16), (6.17) for an elastic body ($k = 1, 2, 3, \ldots, \chi_k \in R, T \equiv 0$). With each fixed value $s_k^{(e)} = i\chi_k$ the eigenvalues $s = s_{km}^{(v)}$ for a viscoelastic body are defined from the equation

$$\beta^2(s) = -\chi_k^2, k = 1, 2, 3, \ldots \tag{6.18}$$

Taking into account the orthogonality of the system of the eigenforms $\{\mathbf{V}^{(k)}(\mathbf{x})\}$ and assuming its completeness in the space of vector functions which are differentiated into Ω twice and satisfying the conditions (6.16), we will be able to write the solution of the problem (6.7)–(6.9) in the form

$$\mathbf{U}(\mathbf{x}, s) = \mathbf{U}^{(0)}(\mathbf{x}, s) + \sum_{k=1}^{\infty} \mathbf{V}^{(k)}(\mathbf{x})\eta_k(s) I_{\Omega}^{(k)},\ \eta_k(s) = -\frac{\beta^2(s)}{\beta^2(s) + \chi_k^2} \tag{6.19}$$

$$I_{\Omega}^{(k)} = \int_{\Omega} \mathbf{U}^{(0)}(\mathbf{x}, s)\mathbf{V}^{(k)}(\mathbf{x})\mathrm{d}\Omega\{\int_{\Omega} [\mathbf{V}^{(k)}(\mathbf{x})]^2 \mathrm{d}\Omega\}^{-1}, \tag{6.20}$$

The work (Pshenichnov 2013) establishes sufficient conditions of the absence of branch points on the complex plane at the components of vector $\mathbf{U}(\mathbf{x}, s)$ without assuming that $\nu = \text{const}$. Here, the essence of this conditions is that the set of eigenvalues of the problem (6.16), (6.17) is no more than countable one and there are no branch points in all components of the vectors $\mathbf{F}(\mathbf{x}, s)$, $\mathbf{P}^{(1)}(\mathbf{x}, s)$, $\mathbf{P}^{(2)}(\mathbf{x}, s)$, as well as function $\Theta(s)$. If these conditions are fulfilled, after application of the methods of contour integration and the theory of residues the value for $\mathbf{u}(\mathbf{x}, t)$ will be simpler if the poles $\mathbf{U}^{(0)}(\mathbf{x}, s)$ and $\eta_k(s)$ have the first order.

If $\mathbf{F}(\mathbf{x}, s)$, $\mathbf{P}^{(1)}(\mathbf{x}, s)$, $\mathbf{P}^{(2)}(\mathbf{x}, s)$, $\Theta(s)$ lead to a situation when no eigenvalue of $s_{km}^{(v)}$ for a viscoelastic body, defined from the Eq. (6.18), is a singularity of $\mathbf{U}^{(0)}(\mathbf{x}, s)$, the order of the pole of vector components $\mathbf{U}(\mathbf{x}, s)$ in points $s = s_{km}^{(v)}$ will be the same as of the function $\eta_k(s)$.

Let us consider a case where the hereditary kernel is written in the form

$$T(t) = ae^{-bt}, 0 < a < b, \Theta(s) = a/(s + b) \tag{6.21}$$

Then, the Eq. (6.18) for all $\chi_k \in R$ is a cubic one with respect to s:

$$s^3 + bs^2 + \chi_k^2 s + \chi_k^2(b - a) = 0, k = 1, 2, 3, \ldots \tag{6.22}$$

It was proved that under the condition $a/b < 8/9$ this equation for any $k = 1, 2, 3, \ldots$ will have exactly one real root $s = z_k$ and two complex conjugate ones $s = \alpha_k \pm i\omega_k$, $\omega_k > 0$, with $z_k < 0$, $\alpha_k < 0$, $z_k \neq -b$, $z_k \neq a - b$. Moreover, it was established that in points z_k and $\alpha_k \pm i\omega_k$ the functions $\eta_k(s)$ have simple poles.

6.4 Example

As an example, let us consider a problem of propagation of a non-stationary longitudinal wave in a cross section of a viscoelastic infinitely long cylinder being initially in a non-perturbed state (Fig. 6.1). The interior surface of the cylinder is rigidly

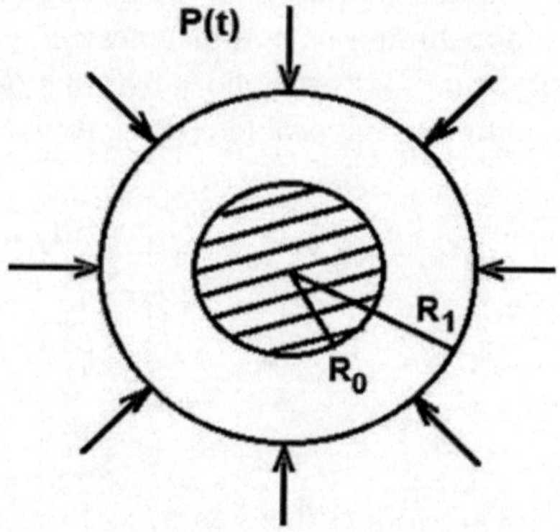

Fig. 6.1 Cylinder cross section

fixed, while the outer surface is exposed to an axisymmetric radial load $P(t)$ evenly distributed along the cylinder's element since the moment $t = 0$.

In a polar coordinate system $R,\ \theta$ with the origin at the center of the cross section, let us introduce dimensionless values

$$r = R/R_1, r_0 = R_0/R_1, \tau = t/t_0, u(r,\tau) = u_R(R,t)/R_1, \sigma_{11}(r,\tau) = \sigma_{RR}(R,t)/2\mu_0$$
$$\sigma_{22}(r,\tau) = \sigma_{\theta\theta}(R,t)/2\mu_0,\ P_0 f(\tau) = P(t)/2\mu_0,\quad \gamma(\tau) = t_0 T(t)$$

where $t_0 = R_1/c$; $R_0,\ R_1$ are interior and outer radiuses of the cylinder, $c = \sqrt{(\lambda_0 + 2\mu_0)/\rho}$ is the longitudinal elastic wave velocity, u_R is the radial displacement, $\sigma_{RR},\ \sigma_{\theta\theta}$ are stresses, P_0 is the dimensionless constant, $T(t)$ is the kernel of relaxation (Poisson's ratio ν_0 is constant). The mathematical statement of the problem consists of equations (prime marks and dots denote derivatives with respect to r and τ correspondingly):

$$(1 - \hat{\gamma})[u'(r,\tau) + u(r,\tau)/r]' = \ddot{u}(r,\tau), \tag{6.23}$$

$$\sigma_{11}(r,\tau) = (1 - \hat{\gamma})[wu'(r,\tau) + (w-1)u(r,\tau)/r], \tag{6.24}$$

and boundary and initial conditions

$$u(r_0,\tau) = 0,\quad \sigma_{11}(1,\tau) = -P_0 f(\tau);\ u(r,0) = 0,\quad \dot{u}\,(r,0) = 0 \tag{6.25}$$

where $w = (1 - \nu_0)/(1 - 2\nu_0)$.

Let us assume that $\gamma(\tau) = ae^{-b\tau}$, $0 < a < b$ and $f(\tau) = h(\tau)$ is Heaviside function. Then, after Laplace transform and inversion operation, the solution will take the form

$$u(r,\tau) = u^{(0)}(r) - 2P_0 \sum_{k=1}^{\infty} Y(r,\chi_k)\{g(z_k)e^{z_k\tau} + 2\mathrm{Re}[g(\alpha_k + i\omega_k)e^{(\alpha_k+i\omega_k)\tau}]\}, \tag{6.26}$$

where

$$u^{(0)}(r) = -P_0 \frac{(1-2\nu_0)r_0}{(1-a/b)[(1-2\nu_0)r_0^2+1]}\left(\frac{r}{r_0} - \frac{r_0}{r}\right), \tag{6.27}$$

$$Y(r,\chi) = \frac{1}{\chi \mathrm{d}X(\chi)/\mathrm{d}\chi}[J_1(r_0\chi)N_1(r\chi) - N_1(r_0\chi)J_1(r\chi)], \tag{6.28}$$

$$X(\chi) = N_1(r_0\chi)[J_1(\chi) - w\chi J_0(\chi)] - J_1(r_0\chi)[N_1(\chi) - w\chi N_0(\chi)], \tag{6.29}$$

$$g(s) = \frac{(s+b)^2}{2s^2 + (4b-3a)s + 2b(b-a)}, \tag{6.30}$$

where J_0, J_1, N_0, N_1 are Bessel functions of the first and second kinds of zero and first indices; χ_k $(k = 1, 2, 3, \ldots)$ are the real roots of the equation

$$X(\chi) = 0 \tag{6.31}$$

while z_k and $\alpha_k + i\omega_k$ $(\omega_k > 0)$ are real and one of two complex conjugate roots of the Eq. (6.22).

6.5 Notes and Comments

With a constant Poisson's ratio in the case of a two-parametrical exponential kernel (6.21), the solution of a non-stationary dynamic problem of linear viscoelasticity will not be hard to obtain if a solution of the corresponding problem of the elasticity theory is already known. In the above-stated example, the solution was presented in a rather simple form and remains true within the whole range of time changing.

Acknowledgements The reported study was funded by Russian Foundation for Basic Research, according to the research projects Nos. 18-08-00471 a, 19-38-70005 mol_a_mos.

References

Amendola, G., Fabrizio, M., Golden, J. M. (2012). *Thermodynamics of materials with memory*, 574 p. Heidelberg: Springer.

Christensen R. M. (1982). *Theory of viscoelasticity: An introduction* (364 p). New York: Academic Press.

Colombaro, I., Giusti, A., & Mainardi, F. (2017). On the propagation of transient waves in a viscoelastic Bessel medium. *Zeitschrift fur Angewandte Mathematik und Physik, 68,* 62. https://doi.org/10.1007/s00033-017-0808-6.

Egorychev, O. A., & Poddayeva, O. I. (2006). Normal impact on end of cylindrical shell. *StroitelnyayaMekhanika I RaschyotSooruzheniy, 1,* 34–36. (in Russian).

Filippov, I. G., & Cheban, V. G. (1988). *The Mathematical theory of vibrations of elastic and viscoelastic plates and rods.* Chisinau: Stiintsa. (in Russian).

Ilyasov M. Kh. (2011). *Non-stationary viscoelastic waves.* Azerbaijan HavaYollary, Baku, 330 p (in Russian).

Jeltkov, V. I., Tolokonnikov, L. A., & Khromova, N. G. (1993). The transient functions in the dynamics of viscoelastic bodies. *Academy of Sciences USSR reports., 329*(6), 718–719. (in Russian).

Kurbanov, N. T., & Nasibzada, V. N. (2015). Investigation of forced oscillations viscoelastic shells. *International Journal of Current research. India., 7*(07), 18356–18360.

Lokshin, A. A., Suvorova, Yu. V. (1982). *Mathematical theory of wave propagation in memory mediums.* Moscow State University, Moscow, 151 p (in Russian).

Lycheva, T. N., & Lychev, S. A. (2016). Spectral decompositions in dynamical viscoelastic problems. *PNRPU Mechanics Bulletin., 4,* 120–150. (in Russian).

Pshenichnov, S. G. (2013). Nonstationary dynamic problems of linear viscoelasticity. *Mechanics of Solids, 48*(1), 68–78.

Chapter 7
Features of Subsonic Stage of Contact Interaction of Viscoelastic Half-Plane and Absolutely Rigid Striker

Leonid Igumnov, Ekaterina A. Korovaytseva, and Dmitrii V. Tarlakovskii

Abstract Plane non-stationary contact problem concerning interaction of symmetric absolutely rigid striker and viscoelastic half-plane at subsonic stage of interaction is considered. The striker motion is considered vertical. Free slipping is taken as contact condition. For half-plane boundary-normal displacement determination, corresponding Green function is constructed. The resolving equation system is formulated, including striker motion equation, relations for the force of contact interaction, the equation connecting half-plane boundary-normal displacement with the striker displacement and integral relation connecting half-plane boundary-normal displacement with contact stresses. For the obtained equation system, solution difference scheme is constructed.

Keywords Viscoelastic half-plane · Non-stationary contact problem · Laplace transform

7.1 Introduction

Only a limited amount of works is devoted to viscoelasticity dynamic contact problems solution in accessible literature, and only static or stationary dynamic processes of contact interaction are considered. Indeed, in the work (Belokon' and Shekhov 1979), contact problem concerning stamp stationary motion along viscoelastic bar is solved. In the work (Grishin 1995), asymptotic solution concerning stationary motion of a stamp with angles along a half-plane hereditary properties of which material are described by a power law of nonlinear creep theory is obtained. In the work (Galin 1980), analytical solutions of the problems concerning rigid stamp pressing viscoelastic half plane, which properties are described by exponential creep kernel,

L. Igumnov · D. V. Tarlakovskii
Research Institute for Mechanics, National Research Lobachevsky State University of Nizhny Novgorod, Nizhny Novgorod, Russia

E. A. Korovaytseva · D. V. Tarlakovskii (✉)
Institute of Mechanics, Lomonosov Moscow State University, Moscow, Russia
e-mail: tdvhome@mail.ru

F. dell'Isola and L. Igumnov (eds.), *Dynamics, Strength of Materials and Durability in Multiscale Mechanics*, Advanced Structured Materials 137,
https://doi.org/10.1007/978-3-030-53755-5_7

and rolling of viscoelastic cylinder along the foundation of the same material are represented. In the work (Grigoryan 1993), solution of the problem concerning body penetration in soil ground is represented, but mathematical model of this medium does not describe its hereditary properties, stresses in contact area, and contact forces which are found using the relations obtained experimentally. Besides, non-stationary regime of motion is not considered.

In accessible literature on the whole, solutions of non-stationary viscoelastic dynamic contact problems are not represented. In this work for the first time, solution of non-stationary dynamic plane problem concerning interaction of absolutely rigid striker with viscoelastic half plane, hereditary properties of which material are described by Koltunov kernel, is obtained.

7.2 Problem Statement

In a rectangular Cartesian coordinate system, we consider viscoelastic half-plane $x_3 \geq 0$. At the initial time $t = 0$ an absolutely rigid striker of mass m starts indenting into the half plane.

Let us introduce the following system of dimensionless variables (they are marked by the primes which are omitted in what follows):

$$x_1 = \frac{x_1'}{L}, x_3 = \frac{x_3'}{L}, \tau = \frac{c_1 t}{L}, \ c_1^2 = \frac{\lambda + 2\mu}{\rho}, \ u_k = \frac{u_k'}{L}, \ \sigma_{kl} = \frac{\sigma_{kl}'}{\lambda + 2\mu},$$

$$M(\tau) = \frac{4L}{3\rho c_1^3} M'(t), \ m = \frac{m'}{\rho L^2}, R_3 = \frac{R_3'}{(\lambda + 2\mu)L}, \beta^2 = \frac{\lambda + \mu}{\lambda + 2\mu},$$

$$\eta = \frac{c_1}{c_2} = \frac{1}{\gamma}, \ \kappa = \frac{\lambda}{\lambda + 2\mu} = 1 - 2\gamma^2, \ c_2^2 = \frac{\mu}{\rho},$$

where L is some characteristic linear dimension; c_1 and c_2 are, respectively, tension–compression and shear wave propagation in elastic medium velocities; t is time; u_k are components of displacement vector; σ_{kl} are stresses; $M(\tau)$ is relaxation kernel; R_3 is contact force; ρ is half-plane material density; λ, μ are elastic Lamé constants.

Dimensionless equations of half-plane motion have the form (Gorshkov et al. 2004)

$$D(\tau) * \left(\beta^2 \frac{\partial \theta}{\partial x_1} + \gamma^2 \Delta u_1\right) = \ddot{u}_1, \ D(\tau) * \left(\beta^2 \frac{\partial \theta}{\partial x_3} + \gamma^2 \Delta u_3\right) = \ddot{u}_3, \tag{7.1}$$

where

$$D(\tau) = \delta(\tau) - M(\tau), \ \theta = \frac{\partial u_1}{\partial x_1} + \frac{\partial u_3}{\partial x_3}, \ \Delta = \frac{\partial^2}{\partial x_1^2} + \frac{\partial^2}{\partial x_3^2}, \ \gamma^2 = 1 - \beta^2.$$

We suppose that at initial time, medium is at rest:

$$u_1|_{\tau=0} = u_3|_{\tau=0} = \dot{u}_1|_{\tau=0} = \dot{u}_3|_{\tau=0} = 0. \tag{7.2}$$

Relations for the stresses have the form:

$$\sigma_{kl} = D(\tau) * T_{kl}(u_1, u_3),$$
$$T_{11}(u_1, u_3) = \frac{\partial u_1}{\partial x_1} + \kappa \frac{\partial u_3}{\partial x_3},\ T_{22}(u_1, u_3) = \kappa\theta,\ T_{33}(u_1, u_3) = \frac{\partial u_3}{\partial x_3} + \kappa \frac{\partial u_1}{\partial x_1},$$
$$T_{12}(u_1, u_3) = T_{23}(u_1, u_3) = 0,\ T_{13}(u_1, u_3) = \gamma^2\left(\frac{\partial u_3}{\partial x_1} + \frac{\partial u_1}{\partial x_3}\right).$$

We suppose that the striker is constrained by a smooth convex cylindrical surface, the equation of which in a coordinate system Cx_1x_3' referred to its mass center C we write down as follows:

$$x_3' = x_3 + l = f(x_1),$$
$$f(-x_1) = f(x_1),\ f'(x_1) < 0\ (x_1 \neq 0),\ f'(0) = 0,\ f''(x_1) < 0. \tag{7.3}$$

Here, l is the distance between mass center and stagnation point.

Striker motion along the axis Ox_3 without external forces action is described by the following initial problem

$$m\ddot{u}_{c3} = R_3,\ R_3(\tau) = \int_{-b(\tau)}^{b(\tau)} \sigma_{330}(x_1, \tau)dx_1,\ \sigma_{330} = \sigma_{33}|_{x_3=0}, \tag{7.4}$$

$$u_{c3}|_{\tau=0} = u_{c30},\ \dot{u}_{c3}|_{\tau=0} = v_{c30},$$

where u_{c3} is striker mass center displacement; $[-b(\tau), b(\tau)]$ is contact area.

We suppose that the half-plane boundary outside contact area is free:

$$\sigma_{13}|_{x_3=0} = \sigma_{33}|_{x_3=0} = 0,\ |x_1| > b(\tau).$$

Interaction of the striker and half-plane is modeled by the condition of free slipping

$$\sigma_{13}|_{x_3=0} = 0,\ u_3|_{x_3=0} = w(x_1, \tau),\ |x_1| \leq b(\tau),$$

where $w(x_1, \tau)$ is half-plane boundary normal displacement.

When writing down these relations, we took into account that in accordance with (7.3) at zero approximation the boundary of the striker at arbitrary time point is determined as follows:

$$x_3 + l - u_{c3} = f(x_1),$$

from which contact area radius is

$$b_2 = f^{-1}(l - u_{c3}) \tag{7.5}$$

Besides, we must require the displacements to be constrained.

Integral equation connecting half-plane boundary normal displacement and contact stresses has the form

$$w(x_1, \tau) = G_{30}(x_1, \tau) * *\sigma_{330}(x_1, \tau), \tag{7.6}$$

where the function $G_{30}(x_1, \tau)$ corresponds to the problem (7.1), (7.2) and boundary conditions

$$\sigma_{13}|_{x_3=0} = 0, \quad \sigma_{33}|_{x_3=0} = \delta(x_1)\delta(\tau) \tag{7.7}$$

On the other hand, the following relation connects half-plane boundary normal displacement with the striker displacement:

$$w = u_{c3} + f(x_1) - l \tag{7.8}$$

So, the resolving equation system has the form (7.4), (7.5), (7.6), (7.8), and for its solution we have to determine Green function $G_{30}(x_1, \tau)$

7.3 Green Function Construction

Equation system (7.1) solution at initial conditions (7.2) and boundary conditions (7.7) is sought in accordance with the statement proved in (Ilyasov 2011) in the form

$$\mathbf{u}(x_1, x_3, \tau) = \int_0^\infty \mathbf{u}_e(x_1, x_3, \alpha) W_\sigma(\alpha, \tau) d\alpha \tag{7.9}$$

Here, $\mathbf{u}_e$ is constrained displacement vector with coordinates u_{e1} and u_{e3}, corresponding to two-dimensional elastic problem

$$\left(\beta^2 \frac{\partial \theta}{\partial x_1} + \gamma^2 \Delta u_{e1}\right) = \ddot{u}_{e1}, \quad \left(\beta^2 \frac{\partial \theta}{\partial x_3} + \gamma^2 \Delta u_{e3}\right) = \ddot{u}_{e3},$$

$$\mathbf{u}_e|_{\alpha=0} = \left.\frac{\partial \mathbf{u}_e}{\partial \alpha}\right|_{\alpha=0} = \mathbf{0},$$

$$\sigma_{13}^{e}\Big|_{x_3=0} = 0,\ \sigma_{33}^{e}\Big|_{x_3=0} = \delta(x_1)\delta(\alpha).$$

The relation (7.9) on the surface $x_3 = 0$ can be written down as follows:

$$G_{30}(x_1, \tau) = \int\limits_0^\infty G_{3e}(x_1, \alpha) W_\sigma(\alpha, \tau)\mathrm{d}\alpha. \tag{7.10}$$

So, the function $G_{3e}(x_1, \alpha)$ in (7.10) represents Lamb's problem solution obtained in (Gorshkov and Tarlakovski 1995):

$$\begin{aligned}
G_{3e}(x_1, \alpha) &= \sum_{k=1}^{2} G_{3e,k}(x_1, \alpha) H(\alpha - \eta_k |x_1|),\\
G_{3e,1}(x_1, \alpha) &= \frac{\eta^4}{\pi} \frac{x_1^2\left(\eta^2 x_1^2 - 2\alpha^2\right)^2}{P_4(x_1^2, \alpha^2)} \sqrt{\alpha^2 - x_1^2},\\
G_{3e,2}(x_1, \alpha) &= \frac{\eta^4}{\pi} \frac{4x_1^2\alpha^2\left(\alpha^2 - x_1^2\right)^2}{P_4(x_1^2, \alpha^2)} \sqrt{\alpha^2 - \eta^2 x_1^2},\\
P_4(x_1, \alpha) &= \left(\eta^2 x_1 - 2\alpha\right)^4 - 16\alpha^2\left(\alpha - \eta^2 x_1\right)(\alpha - x_1),\\
\eta_1 = 1,\ \eta_2 = \eta.
\end{aligned}$$

Then, for Green function for viscoelastic half-plane $G_{30}(x_1, \tau)$ determination, we have to construct the function $W_\sigma(\alpha, \tau)$.

In accordance with (Ilyasov 2011), the function W_σ in (7.9) is constrained solution of one-dimensional viscoelastic problem

$$\begin{aligned}
\ddot{W}_\sigma &= D(\tau) * \frac{\partial^2 W_\sigma}{\partial \alpha^2};\\
W_\sigma\big|_{\tau=0} &= \dot{W}_\sigma\big|_{\tau=0} = 0,\ \alpha \geq 0;\\
[D(\tau) * W_\sigma(\alpha, \tau)]\big|_{\alpha=0} &= \delta(\tau),\ \tau \geq 0.
\end{aligned} \tag{7.11}$$

Applying Laplace transform to (7.11), we obtain

$$\left(W_\sigma^L\right)'' - b^2 \cdot W_\sigma^L = 0;\ b(s) = \frac{s}{\sqrt{1 - M^L(s)}}, \tag{7.12}$$

$$W_\sigma^L(0, s) = \frac{1}{1 - M^L(s)}. \tag{7.13}$$

The function W_σ^L is constrained at $\alpha \to \infty$, $M^L(s)$ is relaxation kernel transform. Boundary problem (7.12)–(7.13) solution has the form:

$$W_\sigma^L(\alpha, s) = \frac{1}{1 - M^L(s)} \mathrm{e}^{-b(s)\alpha}.$$

The function $G^L(\alpha, s) = \mathrm{e}^{-b(s)\alpha}$ original was found in (Korovaytseva et al. 2017) for the case of exponential relaxation kernel $M(\tau) = a\mathrm{e}^{-\vartheta\tau}$. Also restricting ourselves by this case, we represent function $W_\sigma^L(\alpha, s)$ transform as follows:

$$W_\sigma^L(\alpha, s) = F^L(\alpha, s + \vartheta), \quad F^L(\alpha, s) = f(s)\mathrm{e}^{-d(s)\alpha},$$
$$f(s) = \frac{1}{1 - \tilde{M}^L(s)}, \quad d(s) = \frac{s - \vartheta}{\sqrt{1 - \tilde{M}^L(s)}}, \quad \tilde{M}^L(s) = \frac{a}{s}.$$

We expand the function $f(s)$ in series in the vicinity of infinitely distant point s of the plane:

$$f(s) = \frac{1}{1 - \tilde{M}^L(s)} = \sum_{k=0}^{\infty} \left[\tilde{M}^L(s)\right]^k = \sum_{k=0}^{\infty} \frac{a^k}{s^k}.$$

Expansion in series in powers of s of the function $F_0^L(\alpha, s) = e^{-d(s)\alpha}$ was carried out in (Korovaytseva et al. 2017) and had the form

$$F_0^L(\alpha, s) = \mathrm{e}^{-(c_1 a - \vartheta)\alpha} \mathrm{e}^{-s\alpha} Q^L(\alpha, s), \quad Q^L(\alpha, s) = 1$$
$$+ \sum_{l=0}^{\infty} \frac{g_{l+1}(\alpha)}{s^{l+1}} = \sum_{l=0}^{\infty} \frac{g_l(\alpha)}{s^l}, \quad g_0(\alpha) = 1,$$
$$g_l(\alpha) = \sum_{m=1}^{l} \frac{(-1)^m \alpha^m h_{l-m,m}}{m!}, \quad h_{k,m+1} = \sum_{n=0}^{k} h_{k-n,m} e_n \ (m \geq 1), \ h_{k,1} = e_k,$$
$$e_n = d_{n+1} a^{n+1}, \ c_n = \frac{(2n-1)!!}{2^n n!}, \quad d_n = \left[\frac{2n+1}{2(n+1)} a - \vartheta\right] c_n.$$

Then,

$$F^L(\alpha, s) = \mathrm{e}^{-(c_1 a - \vartheta)\alpha} \mathrm{e}^{-s\alpha} \left[1 + \sum_{m=0}^{\infty} \frac{d_{m+1}(\alpha)}{s^{m+1}}\right],$$

where $k + l = m, \ d_m(\alpha) = \sum_{k=0}^{m} a^k g_{m-k}(\alpha).$

So,

$$F(\alpha, \tau) = \mathrm{e}^{-(c_1 a - \vartheta)\alpha} \left[\delta(\tau - \alpha) + \sum_{m=0}^{\infty} \frac{d_{m+1}(\alpha)}{m!} (\tau - \alpha)^m\right] H(\tau - \alpha),$$

and the original of the function W_σ is written down as follows:

$$W_\sigma(\alpha, \tau) = \mathrm{e}^{-\vartheta\tau} F(\alpha, \tau) \tag{7.14}$$

Green function $G_{3e}(x_1, \alpha)$ can be represented in the form (Gorshkov and Tarlakovski 1995)

$$G_f(x_1, \tau) = \left[G_{fr}(x_1, \tau) + G_{fs}(x_1, \tau)\right]H(\tau - |x|), \tag{7.15}$$

where

$$\begin{aligned}
G_{fr}(x_1, \tau) &= G_{r1}(x_1, \tau)H(\eta|x| - \tau) + G_{r3}(x_1, \tau)H(\tau - \eta|x|),\\
G_{rk}(x_1, \tau) &= G_{fk}(x_1, \tau) - G_{fs}(x_1, \tau),\ k = 1, 3\\
G_{f1}(x_1, \tau) &= \frac{\eta^4}{\pi}\frac{x_1^2\left(\eta^2 x_1^2 - 2\tau^2\right)^2}{P_4(x_1^2, \tau^2)}\sqrt{\tau^2 - x_1^2},\\
P_4(x, \tau) &= \left(\eta^2 x - 2\tau\right)^4 - 16\tau^2\left(\tau - \eta^2 x\right)(\tau - x),
\end{aligned}$$

$$\begin{aligned}
G_{f3}(x_1, \tau) &= \frac{\eta^4}{\pi}\frac{x_1^2 R_{22}(x_1^2, \tau^2)}{P_4(x_1^2, \tau^2)}\sqrt{\tau^2 - x_1^2},\\
G_{fs}(x_1, \tau) &= \frac{a_s\tau}{x_1^2 - c_R^2\tau^2},\ a_s = \frac{R_{22}(c_R^2, 1)\sqrt{1 - c_R^2}}{\pi\eta^4 P_2(c_R^2, 1)},\\
R_{22}(x, \tau) &= \left(\eta^2 x - 2\tau\right)^2 + 4\tau\sqrt{\tau - x}\sqrt{\tau - \eta^2 x},\\
P_2(x, \tau) &= x^2 - 2\alpha^2\tau x + \beta^2\tau^2,\\
\alpha^2 &= 2(1 - \kappa) - c_R^2/2,\\
\beta^2 &= 16\left(\eta^2 - 1\right)/\left(\eta^8 c_R^2\right),
\end{aligned}$$

c_R is Rayleigh wave propagation velocity.

Taking into account (7.10), (7.14) and (7.15), we can write down the expression for viscoelastic half-plane Green function $G_{30}(x_1, \tau)$ in the form

$$\begin{aligned}
G_{30}(x_1, \tau) &= G_{30,r} + G_{30,s},\\
G_{30,r} &= \mathrm{e}^{-c_1 a\tau} G_{3e,r}(x_1, \tau)\\
&\quad + \mathrm{e}^{-\vartheta\tau}\int\limits_{|x_1|}^{\tau} G_{3e,r}(x_1, \alpha)\mathrm{e}^{-(c_1 a - \vartheta)\alpha}\sum_{m=0}^{\infty}\frac{d_{m+1}(\alpha)}{m!}(\tau - \alpha)^m \mathrm{d}\alpha,\\
G_{30,s} &= \mathrm{e}^{-c_1 a\tau} G_{3e,s}(x_1, \tau)\\
&\quad + \mathrm{e}^{-\vartheta\tau}\int\limits_{|x_1|}^{\tau} G_{3e,s}(x_1, \alpha)\mathrm{e}^{-(c_1 a - \vartheta)\alpha}\sum_{m=0}^{\infty}\frac{d_{m+1}(\alpha)}{m!}(\tau - \alpha)^m \mathrm{d}\alpha.
\end{aligned} \tag{7.16}$$

7.4 Contact Problem Solution Algorithm

So, finally contact problem for the subsonic stage of interaction is determined by closed equation system

$$\begin{aligned} m\ddot{h} &= R_e + R_3, \\ \dot{h} &= v_{c3},\ h|_{\tau=0} = 0,\ \dot{h}\big|_{\tau=0} = v_{c30}, \\ R_3 &= \int\limits_{-b(\tau)}^{b(\tau)} \sigma_{330}(x_1,\tau)\mathrm{d}x_1, \\ b_2 &= f^{-1}(h), \\ w(x_1,\tau) &= G_{30}(x_1,\tau) * *\sigma_{330}(x_1,\tau), \\ w &= u_{c30} + h + f(x_1). \end{aligned} \tag{7.17}$$

Here, $h = u_{c3} - u_{c30}$ is striker indentation depth, v_{c3} is striker velocity, and v_{c30} is initial striker velocity.

The solution of the problem is constructed using time and coordinate mesh representation of the resolving equations integration area and the following construction of difference schemes for these equations and quadrature formulae for the integrals.

We shall mesh the plane $R^2_{\tau x_1}$ using mesh with time step δ_τ and coordinate step δ_ξ:

$$\tau_i = i\delta_\tau,\ \xi_j = j\delta_\xi (i = 0,\ 1,\ 2,\ \ldots;\ j \in Z).$$

In the general case, $\delta_\tau \neq \delta_\xi$.

We shall assign the functions of one and two variables in the relations (7.17) to mesh functions

$$\begin{aligned} h_i &= h(\tau_i),\ b_i = b(\tau_i),\ v_{c3,i} = v_{c3}(\tau_i),\ f_j = f(x_{1,j}), \\ \sigma_{ij} &= \sigma_{330}(x_{1,j},\tau_i),\ w_{ij} = w\big(x_{1,j},\tau_i\big). \end{aligned}$$

Then, the integral relation for the normal displacement of the half space at the point $\tau = \tau_n, x_1 = \xi_m$ has the form

$$w_{nm} = \iint\limits_{D_{nm}} G_{30}(\xi_m - \xi, \tau_n - \tau)\cdot\sigma(\xi,\tau)\mathrm{d}\tau\mathrm{d}\xi.$$

Representing integration area D_{nm} by a polygon analogous to (Gorshkov and Tarlakovski 1995), we approximately replace the integral by the following sum:

$$w_{nm} \approx I^r_{nm} + I^s_{nm} + \varepsilon_{nm} I^0_{nm},$$

$$\varepsilon_{nm} = 1\,(m < l_n),\ \varepsilon_{nm} = 0\,(m > l_n),\ l_n = \left[\frac{b_n}{\delta_\xi}\right],$$

$$I^r_{nm} = \iint\limits_{B_{nm}} G_{30,r}(\xi_m - \xi, \tau_n - \tau)\sigma(\xi, \tau)\mathrm{d}\tau\mathrm{d}\xi,$$

$$I^s_{nm} = \iint\limits_{B_{nm}} G_{30,s}(\xi_m - \xi, \tau_n - \tau)\sigma(\xi, \tau)\mathrm{d}\tau\mathrm{d}\xi.$$

For integrals calculation, we use rectangular method, taking rectangulars K_{ij} as elementary integration domains and supposing $\sigma(\xi, \tau) \approx \sigma_{nm}$ at $(\xi, \tau) \in K_{i,j} \cup K_{i,j+1}$. Then, the integral I^r_{nm} obtains the form

$$I^r_{nm} \approx \delta_\tau\delta_\xi \sum_{i=1}^{n-1}\sum_{j=p_i}^{q_i} a^r_{n-i,m-j}\sigma_{ij},$$

$$p_i = \max(-l_i, k_{i1}),\ l_i = \left[\frac{b_i}{\delta_\xi}\right],\ k_{i1} = (i-n)\frac{\delta_\tau}{\delta_\xi} + m - 1,$$

$$q_i = \min(l_i, k_{i2}),\ k_{i2} = m - 1 + (n-i)\frac{\delta_\tau}{\delta_\xi}.$$

Here, quadrature formulae coefficients

$$a^r_{nm} = \frac{1}{4}\int\limits_{-1}^{1}\int\limits_{-1}^{1} G_{30,r}\left[\delta_\xi\left(m - \frac{1}{2} - \frac{v}{2}\right), \delta_\tau\left(n - \frac{1}{2} - \frac{u}{2}\right)\right]\mathrm{d}u\mathrm{d}v.$$

Expressions for coefficients a^r_{nm} are obtained using substitution

$$2\tau = (u + 2i - 1)\delta_\tau,$$
$$2\xi = (v + 2j - 1)\delta_\xi.$$

Integral I^s_{nm} is calculated similarly despite the singularity of sub-integral function $G_{30,s}\left[\delta_\xi\left(m - \frac{1}{2} - \frac{v}{2}\right), \delta_\tau\left(n - \frac{1}{2} - \frac{u}{2}\right)\right]$. This is caused by the fact that analytical integral calculation is impossible in this case, and numerical analytic method of singularity extraction (multiplicative method) did not lead to its smoothing.

Integral I^0_{nm} is calculated over the area A_{nm} near the point $\tau = \tau_n, x_1 = \xi_m$:

$$I^0_{nm} = \iint\limits_{A_{nm}} G_{30}(\xi_m - \xi, \tau_n - \tau)\sigma(\xi, \tau)\mathrm{d}\tau\mathrm{d}\xi \approx a_{nm}\sigma_{nm},$$

$$a_{nm} = \int\limits_{(n-1)\delta_\tau}^{n\delta_\tau}\mathrm{d}\tau \int\limits_{\tau+(m-n)\delta_\xi}^{-\tau+(m+n)\delta_\xi} G_{30}(m\delta_\xi - \xi, n\delta_\tau - \tau)\mathrm{d}\xi.$$

The expression for Green function G_{30} (7.16) can be rewritten separating the term containing integral of elastic half-plane Green function:

$$G_{30}(x_1,\tau) = G_{30,1}(x_1,\tau) + G_{30,2}(x_1,\tau),$$
$$G_{30,1}(x_1,\tau) = \mathrm{e}^{-c_1 a\tau} G_{3e}(x_1,\tau),$$
$$G_{30,2}(x_1,\tau) = \mathrm{e}^{-\vartheta\tau} \int\limits_{|x_1|}^{\tau} G_{3e}(x_1,\alpha)\mathrm{e}^{-(c_1 a-\vartheta)\alpha} \sum_{m=0}^{\infty} \frac{d_{m+1}(\alpha)}{m!}(\tau-\alpha)^m \mathrm{d}\alpha.$$

Then, the coefficient a_{nm} is written as follows:

$$a_{nm} = a_{nm,1} + a_{nm,2};$$
$$a_{nm,1} = \int\limits_{(n-1)\delta_\tau}^{n\delta_\tau} \mathrm{e}^{-c_1 a(n\delta_\tau-\tau)}\mathrm{d}\tau \int\limits_{\tau+(m-n)\delta_\xi}^{-\tau+(m+n)\delta_\xi} G_{3e}(m\delta_\xi-\xi, n\delta_\tau-\tau)\mathrm{d}\xi;$$
$$a_{nm,2} = \int\limits_{(n-1)\delta_\tau}^{n\delta_\tau} \mathrm{e}^{-c_1 a(n\delta_\tau-\tau)}\mathrm{d}\tau \int\limits_{\tau+(m-n)\delta_\xi}^{-\tau+(m+n)\delta_\xi} \int\limits_{|m\delta_\xi-\xi|}^{n\delta_\tau-\tau} G_{3e}(m\delta_\xi-\xi, n\delta_\tau-\tau)\mathrm{e}^{-(c_1 a-\vartheta)\alpha}.$$
$$\cdot\sum_{m=0}^{\infty}\frac{d_{m+1}(\alpha)}{m!}(n\delta_\tau-\tau-\alpha)^m \mathrm{d}\alpha\mathrm{d}\xi.$$

So, finally difference scheme for the system (7.17) at the point $\tau=\tau_n$, $x_1=\xi_m$ has the form

$$\begin{aligned}
\sigma_{nm} &= \sum_{i=1}^{n-1}\sum_{j=p_i}^{q_i} a_{n-i,2m-j}\sigma_{ik_j} - \frac{w_{nm}}{a_{nm}}\\
w_{nm} &= h_n + f_m + u_{c0}\\
h_n &= h_{n-1} + V_{n-1}\delta_\tau\\
V_n &= V_{n-1} + R_{n-1}\frac{\delta_\tau}{m}\\
R_n &= 2\delta\left(\sigma_{n_0} + 2\sum_{j=1}^{l_n}\sigma_{nj}\right)\\
b_n &= f^{-1}(-h_n)
\end{aligned}\qquad(7.18)$$

Initial conditions for the system (7.18) have the form

$$u_{c3,0} = u_{c30},\ V_0 = V_{30},\ \sigma_{0m} = -V_{30}.$$

7.5 Example

As an example of the constructed algorithm using, contact problem for parabolic striker is considered. Striker surface equation is $y = x^2$, and parameters of relaxation kernel are $a = 0.1$, $\vartheta = 0.2$. Figures 7.1, 7.2, 7.3, and 7.4 represent time

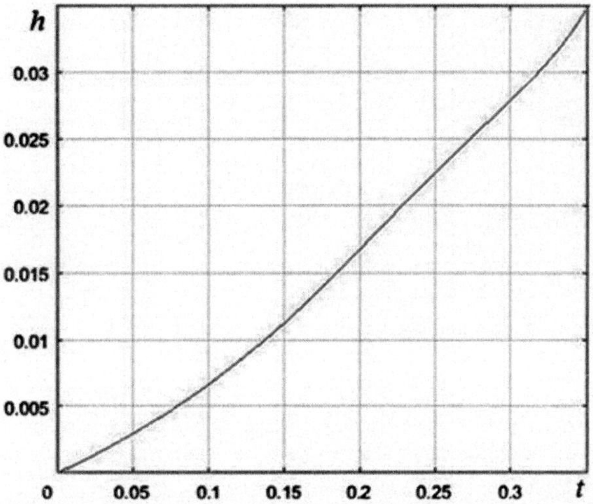

Fig. 7.1 Striker indentation depth time dependence

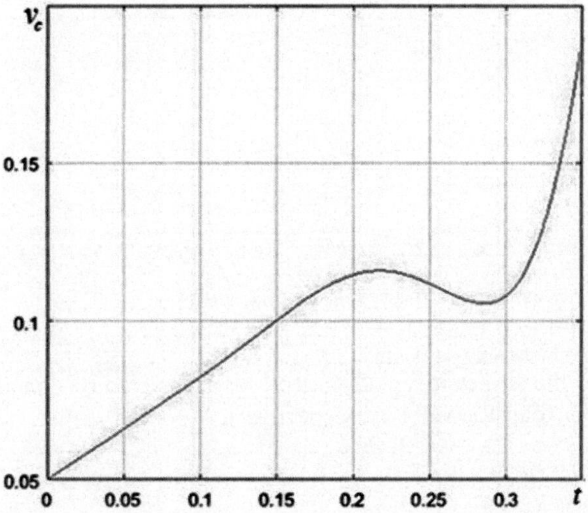

Fig. 7.2 Striker indentation velocity time dependence

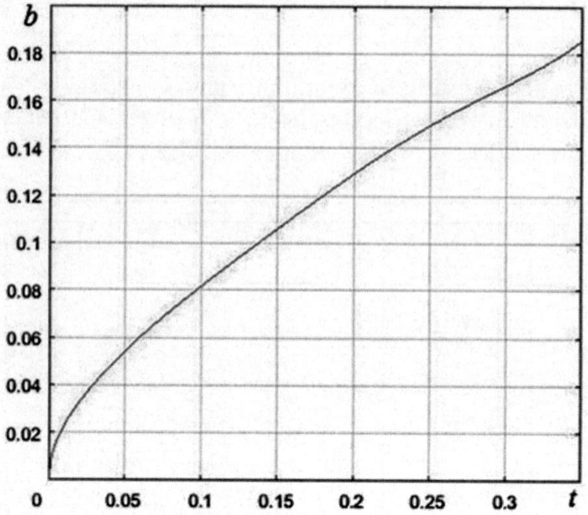

Fig. 7.3 Contact area radius time dependence

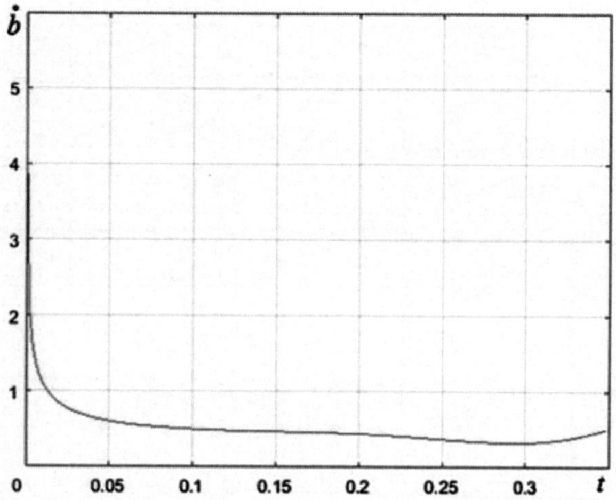

Fig. 7.4 Contact area expansion velocity time dependence

dependencies of the striker indentation depth, striker velocity, contact area radius, and contact area expansion velocity, respectively.

Acknowledgements The reported study was funded by RFBR, according to the research project No. 19-38-70005 mol_a_mos.

References

Belokon', A. V., & Shekhov, V. P. (1979). Problem of stamp under harmonic loading movement along viscoelastic bar. Rostov-on-Don, RSU. 1979. 25p. – Dep. In VINITI 11.12.79, № 4303–79 Dep. [in Russian].

Galin, L. A. (1980). *Contact problems of elasticity and viscoelasticity theory*. Moscow: Nauka, 304p. [in Russian].

Gorshkov, A. G., Medvedski, A. L., Rabinski L. N., & Tarlakovski D. V. (2004). *Waves in continious media*. Moscow: Physmathlit, 2004. 472pp. [in Russian].

Gorshkov, A. G., & Tarlakovski, D. V. (1995). *Dynamic contact problems with moving borders*. Moscow: Physmathlit, 352pp. [in Russian].

Grigoryan, S. S. (1993). Approximate solution of the problem concerning body penetration in soil. *Fluid dynamics, 4*, 18–24. [in Russian].

Grishin, S. A. (1995). Contact problems for power-law creeping body. In *Proceeding of 2nd contact mechanics international symptoms 1994* (pp. 319–326). Carry-le-Rouet, France, New-York; London: Plenum Press.

Ilyasov, M. Kh. (2011). *Non-stationary viscoelastic waves*. Baku, 330pp. [in Russian].

Korovaytseva, E. A., Pshenichnov, S. G., & Tarlakovskii, D. V. (2017). Propagation of one-dimensional non-stationary waves in viscoelastic half space. *Lobachevskii Journal of Mathematics, 38*, 827. https://doi.org/10.1134/S1995080217050237.

Chapter 8
Interaction of Harmonic Waves of Different Types with the Three-Layer Plate Placed in the Soil

Leonid Igumnov, Dmitrii V. Tarlakovskii, Natalia A. Lokteva, and Nguyen Duong Phung

Abstract The solution of two-dimensional task about interaction of harmonic wave with plate with finite length is placed in the soil. Plate's mechanical behavior is described by Pimushin V.N. equations system, and soil mechanical behavior is described by linear propulsion theory equation. Research of vibration-absorbing properties of the plate dependent on frequency and form of harmonic wave acting on the plate was conducted. From practical point of view, this task is connected with protection of underground buildings from vibration impact, formed by moving trains of underground in different distances from object being protected.

Keywords Soil · Plate · Paimushin V.N. Model · Harmonic wave · Oscillation frequency · Vibration absorption · Vibrational acceleration

8.1 Introduction

Due to implementation of new infrastructure in existing city environment, there is more often required to protect already existing buildings and constructions from negative technogenic influence. For residential buildings and public buildings, vibration is the most dangerous event. In buildings engineering equipment, industrial plants and transport (small-scale underground, heavy trucks, cars, trains) are vibrations sources, which are emitting high dynamic loads, which cause spreading vibration in soil and buildings. These vibrations are often are often causing noises inside buildings.

L. Igumnov · D. V. Tarlakovskii
Research Institute for Mechanics, National Research Lobachevsky State University of Nizhny Novgorod, Gagarin ave., 23, 603950 Nizhny Novgorod, Russia

D. V. Tarlakovskii · N. A. Lokteva (✉)
Institute of Mechanics, Lomonosov Moscow State University, Moscow, Russia
e-mail: nlok@rambler.ru

N. A. Lokteva · N. D. Phung
Moscow Aviation Institute (National Research University), Moscow, Russia

F. dell'Isola and L. Igumnov (eds.), *Dynamics, Strength of Materials and Durability in Multiscale Mechanics*, Advanced Structured Materials 137,
https://doi.org/10.1007/978-3-030-53755-5_8

There are various works on vibration absorption and control (Maurini et al. 2004; Andreaus et al. 2004; dell'Isola et al. 2004; Vidoli and dell'Isola 2001). Some new microstructures metamaterials have vibration isolation properties (Barchiesi et al. 2019; Di Cosmo et al. 2018).

More information on metamaterials and one of their mechanical examples—pantographic structures—can be found in (Del Vescovo et al. 2014; Placidi et al. 2016; dell'Isola et al. 2019). And it is also worth to mention that designing metamaterials is based on generalized continuum approaches (Alibert et al. 2003; Auffray et al. 2013; Abali et al. 2017, 2019; dell'Isola et al. 2012, 2015, 2016).

One of the variants of building foundation protection from vibrations appearing in the soil is placing vibration-absorbing plates between source of vibration and the object. Quite a lot of works are dedicated to research of problem of dynamic interaction between object and soil (Umek 1973; Kostrov 1964; Ryl'ko 1977). Homogeneous one-layer sound-absorbing plates in nowadays are well researched. At the same time, the research of multi-layer plates dynamics, including plates filled with filler, requiring proper accounting of transverse shear as well as transverse normal stress in filling layer, in other case the state of dynamic stress can be calculated with severe mistakes. On the other hand, in common, there is a wave stationary influence with different configuration influence on barriers being researched.

In this work, the influence of flat and cylindrical wave influence on barrier with complex structure, accounting all mentioned effects. In practice, this situation can be caused by vibration, caused by single disturbance, appearing in the soil when train in underground passing by in a distance, which appears to be flat, or nearby the barrier which inducting cylindrical wave.

8.2 Problem Statement of Interaction of Harmonic Spherical Wave Propagating in Continuum and a Three-Layered Plate

Three-layered plate placed in the soil is acting like a barrier model. It is separating soil in two parts, «1» and «2». The part «1» is where the wave source is placed. Isotropic propulsive half space described by propulsion theory equations (Rakhmatulin et al. 1983; Berezhnoi et al. 2004) is acting as a soil. Usage of such half space is justified for disturbances with little amplitude, caused for example by train passing through the tunnel. A planar statement of the problem is considered. It is considered that border conditions are equal to the conditions on the simply supported contour of the plate. To describe the plate movement, Paimushin V.N. equation system (Ivanov and Paimushin 1995a, b) is being used. The beginning of coordinates axes in the upper point of the plate. Axis Ox is directed along middle line of the plate toward ground, and axis Oz is directed inside of the half space «2». Undisturbed state of the soil is undeformed. From the negative side of Oz axis («1» environment), there is harmonic wave with amplitude of normal movement p on front is moving toward the barrier.

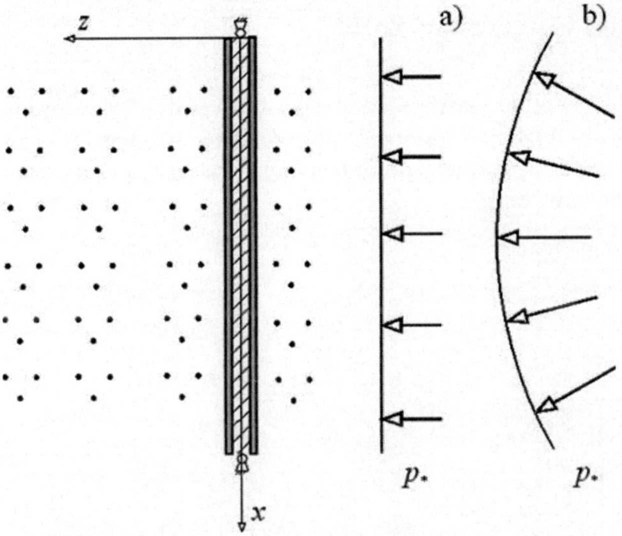

Fig. 8.1 Model of interaction between soil and plate: **a** is the ingoing wave; **b** is the cylindrical ingoing wave

There is two kind of influence being observed—flat wave and cylindrical one. The normal vector of the plane is lying inside the plane *Oxy*, due to this, components of disturbance-deformation state of barrier and soil are independent from axis *Y*, and due to this only normal disturbance is being accounted. The main aim is definition of normal movement in half space «2» after passing through barrier (Fig. 8.1).

The main target is definition of sum of vector field of acceleration a, inducted by passed and emitted waves in half space «2» as frequency function ω and spatial coordinates x and z dependent from plates parameters. Components a_x and a_z, and module a of vector field of accelerations are defined by this formulas:

$$a_x = -\omega^2 u^{(2)}, \quad a_z = -\omega^2 w^{(2)}. \tag{8.1}$$

$$a = \sqrt{a_x^2 + a_z^2}, \tag{8.2}$$

where u and w—movement along axis Ox and Oz.

The mathematical formulation of the problem includes the task of the incident wave, the equations of motion of the soil and the plate, boundary conditions for the plate and the ground, conditions at infinity, as well as the conditions of contact of the soil with an obstacle, where we neglect the adhesion of the plate with the ground.

8.3 Motion Equation of Soil

The model of isotropic elastic medium is used below (Rakhmatulin et al. 1983; Berezhnoi et al. 2004). Its motion is described by the dynamic equations of the elasticity theory (8.3) together with Cauchy relations (8.4) and constitutive equations of the Hookean law (8.5).

– Dynamic equations

$$\rho\ddot{u} = \frac{\partial\sigma_{11}}{\partial x} + \frac{\partial\sigma_{13}}{\partial z},$$
$$\rho\ddot{w} = \frac{\partial\sigma_{31}}{\partial x} + \frac{\partial\sigma_{33}}{\partial z}. \tag{8.3}$$

– Cauchy equations

$$\varepsilon_{11} = \frac{\partial u}{\partial x},\ \varepsilon_{13} = \frac{1}{2}\left(\frac{\partial u}{\partial z} + \frac{\partial w}{\partial x}\right),$$
$$\varepsilon_{33} = \frac{\partial w}{\partial z},\ \theta = \frac{\partial u}{\partial x} + \frac{\partial w}{\partial z}. \tag{8.4}$$

– Constitutive relations

$$\sigma_{11} = \lambda\theta + 2\mu\varepsilon_{11},\ \sigma_{13} = 2\mu\varepsilon_{13},$$
$$\sigma_{33} = \lambda\theta + 2\mu\varepsilon_{33},\ \theta = \varepsilon_{11} + \varepsilon_{33}. \tag{8.5}$$

where: w and u are movements along the axes Ox and Oz; σ_{ij} and ε_{ij} are components of stress and strain tensors; θ is coefficient of volume expansion; ρ and λ, μ are density and Lame's elastic constant soil; dots here and hereinafter denote time derivatives.

There are equivalent equations, describing the motion of the elastic continuum. Let us assume all functions be changing in harmonic manner.

Lame's equation

$$\rho\ddot{u} = (\lambda + \mu)\frac{\partial\theta}{\partial x} + \mu\Delta u,\ \rho\ddot{w} = (\lambda + \mu)\frac{\partial\theta}{\partial z} + \mu\Delta w,$$
$$\Delta = \frac{\partial^2}{\partial x^2} + \frac{\partial^2}{\partial z^2}. \tag{8.6}$$

The equations of dynamics for scalar and vector potentials:

$$\ddot{\varphi} = c_1^2\Delta\varphi,\ \ddot{\psi} = c_2^2\Delta\psi,\ c_1^2 = \frac{\lambda + 2\mu}{\rho},\ c_2^2 = \frac{\mu}{\rho}, \tag{8.7}$$

where:

$$u = \frac{\partial \varphi}{\partial x} - \frac{\partial \psi}{\partial z}, \quad w = \frac{\partial \varphi}{\partial z} + \frac{\partial \psi}{\partial x}. \tag{8.8}$$

Only harmonic waves with the frequency ω are considered

$$\varphi = \varphi_a e^{i\omega t}, \quad \psi = \psi_a e^{i\omega t} \ldots \tag{8.9}$$

Assuming the propagating waves be harmonic only, we obtain the following dynamic equations:

$$\begin{aligned} \rho\omega^2 u + \frac{\partial \sigma_{11}}{\partial x} + + \frac{\partial \sigma_{13}}{\partial z} = 0, \\ \rho\omega^2 w + \frac{\partial \sigma_{31}}{\partial x} + + \frac{\partial \sigma_{33}}{\partial z} = 0, \end{aligned} \tag{8.10}$$

Lame's equation

$$\begin{aligned} \rho\omega^2 u + (\lambda + \mu)\frac{\partial \theta}{\partial x} + \mu \Delta u = 0, \\ \rho\omega^2 w(\lambda + \mu)\frac{\partial \theta}{\partial z} + \mu \Delta w = 0. \end{aligned} \tag{8.11}$$

Finally, for the scalar and vector potentials, we obtain

$$\Delta\varphi + k_1^2\varphi = 0, \; \Delta\psi + k_2^2\psi = 0, \; k_j = \omega/c_j \tag{8.12}$$

Since the medium «2» is not bounded by the coordinate z, then the Somerfield radiation condition acts as a boundary condition (Gorshkov et al. 2004).

$$\frac{\partial \varphi}{\partial r} + ik_1\varphi = o\left(\frac{1}{r}\right), \quad \frac{\partial \psi}{\partial r} + ik_2\psi = o\left(\frac{1}{r}\right), \quad r \to \infty \tag{8.13}$$

8.4 Incoming Wave

To describe plane harmonic wave, the flat one-dimensional stretch-compression wave ($\psi \equiv 0$) (Gorshkov et al. 2004), spreading along positive direction of Oz axis with amplitude of normal pressure, is being reviewed. In this case in first of ten Eq. (8.12), we assume $\varphi = \varphi(z)$. As a result, we have an equation, which depends on potential amplitude, which solution satisfies the conditions (8.13).

$$\varphi_a'' + k_1^2\varphi_a = 0. \tag{8.14}$$

Its solution have a view of (A_{φ}—if arbitrary constant)

$$\varphi_a = A_{\varphi} e^{-ik_1 z}. \tag{8.15}$$

From here (8.5), we get that potential have a type of incoming wave

$$\varphi = A_{\varphi} e^{-ik_1(z-c_1 t)} \tag{8.16}$$

Using this equation consistently in (8.4), (8.5) and (8.8), we get these formulas for movement, deformation, and stresses.

$$\begin{aligned}
&u \equiv 0,\ w = -ik_1 A_{\varphi} e^{-ik_1(z-c_1 t)},\ \varepsilon_{11} = \varepsilon_{13} \equiv 0,\ \varepsilon_{33} = \theta = -k_1^2 A_{\varphi} e^{-ik_1(z-c_1 t)},\\
&\sigma_{33} = -(\lambda + 2\mu) k_1^2 A_{\varphi} e^{-ik_1(z-c_1 t)} = -\rho\omega^2 A_{\varphi} e^{-ik_1(z-c_1 t)},\\
&\sigma_{11} = -\lambda k_1^2 A_{\varphi} e^{-ik_1(z-c_1 t)} = -\kappa\rho\omega^2 A_{\varphi} e^{-ik_1(z-c_1 t)},\ \sigma_{13} \equiv 0,\ \kappa = \frac{\lambda}{\lambda + 2\mu}.
\end{aligned} \tag{8.17}$$

From this, accounting, that $\sigma_{33}|_{t=0,\ z=0} = p_*$ for movement and stress in incoming wave, we get these formulas (only their amplitude values are present)

$$\begin{aligned}
&u = u_* \equiv 0,\ w = w_* = \frac{ik_1 p_*}{\rho\omega^2} e^{-ik_1 z} = \frac{ip_*}{\rho c_1 \omega} e^{-ik_1 z},\ A_{\varphi} = -\frac{p_*}{\rho\omega^2},\\
&\sigma_{11} = \sigma_{11*} = \kappa p_* e^{-ik_1 z},\ \sigma_{33} = \sigma_{33*} = p_* e^{-ik_1 z},\ \sigma_{13} = \sigma_{13*} \equiv 0.
\end{aligned} \tag{8.18}$$

The cylindrical wave is being emitted from source place in point with coordinates $O_l(0,0,-d)$. Setting up cylindrical coordinates system with center in point O_l, parallel to the Oz axis, and with radius

$$r_1 = \sqrt{x^2 + (z+d)^2} \tag{8.19}$$

Assuming, that $\varphi = \varphi(r_1)$, from Eq. (8.12), we get following equations for this function:

$$r_1^{-1}\left(r_1\varphi'\right)' + k_1^2\varphi = 0 \tag{8.20}$$

Its common view will be (Kostrov 1964; Ryl'ko 1977):

$$\varphi = A_{\varphi} H_0^{(2)}(k_1 r_1) + B_{\varphi} H_0^{(1)}(k_1 r_1) \tag{8.21}$$

where $H_{\nu}^{(1)}(\zeta)$ and $H_{\nu}^{(2)}(\zeta)$—Hankel function with order ν, A_{φ} and B_{φ}—are arbitrary constants.

Now, similarly to flat wave defining amplitude values of movements and stresses in incoming wave

$$
\begin{aligned}
u_* &= \frac{p_* x d}{\rho c_1^2 N r_1} H_1^{(2)}(k_1 r_1) e^{i\omega t}, \quad w_* = \frac{p_* d(z+d)}{\rho c_1^2 N r_1} H_1^{(2)}(k_1 r_1) e^{i\omega t}, \\
\sigma_{11*} &= \frac{p_* d}{N r_1^2}\Big[(1+\kappa) r_1 H_1^{(2)}(k_1 r_1) - k_1 r_{11}^2 H_2^{(2)}(k_1 r_1) \Big] e^{i\omega t}, \\
\sigma_{13*} &= -(1-\kappa)\frac{p_* d k_1}{N r_1^2} x(z+d) H_2^{(2)}(k_1 r_1) e^{i\omega t}, \\
\sigma_{33*} &= \frac{p_* d}{N r_1^2}\Big[(1+\kappa) r_1 H_1^{(2)}(k_1 r_1) - k_1 r_{33}^2 H_2^{(2)}(k_1 r_1) \Big] e^{i\omega t}. \\
A_\varphi &= -\frac{p_* d}{\rho \omega c_1 N}, \quad N = k_1 d H_0^{(2)}(k_1 d) - (1-\kappa) H_1^{(2)}(k_1 d).
\end{aligned}
\tag{8.22}
$$

On plane surface in $z = 0$, these values are changing to:

$$
\begin{aligned}
u_*|_{z=0} &= \frac{p_* x d}{\rho c_1^2 N r_{10}} H_1^{(2)}(k_1 r_{10}), \quad w_*|_{z=0} = \frac{d^2 p_*}{\rho c_1^2 N r_{10}} H_1^{(2)}(k_1 r_{10}), \\
\sigma_{11*}|_{z=0} &= \frac{p_* d}{N r_{10}^2}\Big[(1+\kappa) r_{10} H_1^{(2)}(k_1 r_{10}) - k_1 r_{110}^2 H_2^{(2)}(k_1 r_{10}) \Big], \\
\sigma_{13*}|_{z=0} &= -(1-\kappa)\frac{p_* d^2 k_1 x}{N r_{10}^2} H_2^{(2)}(k_1 r_{10}), \\
\sigma_{33*}|_{z=0} &= \frac{p_* d}{N r_{10}^2}\Big[(1+\kappa) r_{10} H_1^{(2)}(k_1 r_{10}) - k_1 r_{330}^2 H_2^{(2)}(k_1 r_{10}) \Big].
\end{aligned}
\tag{8.23}
$$

8.5 The Plate Geometry

The plate consists of three layers, two bearing layers, and the filling one. Bearing layers are isotropic and have thickness t. The filling layer of thickness h has honeycomb structure. The plate is simply supported on the contour.

The plate motion is described by Paimushin V.N. equation system (Ivanov and Paimushin 1995a, b), which takes into account structural features of the plate.

$$
\begin{aligned}
& B\frac{\partial^2}{\partial x^2}\left(u_1^a\right) + 2q^1 + \omega^2 \rho_a u_1^a = 0, \\
& -D\frac{\partial^4}{\partial x^4} w_c + 2k_1 q^1_{,x} + p_1 - p_2 + \omega^2 \rho_c w_c = 0, \\
& -D\frac{\partial^4}{\partial x^4} w_a - 2c_3 w_a + p_1 + p_2 + \omega^2 \rho_{aw} w_a = 0, \\
& u_1^a - k_1 w_{c,x} - k_2\left(q^1_{,x}\right)_{,x} + k_3 q^1 = 0.
\end{aligned}
\tag{8.24}
$$

In the equation system (8.24), the further notation is used: w_c, w_a—deflections; q^1—transverse shear stress in the filling along the x-axis; $u_1^{(k)}$—tangential displacement along the x-axis, respectively, in the *k-th* bearing layer. B и D—tangent and bending stiffness of the tangent plate. $w^{(k)}$—the deflection of k bearing layer

$$u_1^a = u_0^{(1)} - u_0^{(2)};$$
$$w_c = w_0^{(1)} + w_0^{(2)},\ w_a = w_0^{(1)} - w_0^{(2)}.$$

Conditions on the simply supported contour of the plate

$$w_0^{(1)}\Big|_{x=0,l} = w_0^{(2)}\Big|_{x=0,l} = u_{,x}^{(1)}\big|_{x=0,l} = u_{,x}^{(2)}\big|_{x=0,l} = q_{,x}^1\big|_{x=0,l} = 0,$$
$$w_0^{(1)}{}_{,xx}\Big|_{x=0,l_1} = w_0^{(2)}{}_{,xx}\Big|_{x=0,l_1} = 0. \tag{8.25}$$

8.6 Conditions on the Contact Surface

The pressure amplitude of the wave coming from the medium «1» is equal to the sum of the normal stress in the medium and stress, arising as a result of the wave action. In the second medium, the pressure amplitude is the same as the normal stress.

The plane harmonic wave

$$p_1 = \left(\sigma_{33}^{(1)} + \sigma_{33*}\right)\Big|_{z=0},\quad p_2 = -\sigma_{33}^{(2)}\Big|_{z=0}. \tag{8.26}$$

$$\left(w^{(1)} + w_*\right)\big|_{z=0} = w_0^{(1)},\quad w^{(2)}\big|_{z=0} = w_0^{(2)},$$
$$\sigma_{13}^{(1)}\Big|_{z=0} = \sigma_{13}^{(2)}\Big|_{z=0} = 0,\ \sigma_{12}^{(1)}\Big|_{z=0} = \sigma_{12}^{(2)}\Big|_{z=0} = 0,\ \sigma_{23}^{(1)}\Big|_{z=0} = \sigma_{23}^{(2)}\Big|_{z=0} = 0. \tag{8.27}$$

The cylindrical harmonic wave

$$p_{1n} = (\sigma_{33*n} + \sigma_{33n})|_{z=0},\quad p_{2n} = -\sigma_{33n}^{(2)}\Big|_{z=0},$$
$$\sigma_{13n}^{(1)}\Big|_{z=0} = \sigma_{13*}|_{z=0}; \tag{8.28}$$

$$\left(w_n^{(1)} + w_{*n}\right)\big|_{z=0} = w_{0n}^{(1)}\Big|_{z=0},$$
$$w_n^{(2)}\big|_{z=0} = w_{0n}^{(2)}\Big|_{z=0},\ u_{0n}^{(1)}\Big|_{z=0} = u_*|_{z=0}. \tag{8.29}$$

Medium displacements, summed up with ingoing wave displacements, are equal to the displacement of the first plate bearing layer. The displacement of the second medium is the same as the displacement of the second plate bearing layer. Shear stresses are assumed to be vanishing.

8.7 Fourier Decomposition of Unknown Functions

All unknown function, as well for the plate as for both media, are decomposed into trigonometric series and satisfy the boundary conditions of the simple support (Sheddon 1951).

$$K = \sum_{n=1}^{\infty} \Big[w_{0n}^{(1)}, w_{0n}^{(2)}, w_n^{(1)}, w_n^{(2)}, p_n, p_{1n}, p_{2n}, w_{*n}, \sigma_{33*n}, \sigma_{33n}^{(1)}, \sigma_{33n}^{(2)},$$

$$\varphi_n^{(l)}, \psi_{1n}^{(l)}, \varepsilon_{11n}^{(l)}, \varepsilon_{33n}^{(l)}, \sigma_{11n}^{(l)}, \sigma_{13n}^{(l)}, \sigma_{33n}^{(l)}, w_{an}, w_{cn} \Big] \sin(\lambda x),$$

$$L = \sum_{n=1}^{\infty} \Big[u_{0n}^{(1)}, u_{0n}^{(2)}, u_n^{(1)}, u_n^{(2)}, \psi_{3nm}^{(l)}, \varepsilon_{13n}^{(l)}, u_{1n}^{a}, q_n^{1} \Big] \cos(\lambda x),$$

$$\lambda = \frac{\pi n}{l}. \tag{8.30}$$

Dynamic equations for the plate are represented hence through the Fourier coefficients. Then, taking into account (8.30), the equation system for the plate (8.14) can be solved, and for the normal displacements on the surfaces, we have the formulae (8.31).

$$\begin{aligned}
&-\lambda_n^2 B u_{1n}^a + \omega^2 \rho_a u_{1n}^a + 2q_n^1 = 0,\\
&-D\lambda_n^4 w_{cn} + \omega^2 \rho_c w_{cn} - 2\lambda_n k_1 q_n^1 + p_{1n} - p_{2n} = 0,\\
&-D\lambda_n^4 w_{an} + \left(\omega^2 \rho_{aw} - 2c_3\right) w_{an} + p_{1n} + p_{2n} = 0,\\
&u_{1n}^a - k_1 \lambda_n w_{cn} + k_2 \lambda_n^2 q_n^1 + k_3 q_n^1 = 0.
\end{aligned} \tag{8.31}$$

$$\begin{aligned}
\varepsilon_{11n}^{(l)} &= -\lambda_n u_{1n}^{(l)}, \ \varepsilon_{33n}^{(l)} = \frac{\partial w_n^{(l)}}{\partial z},\\
\varepsilon_{13nm}^{(l)} &= \frac{1}{2}\left(\frac{\partial u_{1nm}^{(l)}}{\partial z} + \lambda_{1n} w_{nm}^{(l)} \right),\\
\theta_n^{(l)} &= -\lambda_n u_{1m}^{(l)} + \frac{\partial w_n^{(l)}}{\partial z} \ (n \geq 1).
\end{aligned} \tag{8.32}$$

$$\begin{aligned}
\sigma_{11n}^{(l)} &= \lambda \theta_n^{(l)} + 2\mu \varepsilon_{11n}^{(l)}, \ \sigma_{33n}^{(l)} = \lambda \theta_n^{(l)} + 2\mu \varepsilon_{33n}^{(l)},\\
\sigma_{13n}^{(l)} &= 2\mu \varepsilon_{13n}^{(l)}.
\end{aligned} \tag{8.33}$$

8.8 Computing of the Fourier Coefficients for the Potentials in Ambient Media

The wave equations in the potentials are also represented through the Fourier coefficients (8.30):

$$\frac{\partial^2 \varphi_n^{(l)}}{\partial z^2} + \varphi_n^{(l)}\left(k_1^2 - \lambda_n^2\right) = 0, \tag{8.34}$$

$$\frac{\partial^2 \psi_i^{(l)}{}_n}{\partial z^2} + \psi_{in}^{(l)}\left(k_1^2 - \lambda_n^2\right) = 0, i = 1, 2.$$

Solution of these equations must satisfy the Somerfield condition written in the Fourier coefficients as (8.30).

$$\frac{\partial^2 \varphi_n^{(l)}}{\partial z^2} + \text{sign}(k_1 - \lambda_n)\kappa_{1n}^2 \varphi_n^{(l)} = 0 \ (n \geq 1),$$

$$\frac{\partial^2 \psi_{in}^{(l)}}{\partial z^2} + \text{sign}(k_2 - \lambda_n)\kappa_{2n}^2 \psi_{in}^{(l)} = 0 \ (n \geq 0),$$

$$\kappa_{jn} = \sqrt{\left|k_j^2 - \lambda_n^2\right|}. \tag{8.35}$$

General solution of the wave Eq. (8.34) is given by the following way

$$\varphi_n^{(1)}(z, \omega) = C_{11n}(\omega)\left[e^{i\kappa_{1n}(\omega^2)z} H(k_1 - \lambda_n) + e^{\kappa_{1n}(\omega^2)z} H(\lambda_n - k_1)\right],$$

$$\psi_n^{(1)}(z, \omega) = C_{21n}(\omega)\left[e^{i\kappa_{2n}(\omega^2)z} H(k_2 - \lambda_n) + e^{\kappa_{2n}(\omega^2)z} H(\lambda_n - k_2)\right],$$

$$\varphi_n^{(2)}(z, \omega) = C_{12n}(\omega)\left[e^{-i\kappa_{1n}(\omega^2)z} H(k_1 - \lambda_n) + e^{-\kappa_{1n}(\omega^2)z} H(\lambda_n - k_1)\right],$$

$$\psi_n^{(2)}(z, \omega) = C_{22n}(\omega)\left[e^{-i\kappa_{2n}(\omega^2)z} H(k_2 - \lambda_n) + e^{-\kappa_{2n}(\omega^2)z} H(\lambda_n - k_2)\right], \tag{8.36}$$

To determine the constants $C_{11n}(\omega)$, $C_{21n}(\omega)$, $C_{12n}(\omega)$, $C_{22nm}(\omega)$, we have to use the conditions of the contact between the plate and the medium (8.26), (8.27) for the plane wave and (8.28), (8.29) for the cylindrical wave. This requires that the stresses, deformations, and displacements will be expressed in term of potentials. Then, the normal and tangential displacements of the medium are found based on Eq. (8.8). Further, the vibration acceleration (8.1), (8.2) and displacements (8.8) are determined.

8.9 Example

As an example, we considered a plate with the following dimensions, and here, the plates' parameters are $L = 5$ m, $h = 0.015$ m, $t = 0.004$ m. The bearing layers are made of DIN17100-type steel with density of $\rho_0 = 7859$ kg/m^3, Young's modulus $E_0 = 2 \times 10^5$ MPa , and Poisson's ratio of $\nu_0 = 0.28$. Filler material is aluminum of Al–Mn type with density of $\rho_0 = 2730$ kg/m^3, Young's modulus of $E_0 = 0.71 \times 10^{-5}$ MPa, and Poisson's ratio of $\nu_0 = 0.3$. The materials of the media "1" and "2" are assumed to be fill-up ground compacted under the humidity degree of 0.5, Young's modulus of $E = 10^8$ MPa, and density of $\rho = 1600$ kg/m^3.

The results of calculations of the module of the acceleration field at various points in the medium are presented on Fig. 8.2. The figure shows that a significant decrease in the vibration acceleration module occurs with distance from the plate boundary. Similarly, a cylindrical wave overcomes an obstacle better than a plane wave.

As you know, the greatest danger to the foundations of buildings and structures represent low frequencies. In accordance with (set of rules for the design and construction of the joint venture 23-105-2004), the vibration assessment is carried out in octave bands with geometric mean frequencies of 16, 31.5, and 63 Hz. By changing the plate geometrically, it is possible to avoid coincidence of the resonance with these frequencies

8.10 Notes and Comments

In this article, solution method of couple problem of relation between plane harmonic wave and the simply supported plate is presented. The results depend on the plate's materials and its geometrics. In this way, it becomes possible to choose optimal parameters of the material, the plate is made from, and its geometrics, which is of practical interest.

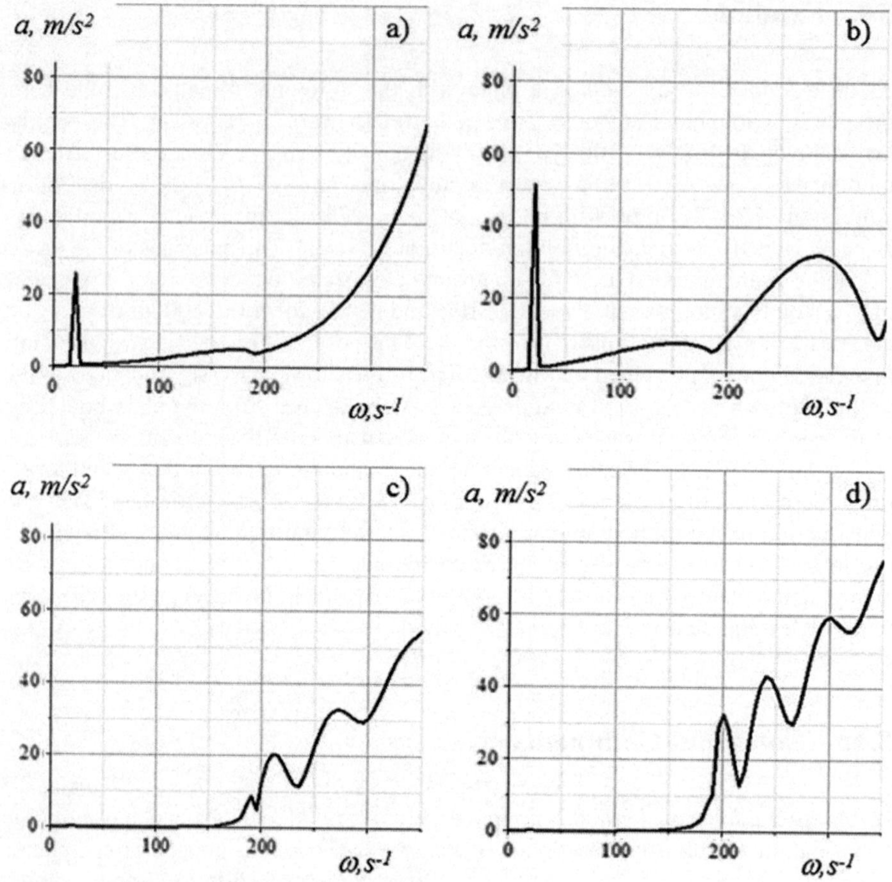

Fig. 8.2 The vibration acceleration module in the different points of medium “2”: **a** on the edge of the plate and medium “2” for $z = 0$ m, **c** on the medium “2” for $z = 10$ m under impact of the plane wave. **b** on the edge of the plate and medium “2” for $z = 0$ m, **d** on the medium “2” for $z = 10$ m under impact of the cylindrical wave

Acknowledgements The reported study was funded by Russian Foundation for Basic Research, according to the research projects Nos. 19-08-00968 A.

References

Abali, B. E., Altenbach, H., dell’Isola, F., Eremeyev, V. A., & Ochsner, A. (Eds). (2019). New achievements in continuum mechanics and thermodynamics. In *A tribute to wolfgang H. muller. Advanced structured materials* (Vol. 108, 564p). Cham. Switzerland.: Springer Nature Switzerland AG. Part of Springer.

Abali, B. E., Muller, W., & dell'Isola, F. (2017). Theory and computation of higher gradient elasticity theories based on action principles. *Archive of Applied Mechanics, 87*(9).

Alibert, J. J., Seppecher, P., dell'Isola, F. (2003). Truss modular beams with deformation energy de-pending on higher displacement gradients. *Mathematics and Mechanics of Solids, 8*(1).

Andreaus, U., dell'Isola, F., & Porfiri, M. (2004). Piezoelectric passive distributed controllers for beam flexural vibrations. *Journal of Vibration and Control, 10*(5), 625–659.

Auffray, N., dell'Isola, F., Eremeyev, V., Madeo, A., & Rossi, G. (2013). Analytical continuum mechanics à la Hamilton-Piola: Least action principle for second gradient continua and capillary fluids. *Mathematics and Mechanics of Solids.*

Barchiesi, E., Spagnuolo, M., & Placidi, L. (2019). Mechanical metamaterials: A state of the art. *Mathematics and Mechanics of Solids, 24*(1), 212–234.

Berezhnoi, D. V., Konoplev, Yu. G., Paimushin, V. N., & Sekaeva, L. R. (2004). Investigation of the interaction between concrete collector and dry and waterlogged grounds. Trudy Vseros. nauch. konf. "Matematicheskoe modelirovanie i kraevye zadachi" [Proc. All-Russ. Sci. Conf. "Mathematical Simulation and Boundary Value Problems"]. Part 1. Mathematical Models of Mechanics, Strength and Reliability of Structures. Samara, SamGTU, pp. 37–39. (In Russian).

Del Vescovo, D., & Giorgio, I. (2014). Dynamic problems for metamaterials: review of existing models and ideas for further research. *International Journal of Engineering Science, 80,* 153–172.

dell'Isola, F., Andreaus, U., & Placidi, L. (2015). At the origins and in the vanguard of peridynamics, non-local and higher-gradient continuum mechanics: An underestimated and still topi-cal contribution of Gabrio Piola. *Mathematics and Mechanics of Solids, 20*(8).

dell'Isola, F., Della Corte, A., & Giorgio, I. (2016). Higher-gradient continua: The legacy of Piola, Mindlin, Sedov and Toupin and some future research perspectives. *Mathematics and Mechanics of Solids.*

dell'Isola, F., Maurini, C., & Porfiri, M. (2004). Passive damping of beam vibrations through distributed electric networks and piezoelectric transducers: prototype design and experimental validation. *Smart Materials and Structures, 13*(2), 299.

dell'Isola, F., Seppecher, P., & Alibert, J. J. (2019). Pantographic metamaterials: an example of math-emphatically driven design and of its technological challenges. *Continuum Mechanics and Thermodynamics, 31*(4), 851–884.

dell'Isola, F., Seppecher, P., & Madeo, A. (2012). How contact interactions may depend on the shape of Cauchy cuts in Nth gradient continua: approach "à la D'Alembert". *Zeitschrift für angewandte Mathematik und Physik, 63*(6).

Di Cosmo, F., Laudato, M., & Spagnuolo, M. (2018). Acoustic metamaterials based on local resonances: Homogenization, optimization and applications. Chapter from a book. In: *Generalized Models and Non-classical Approaches in Complex Materials.* (pp. 247–274) Springer.

Gorshkov, A. G., Medvedskii, A. L., Rabinskii, L. N., & Tarlakovskii, D. V. (2004). *Waves in continuum m*edia (472p.). Moscow: FIZMATLIT (In Russian).

Ivanov, V. A., & Paimushin, V. N. (1995). Refined formulation of dynamic problems of three-layered shells with a transversally soft filler is a numerical-analytical method for solving them. *Applied Mechanics and Technical Physics, 36*(4), 147–151.

Ivanov, V.A., & Paimushin, V. N. (1995). Refinement of the equations of the dynamics of multilayer shells with a transversally soft filler. *Izv. RAS. MTT, 3,* 142–152.

Kostrov, B. V. (1964). Motion of a rigid massive wedge inserted into an elastic medium under the effect of plane wave. *Prikl Mat Mekh, 28*(1), 99–110. (In Russian).

Maurini, C., dell'Isola, F., & Del Vescovo, D. (2004). Comparison of piezoelectronic networks acting as distributed vibration absorbers. *Mechanical Systems and Signal Processing, 18*(5), 1243–1271.

Placidi, L., Barchiesi, E., Turco, E., & Rizzi, N. L. (2016). A review on 2D models for the description of pantographic fabrics. *Zeitschrift für angewandte Mathematik und Physik, 67*(5).

Ryl'ko, M. A. (1977). On the motion of a rigid rectangular insertion under the effect of plane wave. *Mekh Tverd Tela, 1,* 158–164 (In Russian).

Set of rules for the design and construction of the joint venture 23-105-2004 Evaluation of vibration in the design and construction and operation of metro facilities. Moscow: Gosstroy Russia (2014).
Sheddon, I. (1951). *Fourier Transforms* (p. 542). New York: McGraw Hill.
Rakhmatulin, Kh. A., & Sunchalieva, L. M. (1983). Elastic and elastoplastic properties of the ground upon dynamic loads on the foundation. *Department: in VINITI,* 4149–4183 (In Russian).
Umek, A. (1973). Dynamic responses of building foundations to incident elastic waves. PhD Thesis. Illinois, Ill. Inst. Technol.
Vidoli, S., & dell'Isola, F. (2001). Vibration control in plates by uniformly distributed actuators interconnected via electric networks. *European Journal of Mechanics - A/Solids, 20*(3), 435–456.

Chapter 9
Computer Simulation of the Process of Loss of Stability of Composite Cylindrical Shells Under Combined Quasi-static and Dynamic Loads

Nikolaii A. Abrosimov, Aleksandr V. Elesin, and Leonid Igumnov

Abstract Based on the applied theory of shells, an energy-consistent resolving system of equations is constructed, and a complex numerical method is developed which, within the framework of an explicit variational-difference scheme, makes it possible to solve both quasi-static and dynamic problems of nonlinear nonaxisymmetric deformation and loss of stability of composite cylindrical shells. The reliability of the developed method is substantiated by comparing calculation results with experimental data. The characteristic forms and critical buckling loads of GRP cylindrical shells as functions of the level of preloading by a quasi-static internal pressure and of the subsequent dynamic loading by an external pressure are analyzed for various reinforcement patterns in a wide range of loading rate.

Keywords Composite materials · Cylindrical shells · Nonlinear deformation · Stability · Numerical methods · quasi-static and dynamic loads

9.1 Introduction

Due to their high strength and rigidity characteristics, composite materials offer great possibilities for creating rational designs in various areas of modern engineering. When in service, structural elements made from composite materials can be subjected to combined dynamic and static loads, leading to considerable changes in form and buckling instability of the elements. To effectively utilize the potentialities of composite materials, comprehensive studies on the dynamic deformation and buckling instability of structural elements made from materials under combined quasi-static and dynamic effects are necessary.

The current investigations in this direction are devoted, as a rule, to the analysis of the nonlinear behavior and buckling instability of cylindrical shells made from traditional isotropic materials (Manevich et al. 1977; Baskakov et al. 1982;

N. A. Abrosimov · A. V. Elesin · L. Igumnov (✉)
Research Institute for Mechanics, National Research Lobachevsky State University of Nizhny Novgorod, Gagarin ave., 23, 603950 Nizhny Novgorod, Russian Federation
e-mail: igumnov@mech.unn.ru

F. dell'Isola and L. Igumnov (eds.), *Dynamics, Strength of Materials and Durability in Multiscale Mechanics*, Advanced Structured Materials 137,
https://doi.org/10.1007/978-3-030-53755-5_9

Bendyukov and Deryushev 1995; Skurlatov 1972; Dubrovin 2015). In (Manevich et al. 1977), results of an experimental–theoretical study on the buckling instability of a steel cylindrical shell under pulsed loading by external pressure in combination with external (or internal) static pressure are presented. In (Baskakov et al. 1982), the results of experimental investigations into the effect of internal static pressure and loading rate on the stability of aluminum cylindrical shells subjected to pulsed loadings by external pressure are given.

An experimental analysis of buckling of thin-walled cylindrical shells under local pulsed loading at various values of axial static compression is presented in (Bendyukov et al. 1995). In (Skurlatov 1972), the results of investigation into the stability of cylindrical shells subjected to axial static forces and an incident in the longitudinal axial direction pressure wave are given.

Experimental and theoretical investigations into the processes of deformation and loss of stability of composite structural elements under dynamic loading were considered in (Smirnov and Lamzin 2018; Volkov et al. 2018; Jansen 2005; Bisagni 2005; Rahman et al. 2011).

At the same time, the nonlinear 3D problems on the dynamic deformation and loss of stability of preliminary loaded composite cylindrical shells have not been investigated carefully enough (Smirnov and Lamzin 2018). The aim of the present work was to develop a technique of numerical research into nonlinear nonstationary deformation and loss of stability of cylindrical shells of composite materials under combined quasi-static and dynamic loading conditions.

It is also worth to mention about one of the most interesting subjects in modern continuum mechanics and engineering regarding composite materials (Placidi et al. 2015). This is the development of newly (scientifically) conceived materials ('metamaterials') with mechanical properties that cannot be found in nature (Barchiesi and Spagnuolo 2018; Del Del Vescovo and Giorgio 2014). Metamaterials are macroscopic composites whose properties are mainly determined by their periodic cellular microstructure rather than by the chemical and physical properties of the material constituting them.

An example of mechanical metamaterials are pantographic structures (dell'Isola et al. 2016a, b, c, 2017, 2019b; Placidi et al. 2016, 2017; Rahali et al. 2015).

In order to account for multiscale mechanical interactions, which taking place for metamaterials, higher order gradient continuum theories can be a choice (dell'Isola et al. 2012, 2015, 2016, 2019a; Sciarra et al. 2007).

9.2 Problem Formulation and Solution Method

Let us consider a cylindrical shell formed by angle-ply winding of a unidirectional composite material in the orthogonal curvilinear system of coordinates α_i $(i = \overline{1, 3})$, coinciding with lines of the principal curvatures and the external normal to the internal surface of the shell. The Lamé coefficients of the shell are $H_1 = 1, \quad H_2 = 1 +$

$k_2\alpha_3$, $H_3 = 1$, principal curvatures $k_1 = 0$, $k_2 = 1/R$ where R is radius of the internal surface of the shell.

Components of the nonlinear strain tensor in the applied theory of cylindrical shells can be presented in the form (Shapovalov 1997):

$$e_{11} = \frac{1}{H_1}(\varepsilon_{11} + \varepsilon_{13}^2/2 + \alpha_3\chi_{11}),$$
$$e_{12} = \frac{1}{H_1}[\varepsilon_{12} + \varepsilon_{13}\varepsilon_{23}/2 + \alpha_3\chi_{12}] + \frac{1}{H_2}[\varepsilon_{21} + \varepsilon_{13}\varepsilon_{23}/2 + \alpha_3\chi_{21}], (1 \leftrightarrow 2),$$
$$e_{13} = \frac{1}{H_1}(\varphi_1 + \varepsilon_{13}), \tag{9.1}$$

where

$$\varepsilon_{11} = \frac{\partial u_1}{\partial \alpha_1} + k_1 u_3, \chi_{11} = \frac{\partial \varphi_1}{\partial \alpha_1}, \varepsilon_{12} = \frac{\partial u_2}{\partial \alpha_1}, \chi_{12} = \frac{\partial \varphi_2}{\partial \alpha_1},$$
$$\varepsilon_{13} = \frac{\partial u_3}{\partial \alpha_1} - k_1 u_1, (1 \leftrightarrow 2) \tag{9.2}$$

$u_i(\alpha_1, \alpha_2, t)$ $(i = \overline{1, 3})$ are displacements of points of the internal surface of shell in the directions of axes α_i , φ_j $(j = \overline{1, 2})$ are rotation angles of the normal to the internal surface. The symbol $(1 \leftrightarrow 2)$ located on a separate line means that each above-mentioned relation is supplemented by one more relation with the subscript 1 replaced by 2, 2 by 1. The symbol located on one line with a relation means the same operation with this particular relation. The physical relations for the elementary layer of the shell, in view of hypotheses of the applied theory of shells, may be written as (Vasil'yev 1988):

$$\sigma_{11} = \left(A_{11} - \frac{A_{13}^2}{A_{33}}\right)e_{11} + \left(A_{12} - \frac{A_{13}A_{23}}{A_{33}}\right)e_{22}, \sigma_{12} = A_{66}e_{12}, (1 \leftrightarrow 2),$$
$$\sigma_{13} = A_{44}e_{13}, \sigma_{23} = A_{55}e_{23}, \tag{9.3}$$

where A_{mn} are rigidities of the unidirectional layer, which are calculated from the elastic moduli and Poisson ratios of the elementary layer and are step functions of the variable α_3.

Constitutive relations of an isotropic shell are formulated on the basis of the differential theory of plasticity with linear hardening (Abrosimov and Bazhenov 2002)

$$\sigma_{ij} = \frac{\nu E}{(1+\nu)(1-2\nu)}e + \frac{E}{2(1+\nu)}e'_{ij},$$
$$e_{ij} = e'_{ij} + e''_{ij}, \; e = e_{11} + e_{22} + e_{33}, \; \dot{e}_{ij} = \dot{\gamma} S_{ij},$$

$$\sigma=(\sigma_{11}+\sigma_{22}+\sigma_{33})/3,\ \sum_{i,j=1}^{3}S_{ij}S_{ij}={}^{2}\!/_{3}\sigma_{*}^{2},$$

$$S_{ij}=\sigma_{ij}-\sigma\delta_{ij}-\rho_{ij},\ \rho_{ij}=2ge''_{ij},\ e''_{ij}=\int\limits_{0}^{t}\dot{e}''_{ij}dt, \tag{9.4}$$

where E is the elastic modulus; ν is Poisson ratio; e'_{ij}, e''_{ij} are the elastic and plastic components of the strain tensor; δ_{ij} is the Kronecker tensor; σ_*, g are the yield stress and hardening modulus of the material; $\dot{\gamma}$ is a scalar parameter.

An energetically consistent system of equations of motion in the applied theory of cylindrical shells is deduced from the condition of minimum of the functional of full energy of shell, which, for a cylindrical shell loaded by a dynamic internal (external) pressure, can be written as (Abrosimov and Bazhenov 2002):

$$\begin{aligned}
&\iint\limits_{S}\Bigg[N_{11}\frac{\partial(\delta u_1)}{\partial\alpha_1}+N_{21}\frac{\partial(\delta u_1)}{\partial\alpha_2}+N_{22}\frac{\partial(\delta u_2)}{\partial\alpha_2}+N_{12}\frac{\partial(\delta u_2)}{\partial\alpha_1}-N_{23}^{*}k_2\delta u_2\\
&+N_{13}^{*}\frac{\partial(\delta u_3)}{\partial\alpha_1}+N_{23}^{*}\frac{\partial(\delta u_3)}{\partial\alpha_2}+N_{22}k_2\delta u_3+M_{11}\frac{\partial(\delta\varphi_1)}{\partial\alpha_1}+M_{21}\frac{\partial(\delta\varphi_1)}{\partial\alpha_2}\\
&+Q_{13}\delta\varphi_1+M_{22}\frac{\partial(\delta\varphi_2)}{\partial\alpha_2}+M_{12}\frac{\partial(\delta\dot{\varphi}_2)}{\partial\alpha_1}+Q_{23}\delta\dot{\varphi}_2\Bigg]d\alpha_1 d\alpha_2\\
&+\iint\limits_{S}\big[(\bar{B}_{11}\ddot{u}_1+\bar{B}_{12}\ddot{\varphi}_1)\delta u_1+(\bar{B}_{11}\ddot{u}_2+\bar{B}_{12}\ddot{\varphi}_2)\delta u_2+\bar{B}_{11}\ddot{u}_3\delta u_3\\
&+(\bar{B}_{22}\ddot{\varphi}_1+\bar{B}_{21}\ddot{u}_1)\delta\varphi_1\\
&+(\bar{B}_{22}\ddot{\varphi}_2+\bar{B}_{21}\ddot{u}_2)\delta\varphi_2\big]d\alpha_1 d\alpha_2-\sum_{i=1}^{2}\int\limits_{\Gamma_i}N_{11}^{o}\delta u_i^{o}d\alpha_2-\iint\limits_{S}F_3\delta u_3 d\alpha_1 d\alpha_2=0
\end{aligned} \tag{9.5}$$

where

$$(N_{11},N_{12},M_{11},M_{12},Q_{13})=\int\limits_{0}^{h}(\sigma_{11},\sigma_{12},\alpha_3\sigma_{11},\alpha_3\sigma_{12},\sigma_{13})H_2 d\alpha_3\ ,$$

$$N_{13}^{*}=Q_{13}+N_{11}\varepsilon_{13}+N_{12}\varepsilon_{23}\ ,$$

$$(1\leftrightarrow 2)$$

$$\bar{B}_{11}=\rho(h+k_2h^2/2);\qquad \bar{B}_{22}=\rho(h^3/3+k_2h^4/4);$$

$$\bar{B}_{12}=\bar{B}_{21}=\rho(h^2/2+k_2h^3/3),$$

S is the internal surface of the shell, Γ_i is shell boundaries along the lines α_2 ($i =$ 1,2), F_3 is the load along the coordinate axis α_3, index «o» denotes the force applied to the ends of the shell, ρ is the density of shell material, and h is shell thickness.

By minimizing the total energy functional of the shell (9.5), it is possible to obtain a system of equations of motion

$$
\begin{aligned}
L_1(N) &= \bar{B}_{11}\ddot{u}_1 + \bar{B}_{12}\ddot{\varphi}_1\,; \quad L_2(N) + N^*_{13}k_2 = \bar{B}_{11}\ddot{u}_2 + \bar{B}_{12}\ddot{\varphi}_2\,,\\
L_1(M) - Q_{13} &= \bar{B}_{22}\ddot{\varphi}_1 + \bar{B}_{21}\ddot{u}_1\,; \quad L_2(M) - Q_{23} = \bar{B}_{22}\ddot{\varphi}_2 + \bar{B}_{21}\ddot{u}_2\,,\\
L_1(T) &= \frac{\partial T_{11}}{\partial \alpha_1} + \frac{\partial T_{21}}{\partial \alpha_2}\,, \quad (1 \leftrightarrow 2)\,,\\
&\frac{\partial N^*_{13}}{\partial \alpha_1} + \frac{\partial N^*_{23}}{\partial \alpha_2} - k_2 N_{22} + F_3 = \bar{B}_{11}\ddot{u}_3
\end{aligned}
\tag{9.6}
$$

and natural boundary conditions, which have to be assumed in the form

$$
N_{11} = N^o_{11}\,; \quad u_2 = u_3 = \varphi_1 = \varphi_2 = 0. \tag{9.7}
$$

Supplementing relation (9.6) and (9.7) with the necessary number of initial conditions

$$
\begin{aligned}
&u_i(\alpha_1, \alpha_2, 0) = u_i^0(\alpha_1, \alpha_2), \qquad \varphi_j(\alpha_1, \alpha_2, 0) = \varphi_j^0(\alpha_1, \alpha_2),\\
&\dot{u}_i(\alpha_1, \alpha_2, 0) = \dot{u}_i^0(\alpha_1, \alpha_2), \qquad \dot{\varphi}_j(\alpha_1, \alpha_2, 0) = \dot{\varphi}_j^0(\alpha_1, \alpha_2) \qquad (i = \overline{1,3}\,;\ j = 1, 2),
\end{aligned}
$$

we obtain a full system of equations for analyzing nonlinear wave processes of deformation and buckling instability of cylindrical shells made by stacking elementary layers of a unidirectional composite material, loaded with static internal pressure or axial compression and subsequent external dynamic pressure. However, axial loading is carried out through an absolutely rigid ring.

The critical load of buckling instability is determined according to the characteristic kink of the action amplitude–maximum deflection curve.

The numerical method for solving the formulated problem is based on the explicit variational-difference scheme (Abrosimov and Bazhenov 2002; Abrosimov and Elesin 2017). As a result of the transformation of the variational equation of dynamics (9.5) using well-known difference procedures (Abrosimov and Bazhenov 2002), we arrive at the systems of grid equations describing the motion of the internal and boundary nodes:

$$
\begin{aligned}
\bar{B}_{11}\ddot{u}_j + \bar{B}_{12}\ddot{\varphi}_j &= F_{u_j}\,, \qquad \bar{B}_{11}\ddot{u}_3 = F_{u_3}\\
\bar{B}_{22}\ddot{\varphi}_j + \bar{B}_{21}\ddot{u}_j &= F_{\varphi_j}\,, \quad (j = 1, 2)
\end{aligned}
\tag{9.8}
$$

where F_{u_i}, F_{φ_j} are difference analogs of the left parts of the system (9.6).

Solving the system of algebraic Eq. (9.8) relative to generalized accelerations $\ddot{u}_i$, $\ddot{\varphi}_j$, we obtain a system of ordinary differential equations of the second order in time, for the integration of which the explicit "cross" scheme is used (Abrosimov and Bazhenov 2002). As a result, the solution of system (9.8) is reduced to a recurrent calculation using the formulas:

$$\dot{u}_j\left(t^{\kappa+\frac{1}{2}}\right) = \dot{u}_j\left(t^{\kappa-\frac{1}{2}}\right) + \frac{\Delta t}{\bar{B}_{11}\bar{B}_{22} - \bar{B}_{12}\bar{B}_{21}}\left(F_{u_j}\bar{B}_{22} - F_{\varphi_j}\bar{B}_{12}\right),$$

$$\dot{u}_3\left(t^{\kappa+\frac{1}{2}}\right) = \dot{u}_3\left(t^{\kappa-\frac{1}{2}}\right) + \frac{\Delta t}{\bar{B}_{11}}\left(F_{u_3}\right),$$

$$\dot{\varphi}_j\left(t^{\kappa+\frac{1}{2}}\right) = \dot{\varphi}_j\left(t^{\kappa-\frac{1}{2}}\right) + \frac{\Delta t}{\bar{B}_{11}\bar{B}_{22} - \bar{B}_{12}\bar{B}_{21}}\left(F_{\varphi_j}\bar{B}_{11} - F_{u_j}\bar{B}_{21}\right),$$

$$u_i(t^{\kappa+1}) = u_i(t^{\kappa}) + \Delta t\dot{u}_i\left(t^{\kappa+\frac{1}{2}}\right), \qquad (i = \overline{1,3}\ ;\ j = 1, 2), \quad (\kappa = \overline{0,\infty})$$

$$\varphi_j(t^{\kappa+1}) = \varphi_j(t^{\kappa}) + \Delta t\dot{\varphi}_j\left(t^{\kappa+\frac{1}{2}}\right).$$

In this case, the time integration step Δt is determined on the basis of Neumann spectral theorem, which leads to the condition (Abrosimov and Bazhenov 2002).

$\Delta t \leq 2/\omega\text{max}$ (ωmax is maximum eigenfrequency of a semi-discrete system (9.8)).

Herewith, the quasi-static loading mode is modeled by specifying the pressure as a linearly growing function reaching a stationary value during three vibration periods of cylindrical composite shell in the lowest form.

9.3 Results of Research

To validate the reliability and accuracy of the technique suggested, numerical calculations were compared with experimental data (Baskakov et al. 1982) on the dynamic stability of isotropic cylindrical shells loaded by internal pressure and then by dynamic external pressure at different loading rates, with the external pressure distributed uniformly over all shell surface.

The geometrical and physicomechanical parameters of the shell were as follows: $R/h = 104$; $h = 0.0005$ m; $L/R = 1.9$; $E = 73$ GPa; $\nu = 0.3$, $\rho = 2700$ kg/m^3, σ_* $= 0.37$ GPa; $g = 0.6$ GPa, and L is the length of shell generatrix.

The static internal pressure was created by compressed air, and the dynamic pressure by an electrohydraulic discharge caused by the blast of calibrated copper wires (Baskakov et al. 1982). The fastening of shell edges was close to rigid fixation.

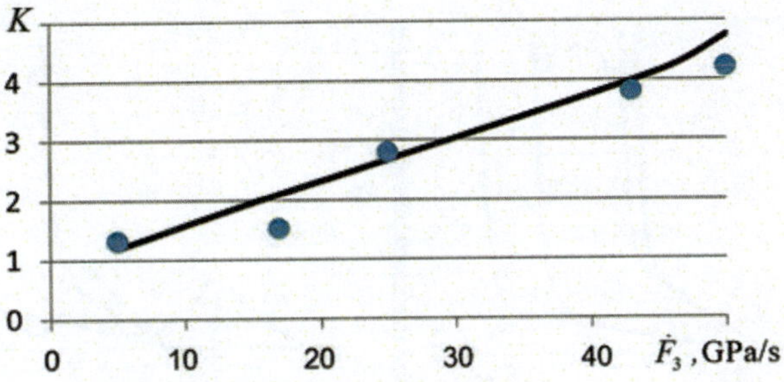

Fig. 9.1 Coefficient of dynamic overload versus loading rate (dots—experiment (Baskakov et al. 1982), curve—calculation by the given technique)

9.3.1 Internal Pressure

Figure 9.1 presents experimental and calculated dynamic coefficients $K = \frac{F_3^*}{F_3^0}$ as function of the rate of loading by the dynamic external pressure $\dot{F}_3^+$ (F_3^*, F_3^0 are the critical loads of buckling instability under the dynamic and static external pressure, respectively). The results presented were obtained at a static internal pressure F_3^-, dimensionless form of which is defined by $\bar{F}_3^- = \frac{F_3^-}{E}\left(\frac{R}{h}\right)^2$ and is equal to $\bar{F}_3^- = 0.07$.

The result obtained points to a good agreement between calculation results and experimental data.

9.3.2 External Pressure

Further, the effect of preliminary axial loading by external pressure on the process of loss of stability of the cylindrical shell was investigated. The shell was made of a composite material with the following geometrical and physicomechanical material parameters: $R = 0072\,\text{m}$; $R/h = 112$; $L/R = 2, 22$; $E_{11} = 200\,\text{GPa}$; $E_{22} = E_{11}/30$; $G_{12} = G_{13} = G_{23} = E_{22}/2$; $\nu_{12} = 0.25$, $\rho = 1800\ \text{kg/m}^3$.

The results of the research into the effect of reinforcement angle and preliminary static loading by external pressure on the process of loss of stability of the shell uniformly distributed over all shell surface are illustrated in Figs. 9.2, 9.3, 9.4, and 9.5.

Figures 9.2 and 9.3 show the absolute values of time-dependent maximum deflections U_3^* of shells with different reinforcement angles and pulse rates of external pressure, preliminary loaded by axial quasi-static compression of various intensity.

The deformed configurations of shells illustrating the effect of reinforcement angle on the process of loss of stability under dynamic loading by external pressure at pulse rates of 5 and 20 GPa/s, both preloaded by quasi-static axial loads of various level and nonpreloaded, are shown in Figs. 9.4 and 9.5.

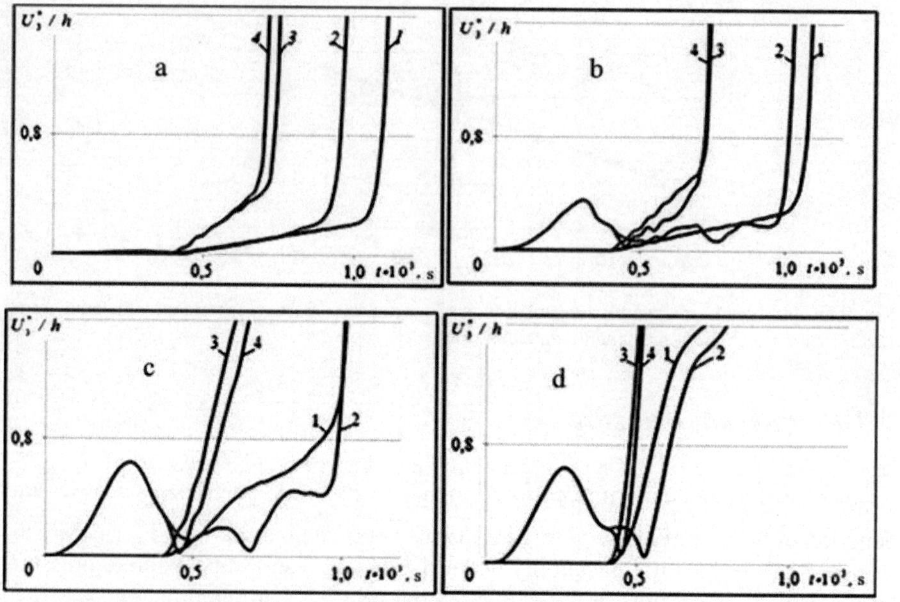

Fig. 9.2 Absolute values of maximum deflection of shell versus time: without considering the axial compression for pulse rate of external pressure 5 GPas (1), 20 GPa/s (3) and considering quasi-static compression $N_{11}^{0} = 0, 9N_{11}^{*}$ (N_{11}^{*} is critical value of quasi-static loss of stability) for the pulse rates of external pressure 5 GPa/s (2), 20 GPa/s (4) reinforcement angles 90° (**a**), 75° (**b**), 60° (**c**), 45° (**d**) relatively the length of shell generatrix, respectively

The analysis of the effect of the rate of the external force in combination with quasi-static compression on the loss of stability testifies that the shells with a reinforcement angle of 90° exhibit the loss of stability as a result of formation of dents elongated in the circular direction, and their intensity being almost the same both under quasi-static and dynamic loading. For shells with a reinforcement angle of 75°, the loss of stability is characterized by decreasing the number of dents and increasing their area when external pressure grows. For shells with a reinforcement angle of 60°, a transformation of a corrugated buckling under quasi-static loading is observed followed by a transition to a configuration with bulges extended along the generatrix. The shells with a reinforcement angle of 45° remain almost cylindrical with the exception of the area adjacent to the ends of the shell.

During the loss of stability of shells with a reinforcement angle close to 90°, loaded by external pressure and axial compression with intensity of $N_{11}^{0} = 0$, $N_{11}^{0} = 0, 45N_{11}^{*}$ and $N_{11}^{0} = 0, 9N_{11}^{*}$, the number of waves is observed to increase in both circular and longitudinal directions. The shells with a reinforcement angle of 75° are characterized by the change in configuration of the loss of stability under growing axial compression from bulges oriented along the generatrix to the formation of local dents. At the loss of stability of shells with a reinforcement angle of 60°, dents are formed along all shell generatrix and their number and depth increase with

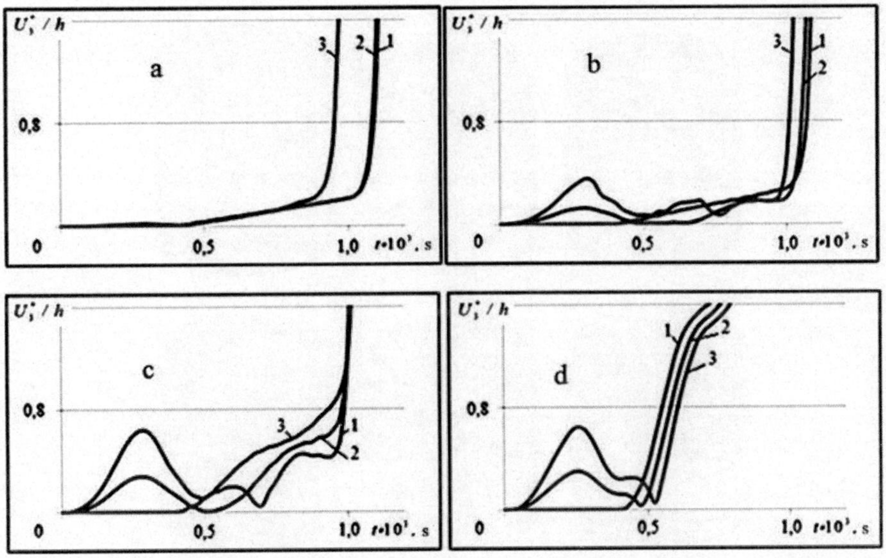

Fig. 9.3 Maximum deflection of shell versus time under loading by external pressure at the rate of 5 GPa/s and various levels of axial compression: $N_{11}^{0} = 0$ (1), $N_{11}^{0} = 0,45N_{11}^{*}$ (2), $N_{11}^{0} = 0,9N_{11}^{*}$ (3) for reinforcement angles 90° (**a**), 75° (**b**), 60° (**c**), 45° (**d**), respectively

increasing axial compression. For shells reinforced with 45° angle, the buckling modes practically retaining the initial cylindrical form, except for the zone of edge effects, are characteristic.

From the results obtained, it follows that the level of preliminary axial loading considerably affects the characteristic forms of the loss of stability of structures reinforced with 60°–90° angles.

With a smaller reinforcement angle, the forms of the buckling instability are almost the same. For all reinforcement angles, precompression has little effect on the critical load of the external pressure pulse, the greatest effect is observed for the shell with a reinforcement angle of 90°.

A similar study was carried out for a cylindrical shell with weakly expressed anisotropy ($E_{22} = E_{11}/2$), which showed a slight effect of pre-axial loading on the critical load of buckling, even for "extreme" reinforcement options (0° and 90°). Herewith, preliminary axial compression has a significant effect on the characteristic buckling forms.

9.4 Conclusion

Precompression for shells with pronounced anisotropy has a significant impact on buckling forms for reinforcement angles of 60°–90°. The values of the critical load

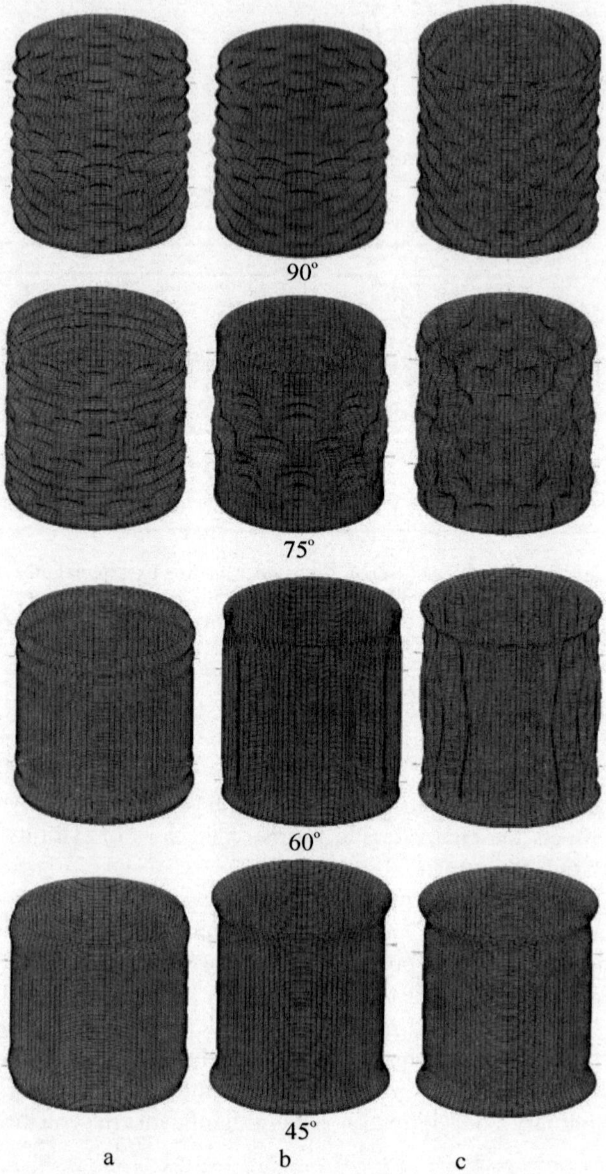

Fig. 9.4 Characteristic buckling forms of composite cylindrical shells for different reinforcement angles: under quasi-static loading (**a**); dynamic loading by external pressure pulse at the rates of 5 GPa/s (**b**) and 20 GPa/s (**c**) in combination with quasi-static compression intensity $N_{11}^0 = 0, 9N_{11}^*$

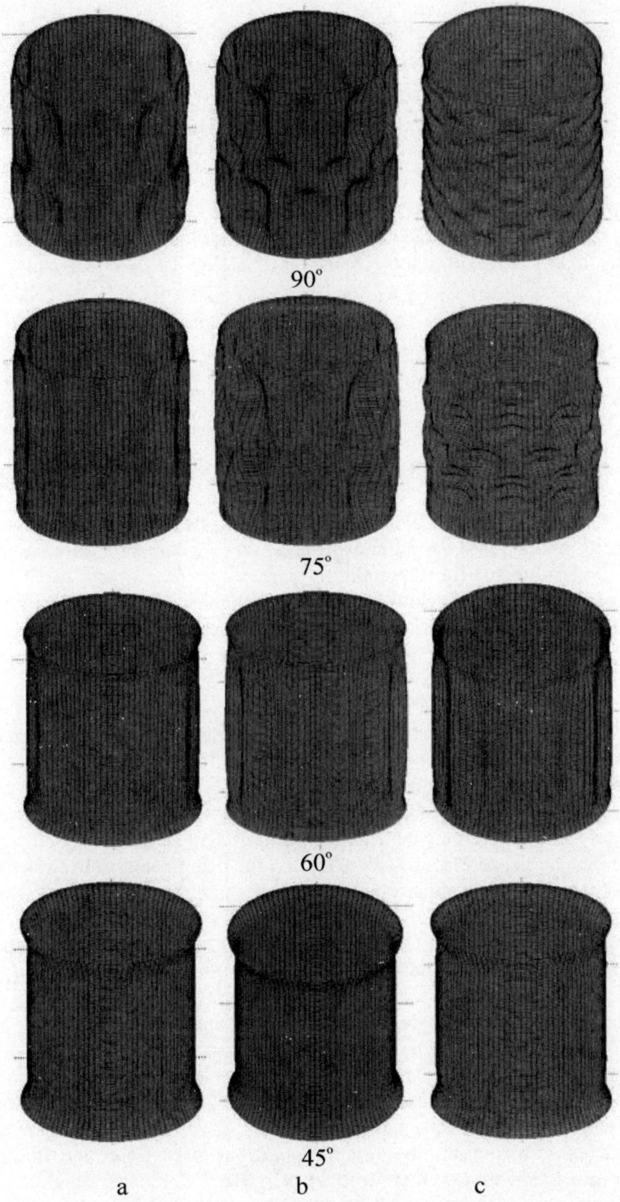

Fig. 9.5 Characteristic buckling forms of composite cylindrical shells under loading by external pulse at a rate of 5 GPa/s for different reinforcement angles and intensity of axial compression: $N_{11}^0 = 0$ (**a**), $N_{11}^0 = 0,45N_{11}^*$ (**b**) and $N_{11}^0 = 0,9N_{11}^*$ (**c**), respectively

vary slightly, and the greatest difference is observed for reinforcement angles of 90°. For shells made from material with a weakly pronounced anisotropy, a slight effect of the preliminary axial loading on the critical value of the dynamic external pressure is observed, though a difference in buckling forms of a shell is noted.

Acknowledgements The method of calculating composite cylindrical shells under combined loads was developed at a financial support of the Ministry of Science and Higher Education of the Russian Federation (task 0729-2020-0054), and numerical analysis of loss of stability of shells was carried out at a financial support of RFBR grants ((№ 18-08-01234, № 19-08-00828).

References

Abrosimov, N. A., & Bazhenov, V. G. (2002). *Nonlinear Problems of dynamics of composite structures*. Nizhni Novgorod: Izd NNGU, 400 [in Russian].

Abrosimov, N. A., & Elesin, A. V. (2017). Numerical analysis of dynamic strength of composite cylindrical shells under multiple-pulse exposures. *Nizhni Novgorod, Problemy prochnosti i plastichnosti, 79*(4), 450–461 [in Russian].

Barchiesi, E., Spagnuolo, M., & Placidi L. (2018). Mechanical metamaterials: a state of the art. *Mathematics and Mechanics of Solids*.

Baskakov, V. N., Kostoglotov, A. I., & Shvetsova, L. A. (1982). Investigation of the dynamic stability of smooth cylindrical shells. *Problemy Prochnosti, 5*, 31–33.

Bendyukov, V. V., & Deryushev, V. V. (1995) Dynamic short-wave instability of thin-walled cylindrical shells at the local action of an external pressure pulse. *Problemy Prochnosti, 4*, 36–43.

Bisagni, C. (2005). Dynamic buckling of fiber composite shells under impulsive axial compression. *Thin-Walled Structure, 43*, 499–514.

Del Vescovo, D., & Giorgio, I. (2014). Dynamic problems for metamaterials: review of existing models and ideas for further research. *International Journal of Engineering Science, 80*, 153–172.

dell'Isola, F., Andreaus, U., & Placidi, L. (2015). At the origins and in the vanguard of peridynamics, non-local and higher-gradient continuum mechanics: An underestimated and still topi-cal contribution of Gabrio Piola. *Mathematics and Mechanics of Solids, 20*(8).

dell'Isola, F., Cuomo, M., Greco, L., Della, Corte, A. (2017). Bias extension test for pantographic sheets: Numerical simulations based on second gradient shear energies. *Journal of Engineering Mathematics*.

dell'Isola, F., Della Corte, A., & Giorgio, I. (2016a). Higher-gradient continua: The legacy of Piola, Mindlin, Sedov and Toupin and some future research perspectives. *Mathematics and Mechanics of Solids*.

dell'Isola, F., Della Corte, A., Greco, L., & Luongo, A. (2016b). Plane bias extension test for a continuum with two inextensible families of fibers: a variational treatment with Lagrange multipliers and a perturbation solution. *International Journal of Solids and Structures*.

dell'Isola, F., Giorgio, I., Pawlikowski, M., & Rizzi, N. (2016c). Large deformations of planar extensible beams and pantographic lattices: Heuristic homogenization, experimental and numerical examples of equilibrium. *Proceedings of The Royal Society A, 472*(2185).

dell'Isola, F., Seppecher, P., & Madeo, A. (2012). How contact interactions may depend on the shape of Cauchy cuts in Nth gradient continua: approach "à la D'Alembert". *Zeitschrift für angewandte Mathematik und Physik, 63*(6).

dell'Isola, F., Seppecher, P., Alibert, J. J., Lekszycki, T., Grygoruk, R., Pawlikowski, M., et al. (2019a). Pantographic metamaterials: An example of mathematically driven design and of its technological challenges. *Continuum Mechanics and Thermodynamics, 31*(4), 851–884.

dell'Isola, F., Seppecher, P., Spagnuolo, M., Barchiesi, E., Hils, F., Lekszycki, T., et al. (2019b). Advances in pantographic structures: design, manufacturing, models, experiments and image analyses. *Continuum Mechanics and Thermodynamics, 31*(4), 1231–1282.

Dubrovin, V. M., & Butina, T. A. (2015). Modeling the dynamic stability of a cylindrical shell at the action of an axial compressing load. *Mechanics of Composite Materials, 6*, 46–57.

Jansen, E. L. (2005). Dynamic stability problems of anisotropic cylindrical shells via a simplified analysis. *Nonlinear Dynamics, 39,* 349–367.

Manevich, L. I., Mikhailov, G. V., Pavlenko, I. D., & Prokopalo, E. F. (1977). Research on the stability of shells at a joint action of static and dynamic loads. *Prikl Mekh XIII, 1*, 27–32.

Placidi, L., Andreaus, U., & Giorgio, I. (2017). Identification of two-dimensional pantographic structure via a linear D4 orthotropic second gradient elastic model. *Journal of Engineering Mathematics.*

Placidi, L., Barchiesi, E., Turco, E., & Rizzi, N. L. (2016). A review on 2D models for the description of pantographic fabrics. *Zeitschrift für angewandte Mathematik und Physik, 67*(5).

Placidi, L., Giorgio, I., Della, Corte A., & Scerrato, D. (2015). Euromech 563 Cisterna di Latina 17–21 March 2014 Generalized continua and their applications to the design of composites and metamaterials: a review of presentations and discussions. *Mathematics & Mechanics of Solids.* https://doi.org/10.1177/1081286515576948.

Rahali, Y., Giorgio, I., Ganghoffer, & J.-F., dell'Isola, F. (2015). Homogenization à la Piola produces second gradient continuum models for linear pantographic lattices. *International Journal of Engineering Science, 97.*

Rahman, T., Jansen, E. L., & Gürdal, Z. (2011). Dynamic buckling analysis of composite cylindrical shells using a finite element based perturbation method. *Nonlinear Dynamics, 66*(3), 389–401.

Sciarra, G., dell'Isola, F., & Coussy, O. (2007). Second gradient poromechanics. *International Journal of Solids and Structures, 44*(20).

Shapovalov, L. A. (1997). Consideration of transverse compression in equations of the nonlinear dynamics of shells. *Izv RAN, Mekh Tverd Tela, 3*, 156–168.

Skurlatov, E. D. (1972). An experimental study on the behaviour of cylindrical shells at dynamic loadings. *Problemy prochnosti, 9*, 79–83.

Smirnov, I. V., Lamzin, D. A., Konstantinov, A. Y., Bragov, A. M., & Lomunov, A. K. (2018). A unified experimental-theoretical approach to predict the critical stress characteristics of failure and yielding under quasi-static and dynamic loading. *Engineering Fracture Mechanics.* https://doi.org/10.1016/j.engfracmech.2018.10.023.

Vasil'yev, V. V. (1988). *Mechanics of structures from composite materials* (Vol. 272). Moscow; Mashinostroenie [in Russian].

Volkov, I. A., Igumnov, L. A., Litvinchuk, S. Y., & Vorobtsov, I. V. (2018). Modeling dynamic deformation and failure of thin-walled structures under explosive loading. *EPJ Web of Conferences, 183*(03016) (2018) DYMAT 2018. https://doi.org/10.1051/epjconf/201818303016.

Chapter 10
The Effect of Preheating on the Thermoelastic Structurally Inhomogeneous Medium Spectral Properties in the Presence of an Initial Strain

Galina Yu. Levi, Leonid Igumnov, and Mikhail O. Levi

Abstract The boundary problem of a prestressed thermoelastic layered half-space oscillations subjected to the action of a mechanical or thermal load is considered. Initial stresses are induced in the body by stretching or compression and by the action of temperature. The two-dimensional Green's function of the medium is constructed. We made the analysis of its real poles behavior, and their distribution is presented graphically. The effect of preheating, pinching, and initial deformation of the first mode phase velocity is studied.

Keywords Thermoelasticity · Half-space · Initial strain · Preheating · Green's function · Phase velocity

10.1 Introduction

Thermoelasticity studies the interaction of deformation and thermal fields. It deals with a dynamic system, the interaction of which with the environment is limited to mechanical work, external forces, and heat transfer. It also includes thermal conductivity, stress, and deformation due to heat flow. In addition, the change in body temperature is caused not only by external and internal sources of heat, but also by the deformation process itself. For this reason, thermoelasticity should be considered as a multifield discipline governed by the interaction of the temperature deformation field. This allows one to determine the stresses created by the temperature field and

G. Yu. Levi · L. Igumnov (✉)
Research Institute for Mechanics of Lobachevsky State University of Nizhny Novgorod, Gagarin ave., 23, 603950 Nizhny Novgorod, Russia
e-mail: igumnov@mech.unn.ru

G. Yu. Levi · M. O. Levi
Southern Scientific Center of Russian Academy of Sciences, st. Chehova, 41, 344006 Rostov-on-Don, Russia

F. dell'Isola and L. Igumnov (eds.), *Dynamics, Strength of Materials and Durability in Multiscale Mechanics*, Advanced Structured Materials 137,
https://doi.org/10.1007/978-3-030-53755-5_10

to calculate the temperature distribution as a result of the action of time-dependent forces and heat sources.

The development of modern technologies causes a considerable interest in the processes of excitation of mechanical vibrations due to the effect of laser radiation (Muratikov 1998). A fairly complete overview of the work in this direction is given in (Achenbach 2003; Xu et al. 2008). In the majority of works, different approaches are used for research, which allow efficiently analyzing various sides of dynamic processes in thermoelastic. These are, first of all, the propagation features of volume (Singh et al. 1985; Sharma et al. 1986; El-Maghraby 2008) and interface (Kumar et al. 2013; Verma 2002) waves, Rayleigh waves (Stan Chirita, 2013), Lamb (Al-Qahtani et al. 2004; Kumar et al. 2008) and cylindrical waves (Sharma 2001). Questions of viscoelastic effects (Elhagary 2013; Levi et al. 2015), as well as the presence of initial stresses (Singh 2010), occupy a special place in the problem of the propagation of thermoelastic waves in semibounded bodies. In article (Sheydakov 2008), following the approach considered in the monograph (Lurie 1980), consecutive linearization of nonlinear equations of a thermoelastic medium was carried out, and equations of motion and determining relationships of the dynamics of a prestressed thermoelastic medium were constructed. The equations are constructed in tensor form, admitting generalization to curve-linear coordinates. The issues of thermoelastic bodies contact interaction are of considerable interest (Levi et al. 2019; Belyankova et al. 1999; Belyankova et al. 2012; Levi et al. 2017; Belyankova et al. 2016). In (Belyankova et al. 1999; Belyankova et al. 2012), the problems of contact interaction of a thermoelastic layer and half-space were investigated. The equations of motion constructed in (Sheydakov 2008) and the defining relations, taking into account the presence of initial stresses, preliminary heating, etc., are generalized to prestressed thermoelectronic stresses.

In the present work, within the framework of the linearized theory of propagation of coupled thermoelastic waves (Sheydakov 2008), the boundary problem of oscillation of a non-uniform half-space under the action of a thermal load given on the surface of the medium is considered. The effect of preheating and initial strain on layered thermoelastic half-space phase velocity is investigated.

10.2 Formulation of the Problem

We consider a structurally inhomogeneous half-space, which is a homogeneous thermoelastic layer rigidly coupled with a homogeneous thermoelastic half-space. The body is subject to prestressing, which is set as deformation and thermal effects. A load acts on a surface in a certain area, imposing mechanical stress or heat flux. Outside this area, the surface is thermally insulated and free from mechanical stress. At the boundary between the layer and the half-space, the conditions of rigid coupling and thermal insulation are assumed.

10.3 The 3D Linear Thermoelasticity Equations

We assumed that the body is in the orthogonal Lagrange coordinate system x_1, x_2, x_3 associated with the natural state of the medium. The axis x_3 is directed vertically upward from the body. Thermoelastic layer $|x_1|,\ |x_2| < \infty,\quad -h \le x_3 \le 0$ is rigidly coupled with a thermoelastic substrate $|x_1|,\ |x_2| < \infty,\ -\infty \le x_3 \le -h$. The governing equations for transversely isotropic thermoelastic materials taking into account the presence of an initially deformed state and preheating are (Levi et al. 2015; Levi et al. 2019; Belyankova et al. 2012; Belyankova et al. 2016):

$$\sigma_1^{*(n)} = c_{11}^{*(n)}\frac{\partial u_1^{(n)}}{\partial x_1} + c_{12}^{*(n)}\frac{\partial u_2^{(n)}}{\partial x_2} + c_{13}^{*(n)}\frac{\partial u_3^{(n)}}{\partial x_3} - \beta_1^{*(n)}u_4^{(n)},$$

$$\sigma_2^{*(n)} = c_{12}^{*(n)}\frac{\partial u_1^{(n)}}{\partial x_1} + c_{22}^{*(n)}\frac{\partial u_2^{(n)}}{\partial x_2} + c_{23}^{*(n)}\frac{\partial u_3^{(n)}}{\partial x_3} - \beta_2^{*(n)}u_4^{(n)},$$

$$\sigma_3^{*(n)} = c_{13}^{*(n)}\frac{\partial u_1^{(n)}}{\partial x_1} + c_{32}^{*(n)}\frac{\partial u_2^{(n)}}{\partial x_2} + c_{33}^{*(n)}\frac{\partial u_3^{(n)}}{\partial x_3} - \beta_3^{*(n)}u_4^{(n)}, \tag{10.1}$$

$$\sigma_{12}^{*(n)} = c_{1212}^{*(n)}\frac{\partial u_1^{(n)}}{\partial x_2} + c_{1221}^{*(n)}\frac{\partial u_2^{(n)}}{\partial x_1},\ \sigma_{21}^{*(n)} = c_{2112}^{*(n)}\frac{\partial u_1^{(n)}}{\partial x_2} + c_{1212}^{*(n)}\frac{\partial u_2^{(n)}}{\partial x_1},$$

$$\sigma_{13}^{*(n)} = c_{1313}^{*(n)}\frac{\partial u_1^{(n)}}{\partial x_3} + c_{1331}^{*(n)}\frac{\partial u_3^{(n)}}{\partial x_1},\ \sigma_{31}^{*(n)} = c_{3113}^{*(n)}\frac{\partial u_1^{(n)}}{\partial x_3} + c_{1313}^{*(n)}\frac{\partial u_3^{(n)}}{\partial x_1},$$

$$\sigma_{23}^{*(n)} = c_{2323}^{*(n)}\frac{\partial u_2^{(n)}}{\partial x_3} + c_{2332}^{*(n)}\frac{\partial u_3^{(n)}}{\partial x_2},\ \sigma_{32}^{*(n)} = c_{3223}^{*(n)}\frac{\partial u_2^{(n)}}{\partial x_3} + c_{2323}^{*(n)}\frac{\partial u_3^{(n)}}{\partial x_2},$$

where $\mathbf{u}^{(n)} = \left\{u_1^{(n)}, u_2^{(n)}, u_3^{(n)}, u_4^{(n)}\right\}$ is the extended displacement vector $u_1^{(n)}, u_2^{(n)}, u_3^{(n)}$ in the direction of the corresponding coordinates and temperature $u_4^{(n)}$ in the body, $n = 0,\ 1$ are the parameters of the substrate and the coating, respectively, $c_{ijkl}^{*(n)}, \beta_{kl}^{*(n)} = \alpha_{ij}^{(n)} c_{ijkl}^{*(n)}$ are the components of the elastic and thermoelastic constant tensors of materials subjected to prestressing conditions.

The oscillations of a prestressed layered thermoelastic medium are described by the equations of motion and thermal conductivity (Belyankova et al. 2012):

$$\frac{\partial \sigma_1^{*(n)}}{\partial x_1} + \frac{\partial \sigma_{12}^{*(n)}}{\partial x_2} + \frac{\partial \sigma_{13}^{*(n)}}{\partial x_3} = \rho_0^{(n)}\frac{\partial^2 u_1^{(n)}}{\partial t^2}$$

$$\frac{\partial \sigma_{21}^{*(n)}}{\partial x_1} + \frac{\partial \sigma_2^{*(n)}}{\partial x_2} + \frac{\partial \sigma_{23}^{*(n)}}{\partial x_3} = \rho_0^{(n)}\frac{\partial^2 u_2^{(n)}}{\partial t^2} \tag{10.2}$$

$$\frac{\partial \sigma_{31}^{*(n)}}{\partial x_1} + \frac{\partial \sigma_{32}^{*(n)}}{\partial x_2} + \frac{\partial \sigma_{3}^{*(n)}}{\partial x_3} = \rho_0^{(n)} \frac{\partial^2 u_3^{(n)}}{\partial t^2}$$

$$\lambda_1^{(n)} \frac{\partial^2 u_4^{(n)}}{\partial x_1^2} + \lambda_2^{(n)} \frac{\partial^2 u_4^{(n)}}{\partial x_2^2} + \lambda_3^{(n)} \frac{\partial^2 u_4^{(n)}}{\partial x_3^2} = \frac{\theta_1^{(n)}}{\theta_0} c_\varepsilon^{(n)} \rho_0^{(n)} \frac{\partial u_4^{(n)}}{\partial t}$$
$$+\theta_1^{(n)} \beta_1^{(n)*} \frac{\partial^2 u_1^{(n)}}{\partial t \partial x_1} + \theta_1^{(n)} \beta_2^{(n)*} \frac{\partial^2 u_2^{(n)}}{\partial t \partial x_2} + \theta_1^{(n)} \beta_3^{(n)*} \frac{\partial^2 u_3^{(n)}}{\partial t \partial x_3}, \tag{10.3}$$

where $\lambda_{ij}^{(n)}$, $\alpha_{ij}^{(n)}$ are the components of the tensors of thermal conductivity coefficients, thermal expansion, $\rho_0^{(n)}$ are the density of the materials in their natural state, $c_\varepsilon^{(n)}$ is the specific heat. θ_0, and $\theta_1^{(n)}$ are accordingly, the body temperature in the undeformed state and the temperature of the nth layer in the initial deformed state.

The constants of the medium with uniform initial deformation and preheating involved in Eqs. (10.1) (10.3) are determined by the expressions:

$$c_{ijkl}^{(n)*} = \frac{\delta_{kj}}{2}{}_{ilmm}^{(n)} \left(\nu_m^{(n)2} - 1\right) +_{ijkl}^{(n)} \nu_j^{(n)} \nu_k^{(n)} - \delta_{kj} \left(\theta_1^{(n)} - \theta_0\right) \beta_{il}^{(n)}, \tag{10.4}$$

$$\beta_j^{(n)*} = \nu_j^{(n)} \beta_j^{(n)}, \tag{10.5}$$

where $\nu_k^{(n)} = 1 + \delta_k^{(n)}$, $\delta_k^{(n)}$ $(k = 1, 2, 3)$ are relative elongation of the fibers.

We introduce an extended stress vector $\mathbf{q}^{(1)\tau} = \left\{\sigma_{31}^{(1)}, \sigma_{32}^{(1)}, \sigma_{3}^{(1)}, -\lambda_3^{(1)} u_{4,3}^{(1)}\right\}$ of a thermoelastic medium. The oscillations in the body are caused by the action of the load $\mathbf{q}_0(x_1, x_2)$ in the region $\Omega = \{|x_1| \le 1, |x_2| \le \infty\}$. Then, the boundary conditions are written:

$$x_3 = 0,\ \mathbf{q}^{(1)\tau} = \begin{cases} \mathbf{q}_0^\tau(x_1, x_2), & (x_1, x_2) \in \Omega, \\ 0, & (x_1, x_2) \notin \Omega, \end{cases}, \tag{10.6}$$

$$x_3 = -h, \begin{cases} \sigma_{31}^{(1)} = \sigma_{31}^{(0)},\ \sigma_{32}^{(1)} = \sigma_{32}^{(0)},\ \sigma_{33}^{(1)} = \sigma_{33}^{(0)}, \\ u_1^{(1)} = u_1^{(0)},\ u_2^{(1)} = u_2^{(0)},\ u_3^{(1)} = u_3^{(0)}, \end{cases} \tag{10.7}$$

$$\begin{cases} -\lambda_{33}^{(1)} u_{4,3}^{(1)} = -\lambda_{33}^{(0)} u_{4,3.}^{(0)} \\ u_4^{(1)} = u_4^{(0)}, \end{cases} \quad \text{(I)} \tag{10.8}$$

$$\begin{cases} -\lambda_{33}^{(0)} u_{4.3}^{(0)} = 0 \\ -\lambda_{33}^{(1)} u_{4.3}^{(1)} = 0, \end{cases} \quad \text{(II)} \tag{10.9}$$

$$x_3 \to -\infty{:}\mathbf{u}^{(0)\tau} \to 0 \tag{10.10}$$

For convenience, the task scaling parameters are introduced [Sharma J.N., 2005]:

$$x_i' = \frac{x_i \cdot \omega *}{V_l}, t' = \omega * t, V_P^{(n)} = \sqrt{c_{11}^{(n)} / \rho_0^{(n)}},$$

$$u_i'^{(n)} = \frac{u_i^{(n)} \cdot \rho_0^{(n)} \omega * V_P^{(n)}}{\beta_1^{(n)} \theta_0}, i = 1 - 3,$$

$$\mathrm{E}^{(n)} = \frac{\theta_0 \beta_1^{(n)2}}{\rho_0^{(n)} c_\varepsilon^{(n)} c_{11}^{(0)}}, u_4'^{(n)} = \frac{u_4^{(n)}}{\theta_0}, c_{ijkl}'^{(n)} = \frac{c_{ijkl}^{(n)}}{c_{11}^{(0)}}, \tag{10.11}$$

$$\beta_i'^{(n)} = \frac{\beta_i^{(n)}}{\beta_1^{(0)}}, \lambda_i^{(n)} = \frac{\lambda_i^{(n)}}{\lambda_1^{(0)}}, \omega' = \frac{\omega}{\omega *}, \omega * = \frac{c_\varepsilon^{(0)} c_{11}^{(0)}}{\lambda_1^{(0)}}$$

10.4 The Plane Surface Waves

The propagation of harmonic plane oscillations is considered. For the purpose of this, we present all the functions in the form:

$$f(x_1, x_2, t) = f(x_1) e^{-i\omega t}$$

Then, the defining expressions (10.1)–(10.3) and the boundary conditions (10.6)–(10.10) are rewritten:

$$c_{11}^{(n)*} u_{1,11}^{(n)} + c_{3113}^{(n)*} u_{1,33}^{(n)} + \omega^2 u_1^{(n)} + c_2^{(n)} u_{3,13}^{(n)} - \beta_1^{(n)*} u_{4,1}^{(n)} = 0,$$

$$c_2^{(n)} u_{1,13}^{(n)} + c_{1331}^{(n)*} u_{3,11}^{(n)} + \omega^2 u_3^{(n)} + c_{33}^{(n)*} u_{3,33}^{(n)} - \beta_3^{(n)*} u_{4,3}^{(n)} = 0, \tag{10.12}$$

$$u_{4,11}^{(n)} + \lambda_3^{(n)} u_{4,33}^{(n)} + i\omega\theta_1 u_4^{(n)} + i\omega\theta_1^{(n)} \mathrm{E}\left(\beta_1^{*(n)} u_{1,1}^{(n)} + \beta_3^{*(n)} u_{3,3}^{(n)}\right) = 0.$$

$$x_3 = 0, \begin{cases} c_{3113}^{(1)*} u_{1,3}^{(1)} + c_{1313}^{(1)*} u_{3,1}^{(1)} = q_1^0, \\ c_{33}^{(1)*} u_{3,3}^{(1)} + c_{13}^{(1)*} u_{1,1}^{(1)} - \beta_3^{(1)*} u_4^{(1)} = q_3^0, \\ -\lambda_{33}^{(1)*} u_{4,3}^{(1)} = q_4^0, \end{cases} \tag{10.13}$$

$$x_3 = -h, \begin{cases} u_1^{(1)} = u_1^{(0)}, \\ u_3^{(1)} = u_3^{(0)}, \\ c_{3113}^{(1)*} u_{1,3}^{(1)} + c_{1313}^{(1)*} u_{3,1}^{(1)} - \\ -c_{3113}^{(0)*} u_{1,3}^{(0)} - c_{1313}^{(0)*} u_{3,1}^{(0)} = 0, \\ c_{33}^{(1)*} u_{3,3}^{(1)} + c_{13}^{(1)*} u_{1,1}^{(1)} - \beta_3^{(1)*} u_4^{(1)} - \\ -c_{33}^{(0)*} u_{3,3}^{(0)} - c_{13}^{(0)*} u_{1,1}^{(0)} + \beta_3^{(0)*} u_4^{(0)} = 0, \end{cases} \tag{10.14}$$

$$\begin{cases} u_4^{(1)} = u_4^{(0)}, \\ -\lambda_3^{(1)} u_{4,3}^{(1)} + \lambda_3^{(0)} u_{4,3}^{(0)} = 0, \end{cases} \quad \text{(I)} \tag{10.15}$$

$$\begin{cases} -\lambda_3^{(1)} u_{4,3}^{(1)} = 0, \\ -\lambda_3^{(0)} u_{4,3}^{(0)} = 0. \end{cases} \quad \text{(II)} \tag{10.16}$$

Here $c_2^{(n)} = c_{13}^{*(n)} + c_{1331}^{*(n)}$, the index after the comma denotes the partial derivatives at the corresponding coordinates.

To solve the boundary problem, we apply a one-dimensional Fourier transform along the coordinate x_1. Thus, we rewrite all the functions in the form [Levi M.O. et al., 2017]: $\{u_j, \sigma_{ij}\} = \{U_j, \Theta_{ij}\}e^{-i\alpha x_1}$.

The equations of motion and heat conduction in the Fourier images are rewritten:

$$-\alpha^2 c_{1111}^{(n)*} U_1^{(n)} + c_{3113}^{(n)*} U_1^{\prime\prime(n)} + \omega^2 U_1^{(n)} - i\alpha c_2^{(n)} U_3^{\prime(n)} + i\alpha\beta_{11}^{(n)*} U_4^{(n)} = 0,$$

$$-i\alpha c_2^{(n)} U_1^{\prime(n)} - \alpha^2 c_{1331}^{(n)*} U_3^{(n)} + \omega^2 U_3^{(n)} + c_{3333}^{(n)*} U_3^{\prime\prime(n)} - \beta_{33}^{(n)*} U_4^{\prime(n)} = 0, \tag{10.17}$$

$$-\alpha^2 U_4^{(n)} + \lambda_{33}^{(n)} U_4^{\prime\prime(n)} + i\omega\theta_1^{(n)} U_4^{(n)} + i\omega \mathrm{E}^{(n)*}\left(-i\alpha\beta_{11}^{(n)*} U_1^{(n)} + \beta_{33}^{(n)*} U_3^{\prime(n)}\right) = 0,$$

Here are the notation: $f' = df/dx_3$, $f'' = \mathrm{d}^2 f/\mathrm{d}x_3^2$. Solutions of Eqs. (10.17) will be sought in the form [Levi G.Yu. et al., 2015]:

$$U_1^{(1)}(\alpha, x_3, \omega) = -i\alpha \sum_{k=1}^{3} f_{1k}^{(1)}\left(C_k sh\sigma_k^{(1)} x_3 + C_{k+3} ch\sigma_k^{(1)} x_3\right),$$

$$U_3^{(1)}(\alpha, x_3, \omega) = \sum_{k=1}^{3} f_{3k}^{(1)}\left(C_k ch\sigma_k^{(1)} x_3 + C_{k+3} sh\sigma_k^{(1)} x_3\right), \tag{10.18}$$

$$U_4^{(1)}(\alpha, x_3, \omega) = \sum_{k=1}^{3} f_{4k}^{(1)}\left(C_k sh\sigma_k^{(1)} x_3 + C_{k+3} ch\sigma_k^{(1)} x_3\right), \quad -h \le x_3 \le 0;$$

$$U_1^{(0)}(\alpha, x_3, \omega) = -i\alpha \sum_{k=1}^{3} f_{1k}^{(0)} C_{k+6} e^{\sigma_k^{(0)} x_3},$$

$$U_3^{(0)}(\alpha, x_3, \omega) = \sum_{k=1}^{3} f_{3k}^{(0)} C_{k+6} e^{\sigma_k^{(0)} x_3}, \tag{10.19}$$

$$U_4^{(0)}(\alpha, x_3, \omega) = \sum_{k=1}^{3} f_{4k}^{(0)} C_{k+6} e^{\sigma_k^{(0)} x_3}, \; x_3 \le -h.$$

We substitute the representation of the solution (10.18)–(10.19) to the boundary conditions (10.13) (10.16) written in the images of the Fourier space [Levi G.Yu. et al., 2015], and linear algebraic equations system is obtained. It can be written in the matrix form:

$$\mathbf{LC} = \mathbf{Q}^{\tau}, \tag{10.20}$$

where $\mathbf{Q} = \left\{ Q_1 \; Q_3 \; Q_4 \; 0\;0\;0\;0\;0\;0 \right\}$ is the Fourier transform of $\mathbf{q}_0$,

$$\mathbf{L} = (\mathbf{L}_1, \mathbf{L}_2, \mathbf{L}_3), \tag{10.21}$$

$$\mathbf{L}_1 = \begin{pmatrix} l_{11}^{(1)} & l_{13}^{(1)} & l_{14}^{(1)} \\ 0 & 0 & l_{14}^{(1)} \\ l_{41}^{(1)} & l_{43}^{(1)} & l_{44}^{(1)} \\ u_1^{(1)} s_1^{(1)} & u_1^{(1)} s_3^{(1)} & u_1^{(1)} s_4^{(1)} \\ u_3^{(1)} c_1^{(1)} & u_3^{(1)} c_3^{(1)} & u_3^{(1)} c_4^{(1)} \\ u_4^{(1)} s_1^{(1)} & u_4^{(1)} s_3^{(1)} & u_4^{(1)} s_4^{(1)} \\ l_{11}^{(1)} c_1^{(1)} & l_{13}^{(1)} c_3^{(1)} & l_{14}^{(1)} c_4^{(1)} \\ l_{31}^{(1)} s_1^{(1)} & l_{33}^{(1)} s_3^{(1)} & l_{34}^{(1)} s_4^{(1)} \\ l_{41}^{(1)} c_1^{(1)} & l_{43}^{(1)} c_3^{(1)} & l_{44}^{(1)} c_4^{(1)} \end{pmatrix},$$

$$\mathbf{L}_2 = \begin{pmatrix} 0 & 0 & 0 \\ l_{31}^{(1)} & l_{33}^{(1)} & l_{34}^{(1)} \\ 0 & 0 & 0 \\ u_1^{(1)} c_1^{(1)} & u_1^{(1)} c_3^{(1)} & u_1^{(1)} c_4^{(1)} \\ u_3^{(1)} s_1^{(1)} & u_3^{(1)} s_3^{(1)} & u_3^{(1)} s_4^{(1)} \\ u_4^{(1)} c_1^{(1)} & u_4^{(1)} c_3^{(1)} & u_4^{(1)} c_4^{(1)} \\ l_{11}^{(1)} s_1^{(1)} & l_{13}^{(1)} s_3^{(1)} & l_{14}^{(1)} s_4^{(1)} \\ l_{31}^{(1)} c_1^{(1)} & l_{33}^{(1)} c_3^{(1)} & l_{34}^{(1)} c_4^{(1)} \\ l_{41}^{(1)} s_1^{(1)} & l_{43}^{(1)} s_3^{(1)} & l_{44}^{(1)} s_4^{(1)} \end{pmatrix},$$

$$\mathbf{L}_3 = \begin{pmatrix} 0 & 0 & 0 \\ 0 & 0 & 0 \\ 0 & 0 & 0 \\ -u_1^{(0)} e_1^{(0)} & -u_1^{(0)} e_3^{(0)} & -u_1^{(0)} e_4^{(0)} \\ -u_3^{(0)} e_1^{(0)} & -u_3^{(0)} e_3^{(0)} & -u_3^{(0)} e_4^{(0)} \\ -u_4^{(0)} e_1^{(0)} & -u_4^{(0)} e_3^{(0)} & -u_4^{(0)} e_4^{(0)} \\ -l_{11}^{(0)} e_1^{(0)} & -l_{13}^{(0)} e_3^{(0)} & -l_{14}^{(0)} e_4^{(0)} \\ -l_{31}^{(0)} e_1^{(0)} & -l_{33}^{(0)} e_3^{(0)} & -l_{34}^{(0)} e_4^{(0)} \\ -l_{41}^{(0)} e_1^{(0)} & -l_{43}^{(0)} e_3^{(0)} & -l_{44}^{(0)} e_4^{(0)} \end{pmatrix}$$

$$l_{1k}^{(n)} = -i\alpha(c_{3113}^{*(n)}\sigma_k^{(n)} f_{1k}^{(n)} + c_{1313}^{*(n)} f_{3k}^{(n)}),$$

$$l_{3k}^{(n)} = -\alpha^2 c_{1133}^{*(n)} f_{1k}^{(n)} + \sigma_k^{(n)} c_{3333}^{*(n)} f_{3k}^{(n)} - \beta_3^{*(n)} f_{4k}^{(n)},$$

$$l_{4k}^{(n)} = \sigma_k^{(n)} f_{4k},\ u_p^{(n)} s_k^{(n)} = f_{pk}^{(n)} sh(-h\sigma_k^{(n)}),\ u_p^{(n)} c_k^{(n)} = f_{pk}^{(n)} ch(-h\sigma_k^{(n)}),$$

$$u_p^{(n)} e_k^{(n)} = f_{pk}^{(n)} e^{(-h\sigma_k^{(n)})},\ c_k^{(1)} = ch(-h\sigma_k^{(1)}),$$

$$s_k^{(1)} = sh(-h\sigma_k^{(1)}),\ e_k^{(0)} = e^{(-h\sigma_k^{(0)})}.$$

The system's (10.20) non-trivial solution is insured by the equation:

$$\det \mathbf{L} = 0 \tag{10.22}$$

After finding C_k from Eqs. (10.20), we construct the solution of the boundary value problem in the Fourier form. Applying the inverse Fourier transform to (10.18)–(10.19), the solution of the initial boundary value problem is obtained in the form:

$$u_i^{(n)}(x_1, x_3) = \frac{1}{2\pi} \int_{-1}^{1} k_{ij}^{(n)}(x_1 - \xi, x_3, \omega)\, q_{j0}(\xi)\, d\xi,\ i = 1, 3, 4 \tag{10.23}$$

$$k_{ij}^{(n)}(s, x_3, \omega) = \int_{\Gamma} K_{ij}^{(n)}(\alpha, x_3, \omega) e^{-i\alpha s} d\alpha, \tag{10.24}$$

where $K_{ij}^{(n)}(\alpha, x_3, \omega)$ are elements of the Green function matrix, which are calculated by the following relations:

$$K_{1j}^{(1)}(\alpha, x_3) = -i\alpha \frac{1}{\Delta_0} \sum_{k=1}^{3} f_{1k}^{(1)}(\Delta_{jk} sh\sigma_k^{(1)} x_3 + \Delta_{jk+3} ch\sigma_k^{(1)} x_3),$$

$$K_{3j}^{(1)}(\alpha, x_3) = \frac{1}{\Delta_0} \sum_{k=1}^{3} f_{3k}^{(1)} \sum_{k=1}^{3} f_{3k}^{(1)}(\Delta_{jk} ch\sigma_k^{(1)} x_3 + \Delta_{jk+3} sh\sigma_k^{(1)} x_3), \tag{10.25}$$

$$K_{4j}^{(1)}(\alpha, x_3) = \frac{1}{\Delta_0} \sum_{k=1}^{3} f_{4k}^{(1)} \sum_{k=1}^{3} f_{4k}^{(1)}(\Delta_{jk} sh\sigma_k^{(1)} x_3 + \Delta_{jk+3} ch\sigma_k^{(1)} x_3),$$

$$K_{1j}^{(0)}(\alpha, x_3) = -i\alpha \frac{1}{\Delta_0} \sum_{k=1}^{3} f_{1k}^{(0)} \Delta_{jk+6} e^{\sigma_k^{(0)} x_3},$$

$$K_{3j}^{(0)}(\alpha, x_3) = \frac{1}{\Delta_0} \sum_{k=1}^{3} f_{3k}^{(0)} \Delta_{jk+6} e^{\sigma_k^{(0)} x_3}, \tag{10.26}$$

$$K_{4j}^{(0)}(\alpha, x_3) = \frac{1}{\Delta_0} \sum_{k=1}^{3} s f_{4k}^{(0)} \Delta_{jk+6} e^{\sigma_k^{(0)} x_3}.$$

In Eqs. (10.25)–(10.26) the notations:Δ_0 is the matrix **L** determinant, $\Delta_{j,k}$ is the algebraic complement of the element with the index jk of the matrix **L**. The indexes $j = 1, 3, 4$ are surface load, respectively.

Thus, relations (10.23)–(10.26) determine the displacement ($i = 1, 3$) or temperature ($i = 4$) of a thermoelastic structurally inhomogeneous half-space arbitrary point under the action of a given in the region oscillating load $q_{j0}(x_1)$ taking into account the presence of initial deformations and preheating.

10.5 Results Discussion

To study the influence of the initial deformed states, we consider the problem of vibrations in cadmium sulfide (CdS) layer rigidly coupled with magnesium oxide (MgO) half-space. The initial prestress state created by using initial deformation and preheating conditions. There is a mechanical and thermal load $\mathbf{q}_0(x_1)e^{-i\omega t}$ on the top of layer surface distributed in the area $x_1 \in [-1, 1]$. The following are their physical characteristics:

$c_{1111}^{(1)} = 9.07 \times 10^{10}$ N/m², $c_{1133}^{(1)} = 5.1 \times 10^{10}$ N/m², $c_{3333}^{(1)} = 9.38 \times 10^{10}$ N/m², $c_{1331}^{(1)} = c_{3113}^{(1)} = 1.50 \times 10^{10}$ N/m², $\beta_{11}^{(1)} = 7.09 \times 10^5$ N/K/m², $\beta_{33}^{(1)} = 7.24 \times 10^5$ N/K/m², $c_\varepsilon^{(1)} = 329$ J/kg/K, $\lambda_{11}^{(1)} = \lambda_{33}^{(1)} = 20.0$ W/m/K, $\rho^{(1)} = 4820$ kg/m³, $\tau_0 = 280$ K,
$c_{1111}^{(0)} = 30.0 \times 10^{10}$ N/m², $c_{1133}^{(0)} = 10.1 \times 1\text{–}0^{10}$ N/m², $c_{3333}^{(0)} = 30.0 \times 10^{10}$ N/m², $c_{1331}^{(0)} = c_{3113}^{(0)} = 15.75 \times 10^{10}$ N/m², $\beta_{11}^{(0)} = \beta_{33}^{(0)} = 4.4 \times 10^6$ N/K/m², $c_\varepsilon^{(0)} = 875$ J/kg/K, $\lambda_{11}^{(0)} = \lambda_{33}^{(0)} = 58.0$ W/m/K, $\rho^{(0)} = 3576$ kg/m³.

Figure 10.1 shows the Green's function **K** poles distribution of a layered thermoelastic half-space, in the absence of initial stresses. A special feature of problems for

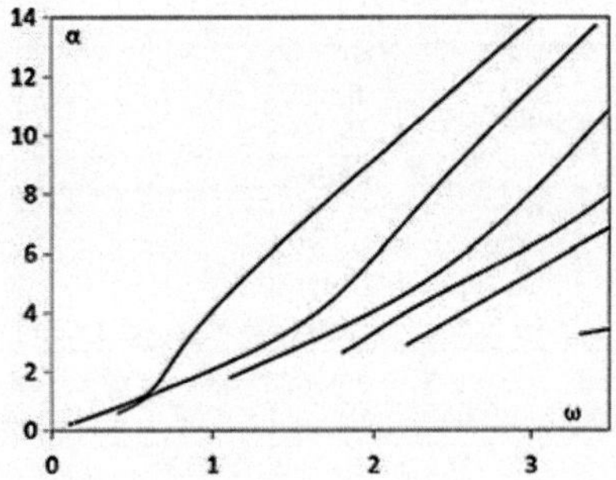

Fig. 10.1 Real poles of Green's function K

thermoelastic bodies is that the elements of function (10.25)–(10.26) have a countable set of complex zeros and poles, some of which have a small imaginary part (Belyankova et al. 2012). The analysis showed that, along with the complex poles with a small imaginary part, the function **K** has many significantly complex poles with a large imaginary part.

Figures 10.2 show the effect of preheating on the phase velocities differences $dV_f = V_f^{\nu\theta} - V_f$ depending on the pinching of the thermoelastic layered half-space. There $V_f^{\nu\theta}$ and V_f are phase velocities of the first mode in the presence of prestressing and in the natural state, respectively. Hereinafter, the following designations of preheating are introduced: a solid line $d\theta = 0.5$ (150 K), a dashed line $d\theta = 0.3$ (100 K), and a doted dashed line $d\theta = 0.1$ (30 K).

From Fig. 10.2, it can be seen that for all pinching modes there is a frequency in which the effect of preheating on the phase velocity increases significantly. In addition, with biaxial pinching (b), there is a frequency in which the phase velocity does not depend on preheating. For uniaxial pinching along the x_1 axis (c), the phase velocity changes only in the range of a certain resonant frequency. With uniaxial pinching along the x_2 axis (d) and the absence of pinching (a), preheating increases the phase velocity at all frequencies. It should also be noted that pinching in the x_1 direction significantly reduces the effect of preheating, while pinching in the x_2 direction increases one.

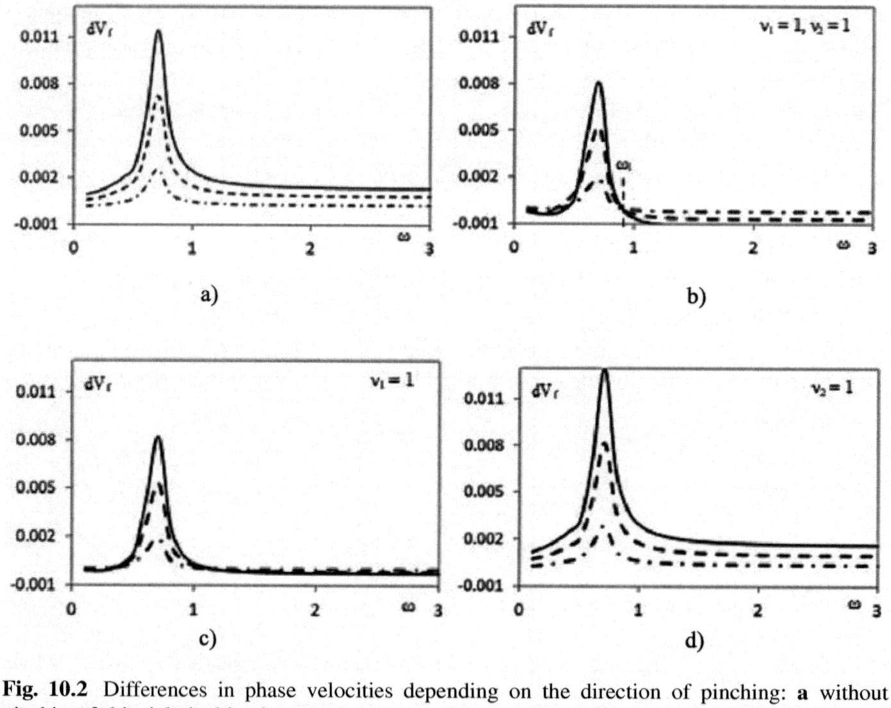

Fig. 10.2 Differences in phase velocities depending on the direction of pinching: **a** without pinching; **b** biaxial pinching in x_1 and x_2; **c** uniaxial pinching on x_1; **d** uniaxial pinching on x_2

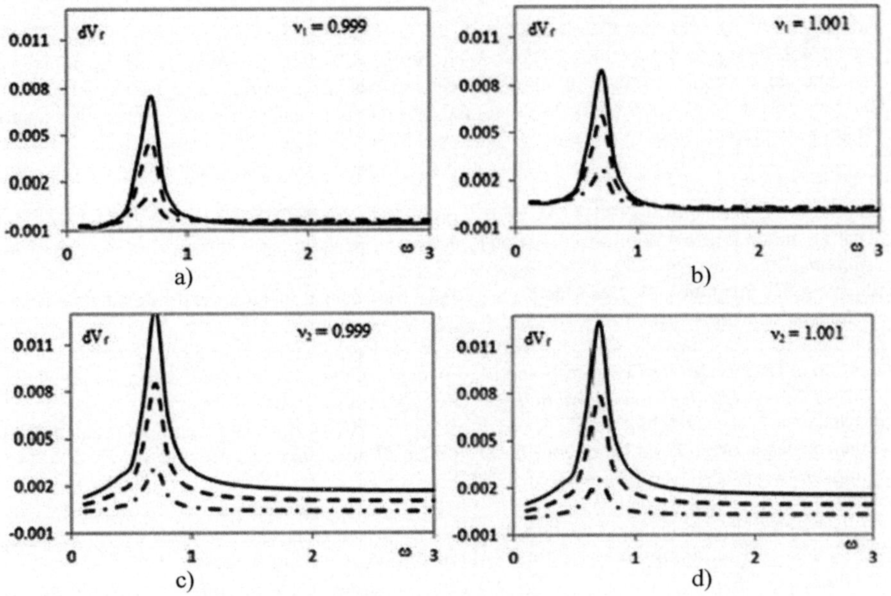

Fig. 10.3 Influence of the initial deformation on the phase velocity: **a** compression in the direction of x_1; **b** stretching in the direction of x_1; **c** compression in the direction of x_2; **d** stretching in the direction of x_2

Figures 10.3 shows the effect of preheating on the phase velocities difference depending on the initial deformations of the thermoelastic layered half-space. Compression in the direction of the x_1 axis reduces the maximum difference in phase velocities. Stretching also affects the opposite direction, and moreover, the difference values even at high frequencies become nonzero. Stretching in the x_2 direction acts opposite and reduces the effect of preheating on phase velocities.

Acknowledgements The reported study was performed as part of the implementation of the state assignment of the Southern Scientific Center of the Russian Academy of Sciences, project 01201354242 and with partial financial support from the Russian Foundation for Basic Research, grants № 19-01-00719, 19-08-01051.

References

Achenbach, J. D. (2003). Laser excitation of surface wave motion. *Journal of the Mechanics and Physics of Solids, 51,* 1885–1902. https://doi.org/10.1016/j.jmps.2003.09.021

Al-Qahtani, H., & Datta, S. K. (2004). Thermoelastic waves in an anisotropic infinite plate. *Journal of Applied Physics, 96,* 3645–3657. https://doi.org/10.1063/1.1776323

Alibert, J. J., Seppecher, P., & Dell'Isola, F. (2003). Truss modular beams with deformation energy depending on higher displacement gradients. *Mathematics and Mechanics of Solids, 8*(1), 51–73.

Bao, H., Bielak, J., Ghattas, O., Kallivokas, L. F., O'Hallaron, D. R., Shewchuk, J. R., & Xu, J. (1998). Large-scale simulation of elastic wave propagation in heterogeneous media on parallel computers. *Computer Methods in Applied Mechanics and Engineering, 152*(1–2), 85–102.

Barchiesi, E., & Khakalo, S. (2019). Variational asymptotic homogenization of beam-like square lattice structures. *Mathematics and Mechanics of Solids, 24*(10), 3295–3318.

Barchiesi, E., Laudato, M., & Di Cosmo, F. (2018). Wave dispersion in non-linear pantographic beams. *Mechanics Research Communications, 94,* 128–132.

Barchiesi, E., & Placidi, L. (2017). A review on models for the 3D statics and 2D dynamics of pantographic fabrics. *Wave dynamics and composite mechanics for microstructured materials and metamaterials* (pp. 239–258). Singapore: Springer.

Barchiesi, E., Spagnuolo, M., & Placidi, L. (2019). Mechanical metamaterials: A state of the art. *Mathematics and Mechanics of Solids, 24*(1), 212–234.

Belyankova, T. I., Vorovich, E. I., Kalinchuk, V. V., & Puzanov, Yu. E. (1999). Dynamic contact problem for thermo-elastic layer. Scientific-Educational and Applied Journal University News. *North-Caucasian Region. Natural Sciences Series, 4,* 109–110. ((In Russian)).

Belyankova, T. I., Kalinchuk, V. V., & Suvorova, G. Y. (2012). A dynamic contact problem for a thermoelastic prestressed layer. *Journal of Applied Mathematics and Mechanics, 75,* 537–546. https://doi.org/10.1016/j.jappmathmech.2012.11.013

Belyankova, T. I., & Kalinchuk, V. V. (2016). Green's function for a prestressed thermoelastic half-space with an inhomogeneous coating. *Journal of Applied Mechanics and Technical Physics, 57,* 828–840. https://doi.org/10.1134/S0021894416050096

Boutin, C., Giorgio, I., & Placidi, L. (2017). Linear pantographic sheets: Asymptotic micro-macro models identification. *Mathematics and Mechanics of Complex Systems, 5*(2), 127–162.

Dell'Isola, F., Andreaus, U., & Placidi, L. (2015). At the origins and in the vanguard of peridynamics, non-local and higher-gradient continuum mechanics: An underestimated and still topical contribution of Gabrio Piola. *Mathematics and Mechanics of Solids, 20*(8), 887–928.

Dell'Isola, F., Corte, A. D., & Giorgio, I. (2017). Higher-gradient continua: The legacy of Piola, Mindlin, Sedov and Toupin and some future research perspectives. *Mathematics and Mechanics of Solids, 22*(4), 852–872.

dell'Isola, F., Giorgio, I., & Andreaus, U. (2015). Elastic pantographic 2D lattices: A numerical analysis on the static response and wave propagation. *Proceedings of the Estonian Academy of Sciences, 64*(3), 219.

Dell'Isola, F., Seppecher, P., Alibert, J. J., Lekszycki, T., Grygoruk, R., Pawlikowski, M., & Gołaszewski, M. (2019). Pantographic metamaterials: An example of mathematically driven design and of its technological challenges. *Continuum Mechanics and Thermodynamics, 31*(4), 851–884.

Dell'Isola, F., Seppecher, P., & Madeo, A. (2012). How contact interactions may depend on the shape of Cauchy cuts in Nth gradient continua: Approach "à la D'Alembert." *Zeitschrift für angewandte Mathematik und Physik, 63*(6), 1119–1141.

dell'Isola, F., Seppecher, P., Spagnuolo, M., Barchiesi, E., Hild, F., Lekszycki, T., & Eugster, S. R. (2019). Advances in pantographic structures: Design, manufacturing, models, experiments and image analyses. *Continuum Mechanics and Thermodynamics, 31*(4), 1231–1282.

Elhagary, M. (2013). A two-dimensional generalized thermoelastic diffusion problem for a half-space subjected to harmonically varying heating. *Acta Mechanica, 224,* 3057–3069. https://doi.org/10.1007/s00707-013-0902-6

El-Maghraby, N. M. (2008). A two-dimensional generalized thermoelasticity problem for a half-space under the action of a body force. *Journal of Thermal Stresses, 31,* 557–568. https://doi.org/10.1080/01495730801978281

Eremeyev, V. A., Alzahrani, F. S., Cazzani, A., dell'Isola, F., Hayat, T., Turco, E., & Konopińska-Zmysłowska, V. (2019). On existence and uniqueness of weak solutions for linear pantographic beam lattices models. *Continuum Mechanics and Thermodynamics, 31*(6), 1843–1861.

Eugster, S., & Steigmann, D. (2019). Continuum theory for mechanical metamaterials with a cubic lattice substructure. *Mathematics and Mechanics of Complex Systems, 7*(1), 75–98.

Kumar, R., & Gupta, V. (2013). Reflection and transmission of plane waves at the interface of an elastic half-space and a fractional order thermoelastic half-space. *Archive of Applied Mechanics, 83,* 1109–1128. https://doi.org/10.1007/s00419-013-0737-6

Kumar, R., & Kansal, T. (2008). Propagation of Lamb waves in transversely isotropic thermoelastic diffusive plate. *International Journal of Solids and Structures, 45,* 5890–5913. https://doi.org/10.1016/S0020768308002710

Levi, G. Yu., & Belyankova, T. I. (2019). Some properties of the transversely isotropic thermoelastic layer under initial stress. *Applied Mechanics and Systems Dynamics. Journal of Physics: Conference Series* 1210. doi:10.1088/1742-6596/1210/1/012080.

Levi, GYu., & Igumnov, L. A. (2015). Some properties of the thermoelastic prestressed medium Green function. *Materials Physics and Mechanics, 23,* 42–46.

Levi, M. O., Levi, GYu., & Lyzhov, V. A. (2017). Some features of the dynamics of ferroelectric (ferromagnetic) heterostructures. *Journal of Applied Mechanics and Technical Physics, 58,* 47–53.

Lurie, A. I. (1980). *Nelinejnaja teorija uprugosti [Nonlinear Theory of Elasticity].* Moscow: Nauka Publishers, 512 p (In Russian).

Madeo, A., Della Corte, A., Greco, L., & Neff, P. (2014). *Wave Propagation in Pantographic 2D Lattices with Internal Discontinuities.* arXiv preprint arXiv:1412.3926.

Muratikov, K. L. (1998). On the theory of oscillations generation by laser radiation in solids with internal stresses by the thermoelastic method. *Pisma v zhurnal tekhnicheskoi fiziki, 24,* 82–88. ((In Russian)).

Placidi, L., Dell'Isola, F., Ianiro, N., & Sciarra, G. (2008). Variational formulation of prestressed solid-fluid mixture theory, with an application to wave phenomena. *European Journal of Mechanics-A/Solids, 27*(4), 582–606.

Sharma, J. N. (2001). Three-dimensional vibration analysis of a homogeneous transversely isotropic thermoelastic cylindrical panel. *The Journal of the Acoustical Society of America, 110,* 254–259. https://doi.org/10.1121/1.1378350

Sharma, J. N., Pal, M., & Chand, D. (2005). Propagation characteristics of Rayleigh waves in transversely isotropic piezothermoelastic materials. *Journal of Sound and Vibration, 284,* 227–248. https://doi.org/10.1016/j.jsv.2004.06.036

Sharma, J. N., & Sidhu, R. S. (1986). On the propagation of plane harmonic waves in anisotropic generalized thermoelasticity. *International Journal of Engineering Science, 24,* 1511–1516. https://doi.org/10.1016/0020-7225(86)90160-6

Sheydakov, D. N., Belyankova, T. I., Sheydakov, N. E., & Kalinchuk, V. V. (2008). Dynamics equations for prestressed thermo-elastic medium. *Vestnik Yuzhnogo Nauchnogo Tsentra, 4,* 3–8. ((In Russian)).

Singh, B. (2010). Wave propagation in an initially stressed transversely isotropic thermoelastic solid half-space. *Applied Mathematics and Computation, 217,* 705–715. https://doi.org/10.1016/j.amc.2010.06.008

Singh, H., & Sharma, J. N. (1985). Generalised thermoelastic waves in transversely isotropic media. *The Journal of the Acoustical Society of America, 77,* 1046–1053. https://doi.org/10.1121/1.392391

Spagnuolo, M., & Andreaus, U. (2019). A targeted review on large deformations of planar elastic beams: Extensibility, distributed loads, buckling and post-buckling. *Mathematics and Mechanics of Solids, 24*(1), 258–280.

Chirita, S. (2013). On the Rayleigh surface waves on an anisotropic homogeneous thermoelastic half space. *Acta Mechanica, 224,* 657–674. https://doi.org/10.1007/s00707-012-0776-z

Verma, K. L. (2002). On the propagation of waves in layered anisotropic media in generalized thermoelasticity. *International Journal of Engineering Science, 40,* 2077–2096. https://doi.org/10.1016/S0020-7225(02)00030-7

Xu, B. Q., Feng, J., & Xu, G. D. (2008). Laser-generated thermoelastic acoustic sources and Lamb waves in anisotropic plates. *Applied Physics and-Materials Science & Processing, 91,* 173–179. https://doi.org/10.1007/s11431-009-0065-9

Chapter 11
Numerical Evaluation of Integrals in Laplace Domain Anisotropic Elastic Fundamental Solutions for High Frequencies

Ivan P. Markov and Marina V. Markina

Abstract In this paper, we address a problem of computing the integrals appearing in integral expressions of Laplace domain anisotropic elastic displacement fundamental solutions. The essence of the problem is that these integrals can become highly oscillatory for high values of frequency or large distance between source and observation points. The modified integral expressions for displacement fundamental solutions and their first derivative are given. We propose a procedure based on the quadrature rule developed by Evans and Webster for the evaluation of rapidly oscillatory integrals. For a triclinic anisotropic elastic material, we consider an illustrative numerical example which involves phase functions with stationary points.

Keywords Anisotropic elasticity · Laplace transform · Fundamental solutions · Oscillatory integrals · Evans-Webster quadrature rule · Boundary element method

11.1 Introduction

In modern science, computational work is an important complement to both experiments and theory, and nowadays, a vast majority of both experimental and theoretical papers involve some numerical calculations, simulations, or computer modeling. One of the most interesting subjects in modern continuum mechanics and engineering worth mentioning is the development of newly (scientifically) conceived materials ('metamaterials') with mechanical properties that cannot be found in nature (Del Vescovo and Giorgio 2014; Barchiesi et al. 2018). These (macroscopic) properties are mainly determined by the micro- or nanostructure of the considered metamaterial rather than by the chemical and physical properties of the materials constituting it at the microscopic level. Designing of such metamaterials is based on high gradient continuum approaches (Alibert et al. 2003; Sciarra et al. 2007; dell'Isola et al. 2012,

I. P. Markov (✉) · M. V. Markina
Research Institute for Mechanics, National Research Lobachevsky State University of Nizhny Novgorod, 23, Bldg. 6, Prospekt Gagarina (Gagarin Avenue), Nizhny Novgorod 603950, Russian Federation
e-mail: markov@mech.unn.ru

F. dell'Isola and L. Igumnov (eds.), *Dynamics, Strength of Materials and Durability in Multiscale Mechanics*, Advanced Structured Materials 137,
https://doi.org/10.1007/978-3-030-53755-5_11

2016a, b, c; Auffray et al. 2013; dell'Isola et al. 2015; Rahali et al. 2015). Pantographic structures are an example of the mechanical metamaterials (Placidi et al. 2016; dell'Isola et al. 2019a, b, c). Identification of constitutive parameters and validation of theoretical predictions of built models are only a few subjects where numerical simulations are used on everyday basis (dell'Isola et al. 2016a, b, c, 2017; Placidi et al. 2017).

Green's functions or fundamental solutions are extensively used in the solution of boundary value problems, and they have many applications in different fields, such as acoustics, earthquake engineering, nondestructive evaluation of the materials, studying propagation and scattering of elastic waves in elastic media and so on. They are essential for the development of various formulations of Boundary Element Method (BEM), which is a very powerful numerical tool for engineering analysis (dell'Isola et al. 2019a). However, when the elastic media with arbitrary degree of anisotropy is considered, implementation of any method based on integral equations is somewhat hindered because static and dynamic anisotropic elastic fundamental solutions are not available in the explicit and closed-form analytical expressions. Efficiency of any BEM formulation utilizing the anisotropic elastic fundamental solutions heavily depends on computation efficiency of the method employed to evaluate said fundamental solutions.

Frequency and Laplace transformed dynamic anisotropic elastic fundamental solutions can be represented as a sum of static (singular) and dynamic (regular) parts. Static part does not depend on frequency and was intensely studied starting with Fredholm (1900). Over the years, several different approaches to calculate the static anisotropic elastic fundamental solutions and their derivatives were proposed: using integral expressions (Lifshitz and Rozenzweig 1947; Barnett 1972), employing the residue calculus (Sales and Gray 1998; Phan et al. 2004, 2005), expressions in terms of the Stroh eigenvalues (Ting 1997; Lee 2009; Shiah et al. 2010; Buroni and Sáez 2013; Xie et al. 2016) or eigenvectors (Malén 1971; Nakamura and Tanuma 1997), solution in form of uniformly convergent series (Mura and Kinoshita 1971), a double Fourier-series expansion (Shiah et al. 2012; Tan et al. 2013), an interpolation scheme using stored values (Wilson and Cruse 1978). A detailed review of works on static anisotropic elastic Greens' functions is provided by Pan and Chen (2015).

Regarding the dynamic parts of anisotropic elastic fundamental solutions, practically important results belong to Wang and Achenbach (1994, 1995) who used Radon transform and obtained expressions in terms of integrals over the surface of a half of a unit sphere. However, due to the oscillatory nature of the integrands of these integrals, they cannot be efficiently calculated using traditional quadrature rules (e.g., Newton–Cotes, Gauss–Legendre, Clenshaw–Curtis, etc.) for high frequencies or large distances between source and observation points. Asymptotic high-frequency anisotropic fundamental solutions in the far field were obtained by Vavryčuk (2007). For the particular case of transversely isotropic materials, a number of works are dedicated to somewhat alleviate this problem: Sáez and Domínguez (1999) derived expressions more suitable for numerical evaluation and later presented far field fundamental solutions (Sáez and Domínguez 2000), and Fooladi and Kundu (2019a, b) were able to reduce integration domain from half of a unit sphere to a quarter.

Overall excellent review of works on three-dimensional dynamic elastic fundamental solutions is provided by Dineva et al. (2019).

Evaluating highly oscillatory integrals even in one dimension is very expensive with traditional methods and nowadays there are several different approaches available that are specifically designed to deal with this type of integrals (Iserles et al. 2006). In this paper we propose a technique for numerical evaluation of one-dimensional integrals in Laplace domain anisotropic elastic fundamental solutions aimed for high frequencies and/or large distances between source and observation points. Our approach is based on quadrature rule developed by Evans and Webster (1997), which is a variation of Levin's method (Levin 1982).

11.2 Fundamental Solutions

Laplace-transformed full-space displacement fundamental solutions for anisotropic and linearly elastic media can be represented as a sum of static (singular) and dynamic (regular) terms as

$$\overline{g}_{jp}(\mathbf{y},\mathbf{x},s)=\overline{g}_{jp}(\mathbf{r},s)=g^{\mathrm{S}}_{jp}(\mathbf{r})+\overline{g}^{\mathrm{R}}_{jp}(\mathbf{r},s),\quad j,p=\overline{1,3}, \tag{11.1}$$

$$\mathbf{r}=\mathbf{y}-\mathbf{x},\quad r=|\mathbf{r}|, \tag{11.2}$$

where g^{S}_{jp} is static (singular) term, $\overline{g}^{\mathrm{R}}_{jp}$ is dynamic (regular) term, s is the complex Laplace transform parameter, $\mathbf{x}$ is the position vector of the source point, and $\mathbf{y}$ is the position vector of the observation point.

Employing the Radon transform according to Wang and Achenbach (1994, 1995), the dynamic term of the fundamental solution and its derivative can be expressed as the following integrals

$$\overline{g}^{\mathrm{R}}_{ij}(\mathbf{r},s)=-\frac{1}{8\pi^2}\int_0^{2\pi}\int_0^1\sum_{m=1}^{3}\frac{k_m E_{im}E_{jm}}{\rho c_m^2}\mathrm{e}^{-k_m br}\,\mathrm{d}\varphi\mathrm{d}b, \tag{11.3}$$

$$\overline{g}^{\mathrm{R}}_{ij,p}(\mathbf{r},s)=\frac{1}{8\pi^2}\int_0^{2\pi}\int_0^1\sum_{m=1}^{3}\frac{n_p(\varphi,b)k_m^2 E_{im}E_{jm}}{\rho c_m^2}\mathrm{e}^{-k_m br}\,\mathrm{d}\varphi\mathrm{d}b, \tag{11.4}$$

$$c_m=\sqrt{\lambda_m/\rho},\quad k_m=s/c_m, \tag{11.5}$$

where λ_m are the eigenvalues of the matrix $\Gamma_{ij}(\mathbf{n}(\varphi,b))=C_{kijl}n_k n_l$ and E_{jm} are the corresponding eigenvectors, C_{kijl} denote the fourth-order elasticity tensor, and ρ is mass density. Vector $\mathbf{n}(\varphi,b)$ which specifies the direction of the plane wave propagation, we define as (see Figs. 11.1 and 11.2)

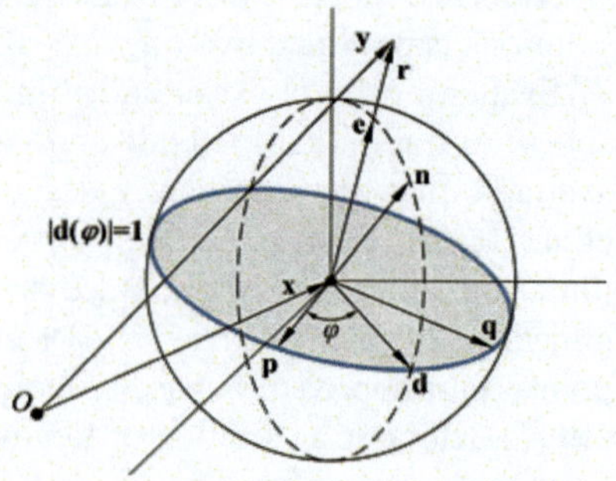

Fig. 11.1 Geometry of vectors **p**, **q**, **d**, **n** and **e**

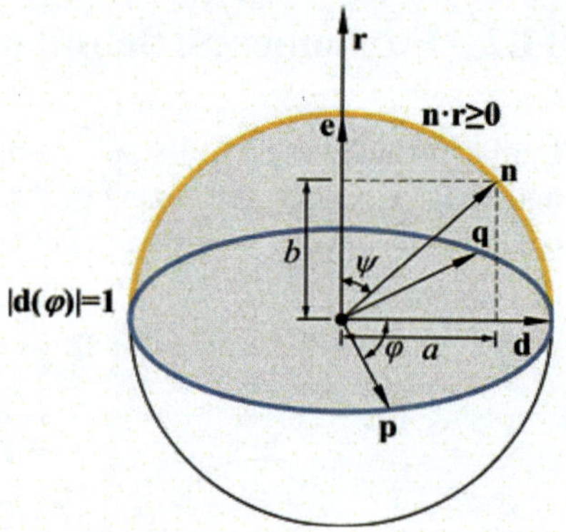

Fig. 11.2 Relations between a, b, φ, ψ and vectors **e, p, q, d, n**

$$\mathbf{n}(\varphi, b) = a\mathbf{d}(\varphi) + b\mathbf{e} = [n_1, n_2, n_3]^{\mathrm{T}}, \tag{11.6}$$

$$a = \sqrt{1 - b^2}, \quad \mathbf{d}(\varphi) \cdot \mathbf{r} = 0, \quad \mathbf{e} = \frac{\mathbf{r}}{|r|}, \quad \mathbf{e} = [e_1, e_2, e_3]^{\mathrm{T}}. \tag{11.7}$$

To define vector $\mathbf{d}(\varphi)$, we introduce a set of orthonormal vectors **p** and **q**

$$\mathbf{p} = \frac{[e_2, -e_1, 0]^{\mathrm{T}}}{\sqrt{1 - e_3^2}}, \quad \mathbf{q} = \mathbf{e} \times \mathbf{p} = \frac{[e_1 e_3, e_2 e_3, -(1 - e_3^2)]^{\mathrm{T}}}{\sqrt{1 - e_3^2}}, \tag{11.8}$$

$$\mathbf{p} \cdot \mathbf{e} = 0, \quad \mathbf{q} \cdot \mathbf{e} = 0, \quad \mathbf{p} \cdot \mathbf{q} = 0. \tag{11.9}$$

That way, we can express vector $\mathbf{d}(\varphi)$ as

$$\mathbf{d}(\varphi) = \mathbf{p} \cos\varphi + \mathbf{q} \sin\varphi, \quad 0 \le \varphi \le 2\pi, \tag{11.10}$$

$$\mathbf{d}(\varphi) = \frac{\left[e_2 \cos\varphi + e_1 e_3 \sin\varphi, -e_1 \cos\varphi + e_2 e_3 \sin\varphi, -(1 - e_3^2) sin\varphi\right]^{\mathrm{T}}}{\sqrt{1 - e_3^2}}. \tag{11.11}$$

Now in integrals (11.3) and (11.4), which define $\overline{g}^{\mathrm{R}}_{ij}(\mathbf{r}, s)$ and $\overline{g}^{\mathrm{R}}_{ij,p}(\mathbf{r}, s)$, we make a change of integration variable $b = \cos\psi$, where ψ is the angle between vectors $\mathbf{n}$ and $\mathbf{r}$ (see Fig. 11.2)

$$\mathrm{d}b = -\sin\psi \mathrm{d}\psi, \;\; \mathbf{n}(\varphi, \psi) = \mathbf{d}(\varphi)\sin\psi + \mathbf{e}\cos\psi = [n_1, n_2, n_3]^{\mathrm{T}}, \tag{11.12}$$

$$b = 0 \Rightarrow \psi = \frac{\pi}{2}, \quad b = 1 \Rightarrow \psi = 0, \tag{11.13}$$

$$\int_0^1 \ldots \mathrm{d}b = -\int_{\pi/2}^{0} \ldots \sin\psi \mathrm{d}\psi = \int_0^{\pi/2} \ldots \sin\psi \mathrm{d}\psi. \tag{11.14}$$

After some simple algebraic transformations, we get the following new expressions for dynamic term and its derivative

$$\overline{g}^{\mathrm{R}}_{ij}(\mathbf{r}, s) = -\frac{s\sqrt{\rho}}{8\pi^2}\int_0^{2\pi}\int_0^{\pi/2}\sum_{m=1}^{3}\frac{\sin\psi E_{im}E_{jm}}{\lambda_m^{3/2}}\mathrm{e}^{sr\sqrt{\rho}\left(-\frac{\cos\psi}{\sqrt{\lambda_m}}\right)}\mathrm{d}\psi\mathrm{d}\varphi, \tag{11.15}$$

$$\overline{g}^{\mathrm{R}}_{ij,p}(\mathbf{r}, s) = \frac{s^2\rho}{8\pi^2}\int_0^{2\pi}\int_0^{\pi/2}\sum_{m=1}^{3}\frac{\sin\psi n_p(\varphi, \psi)E_{im}E_{jm}}{\lambda_m^{2}}\mathrm{e}^{sr\sqrt{\rho}\left(-\frac{\cos\psi}{\sqrt{\lambda_m}}\right)}\mathrm{d}\psi\mathrm{d}\varphi. \tag{11.16}$$

When calculating integrands in (11.15) and (11.16) for each value of integration variables, it is required to find all eigenvalues and corresponding eigenvectors of matrix $\Gamma_{ij}(\mathbf{n}(\varphi, b)) = C_{kijl}n_k n_l$. Though Wang and Achenbach (1994, 1995) provided an alternative approach which requires calculation only of eigenvalues of matrix $\Gamma_{ij}(\mathbf{n}(\varphi, b))$ and does not involve direct calculations of the corresponding eigenvectors, it still remains a computationally intensive task. For a reasonable implementation of dynamic anisotropic elastic fundamental solutions in any boundary element formulation, it is necessary to minimize computation time of fundamental solutions and, therefore, to minimize the number of points at which the integrands are calculated.

In all further considerations, we will deal with inner integrals with respect to ψ for a fixed value of φ

$$I_1^{ij}[\varphi, r, s] = I_1[\varphi, r, s] = \int_0^{\pi/2}\sum_{m=1}^{3}\frac{\sin\psi E_{im}E_{jm}}{\lambda_m^{3/2}}\mathrm{e}^{sr\sqrt{\rho}\left(-\frac{\cos\psi}{\sqrt{\lambda_m}}\right)}\mathrm{d}\psi, \tag{11.17}$$

$$I_2^{ij,k}[\varphi, r, s] = I_2[\varphi, r, s] = \int_0^{\pi/2}\sum_{m=1}^{3}\frac{n_k(\varphi, \psi)\sin\psi E_{im}E_{jm}}{\lambda_m^{2}}\mathrm{e}^{sr\sqrt{\rho}\left(-\frac{\cos\psi}{\sqrt{\lambda_m}}\right)}\mathrm{d}\psi. \tag{11.18}$$

We introduce the following variable

$$\tau = sr\sqrt{\rho} = \langle s = \alpha + i\omega \rangle = \alpha r\sqrt{\rho} + i\omega r\sqrt{\rho}. \tag{11.19}$$

Taking (11.19) into account, we rewrite (11.17) and (11.18) as

$$I_1^{ij}[\varphi, \tau] = I_1[\varphi, \tau] = \int_0^{\pi/2} \sum_{m=1}^{3} \frac{\sin\psi E_{im} E_{jm}}{\lambda_m^{3/2}} e^{\tau\left(-\frac{\cos\psi}{\sqrt{\lambda_m}}\right)} d\psi, \tag{11.20}$$

$$I_2^{ij,k}[\varphi, \tau] = I_2[\varphi, \tau] = \int_0^{\pi/2} \sum_{m=1}^{3} \frac{\sin\psi n_k(\varphi, \psi) E_{im} E_{jm}}{\lambda_m^2} e^{\tau\left(-\frac{\cos\psi}{\sqrt{\lambda_m}}\right)} d\psi. \tag{11.21}$$

From the structure of integrands of $I_1[\varphi, \tau]$ and $I_2[\varphi, \tau]$, we notice that for the large values of the imaginary part of the complex frequency τ these integrals become highly oscillatory, and to efficiently compute them, we need to use specialized methods.

11.3 Evaluation of Integrals $I_1[\varphi, \tau]$ and $I_2[\varphi, \tau]$

Integrals of the form

$$I = \int_{-1}^{1} f(x) e^{i\gamma q(x)} dx, \tag{11.22}$$

are often called *highly oscillatory integrals*, where amplitude function $f(x)$ and phase function $q(x)$ are smooth. Let us recall two well-known approaches for calculation of highly oscillatory integrals: Filon method and Levin's collocation method.

In Filon method, amplitude function is replaced by suitable polynomial interpolating function, and then, oscillatory integral can be evaluated by computing moments $\int_{-1}^{1} x^k e^{i\gamma q(x)} dx$, $k = 0, 1, 2, \ldots$. This strategy if very effective for the integrals with linear phase functions or for the integrals in which the following change of variables is possible

$$I = \int_{q(-1)}^{q(1)} \frac{f\left(q^{-1}(y)\right)}{q'\left(q^{-1}(y)\right)} e^{i\gamma y} dy. \tag{11.23}$$

For such change of variables, we need to be able to obtain inverse function $q^{-1}(y)$, and phase function must not have any stationary points, so $q'(x) \neq 0$ in interval of integration. In our case in integrals $I_1[\varphi, \tau]$ and $I_2[\varphi, \tau]$, amplitude and phase function are not available in explicit closed form, and we need to evaluate them numerically for every value of integration variable. So to apply Filon-type method, we need to numerically calculate inverse of the phase function. Also, we do not know beforehand if phase function has any stationary points.

Levin's method (Levin 1982) is entirely different method suitable for solving highly oscillatory integrals with more complicated phase functions and does not require computation of moments. Levin's method transforms integration problem into the problem of finding a certain function $F(x)$ to satisfy the following ordinary differential equation (ODE)

$$F'(x) + i\gamma q'(x)F(x) = f(x). \tag{11.24}$$

If we solve this ODE, then we can evaluate the integral (11.22) trivially by substituting (11.24) into (11.22)

$$\begin{aligned} I &= \int_{-1}^{1} \left(F'(x) + i\gamma q'(x)F(x)\right)\mathrm{e}^{i\gamma q(x)}\mathrm{d}x \\ &= \int_{-1}^{1} \frac{\mathrm{d}}{\mathrm{d}x}\left[F(x)\mathrm{e}^{i\gamma q(x)}\right]\mathrm{d}x = F(1)\mathrm{e}^{i\gamma q(1)} - F(-1)\mathrm{e}^{i\gamma q(-1)}. \end{aligned} \tag{11.25}$$

To obtain $F(x)$, it is approximated by $F_n(x)$ that is a combination of some linearly independent basis functions

$$F(x) \approx F_n(x) = \sum_{k=1}^{n} \beta_k v_k(x). \tag{11.26}$$

To determine the coefficients β_k, the following system of collocation equations is solved on a sequence of collocation points $\{x_j\}$, $j = \overline{1, n}$:

$$\sum_{k=1}^{n} \beta_k v_k'(x_j) + i\gamma q'(x_j) \sum_{k=1}^{n} \beta_k v_k(x_j) = f(x_j). \tag{11.27}$$

Equation (11.27) is a complex-valued linear algebraic system for coefficients β_k After solving it, we can approximate the integral (11.22) as

$$I \approx I_n = \sum_{k=1}^{n} \beta_k v_k(1)\mathrm{e}^{i\gamma q(1)} - \sum_{k=1}^{n} \beta_k v_k(-1)\mathrm{e}^{i\gamma q(-1)}. \tag{11.28}$$

We notice that right-hand side in a complex-valued linear algebraic system (11.27) is defined by values of amplitude function $f(x)$. For our case, it means that to calculate integrals $I_1[\varphi, \tau]$ and $I_2[\varphi, \tau]$, we need to solve three linear systems, one for each phase function q_m, $m = 1, 2, 3$, with 24 different right-hand sides each: six different amplitude functions in I_1 and 18 different amplitude functions in I_2.

To overcome this drawback, we will consider a variation of Levin's method. Evans and Webster (1997) proposed to form quadrature rule

$$I = \int_{-1}^{1} f(x) \mathrm{e}^{i\tau q(x)} \mathrm{d}x \approx \sum_{j=0}^{N} w_j f(x_j), \tag{11.29}$$

where the weights w_j are chosen, so that formula (11.29) is exact for the functions

$$h_k(x) = i\tau q'(x) p_k(x) + p_k'(x), \quad k = \overline{0, N}. \tag{11.30}$$

Substituting the functions $h_k(x_j)$ on a set of collocations points $\{x_j\}$ into the quadrature formula, we obtain a complex-valued linear algebraic system for weights w_j

$$a_{kj} w_j = b_k, \tag{11.31}$$

$$a_{kj} = i\tau q'(x_j) p_k(x_j) + p_k'(x_j), \quad k, j = \overline{0, N}, \tag{11.32}$$

$$b_k = p_k(1) \mathrm{e}^{i\tau q(1)} - p_k(-1) \mathrm{e}^{i\tau q(-1)}, \quad k = \overline{0, N}. \tag{11.33}$$

This approach provides freedom of choice of collocation points $\{x_j\}$ and functions p_k. Originally Evans and Webster suggested the following

$$x_j = \cos(j\pi/N), \quad p_j(x) = T_j(x), \quad j = \overline{0, N}, \tag{11.34}$$

where $T_j(x)$ is the j-th Chebyshev polynomial of the first kind.

In our case to compute integrals $I_1[\varphi, \tau]$ and $I_2[\varphi, \tau]$ with Evans-Webster method, we need to solve only three complex-valued linear algebraic systems, one for each phase function q_m, $m = 1, 2, 3$.

We start with defining phase functions and obtaining their derivatives

$$q_m(\psi) = -\frac{\cos\psi}{\sqrt{\lambda_m}}, \quad q_m'(\psi) = \frac{1}{2} \frac{2\lambda_m \sin\psi + \lambda_m' \cos\psi}{\lambda_m^{3/2}}, \tag{11.35}$$

$$\lambda_m'(\psi) = \sum_{i=1}^{3} \sum_{j=1}^{3} E_{im}(\mathbf{n}) E_{jm}(\mathbf{n}) \Lambda_{ij}(\mathbf{n}, \boldsymbol{\beta}), \tag{11.36}$$

$$\boldsymbol{\beta}(\varphi,\psi) = \mathbf{d}(\varphi)\cos\psi - \mathbf{e}\sin\psi = [\beta_1, \beta_2, \beta_3]^{\mathrm{T}}, \tag{11.37}$$

$$\Lambda_{jk}(\mathbf{n},\boldsymbol{\beta}) = C_{ijkl}(\beta_i n_l + n_i \beta_l), \quad i, j, k, l = \overline{1, 3}, \tag{11.38}$$

where it should be noted that there is no summation in Eq. (11.36) over m.

We propose the following simple procedure for computing integrals $I_1[\varphi, \tau]$ and $I_2[\varphi, \tau]$.

1. Subdivide an integration interval $[0, \pi/2]$ into *sufficient* number of subintervals. Perform integration with Evans-Webster quadrature rule on each subinterval and then sum all partial results.
2. We assume that for a relatively large number of subintervals, phase function may contain only one stationary point in each subinterval. Hence, to determine if phase function has stationary point in a current subinterval, we simply have to check the signs of the phase function derivative at the end points of subinterval. If signs are different, then the phase function has stationary point in the subinterval.
3. If there is a stationary point in the current subinterval, a complex-valued linear algebraic system for weights w_j (11.31) becomes ill-conditioned or even singular. In order to deal with such system, we use truncated singular value decomposition.

11.4 Numerical Example and Discussions

11.4.1 Numerical Example Explanation

For computation of integrals $I_1[\varphi, \tau]$ and $I_2[\varphi, \tau]$ with the proposed procedure, we consider anisotropic elastic material with density $\rho = 2216$ kg/m^3 and

$$\mathbf{C} = \begin{bmatrix} 17.77 & 3.78 & 3.76 & 0.24 & -0.28 & 0.03 \\ & 19.45 & 4.13 & -0.41 & 0.07 & 1.13 \\ & & 21.79 & -0.12 & 0.01 & 0.38 \\ & & & 8.30 & 0.66 & 0.06 \\ & \text{symm.} & & & 7.62 & 0.52 \\ & & & & & 7.77 \end{bmatrix} \text{GPa},$$

where $\mathbf{C}$ is the elasticity tensor C_{ijkl} given in Voigt notation.

The integration problem is considered in dimensionless quantities and variables

$$\widetilde{\mathbf{C}} = \mathbf{C}/p_{\max},\ \tilde{\rho} = \rho l_{\max}^2/p_{\max}/t_{\max}^2,\ \tilde{s} = 2.0 + i\omega,\ \tau = \tilde{s}\tilde{r}\sqrt{\tilde{\rho}},$$
$$t_{\max} = 0.005\ \text{s},\ p_{\max} = 1.0 \times 10^{10}\ \text{Pa},\ l_{\max} = 200\ \text{m}.$$

The value of φ is fixed and equal $\pi/5$. The dimensionless position vectors of the source point $\mathbf{x}$ and observation point $\mathbf{y}$ are

$$\tilde{\mathbf{x}} = [0, 0, 0]^{\mathrm{T}},$$
$$\tilde{\mathbf{y}} = [\sin(\pi/3)\cos(\pi/4), \sin(\pi/3)\sin(\pi/4), \cos(\pi/3)]^{\mathrm{T}}, \quad \tilde{r} = |\tilde{\mathbf{y}} - \tilde{\mathbf{x}}| = 1.$$

To measure the accuracy of the calculations, the errors are defined as

$$\mathrm{err}(I_1) = \frac{\left\| I_1^{\mathrm{ref}} - I_1 \right\|_2}{\left\| I_1^{\mathrm{ref}} \right\|_2}, \quad \mathrm{err}(I_2) = \max_{1 \le k \le 3} \left(\frac{\left\| I_2^{ij,k;\mathrm{ref}} - I_2^{ij,k} \right\|_2}{\left\| I_2^{ij,k;\mathrm{ref}} \right\|_2} \right),$$

where $\|\|_2$ denotes the 2-norm (spectral norm) of a complex-valued matrix and superscript "ref" represents converged value of I_k obtained with high-order Gauss–Legendre (GL) quadrature rule.

For clarity and further convenience, we expand the expressions for $I_1[\varphi, \tau]$ and $I_2[\varphi, \tau]$ here

$$I_1^{ij}[\varphi, \tau] = I_1[\varphi, \tau] = \int_0^{\pi/2} \sum_{m=1}^{3} \frac{\sin\psi\, E_{im} E_{jm}}{\lambda_m^{3/2}} \mathrm{e}^{\tau\left(-\frac{\cos\psi}{\sqrt{\lambda_m}}\right)} \mathrm{d}\psi = \int_0^{\pi/2} \sum_{m=1}^{3} f_{ij}^m(\psi) \mathrm{e}^{\tau q_m(\psi)} \mathrm{d}\psi, \tag{11.39}$$

$$\begin{aligned} I_2^{ij,k}[\varphi, \tau] = I_2[\varphi, \tau] &= \int_0^{\pi/2} \sum_{m=1}^{3} \frac{n_k(\varphi, \psi)\sin\psi\, E_{im} E_{jm}}{\lambda_m^2} \mathrm{e}^{\tau\left(-\frac{\cos\psi}{\sqrt{\lambda_m}}\right)} \mathrm{d}\psi \\ &= \int_0^{\pi/2} \sum_{m=1}^{3} n_k(\varphi, \psi) f_{ij}^m(\psi) \mathrm{e}^{\tau q_m(\psi)} \mathrm{d}\psi, \end{aligned} \tag{11.40}$$

$$f_{ij}^m(\psi) = \frac{\sin\psi\, E_{im} E_{jm}}{\lambda_m^{3/2}}, \quad m = \overline{1, 3}, \tag{11.41}$$

where $f_{ij}^m(\psi)$ are amplitude functions corresponding to phase functions $q_m(\psi)$ defined in Eq. (11.35).

11.4.2 Computations

We start with displaying amplitude functions $f_{ij}^m(\psi)$ in Figs. 11.3, 11.4, and 11.5. It can be observed that amplitude functions are rather smooth and non-oscillatory.

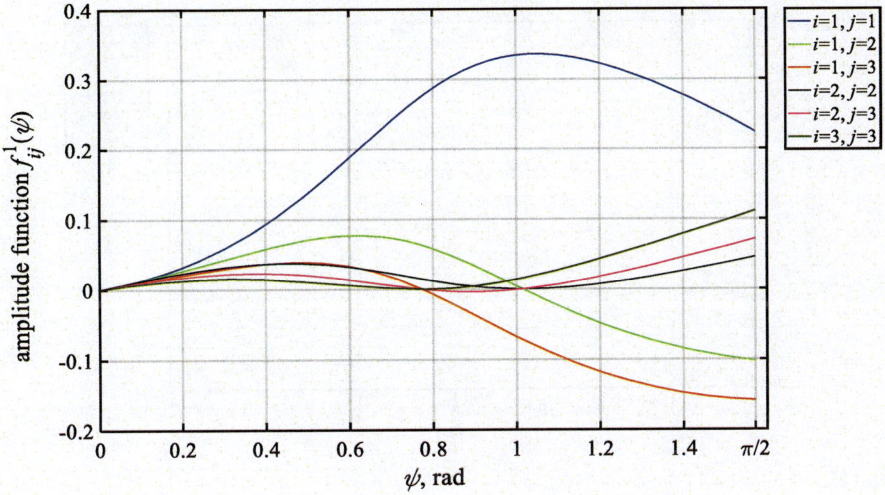

Fig. 11.3 Amplitude function $f^1_{ij}(\psi)$

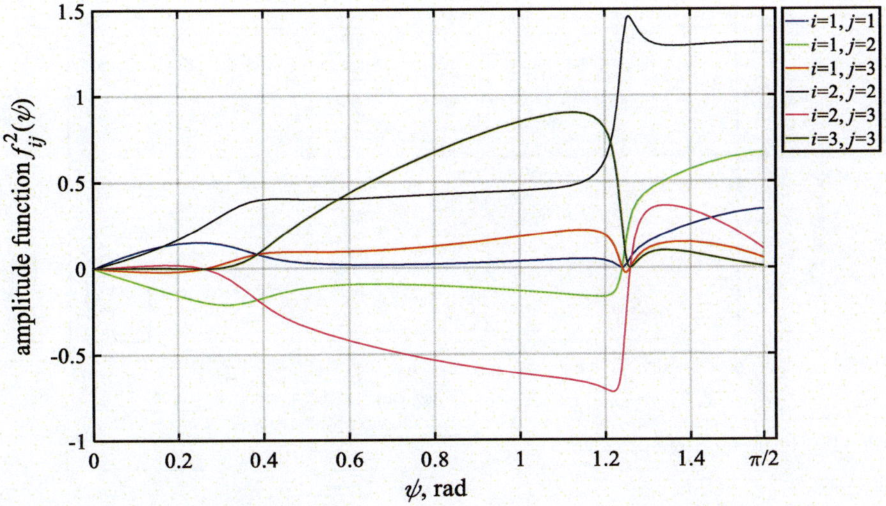

Fig. 11.4 Amplitude function $f^2_{ij}(\psi)$

In Fig. 11.6, phase functions $q_m(\psi)$, $m = \overline{1, 3}$, are shown. We make a special note that $q_1(\psi)$ and $q_3(\psi)$ have stationary points. The positions of stationary points are clearly seen as zeros of derivatives of phase functions $q'_m(\psi)$ as depicted in Fig. 11.7.

In Figs. 11.8 and 11.9, errors $\mathrm{err}(I_1)$ and $\mathrm{err}(I_2)$ are depicted versus total number of integrand evaluations for fixed value $\omega = 500$ for the following orders of Evans-Webster method (EW) uniformly applied on subintervals of integration: 5, 6, 7

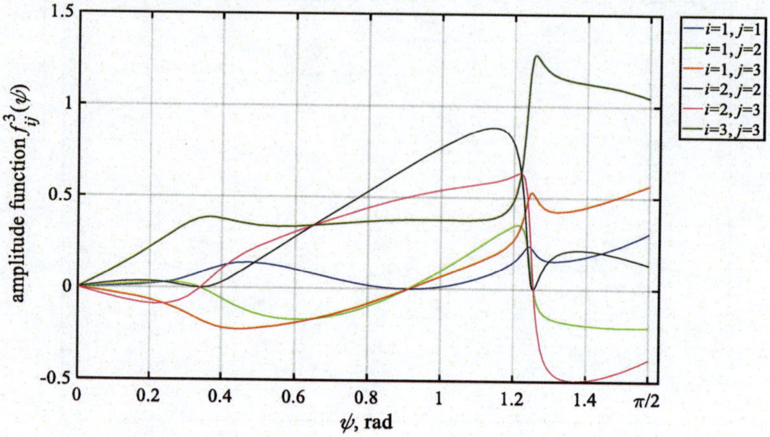

Fig. 11.5 Amplitude function $f_{ij}^3(\psi)$

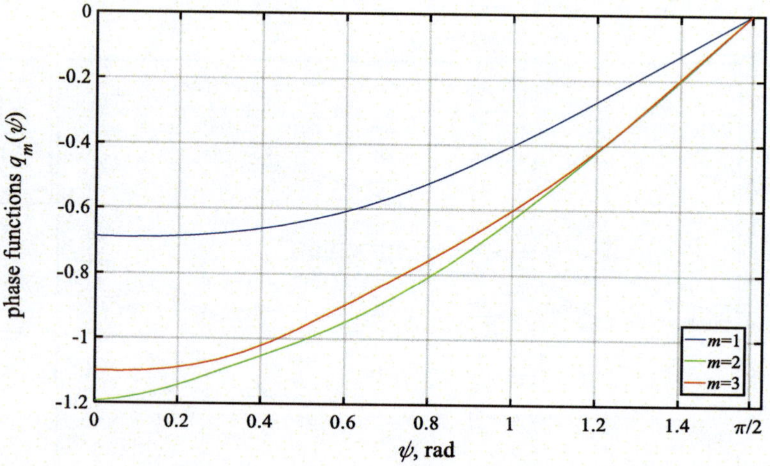

Fig. 11.6 Phase functions $q_m(\psi)$

(Fig. 11.8), 8, 9, and 10 (Fig. 11.9). Errors err(I_1) and err(I_2) are very close to each other for every order of EW method, and err(I_2) seems to be always slightly bigger than err(I_1). As the total number of integrand evaluations increases, differences between errors from different orders of EW method are getting smaller and smaller, suggesting that the proposed composite integration procedure converges in number of subintervals and in order of underlining EW method. We point out that already for total number of integrand evaluations ~400 both errors for all cases are less than 10^{-10}.

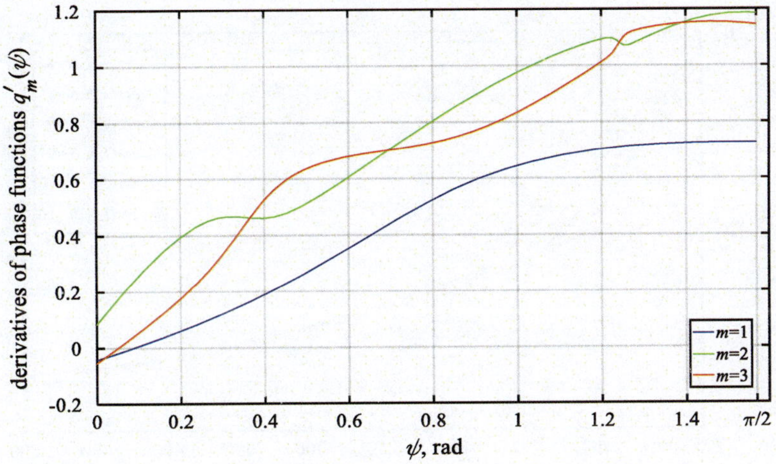

Fig. 11.7 Derivatives of phase functions $q'_m(\psi)$

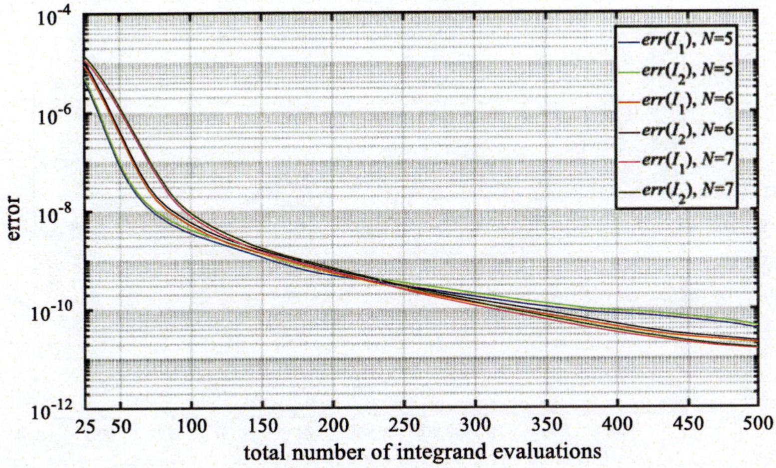

Fig. 11.8 $\mathrm{err}(I_1)$ and $\mathrm{err}(I_2)$ for $\omega = 500$, $N = 5, 6, 7$

Now, we fix total number of integrand evaluations as 300 and for $1 \leq \omega \leq 1000$ in Figs. 11.10 and 11.11 we compare errors $\mathrm{err}(I_2)$ for different orders of EW method: $N = 5, 6, 7$ (Fig. 11.10), 8, 9, and 10 (Fig. 11.11). Results indicate that error decreases as ω increases which attributes to asymptotic property of Levin's method.

In Fig. 11.12, errors $\mathrm{err}(I_2)$ are depicted for $1 \leq \omega \leq 1000$ order of EW method $N = 5$ and different number M of subintervals: $M = 20, 40, 60$, and 80. For four times more of integrand evaluations (from $M = 20$ to $M = 80$), error $\mathrm{err}(I_2)$ decreased roughly by 10^2 from 10^{-8} to 10^{-10}.

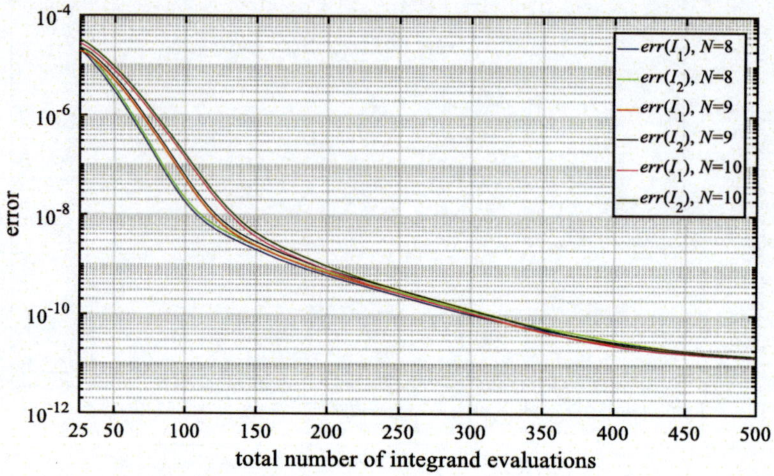

Fig. 11.9 $\mathrm{err}(I_1)$ and $\mathrm{err}(I_2)$ for $\omega = 500$, $N = 8, 9, 10$

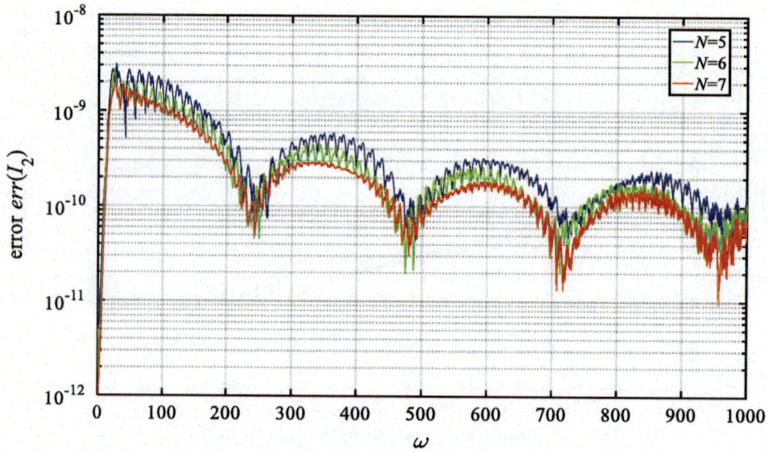

Fig. 11.10 $\mathrm{err}(I_2)$ for $1 \leq \omega \leq 1000$, $N = 5, 6, 7$, and 300 integrand evaluations

11.4.3 Discussion

Now we turn to most important practical question—how much integrand evaluations does our procedure require for a given accuracy and how it performs against Gauss–Legendre quadrature rule. First, we find that high-order Gauss–Legendre formula for whole integration interval works much better than a composite GL rule with fixed low order quadrature on each subinterval (as implemented in, e.g., Dravinski and Niu 2000, 2001). And nowadays, computing even millions of Gauss–Legendre quadrature nodes and weights is not a problem anymore (Bogaert 2014). In Fig. 11.13, total

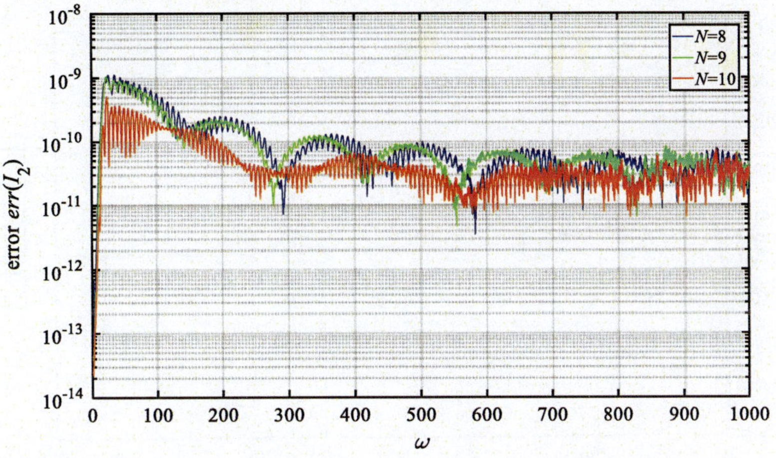

Fig. 11.11 err(I_2) for $1 \le \omega \le 1000$, $N = 8, 9, 10$, and 300 integrand evaluations

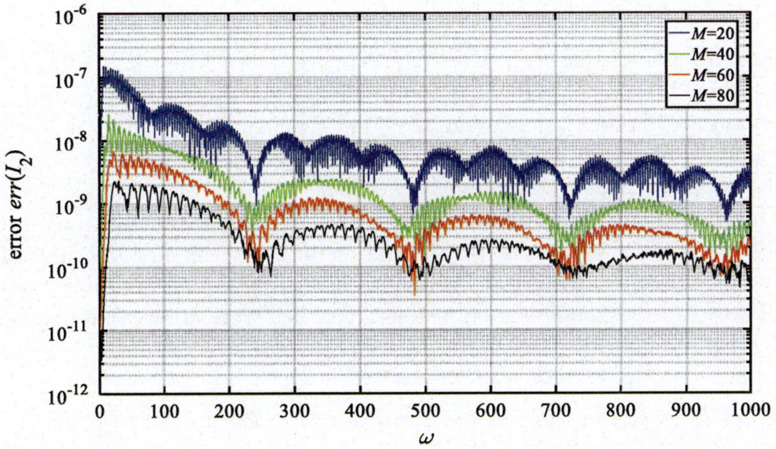

Fig. 11.12 err(I_2) for $1 \le \omega \le 1000$ and different number of subintervals of integration

number of integrand evaluations required for err(I_2) $\le 1 \times 10^{-10}$ is depicted versus frequency $1 \le \omega \le 1000$ for GL and our procedure with EW method orders $N = 5$, 6, 7, 8, 9, and 10. As expected, GL performs better for lower frequencies $\omega \le 100$. Again, as ω increases, the required number of integrand evaluations decreases for our procedure.

We introduce a dimensionless performance coefficient

$$\eta = \frac{\text{number of integrand evaluations for GL}}{\text{number of integrand evaluations for EW} - \text{based procedure}}$$

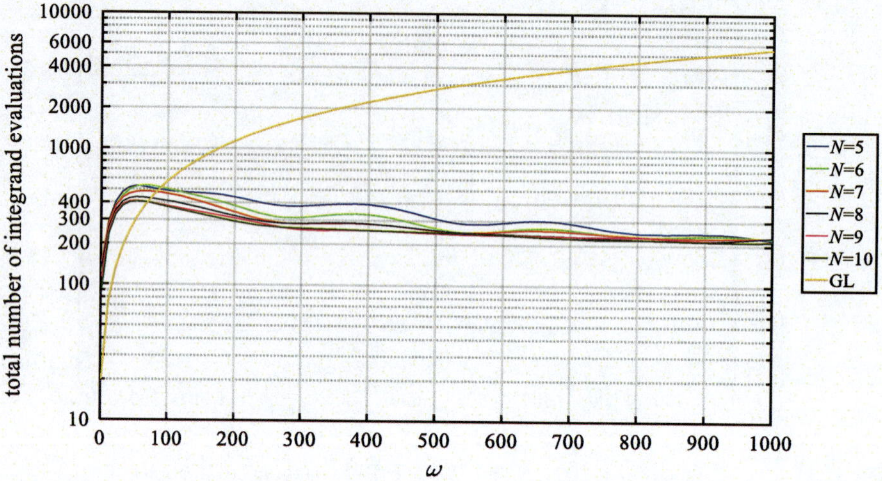

Fig. 11.13 Total number of integrand evaluations for $1 \le \omega \le 1000$ and $\mathrm{err}(I_2) \le 1 \times 10^{-10}$

and depict it versus frequency $1 \le \omega \le 1000$ in Fig. 11.14. From $\omega \ge 100$ η steadily increases from 1 to almost 25 for $\omega \approx 1000$, meaning that for $\omega \approx 1000$ our procedure requires almost 25 times *less* integrand evaluations than GL for $\mathrm{err}(I_2) \le 1 \times 10^{-10}$. Total numbers of integrand evaluations with our procedure are relatively close for all considered orders of underlining EW method.

We find it is advantageous to use low order ($N = 5$ or 6, for example) of EW method and larger number of subintervals of integration than high order of EW method with lesser number of subintervals since solving our $(N+1) \times (N+1)$ complex-valued general linear system requires no less than $O((N+1)^3)$ floating-point operations.

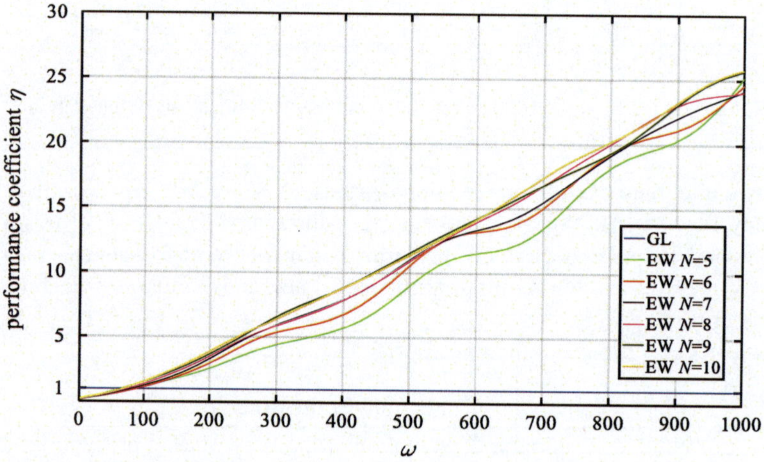

Fig. 11.14 Performance coefficient η for $1 \le \omega \le 1000$ and $\mathrm{err}(I_2) \le 1 \times 10^{-10}$

The question of selecting required number of subintervals to attain desirable accuracy before performing the integration still remains open. For evaluations of dynamic anisotropic elastic fundamental solutions in practical applications, it is possible to employ an approach similar to the one presented by Fooladi and Kundu (2019a, b). For orthotropic materials, they proposed a calibration strategy for choosing the number of integration points for the dynamic part of the anisotropic Green's function. Their technique involves determining an isotropic equivalent of the original anisotropic elasticity tensor and then using it to find an optimum number of integration points for a user-defined set of distances between source and observation points.

Acknowledgements The work is financially supported by the Russian Science Foundation under grant No. 18-79-00082.

References

Alibert, J., Seppecher, P., & dell'Isola, F. (2003). Truss modular beams with deformation energy depending on higher displacement gradients. *Mathematics and Mechanics of Solids, 8,* 51–73. https://doi.org/10.1177/1081286503008001658.

Auffray, N., dell'Isola, F., Eremeyev, V., Madeo, A., & Rossi, G. (2013). Analytical continuum mechanics à la Hamilton-Piola: Least action principle for second gradient continua and capillary fluids. *Mathematics and Mechanics of Solids, 20,* 375–417. https://doi.org/10.1177/1081286513497616.

Barchiesi, E., Spagnuolo, M., & Placidi, L. (2018). Mechanical metamaterials: A state of the art. *Mathematics and Mechanics of Solids, 24,* 212–234. https://doi.org/10.1177/1081286517735695.

Barnett, D. (1972). The precise evaluation of derivatives of the anisotropic elastic Green's functions. *Physica Status Solidi (B), 49,* 741–748. https://doi.org/10.1002/pssb.2220490238.

Bogaert, I. (2014). Iteration-free computation of Gauss-Legendre quadrature nodes and weights. *SIAM Journal on Scientific Computing, 36,* A1008–A1026. https://doi.org/10.1137/140954969.

Buroni, F., & Sáez, A. (2013). Unique and explicit formulas for green's function in three-dimensional anisotropic linear elasticity. *Journal of Applied Mechanics, 80,* 051018. https://doi.org/10.1115/1.4023627.

Del Vescovo, D., & Giorgio, I. (2014). Dynamic problems for metamaterials: Review of existing models and ideas for further research. *International Journal of Engineering Science, 80,* 153–172. https://doi.org/10.1016/j.ijengsci.2014.02.022.

dell'Isola, F., Andreaus, U., & Placidi, L. (2015). At the origins and in the vanguard of peridynamics, non-local and higher-gradient continuum mechanics: An underestimated and still topical contribution of Gabrio Piola. *Mathematics and Mechanics of Solids, 20*(8), 887–928. https://doi.org/10.1177/1081286513509811.

dell'Isola, F., Cuomo, M., Greco, L., & Della Corte, A. (2017). Bias extension test for pantographic sheets: Numerical simulations based on second gradient shear energies. *Journal of Engineering Mathematics, 103*(1), 127–157. https://doi.org/10.1007/s10665-016-9865-7.

dell'Isola, F., Della Corte, A., & Giorgio, I. (2016a). Higher-gradient continua: The legacy of Piola, Mindlin, Sedov and Toupin and some future research perspectives. *Mathematics and Mechanics of Solids, 22*(4), 852–872. https://doi.org/10.1177/1081286515616034.

dell'Isola, F., Della Corte, A., Greco, L., & Luongo, A. (2016b). Plane bias extension test for a continuum with two inextensible families of fibers: A variational treatment with Lagrange

multipliers and a perturbation solution. *International Journal of Solids and Structures, 81,* 1–12. https://doi.org/10.1016/j.ijsolstr.2015.08.029.
dell'Isola, F., Giorgio, I., Pawlikowski, M., & Rizzi, N. (2016c). Large deformations of planar extensible beams and pantographic lattices: Heuristic homogenization, experimental and numerical examples of equilibrium. *Proceedings of the Royal Society A: Mathematical, Physical and Engineering Sciences, 472,* 20150790. https://doi.org/10.1098/rspa.2015.0790.
dell'Isola, F., Igumnov, L., Litvinchuk, S., Ipatov, A., Petrov, A., & Modin, I. (2019a). Surface waves in dissipative poroviscoelastic layered half space: Boundary element analyses. In H. Altenbach, A. Belyaev, V. Eremeyev, A. Krivtsov, & A. Porubov (Eds.), *Advanced Structured Materials. Dynamical processes in generalized continua and structures* (Vol. 103). Cham: Springer.
dell'Isola, F., Seppecher, P., Alibert, J., Lekszycki, T., Grygoruk, R., Pawlikowski, M., et al. (2019b). Pantographic metamaterials: An example of mathematically driven design and of its technological challenges. *Continuum Mechanics and Thermodynamics, 31*(4), 851–884. https://doi.org/10.1007/s00161-018-0689-8.
dell'Isola, F., Seppecher, P., & Madeo, A. (2012). How contact interactions may depend on the shape of Cauchy cuts in Nth gradient continua: Approach "à la D'Alembert." *Zeitschrift für Angewandte Mathematik und Physik, 63*(6), 1119–1141. https://doi.org/10.1007/s00033-012-0197-9.
dell'Isola, F., Seppecher, P., Spagnuolo, M., Barchiesi, E., Hils, F., Lekszycki, T., et al. (2019c). Advances in pantographic structures: Design, manufacturing, models, experiments and image analyses. *Continuum Mechanics and Thermodynamics, 31*(4), 1231–1282. https://doi.org/10.1007/s00161-019-00806-x.
Dineva, P., Manolis, G., & Wuttke, F. (2019). Fundamental solutions in 3D elastodynamics for the BEM: A review. *Engineering Analysis with Boundary Elements, 105,* 47–69. https://doi.org/10.1016/j.enganabound.2019.04.003.
Dravinski, M., & Niu, Y. (2001). Three-dimensional time-harmonic Green's functions for a triclinic full-space using a symbolic computation system. *International Journal for Numerical Methods in Engineering, 53,* 445–472. https://doi.org/10.1002/nme.292.
Dravinski, M., & Zheng, T. (2000). Numerical evaluation of three-dimensional time-harmonic Green's functions for a nonisotropic full-space. *Wave Motion, 32,* 141–151. https://doi.org/10.1016/s0165-2125(00)00034-2.
Evans, G., & Webster, J. (1997). A high order, progressive method for the evaluation of irregular oscillatory integrals. *Applied Numerical Mathematics, 23,* 205–218. https://doi.org/10.1016/s0168-9274(96)00058-x.
Fooladi, S., & Kundu, T. (2019a). An improved technique for elastodynamic Green's function computation for transversely isotropic solids. *Journal of Nondestructive Evaluation, Diagnostics and Prognostics of Engineering Systems, 2,* 021005. https://doi.org/10.1115/1.4043605.
Fooladi, S., & Kundu, T. (2019b). Distributed point source modeling of the scattering of elastic waves by a circular cavity in an anisotropic half-space. *Ultrasonics, 94,* 264–280. https://doi.org/10.1016/j.ultras.2018.09.002.
Fredholm, I. (1900). Sur les équations de l'équilibre d'un corps solide élastique. *Acta Mathematica, 23,* 1–42. https://doi.org/10.1007/bf02418668.
Iserles, A., Nørsett, S., & Olver, S. (2006) Highly oscillatory quadrature: The story so far. *Numerical Mathematics and Advanced Applications*, 97–118. https://doi.org/10.1007/978-3-540-34288-5_6.
Lee, V. (2009). Derivatives of the three-dimensional Green's functions for anisotropic materials. *International Journal of Solids and Structures, 46,* 3471–3479. https://doi.org/10.1016/j.ijsolstr.2009.06.002.
Levin, D. (1982). Procedures for computing one- and two-dimensional integrals of functions with rapid irregular oscillations. *Mathematics of Computation, 38,* 531. https://doi.org/10.2307/2007287.
Lifshitz, I., & Rozenzweig, L. (1947). Construction of the Green tensor for the fundamental equation of elasticity theory in the case of unbounded elastically anisotropic medium. *Zhurnal Eksperimental'noi i Teoreticheskoi Fiziki, 17,* 783–791.

Malén, K. (1971). A unified six-dimensional treatment of elastic green's functions and dislocations. *Physica Status Solidi (B), 44,* 661–672. https://doi.org/10.1002/pssb.2220440224.

Mura, T., & Kinoshita, N. (1971). Green's functions for anisotropic elasticity. *Physica Status Solidi (B), 47,* 607–618. https://doi.org/10.1002/pssb.2220470226.

Nakamura, G., & Tanuma, K. (1997). A formula for the fundamental solution of anisotropic elasticity. *The Quarterly Journal of Mechanics and Applied Mathematics, 50,* 179–194. https://doi.org/10.1093/qjmam/50.2.179.

Pan, E., & Chen, W. (2015). *Static Green's functions in anisotropic media.* Cambridge: Cambridge University Press.

Phan, A., Gray, L., & Kaplan, T. (2004). On the residue calculus evaluation of the 3-D anisotropic elastic Green's function. *Communications in Numerical Methods in Engineering, 20,* 335–341. https://doi.org/10.1002/cnm.675.

Phan, A., Gray, L., & Kaplan, T. (2005). Residue approach for evaluating the 3D anisotropic elastic Green's function: Multiple roots. *Engineering Analysis with Boundary Elements, 29,* 570–576. https://doi.org/10.1016/j.enganabound.2004.12.012.

Placidi, L., Andreaus, U., & Giorgio, I. (2017). Identification of two-dimensional pantographic structure via a linear D4 orthotropic second gradient elastic model. *Journal of Engineering Mathematics, 103,* 1–21. https://doi.org/10.1007/s10665-016-9856-8.

Placidi, L., Barchiesi, E., Turco, E., & Rizzi, N. (2016). A review on 2D models for the description of pantographic fabrics. *Zeitschrift für Angewandte Mathematik und Physik, 67,* 121. https://doi.org/10.1007/s00033-016-0716-1.

Rahali, Y., Giorgio, I., Ganghoffer, J.-F., & dell'Isola, F. (2015). Homogenization à la Piola produces second gradient continuum models for linear pantographic lattices. *International Journal of Engineering Science, 97,* 148–172. https://doi.org/10.1016/j.ijengsci.2015.10.003.

Sáez, A., & Domínguez, J. (1999). BEM analysis of wave scattering in transversely isotropic solids. *International Journal for Numerical Methods in Engineering, 44,* 1283–1300. https://doi.org/10.1002/(sici)1097-0207(19990330)44:9<1283::aid-nme544>3.0.co;2-o.

Sáez, A., & Domínguez, J. (2000). Far field dynamic Green's functions for BEM in transversely isotropic solids. *Wave Motion, 32,* 113–123. https://doi.org/10.1016/s0165-2125(00)00032-9.

Sales, M., & Gray, L. (1998). Evaluation of the anisotropic Green's function and its derivatives. *Computers & Structures, 69,* 247–254. https://doi.org/10.1016/s0045-7949(97)00115-6.

Sciarra, G., dell'Isola, F., & Coussy, O. (2007). Second gradient poromechanics. *International Journal of Solids and Structures, 44*(20), 6607–6629. https://doi.org/10.1016/j.ijsolstr.2007.03.003.

Shiah, Y., Tan, C., & Lee, R. (2010). Internal point solutions for displacements and stresses in 3D anisotropic elastic solids using the boundary element method. *Computer Modeling in Engineering & Sciences, 69,* 167–179. https://doi.org/10.3970/cmes.2010.069.167.

Shiah, Y., Tan, C., & Wang, C. (2012). Efficient computation of the Green's function and its derivatives for three-dimensional anisotropic elasticity in BEM analysis. *Engineering Analysis with Boundary Elements, 36,* 1746–1755. https://doi.org/10.1016/j.enganabound.2012.05.008.

Tan, C., Shiah, Y., & Wang, C. (2013). Boundary element elastic stress analysis of 3D generally anisotropic solids using fundamental solutions based on Fourier series. *International Journal of Solids and Structures, 50,* 2701–2711. https://doi.org/10.1016/j.ijsolstr.2013.04.026.

Ting, T. (1997). The three-dimensional elastostatic Green's function for general anisotropic linear elastic solids. *The Quarterly Journal of Mechanics and Applied Mathematics, 50,* 407–426. https://doi.org/10.1093/qjmam/50.3.407.

Vavryčuk, V. (2007). Asymptotic Green's function in homogeneous anisotropic viscoelastic media. *Proceedings of the Royal Society A: Mathematical, Physical and Engineering Sciences, 463,* 2689–2707. https://doi.org/10.1098/rspa.2007.1862.

Wang, C., & Achenbach, J. (1994). Elastodynamic fundamental solutions for anisotropic solids. *Geophysical Journal International, 118,* 384–392. https://doi.org/10.1111/j.1365-246x.1994.tb03970.x.

Wang, C., & Achenbach, J. (1995). Three-dimensional time-harmonic elastodynamic Green's functions for anisotropic solids. *Proceedings of the Royal Society A: Mathematical, Physical and Engineering Sciences, 449,* 441–458. https://doi.org/10.1098/rspa.1995.0052.

Wilson, R., & Cruse, T. (1978). Efficient implementation of anisotropic three dimensional boundary-integral equation stress analysis. *International Journal for Numerical Methods in Engineering, 12,* 1383–1397. https://doi.org/10.1002/nme.1620120907.

Xie, L., Zhang, C., Sladek, J., & Sladek, V. (2016). Unified analytical expressions of the three-dimensional fundamental solutions and their derivatives for linear elastic anisotropic materials. *Proceedings of the Royal Society A: Mathematical, Physical and Engineering Science, 472,* 20150272. https://doi.org/10.1098/rspa.2015.0272.

Chapter 12
The Dynamics of Eccentric Vibration Mechanism (Part 2)

Leonid Igumnov, S. Vladimir Metrikin, V. Irina Nikiforova, and Lubov N. Fevral'skikh

Abstract This paper presents a detailed study of the dynamics of the two-piston vibro-impact mechanism with a crank vibration exciter. The mechanism has quite a wide range of applications: It is effective for vibro-impact compaction of various types of soil, sand and concrete in strained industrial conditions; for breaking the ice and other harder objects; driving piles and structures, etc. The presented mathematical model (MM) is a substantially nonlinear dynamic system with a variable structure. An original method for numerical–analytical study of dynamic characteristics, such as periodic motion modes with alternating impact interaction of pistons, as well as complex motion modes with an arbitrary number of impacts, including chaotic ones, has been elaborated on the basis of the point mapping method. Thus, it was possible to obtain for the first time the engineering formulas for tuning the mechanism into required mode of operation, at the stage of preliminary MM dynamic analysis, using a special coordinate transformation and the geometry of the Poincaré surface of section. The obtained engineering formulas allow us to indicate the regions of various qualitative behavior of the mechanism in the parameter space. The bifurcation diagrams make it possible to determine the influence of the main parameters on the processes of reorganization of motion modes, starting from the simplest to the most complex modes, including chaotic ones. The obtained results and proposed numerical–analytical approaches for investigating the dynamic characteristics of crank-type vibro-impact mechanisms enable practitioners to use them for tuning and analyzing the operational regimes of specific mechanisms.

Keywords Mathematical model · Point mapping · Stability · Bifurcation diagrams · Chaos

---

L. Igumnov · S. V. Metrikin (✉)
Research Institute for Mechanics, National Research Lobachevsky State University of Nizhny Novgorod, Nizhny Novgorod, Russia
e-mail: v.s.metrikin@mail.ru

V. I. Nikiforova · L. N. Fevral'skikh
National Research Lobachevsky State University of Nizhny Novgorod, Nizhny Novgorod, Russia

F. dell'Isola and L. Igumnov (eds.), *Dynamics, Strength of Materials and Durability in Multiscale Mechanics*, Advanced Structured Materials 137,
https://doi.org/10.1007/978-3-030-53755-5_12

12.1 Introduction

The present work is a direct continuation of investigations carried out in (Igumnov et al. 2017), relating to the study of nonlinear dynamics of new designs of eccentric shock–vibration mechanisms (ESVM) with a crank-sliding bar vibration exciter (CSVE) (Shilkov et al. 2005; Nagaev 1978; Kobrinskiy and Kobrinskiy 1973; Babitskiy and Krupenin 1985; Biderman 1972; Bogodukhov et al. 2005; Vagapov et al. 2008; Zakrzhevskiy 1980; Babitsky 1998; Cveticanin 2002; Pavloaskaia and Wiercigroch 2003; Vagapov et al. 2007; Luo and Ma 2008; Leine and Heimsch 2012; Goebel et al. 2008; Zheleztsov 1949).

The mechanism has quite a wide range of applications: It is effective for vi-bro-impact compaction of various types of soil, sand and concrete in strained industrial conditions; for breaking the ice and other harder objects; driving piles and structures, etc. It also can be used for performing experiments: testing new artificial microstructured metamaterials (Del Vescovo and Giorgio 2014; Barchiesi et al. 2018) and investigating their mechanical response under heavy periodic loads. An example of mechanical metamaterials is pantographic structure (Placidi et al. 2016; dell'Isola et al. 2019). Designing of such metamaterials is based on generalized continuum approaches (Alibert et al. 2003; dell'Isola et al. 2012; Auffray et al. 2013; Rahali et al. 2015; dell'Isola et al. 2015, 2016a; b, c) and widely uses numerical simulations for identification of constitutive parameters and validation of theoretical predictions (dell'Isola et al. 2016a, b, c; Giorgio 2016; dell'Isola et al. 2017; Placidi et al. 2017).

In (Igumnov et al. 2017), a model of ESVM with a CSVE and an arbitrary number of piston-strikers is presented. Stability regions of periodic motion and a bifurcation diagram for specific set of parameters are given for the simplest mechanism with one piston-striker (PS). In the case of an absolutely inelastic impact interaction of PS mechanism with a fixed stopper, new results of investigation of the point mapping of a circle into itself are presented. In particular, analytical equations of inaccessible boundaries are given, in the neighborhood of which there is a countable set of unstable periodic motions. The present work gives a detailed study of nonlinear dynamics of a vibro-impact mechanism with CSVE and two PSs. To study complex modes of motion with an arbitrary number of impacts, including chaotic ones, an original numerical–analytical method of the research based on the point mapping method has been developed. As a result, it was possible to propose simple engineering formulas for finding in the parameter space the region with different qualitative mechanism behavior, convenient for tuning the mechanism for the required mode of operation. The created bifurcation diagrams allow us to distinguish in the parameter space the regions of arbitrarily complex motion modes, including chaotic ones. It is clearly shown that dynamic systems with vibro-impact interaction belong to the class of highly nonlinear systems that are used in various mechanisms (Masri and Caughey 1966; Babitsky 2013). The difficulty in studying the nonlinear dynamics of these systems is due to their very rich, complex and highly nonlinear behavior under deterministic loading (Pavlovskaia et al. 2015; Bernardini and Litak 2016; Tusset et al. 2017; Ing et al. 2010; Liu et al. 2015; Neimark 2010; Feigin 1994). Thus, the dynamics of such systems should be studied with the help of numerical–analytical methods.

12.2 Problem Setting

The new constructive solution is based on the “inverted vibrator” principle (Igumnov et al. 2017; Shilkov et al. 2005). The power pulse, transmitted to the processed medium (fixed obstacle, soil, piles, structures, etc.), arises both due to the thrust against the shoulder of the eccentric shaft and to the drop in the kinetic energy of the operating element. It is evident (Nagaev 1978) that a dense and, at the same time, strong structure of the processed medium can be achieved only when specific pressure on the contact surface of an operating element with the processed medium increases gradually. Its lower limit is ditial state (before competermined by the physical properties of the medium in its inaction process), while the upper one by the ultimate strength of the medium or technological conditions. Thus, the parameters of such machines and mechanisms must be determined from the conditions close to quasi-plastic interaction (Nagaev 1978). Such multi-pulse loading mode can be realized using multi-impact ESVM with CSVE, designed to easily regulate working modes by changing the geometry of kinematic connections.

This paper presents a detailed study of nonlinear dynamics of the mechanism with two PSs. It is shown that the used mathematical apparatus of the point mapping method for studying the dynamics of the mechanism allows us to obtain rather simple analytical relations for geometrical parameters of the mechanism, which indicate in the parameter space, the regions of existence of impact motions of the mechanism after the stroke either by just one PS or alternately by two PSs. Multidimensional parameter space of the model is divided into regions of stable periodic motion modes with both two impacts of PS and a large number of PS impacts. The developed software written in high-level language makes it possible to calculate bifurcation diagrams for the main parameters.

Figure 12.1 presents the scheme of the mechanism under consideration, where 1 is the frame of the mechanism, 2 is a flywheel, 3 are cranks with adjustable eccentricities r_i, 4 are connecting rods (of the length l_i) with a stationary phase shift ϕ, 5 are piston-strikers, 6 is a fixed stopper, 7 is eccentric shaft, 8 is a guide sleeve, 9 is a frame guide rod, 10 is an anvil.

Neglecting the masses of PSs, rods and cranks, the equation of free (without impact) motion of the system for $y_{pi} > 0$ can be written in the following form:

$$M\frac{\mathrm{d}^2 y}{\mathrm{d}t^2} = -Mg,$$

where y is coordinate of the mass center of the body, counted from fixed anvil block, g is free-fall acceleration. Let y_{pi} be the distance of the base ith PS from an anvil ($i = 1, 2$). When one of the PSs contacts the anvil, a momentary impact interaction

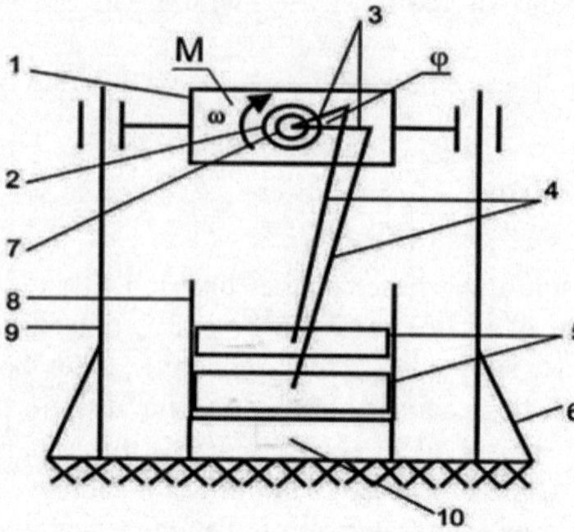

Fig. 12.1 Scheme of two-piston shock-vibration mechanism

takes place, described using the Newton hypothesis in the form $\dot{y}^{+}_{pi} = -R\dot{y}^{-}_{pi}$, where $\dot{y}^{-}_{pi}$ and $\dot{y}^{+}_{pi}$ are velocities of the ith PS immediately before and after the impact interaction, respectively.

The position of the eccentricities of the length r_i will be measured by the angles $\theta_1 = \omega t,\ \theta_2 = \omega t - \phi$, counted off the vertical axis. Then, it follows that:

$$y_{p1} = h_1 + y - s_1 + r_1 \cos \omega t - \sqrt{l_1^2 - r_1^2 \sin^2 \omega t},\ y_{p2} = h_2 + y - s_2 \\ + r_2 \cos(\omega t - \phi) - \sqrt{l_2^2 - r_2^2 \sin^2(\omega t - \phi)},$$

where $s_1,\ s_2$ are distance from the fixture point of the rod to the PS base, $\omega = 2\pi n$ is cyclic frequency of a flywheel rotation, n rotational speed, h_i is anvil height. Thus, equations of motion have the form:

$$M\frac{\mathrm{d}^2 y}{\mathrm{d}t^2} = -Mg,\ y_{pi} > h_i,\ \text{only when} \tag{12.1}$$

$$\dot{y}^{+}_{pi} = -R\dot{y}^{-}_{pi},\ y_{pi} = h_i, i = 1, 2. \tag{12.2}$$

When changing in system (12.1)–(12.2) to dimensionless: time $\tau = \omega t$, coordinate $x = (y - s_2 - l)/l$, parameter $\mu = \frac{r_1}{l},\ \gamma = \frac{r_2}{r_1},\ \varepsilon = \frac{s_1 - s_2}{l}, p = \frac{g}{\omega^2 l}, k_1 = \frac{h_1}{l_1},\ k_2 = \frac{h_2}{l_1},\ \Delta k = k_2 - k_1$, accounting for $r_i \ll l_i\ (l_i \approx l)$ and by introducing functions $f_1(\tau) = k_1 + \varepsilon - \mu \cos \tau,\ f_2(\tau) = k_1 + \Delta k - \mu\gamma \cos(\tau - \phi)$, we obtain the equations, describing impact-oscillatory motions of the mechanism in the form:

$$\frac{\mathrm{d}^2 x}{\mathrm{d}\tau^2} = -p,\ x > f(\tau). \tag{12.3}$$

$$\left.\frac{dx}{d\tau}\right|_+ = -R\left.\frac{dx}{d\tau}\right|_- + (1+R)\frac{df(\tau)}{d\tau}, \; x = f(\tau), \tag{12.4}$$

where $f(\tau) = \max_\tau\{f_1(\tau), f_2(\tau)\}$.

12.3 The Phase Space

The phase space of the system (12.3)–(12.4) $\Phi(x \geq f(\tau),\; \dot{x} < +\infty)$ in coordinates $x,\; \dot{x},\; \tau$ is truncated along x. $S(x = f(\tau))$ is cylindrical along $\dot{x}$ surface, formed by the intersection of two surfaces $x = f_1(\tau),\;\; x = f_2(\tau)$, the qualitative view of which, together with the phase trajectories, is given in Fig. 12.2.

Obviously, the kind of the surface S, shown in Fig. 12.2, is preserved only when surfaces $x = f_1(\tau)$ and $x = f_2(\tau)$ intersect. This condition allows us to obtain ratios for parameters of the mechanism in the form:

$$(\varepsilon - \Delta k/\mu)^2 = (1 - \gamma\cos\phi)^2 + \gamma^2\sin^2\phi.$$

Using this ratio, it is possible to indicate the regions, where the motion mode of the mechanism is possible after the impact either by one PS or alternately by two PSs. These regions are shown in Fig. 12.3 on the plane $\left(\left(\gamma, (\varepsilon - \Delta k/\mu)^2\right)\right.$ at different values of the phase shift between eccentricities ϕ. Only in shaded areas, the motion modes of the mechanism with impacts by each PS are possible.

It is clearly seen from Fig. 12.3 that with increasing ϕ, there is an increase in regions of motion modes with impacts by each PS.

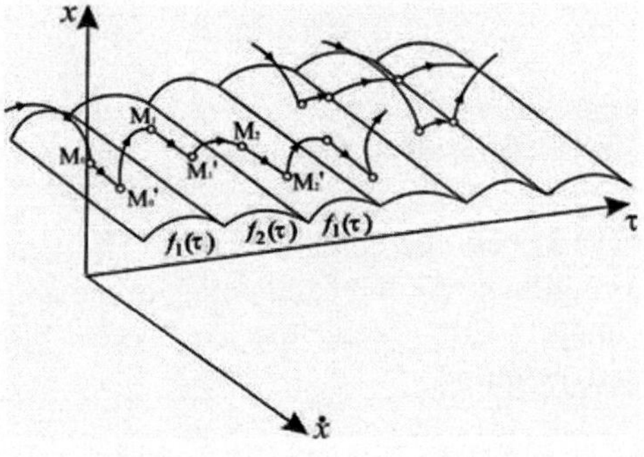

Fig. 12.2 Qualitative view of the phase space

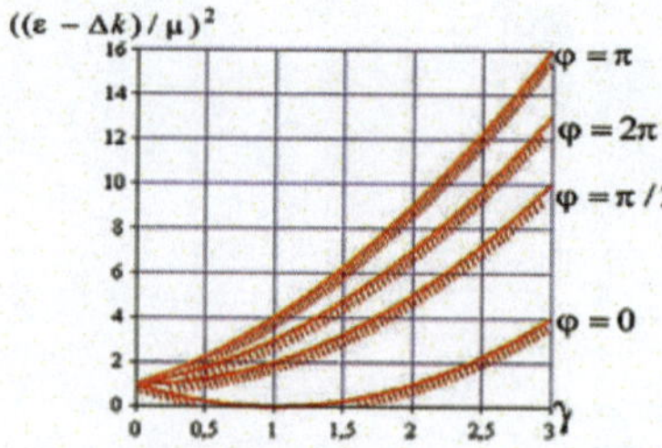

Fig. 12.3 Boundary regions (in the direction of shading) of motion modes with alternate impacts of each PS

12.4 Investigation of Nonlinear Dynamics of the Mechanism Using the Method of Point Transformations.

The dynamics of the mechanism, described by Eqs. (12.3) and (12.4), was studied by the numerical–analytical method using the point transformation method (Neimark 2010) of the surface S into itself.

Let us suppose:
$M_0(\tau = \tau_0, x_0 = f_1(\tau_0), \dot{x} = \dot{x}_0) \in S_1(x = f_1(\tau))$,
$M_1(\tau = \tau_1, x_1 = f_2(\tau_1), \dot{x} = \dot{x}_1) \in S_2(x = f_2(\tau))$,
$M_2(\tau = \tau_2, x = f_1(\tau_2), \dot{x} = \dot{x}_2) \in S_1$, then using (12.3) and (12.4), point transformation $T = T_2 T_1$ points $M_0 \overset{T_1}{\rightarrow} M_1 \overset{T_2}{\rightarrow} M_2$ can be written in the following form:

$$\begin{cases} \Delta k - \mu\gamma\cos(\tau_1 - \phi) = -p\frac{(\tau_1-\tau_0)^2}{2} + \dot{x}_0(\tau_1 - \tau_0) + \varepsilon - \mu\cos\tau_0, \\ \dot{x}_1 = R(p(\tau_1 - \tau_0) - \dot{x}_0) + (1+R)\mu\gamma\sin(\tau_1 - \phi), \end{cases}$$
$$\begin{cases} \varepsilon - \mu\cos\tau_2 = -p\frac{(\tau_2-\tau_1)^2}{2} + \dot{x}_1(\tau_2 - \tau_1) + \Delta k - \mu\gamma\cos(\tau_1 - \phi), \\ \dot{x}_2 = R(p(\tau_2 - \tau_1) - \dot{x}_1) + (1+R)\mu\sin\tau_2. \end{cases} \quad (12.5)$$

By adding to Eq. (12.5) periodic conditions $\dot{x}_2 = \dot{x}_0 = \dot{x}, \ \tau_2 = \tau_0 + 2\pi n$, we can determine the coordinates of the fixed points of the point transformation T, corresponding to two-impact (with successive strokes by each piston-striker) periodic motions (main mode) in the form:

$$\dot{x}_0 = \frac{Rp(2\pi n - \xi(1+R)) + (1+R)\mu(\sin\tau_0 - R\gamma\sin(\tau_0 + \alpha))}{1 - R^2},$$
$$\dot{x}_1 = \frac{Rp(\xi - R(2\pi n - \xi)) + (1+R)\mu(\gamma\sin(\tau_0 + \alpha) - R\sin\tau_0)}{1 - R^2}, \quad (12.6)$$

$$\mu a\cos\tau_0 - \mu b\sin\tau_0 = A, \quad -\mu c\cos\tau_0 + \mu d\sin\tau_0 = -B,$$

where

$$a = (1 - R)(1 - \gamma \cos \alpha) + R\gamma\xi \sin \alpha,$$
$$b = \xi(1 - R\gamma \cos \alpha) - (1 - R)\gamma \sin \alpha,$$
$$c = (1 - R)(1 - \gamma \cos \alpha) + (2\pi n - \xi)\gamma \sin \alpha,$$
$$d = (2\pi n - \xi)(R - \gamma \cos \alpha) - (1 - R)\gamma \sin \alpha,$$

$$A = (1 - R)(\varepsilon - \Delta k) + \frac{p\xi}{(1 + R)}\left(2\pi nR - \frac{\xi}{2}(1 + R)^2\right),$$
$$B = (1 - R)(\varepsilon - \Delta k) + \frac{p(2\pi n - \xi)}{2(1 + R)}\big((2\pi n - \xi)(1 + R^2) - 2R\xi\big),$$

$$\xi = \tau_1 - \tau_0, \ \alpha = \xi - \varphi,$$

τ_0, τ_1, are determined by the last two equations of the system (12.6), while post-impact velocities $\dot{x}_0.\dot{x}_1$ are defined by the first two.

The stability in the small of the found fixed points, corresponding to the periodic modes of motion of the mechanism with alternate impacts by each PS, is determined in accordance with the roots of the characteristic equation $\chi(z) = 0$ (Feigin 1994), which after the linearization of the point transformation Eq. (12.5), is written as:

$$\chi(z) = b_{11}b_{22}z^2 + (a_{11}b_{22} + a_{22}b_{11} - a_{21}b_{12})z + a_{11}a_{22} - a_{12}a_{21}. \qquad (12.7)$$

In (12.7), the following definitions are introduced

$$a_{11} = R(1 + R)\mu\gamma \cos(\tau_0 + \alpha) + Rp + \frac{R^2}{\xi}(\dot{x} - \mu\gamma \sin(\tau_0 + \alpha)),$$
$$a_{12} = \frac{R^2}{\xi}(\mu \sin \tau_0 - \dot{x}),$$

$$a_{21} = (2\pi n - \xi)\frac{R}{\xi}(\dot{x} - \mu\gamma \sin(\tau_0 + \alpha))$$
$$+ (2\pi n - \xi)(1 + R)\mu\gamma \cos(\tau_0 + \alpha)$$
$$+ (2\pi n - \xi)p - \dot{x}_1 + \mu\gamma \sin(\tau_0 + \alpha),$$
$$a_{22} = (2\pi n - \xi)\frac{R}{\xi}(\mu \sin \tau_0 - \dot{x}), \quad b_{11} = \frac{1}{\xi}(p\xi - \dot{x} + \mu\gamma \sin(\tau_0 + \alpha)),$$
$$b_{12} = \frac{1}{\xi}(\dot{x} - p\xi - \mu \sin \tau_0) - (Rp + (1 + R)\mu \cos \tau_0),$$
$$b_{22} = \dot{x}_1 - (2\pi n - \xi)p - \mu \sin \tau_0.$$

The periodic solution is stable if all the roots (12.7) lie within the unit circle; i.e., the inequalities $|z_{1,2}| < 1$ are satisfied. The boundaries of the stability region of the considered periodic motion, denoted by N_+, N_-, N_ϕ, satisfy the equations (Feigin 1994):

$$\chi(1) = 0,\ \chi(-1) = 0,\ \chi(e^{\pm j\phi}) = 0,\ 0 \le \phi \le \pi.$$

Besides, it is known (Feigin 1994) that with a continuous change in parameters, the periodic motion mode disappears either due to the loss of stability conditions or to the exit of the phase trajectory from the region of the corresponding point transformation (bifurcation surface N_C). Thus, the existence and stability regions of periodic motions of the system under consideration are limited by the surfaces N_+, N_-, N_ϕ and N_C.

The boundaries of the stability regions N_+, N_-, N_ϕ have the form:

$$\begin{aligned}
N_+ :&\quad b_{11}b_{22} + (a_{11}b_{22} + a_{22}b_{11} - a_{21}b_{12}) + (a_{11}a_{22} - a_{12}a_{21}) = 0,\\
N_- :&\quad b_{11}b_{22} - (a_{11}b_{22} + a_{22}b_{11} - a_{21}b_{12}) + (a_{11}a_{22} - a_{12}a_{21}) = 0,\\
N_\phi :&\quad b_{11}b_{22} - (a_{11}a_{22} - a_{12}a_{21}) = 0.
\end{aligned}$$

The bifurcation boundary N_C was calculated numerically as follows: The values of system parameters $\varepsilon, \mu, \gamma, \phi, p, R, k, \Delta k$ are being selected, which belong to the stability region, limited by N_+, N_-, N_ϕ surfaces;

- coordinates of stable fixed point $\tau_0, \tau_1, \dot{x}, \dot{x}_1$, corresponding two-stroke double-piston periodic motion mode, are determined from the relation (12.6);
- using the coordinates of the fixed point as initial conditions to find the solution of the system (12.3)–(12.4) on time intervals $[\tau_0, \tau_1)$ and $[\tau_1, \tau_2)$, the conditions

$$1.\ x(\tau) > f(\tau), \tau \ne \tau_0, \tau \ne \tau_1, \tau \ne \tau_2, \tag{12.8}$$

$$2.\ x(\tau) = f(\tau), \tau = \tau_0, \tau = \tau_1, \tau = \tau_2$$

are verified at every instant $\tau \in [\tau_0, \tau_2]$.

When the conditions (12.8) are satisfied, this means that the selected values of the parameters belong to the regions of existence and stability of periodic motion mode of alternate impact interaction with anvil by each PS. If, at some instant, the conditions (12.8) are not satisfied, then the loss of stability of periodic motions with alternate impacts by each PS occurs.

12.5 The Numerical Study of the Dynamics of the Mechanism

The complex dynamics of the two-piston mechanism was studied using numerical–analytical computations with the help of a software complex developed in the Borland Developer Studio 2006 (Arkhangelskiy 2002; Kernighan and Ritchie 2016; Stroustrup 2011). This complex allows us to compute phase trajectories, bifurcation diagrams and regions of existence and stability of the main periodic motion modes of the mechanism.

12.5.1 *The Region of Existence and Stability of Periodic Motions*

Let us denote the region of existence and stability of periodic motions in the parameter space through $D(m_1, m_2)$, where m_1 is the number of strokes by the first piston against the rod,m_2 is the number of strokes by the second piston against the rod.

Figure 12.4 depicts in the plane (p, R) the region $D(1,1)$ (shaded) for parameters $\gamma = 4, \phi = 0.52, \mu = 0.1, k = 0, \Delta k = 0$ and two values ε (relative difference of distances from the connecting points of rods to the PS bases). In Fig. 12.4a $\varepsilon = 0.02$, in Fig. 12.4b $\varepsilon = 0.15$.

The bifurcation boundary N_c cuts off the part $D(m_1, m_2)$(not shaded) from the stability region, formed by the surfaces N_+, N_-, where m_1 and m_2 are not equal to 1 at the same time. It should be noted that in the region $D(1,1)$ for one set of parameters, there always exist two fixed points, the first one being stable, whereas the second one unstable.

Comparing Fig. 12.5a, b, it can be seen that an increase in parameter ε leads to a significant decrease in the size of the existence and stability region of the periodic

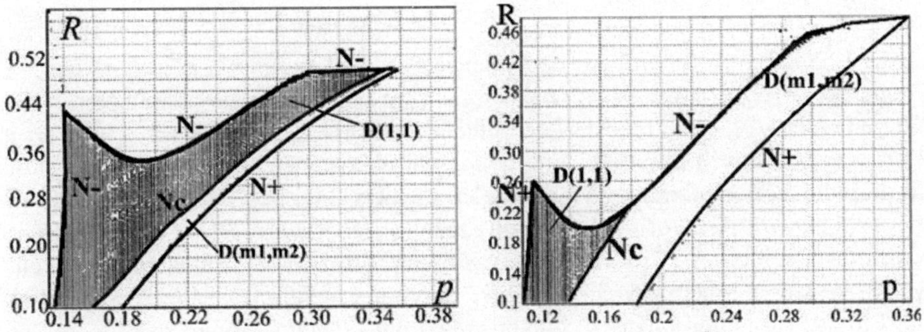

Fig. 12.4 Regions $D(1,1)$ (shaded) of existence of stable periodic motions of alternate PS impacts $\varepsilon = 0.02$ (**a**), $\varepsilon = 0.15$(**b**)

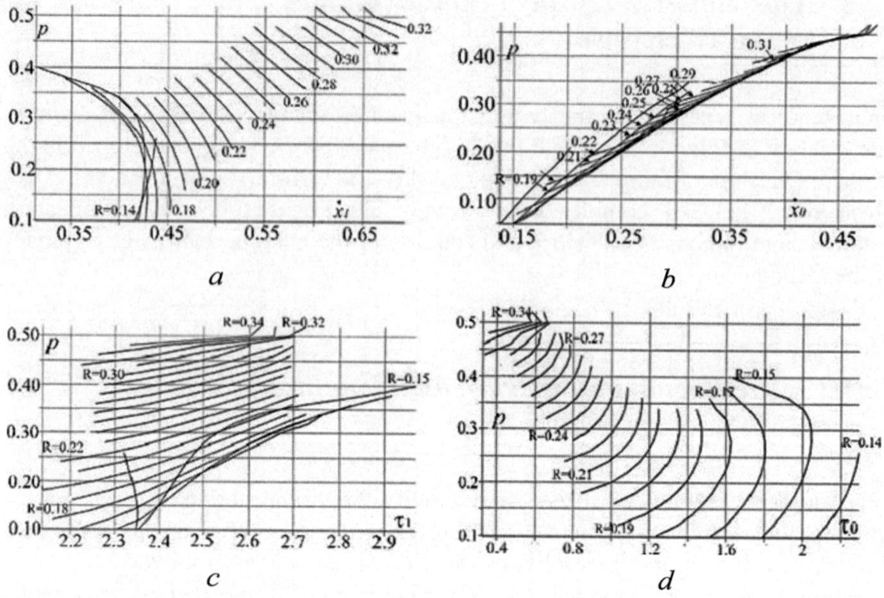

Fig. 12.5 Coordinate dependencies of fixed points $6a \rightarrow \dot{x}_1^+(p)$, $6b \rightarrow \dot{x}_0^+(p)$, $6c \rightarrow \tau_1^*(p)$, $6d \rightarrow \tau_0^*(p)$ on frequency parameter p for different values of the velocity recovery coefficient R

motion modes of the mechanism and to the displacement of the region itself in the direction of decreasing the frequency parameter p. Hence, it is better to connect the rods to PS at equal distances from PS bases.

12.5.2 Coordinate Dependencies of Fixed Points on Frequency Parameter

Figure 12.5a–d shows coordinate dependencies of fixed points $\dot{x}_1^+$, $\dot{x}_0^+$, τ_1^*, τ_0^* ($6a \rightarrow \dot{x}_1^+(p)$, $6b \rightarrow \dot{x}_0^+(p)$, $6c \rightarrow \tau_1^*(p)$, $6d \rightarrow \tau_0^*(p)$, on frequency parameter p for different values of speed recovery coefficient R for a set of parameters $\gamma = 4$, $\varphi = 0.52$, $\mu = 0.1$, $k = 0$, $\Delta k = 0$ from the region of existence and stability $D(1.1)$ of the periodic motion modes with alternate PS impacts (Fig. 12.4). It can be seen from Fig. 12.5 that with increasing the velocity recovery coefficient R, the coordinates of post-impact velocities $\dot{x}_1^+$, $\dot{x}_0^+$ increase, while impact times τ_1^*, τ_0^* decrease. Herein with increasing frequency parameter p, $\dot{x}_1^+$ decreases, herein $\dot{x}_0^+$, τ_1^* increase, and τ_0^* starts to decrease after having reached its maximum. The maximum value increases with decreasing the velocity recovery coefficient R. The above dependencies can be successfully used for tuning the parameters of the mechanism for a particular motion mode.

12.5.3 Bifurcation Diagrams

Figures 12.7 and 12.8 present bifurcation diagrams plotted for the same set of parameters as in Fig. 12.4, but at $R = 0.22$ and for two parameter values ε, in which the values of the frequency parameter p are given on x-axis, while the values of post-impact velocities $\dot{x}^{+}$ are given on the y-axis. Figures 12.7a and 12.8a correspond to the diagram with impacts by the first piston, while Figs. 12.7b and 12.8b to the diagram with impacts by the second piston.

It is clearly seen from Figs. 12.7 and 12.8 that with increasing ε, periodic stable motion modes with alternate PS impacts are observed at lower values of the frequency parameter. Thus, the main periodic motion mode is observed at $0.145 \leq p \leq 0.18$ Fig. 12.6, while in Fig. 12.7 at $0.121 \leq p \leq 0.13$.

Figure 12.8a, b presents the region $D(1,1)$ in the plane (p, R) for $\varepsilon = 0.02$, $\gamma = 4$, $\mu = 0.1$, $k = 0$, $\Delta k = 0$ and different angle values ϕ : $\phi = 0.3$, Fig. 12.8b, $\phi = 3$ Fig. 12.8b.

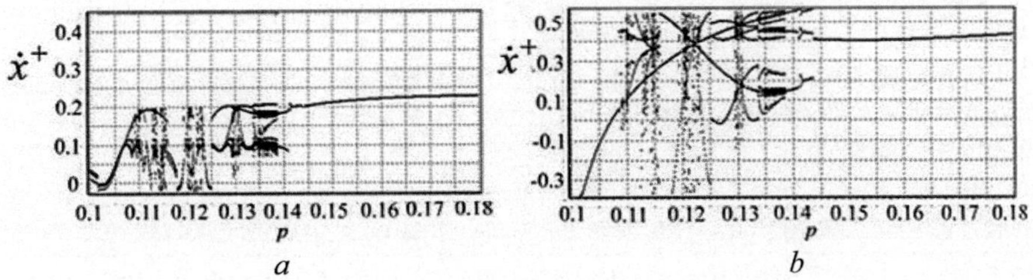

Fig. 12.6 Bifurcation diagrams with frequency parameter p at $R = 0.22$, $\varepsilon = 0.02$, $\gamma = 4$, $\phi = 0.52$, $\mu = 0.1$, $k = 0$, $\Delta k = 0$

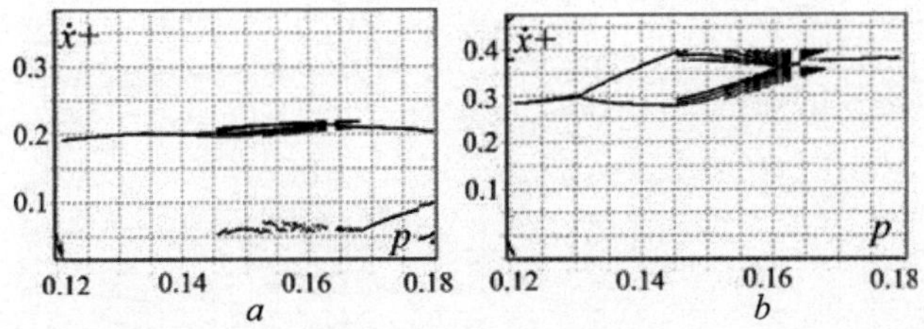

Fig. 12.7 Bifurcation diagrams with frequency parameter p at $R = 0.22$, $\varepsilon = 0.15$, $\gamma = 4$, $\phi = 0.52$, $\mu = 0.1$, $k = 0$, $\Delta k = 0$

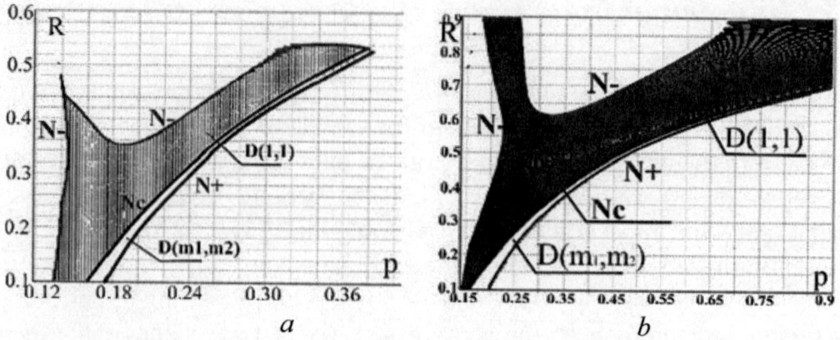

Fig. 12.8 Regions $D(1,1)$ of existence and stability of periodic motions (shaded) with alternate PS impacts a ($\phi = 0.3,$), $b(\phi = 3)$

12.5.4 The Analysis of the Diagrams and Stability Regions

It is seen from Fig. 12.8 that with increasing eccentricity angle, the region of stability of periodic motions with alternate PS impacts increases.

Figures 12.10 and 12.11 show the bifurcation diagrams with frequency parameter p for the same set of parameters as in Fig. 12.8a, b respectively, but at $R = 0.3$ for different values φ.

It is seen from Fig. 12.9 that the main mode ($m_1 = 1, m_2 = 1$) is observed in the range of $0.15 \le p \le 0.215$, while in the range of $0.12 \le p \le 0.13$, a chaotic mode occurs.

Figure 12.10 shows that at $0.18 \le p \le 0.24$ there exists the main mode $m_1 = 1, m_2 = 1$, and at $0.1 \le p \le 0.12$ and $0.122 \le p \le 0.16$, a chaotic mode is observed.

The analysis of the presented stability regions and bifurcation diagrams proved that an increase in ϕ (phase shift between eccentricities) leads to an increase in the regions of existence of stability of periodic motion modes; i.e., the range of frequency parameter value p for which periodic motions are known to exist becomes wider.

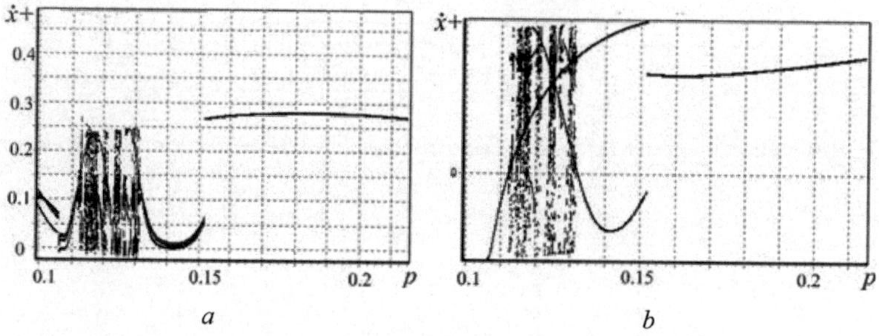

Fig. 12.9 Bifurcation diagrams with frequency parameter p for $\phi = 0.3$

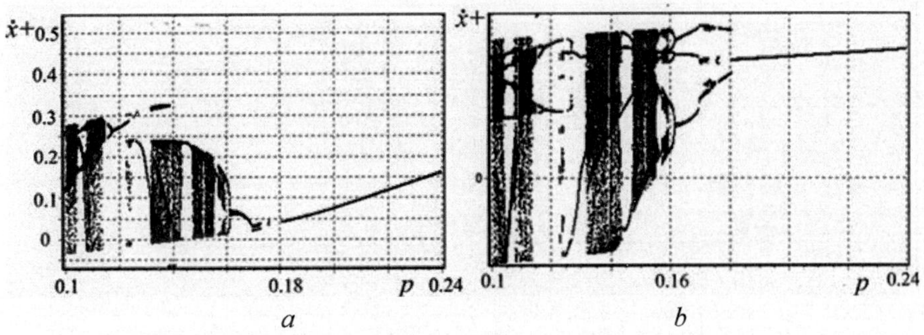

Fig. 12.10 Bifurcation diagrams with frequency parameter p for $\phi = 3$

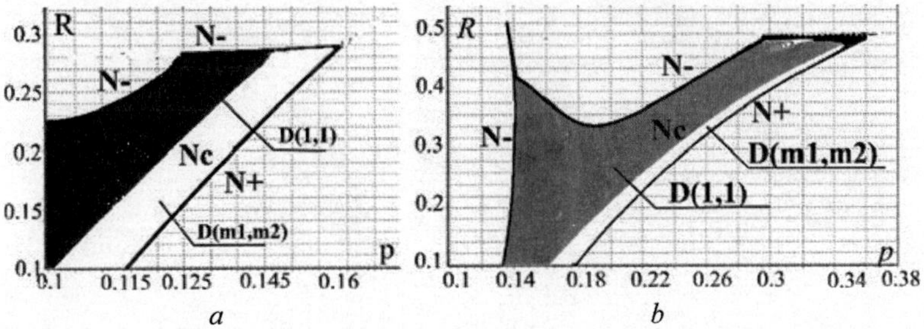

Fig. 12.11 Region $D(1,1)$ of existence and stability of periodic motions with alternate PS impacts $a(\gamma = 2.5)$, $b(\gamma = 4)$

Figure 12.11 depicts the region $D(1,1)$ in the plane (p, R) with a set of parameters $\varepsilon = 0.02, \gamma = 2.5, \phi = 0.52, \mu = 0.1, k = 0, \Delta k = 0$, and different values $\gamma = 2.5$(Fig. 12.11a) $\gamma = 4$(Fig. 12.11b).

Figures 12.13 and 12.14 present bifurcation diagrams for $R = 0.25$, plotted with the same set of parameters as in Fig. 12.11.

Figure 12.12 shows that the main mode with $m_1 = 1, m_2 = 1$ exists at $0.116 \leq p \leq 0.125$; at $0.112 \leq p \leq 0.116 m_1 = 2, m_2 = 2$, $p = 0.116$ is bifurcation

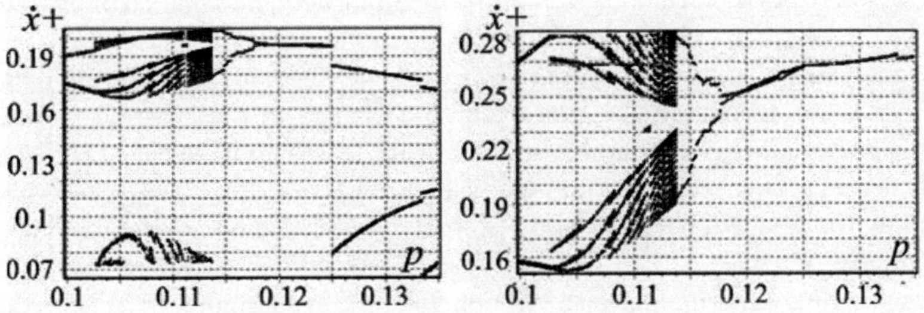

Fig. 12.12 Bifurcation diagrams with frequency parameter p for $\gamma = 2.5$

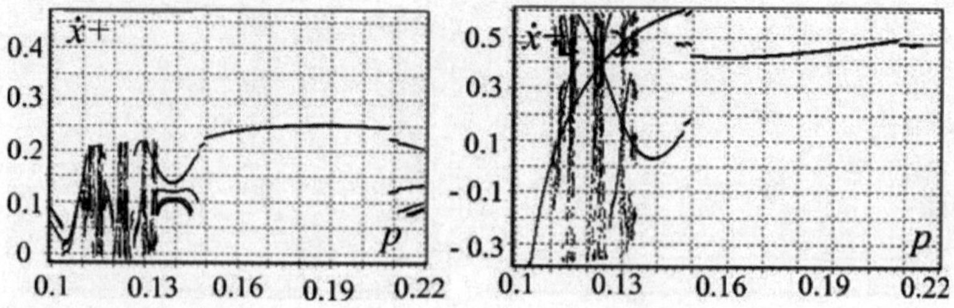

Fig. 12.13 Bifurcation diagrams with frequency parameter p for $\gamma = 4$

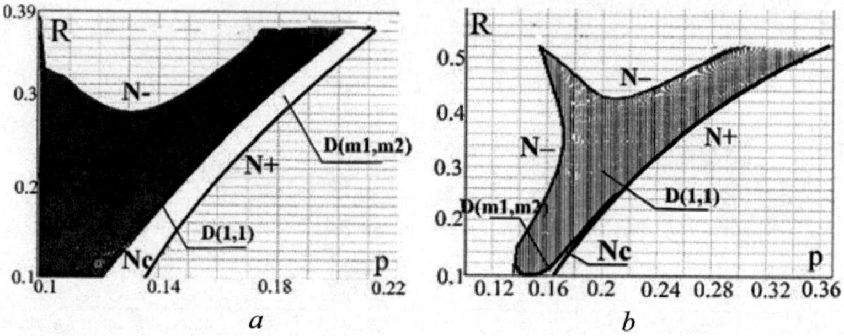

Fig. 12.14 The existence and stability region $D(1,1)$ of periodic motions with alternate PS impacts $a(k_1 = 0, \Delta k = 0)$, $b(\ k_1 = 0.1, \Delta k = 0.1)$

parameter value; at $0.104 \leq p \leq 0.112$ a chaotic mode is observed; at $0.1 \leq p \leq 0.104 m_1 = 2, m_2 = 2$.

Figure 12.13 shows that the main mode is observed at $0.15 \leq p \leq 0.21$, $0.13 \leq p \leq 0.15 m_1 = 3, m_2 = 2$, $0.21 \leq p \leq 0.22 m_1 = 4, m_2 = 1$, while at $0.1 \leq p \leq 0.13$, a chaotic mode exists.

Figure 12.14 presents the stability region in the plane with parameters (p, R), where $\mu = 0.1$, $\varepsilon = 0.02, \gamma = 4$, $\phi = 0, 52$, $\lambda_1 = 0$, $\lambda_2 = 0$, while in Fig. 12.14a $k_1 = 0, \Delta k = 0$(without accounting for anvil heights), in Fig. 12.14b $k_1 = 0.1, \Delta k = 0.1$(with accounting for anvil heights).

Figures 12.16a, b present the bifurcation diagrams with the same set of parameters, as in Fig. 12.14a, b but for $R = 0.25$.

The analysis of the diagrams and stability regions showed that if the anvil heights $k_1, k_1 + \Delta k$ are accounted for, this leads to a change in the configuration of the existence region of periodic motion modes. Thus, it is seen from Fig. 12.15 that the periodic mode for the selected parameter values exists at $0.15 \leq p \leq 0.205$ (without accounting for the anvil heights), and Fig. 12.16 shows that the periodic mode exists at $0.16 \leq p \leq 0.195$.

The numerical experiments led to the conclusion that parameters $\varepsilon, \phi, k_1, \Delta k$ have the most significant effect on the dynamics of the mechanism.

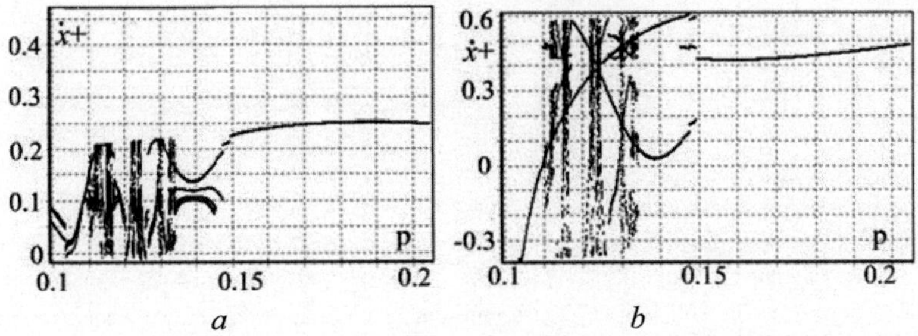

Fig. 12.15 Bifurcation diagrams with frequency parameter p for $k = 0$, $\Delta k = 0$

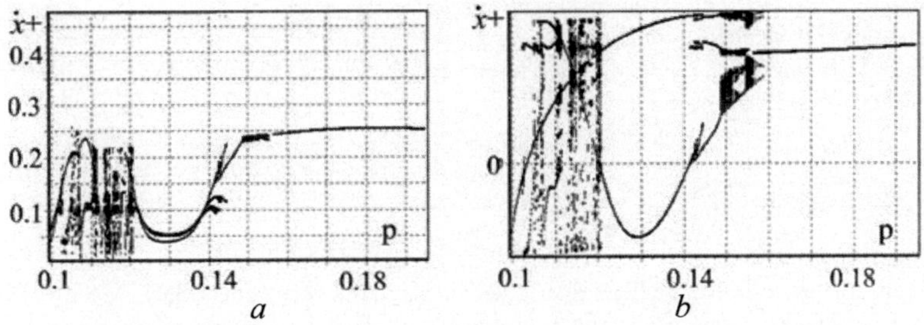

Fig. 12.16 Bifurcation diagrams with frequency parameter p for $k = 0.1$, $\Delta k = 0.1$

12.6 Conclusion

The main results of the present work are as follows:

- a new mathematical model of a two-piston vibro-impact mechanism with a crank vibration exciter is presented;
- analytical relations for the mechanism parameters are given, with the help of which one can indicate regions where mechanism motions are possible after impacts either by one PS or alternately by two PSs;
- bifurcation diagrams for all parameters of the mechanism allowed us to identify existence regions of various motion modes, including chaotic ones;
- two periodic motion modes are proved to exist in the stability regions, formed by surfaces N_+, N_-, N_ϕ and N_C with one set of parameters, one periodic motion being always stable, the other unstable;
- it has been numerically computed, that the parameters ε (distance from the connecting point of the rod to the lower base of PS), ϕ (phase shift between eccentricities) and the height of the anvils $k_1, k_1 + \Delta k$ are the most significant parameters that affect the dynamics of the mechanism.

Acknowledgements This work was supported by a grant of the Russian Science Foundation (16-19-10237-P).

References

Alibert J.J., Seppecher P., dell'Isola F. Truss modular beams with deformation energy depending on higher displacement gradients. Mathematics and Mechanics of Solids, 2003, 8(1)

Arkhangelskiy A. Ya. C++Builder 6. Spravochnoe posobiye. M.:Binom-Press, 2002. – p.544 (in Russian).

Auffray N., dell'Isola F., Eremeyev V., Madeo A., Rossi G. Analytical continuum mechanics à la Hamilton-Piola: least action principle for second gradient continua and capillary fluids. Mathematics and Mechanics of Solids, 2013

Babitskiy V. I., Krupenin V. L. Oscillations in Strongly Nonlinear Systems. Science, 1985, p. 320, (in Russian).

Babitsky, V. I. (1998). *Theory of Vibro-Impact Systems.* New York: Springer. ((in Russian)).

Babitsky, V. I. Theory of Vibro-impact Systems and Applications, Springer Science & Business Media, 2013.

Bakhvalov N.S., Zhidkov N.P., Kobelkov G.M. Numerical methods. Binom. Laboratoriya znaniy, 2003 – p. 640 (in Russian).

Barchiesi, E., Spagnuolo, M., & Placidi, L. (2018). Mechanical metamaterials: A state of the art. *Mathematics and Mechanics of Solids.*

Bernardini D., & Litak, G. An overview of 0–1 test for chaos, Journal of the Brazilian Society of Mechanical Sciences and Engineering, vol. 38, no. 5. 2016, p. 1433–1450.

Biderman V. L. The Applied Theory of Mechanical Oscillations. High School, 1972, p. 416, (in Russian).

Bogodukhov S. I., Grebenyuk V. F., Proskurin A. D. Processing of strengthened surfaces in mechanical engineering a maintenance. Mechanical Engineering, 2005, p. 256, (in Russian).

Cveticanin, L. (2002). The motion of a two-mass system with non-linear connection. *Journal of Sound and Vibration, 252*(2), 361–369.

Del Vescovo, D., & Giorgio, I. (2014). Dynamic problems for metamaterials: Review of existing models and ideas for further research. *International Journal of Engineering Science, 80,* 153–172.

dell'Isola F., Andreaus U, Placidi L. At the origins and in the vanguard of peridynamics, non-local and higher-gradient continuum mechanics: An underestimated and still topical contribution of Gabrio Piola. Mathematics and Mechanics of Solids, 2015, 20(8)

dell'Isola F., Cuomo M., Greco L., Della Corte A. Bias extension test for pantographic sheets: numerical simulations based on second gradient shear energies. Journal of Engineering Mathematics, 2017

dell'Isola, F., Della, C. A., & Giorgio, I. (2016). *Higher-gradient continua: The legacy of Piola, Mindlin.* Mathematics and Mechanics of Solids: Sedov and Toupin and some future research perspectives.

dell'Isola F., Della Corte A., Greco L., Luongo A. Plane bias extension test for a continuum with two inextensible families of fibers: a variational treatment with Lagrange multipliers and a perturbation solution. International Journal of Solids and Structures, 2016

dell'Isola F., Giorgio I., Pawlikowski M., Rizzi N. Large deformations of planar extensible beams and pantographic lattices: heuristic homogenization, experimental and numerical examples of equilibrium. Proceedings of The Royal Society A, 2016, 472(2185)

dell'Isola, F., Seppecher, P., & Alibert, J. J. (2019). Pantographic metamaterials: An example of mathematically driven design and of its technological challenges. *Continuum Mechanics and Thermodynamics, 31*(4), 851–884.

dell'Isola F., Seppecher P., Madeo A. How contact interactions may depend on the shape of Cauchy cuts in Nth gradient continua: approach "à la D'Alembert". Zeitschrift für angewandte Mathematik und Physik, 2012, 63(6)
Feigin M.I. Forced oscillations of systems with discontinuous nonlinearities. M.:Nauka, 1994. –p. 288 (in Russian).
Giorgio I. Numerical identification procedure between a micro-Cauchy model and a macro-second gradient model for planar pantographic structures. Zeitschrift für angewandte Mathematik und Physik, 2016
Goebel, R., Sanfelice, R. G., & Teel, A. R. (2008). Invariance principles for switching systems via hybrid systems techniques. *Systems and Control Letters, 57,* 980–986.
Igumnov, L. A., Metrikin, V. S., & Nikiforova, I. V. (2017). The dynamics of eccentric vibration mechanism (Part 1). *JVE Journal of Vibroengineering, 19,* 4816–5656.
Ing, J., Pavlovskaia, E. E., Wiercigroch, M., & Banerjee, S. (2010). Bifurcation Analysis of Impact Oscillator with One-Sided Elastic Constraint. *Physica D: Nonlinear Phenomena, 239*(3), 12–321.
Kernighan B.W., Ritchie D.M. The C programming language. Wilyams, 2016. – p.288.
Kobrinskiy A. A., Kobrinskiy A. E.Vibro-Impact Systems. Science, 1973, p. 592, (in Russian).
Leine, R. I., & Heimsch, T. F. (2012). Global uniform symptotic attractive stability of the non-autonomous bouning ball system. *Journal Physica, 241,* 2029–2041.
Liu, Y., Pavlovskaia, E. E., Wiercigroch, M., & Peng, Z. K. (2015). Forward and Backward Motion Control of a Vibro-Impact Capsule System. *Int. J. Non-LinearMech., 70,* 30–46.
Luo, G. W., & Ma, X. H. L. (2008). Periodic – impact motions and bifurcations in dynamics of a plastic impact oscillator with a frictional slider. *Journal Mechanics a/Solids, 27,* 1088–1107.
Masri, S. F., & Caughey, T. K. (1966). On the stability of the impact damper. *Journal of Applied Mechanics, 33*(3), 586–592.
Nagaev R.F. Periodic modes of vibration motion. M.: Nauka, 1978. – p. 160, (in Russian).
Neimark Yu. I. The point mapping method in the theory of nonlinear oscillations. M.: LIBROKOM, 2010. – p. 472, (in Russian).
Pavloaskaia, E., & Wiercigroch, M. (2003). Periodic solution finder for an impact oscillator with a drift. *Journal of Sound and Vibration, 267*(4), 893–911.
Pavlovskaia E., Hendry D.C., WiercigrochM. Modelling of high frequency vibro-impact drilling, Int. J. Mech. Sci. 91. 2015, p. 110–119.
Placidi L., Andreaus U., Giorgio I., Identification of two-dimensional pantographic structure via a linear D4 orthotropic second gradient elastic model. Journal of Engineering Mathematics, 2017
Placidi L., Barchiesi E., Turco E., Rizzi N.L. A review on 2D models for the description of pantographic fabrics. Zeitschrift für angewandte Mathematik und Physik, 2016, 67(5)
Rahali Y., Giorgio I., Ganghoffer J.-F., dell'Isola F. Homogenization à la Piola produces second gradient continuum models for linear pantographic lattices. International Journal of Engineering Science, 2015, 97
Shilkov V. A., Savalyuk A. D., Metrikin V. S., Polyakov A. A., Shabardin A. K., Alyokhin A. I., OmenenkoiYaA.c. 1020479 USSR, A Vibro-ram, No. 3376593/29–33, 2005, (in Russian)
Stroustrup B. The C++ programming language. Binom, 2011. – p.1136.
Turco E., Dell'Iola F., Cazzani A., Rizzi N.L. Hencky-type discrete model for pantographic structures: numerical comparison with second gradient continuum models. Zeitschrift für angewandte Mathematik und Physik, 2016
Tusset, A. M., Janzen, F. C., Piccirillo, V., Rocha, R. T., Balthazar, J. M. and Litak, G. (2017). On nonlinear dynamics of a parametrically excited pendulum using both active control and passive rotational (MR) damper. *Journal of Vibration and Control, 24*(9), 1587–1599,
Vagapov, I. K., Ganiev, M. M., & Shinkarev, A. S. (2007). Forced vibrations of two nonlinearly connected solid waveguides under static load. *Journal of Sound and Vibration, 302,* 425–141. ((in Russian)).
Vagapov, I. K., Ganiyev, M. M., & Shinkaryov, A. S. (2008). Theoretically and experimentally investigating the dynamics of an ultrasonic vibro-impact system with an intermediate striker. *Mechanical Engineering, 5,* 3–24. ((in Russian)).

Zakrzhevskiy, M. V. (1980). *Oscillations of a Substantially Nonlinear Substantially-Nonlinear Mechanical Systems* (p. 190). Riga: Zinatne.

Zheleztsov N. A. The point transform method and the problem of forced oscillations of an oscillator with "combined" friction. PMM, Vol. 1949, Issue 13, 1, p. 3–40, (in sRussian).

Chapter 13
High Strain Rate Tension Experiments Features for Visco-Plastic Materials

Artem V. Basalin, Anatolii M. Bragov, Aleksandr Yu. Konstantinov, Andrey K. Lomunov, and Aleksandr V. Zhidkov

Abstract The paper considers distinguishing features of the experiment on visco-plastic materials subjected to high strain rate tension, namely non-uniformity of stress–strain state in the working part of a specimen due to the existence of fixing parts and plastic strain localization. The modification of the Kolsky method (or Split Hopkinson Pressure Bar method) is used as experimental technique. The main experimental setup configurations for testing specimens under high strain rate tension are reviewed. We present mathematical models used for assessment of stress components distributed in a neck. The numerical modeling of high rate tension of a visco-plastic axisymmetric specimen is performed, allowing the accuracy of above models to be estimated. The experimental–numerical procedure used to construct a true stress–strain curve on the basis of high rate tensile experiment is described.

13.1 Introduction

Tension experiments (including high strain rate tests) hold a special place in the system of basic experiments used for identification of material behavior. This kind of tests allows determining deformation diagrams as well as ultimate fracture characteristics, which are necessary for prediction of strength of the structures (Volkov et al. 2018).

Nowadays, fascinating research direction is the development of newly (scientifically) conceived materials ("metamaterials") with mechanical properties that cannot be found in nature (Barchiesi et al. 2018; Del Vescovo and Giorgio 2014; Placidi et al. 2016; dell'Isola et al. 2019). These (macroscopic) properties are mainly determined by the microstructure or nanostructure of the considered metamaterial rather than by the chemical and physical properties of the materials constituting it at the microscopic level. Designing of such metamaterials based on high gradient continuum

A. V. Basalin · A. M. Bragov (✉) · A. Yu. Konstantinov · A. K. Lomunov · A. V. Zhidkov
Research Institute for Mechanics, National Research Lobachevsky State University of Nizhny Novgorod, Gagarin ave. 23, 603950 Nizhny Novgorod, Russia
e-mail: bragov@mech.unn.ru

F. dell'Isola and L. Igumnov (eds.), *Dynamics, Strength of Materials and Durability in Multiscale Mechanics*, Advanced Structured Materials 137,
https://doi.org/10.1007/978-3-030-53755-5_13

models (Alibert et al. 2003; Sciarra et al. 2007; dell'Isola et al. 2012; Auffray et al. 2013; dell'Isola et al. 2015; Rahali et al. 2015; dell'Isola et al. 2016a, b, c). Tensile experiments (among the others) are widely used for validation of theoretical (and numerical) predictions (dell'Isola et al. 2016a, b, c; dell'Isola 2017; Giorgio 2016, Turco et al. 2016; Placidi et al. 2017).

However, tensile tests have a number of features, which have to be taken into account while planning an experiment and interpreting its results. A specially shaped specimen is usually used. Specimen should have a gauge area and areas for their fixing in experimental setup. Areas for fixing might affect stress and strain fields in gauge area. These effects can be minimized by increasing the length of gauge area. This method is employed for static experiments. In case of dynamic loading, there are some length limitations for gauge area due to the wave effects. Therefore, influence of fixing areas can be significant. Moreover, strain localization and necking process cause difficulties in obtaining true strength and strain characteristics of material.

This paper considers some distinguishing features of high strain rate tension experiments within the strain rate range from 500 to 5000 1/s.

13.2 Split Hopkinson bar Technique

One of the most widely used loading techniques at high strain rates is the split Hopkinson pressure bar (SHPB) or the Kolsky method developed by Kolsky (1949). There are some review works (Field et al. 1994; Bacon and Lataillade 2001; Gama et al. 2004) that describe a historical background of SHPB method. This method can be easily implemented and well justified and realized theoretically.

Davies (1948) and Kolsky (1949) proposed to use SHPB for compression tests almost at the same time. SHPB consists of two stiff long bars (incident bar *2* and transmission bar *5*) and a short soft specimen *4* sandwiched between them (Fig. 13.1). The striker *1* impacts the incident bar *2* and generates compressive pulse $\varepsilon^I(t)$, which propagates toward the specimen at the speed *C*. Upon reaching the interface between the incident bar and the specimen, a portion of the stress pulse $\varepsilon^T(t)$ travels through the specimen into transmission bar, while the remaining portion is reflected back into the incident bar as a tension pulse $\varepsilon^R(t)$. The ratio of transmitted and reflected

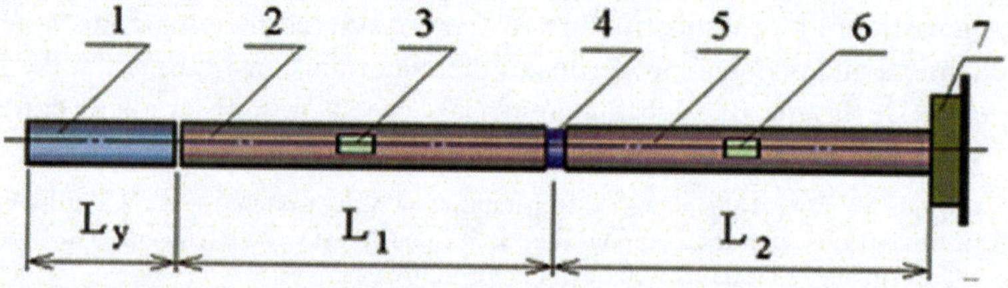

Fig. 13.1 SHPB installation configuration: 1—striker, 2—incident (or input) bar, 3, 6—strain gauges, 4—specimen, 5—transmission (or output) bar, 7—dumper

pulses is determined by material properties of a specimen. The specimen exhibits elastic–plastic deformation while the measuring bars elastic one. The data from the strain gauges *3* and *4* allows us to calculate time histories of strain rate, strain, and stress in the specimen by using the formulas proposed by Kolsky.

The main relations which associate incident $\varepsilon^I(t)$, transmitted $\varepsilon^T(t)$ and reflected $\varepsilon^R(t)$ pulses in measuring bars with strain rate, strain $\varepsilon_s(t)$, and stress $\sigma_s(t)$ in specimen are given below (Kolsky 1949):

$$\dot{\varepsilon}_s(t) = \frac{1}{L_0}\left(C^I\left(\varepsilon^I(t) - \varepsilon^R(t)\right) - C^T\varepsilon^T(t)\right)$$

$$\varepsilon_s(t) = \int\limits_0^t \dot{\varepsilon}_s(\tau)\mathrm{d}\tau$$

$$\sigma_s(t) = \frac{1}{A_0}E^T A^T \varepsilon^T(t)$$

where A_0, L_0 are the initial cross-section area and the length of the specimen, C^I, E^I, A^I, C^T, E^T, A^T are bar sound speeds, Young's modules, and cross-section areas of the incident (superscript *I*) and transmission (superscript *T*) bars, respectively.

By excluding time parameter from the relations $\sigma_s(t)$, $\varepsilon_s(t)$ and we obtain deformation diagram $\sigma_s \sim \varepsilon_s$ under known strain rate history $\dot{\varepsilon}_s \sim \varepsilon_s$. It should be noted that these values are technical (engineering) values as they do not account for the change in size of the specimen under loading. To calculate stresses in a specimen, the force is divided by the initial cross-section area of the specimen. The strain is calculated as ratio of the specimen shortening to its initial length. The stresses and strains have to be corrected to obtain the true stress–strain diagram at large strains. In case of homogeneity of deformation in working part of the specimen, the true stress is calculated as follows:

$$\sigma_S^T(t) = \sigma_S(t) \cdot (1 \mp \varepsilon_S(t))$$

The true (logarithmic) strain is determined by formula:

$$\varepsilon_S^T(t) = \ln(1 \mp \varepsilon_S(t))$$

The true strain rate is calculated as follows:

$$\dot{\varepsilon}_S^T(t) = \frac{\dot{\varepsilon}_S}{1 \mp \varepsilon_S(t)}$$

In above formulas, the «−» sign corresponds to compression, while the «+» sign corresponds to tension. Stresses and strains are considered to be positive both for compression and tension.

13.3 High Strain Rate Tension Based on Measuring Bars Techniques

Classical compressive SHPB method has good theoretical justification and it is fairly well explored. But some practically important characteristics of materials cannot be determined from this type of experiment, namely ultimate strength and ultimate plastic strain. Moreover, deformation diagram may be stress state dependent, that is the stress–strain curve obtained in tensile experiments may differ from compression one. Therefore, determination of deformation diagram in tension is an important independent problem.

Some schemes of high rate tensile testing which use measuring bars technique can be found in works (Hauser 1966; Eskandari and Nemes 2000; Lindholm and Yeakley 1968; Nicholas 1981; Caverzan et al. 2012; Jiang and Zhang 2006; Bragov et al. 2018). The main difference of these schemes is the way they generate tension pulse.

13.3.1 *Schemes for Dynamic Tension Experiments*

A number of schemes for dynamic tension experiments are proposed on the basis of measuring bar technique.

Hauser (1966) used the scheme of direct tension loading of a specimen through a special transmitting system as shown in Fig. 13.2. Tensile loading was generated by a striker and transmitted to an input bar using hollow cylinder, which houses the measuring bars system.

Later, similar scheme with the special transmitting bars system was realized by Eskandari and Nemes (2000) (Fig. 13.3).

In Lindholm's work (Lindholm and Yeakley 1968), tension in the working part of the specimen was obtained by using a specimen of special form (Fig. 13.4). The specimen was sandwiched between an input bar and an output tube. Compressive loading pulse was generated by the impact of the striker. The complex shape of the

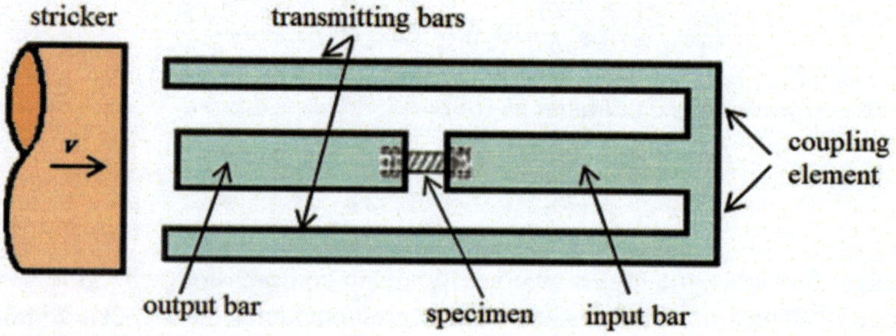

Fig. 13.2 Experimental setup (Hauser 1966)

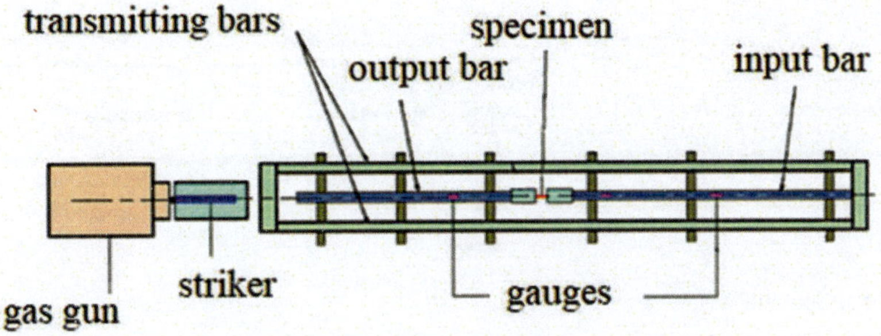

Fig. 13.3 Experimental setup (Eskandari and Nemes 2000)

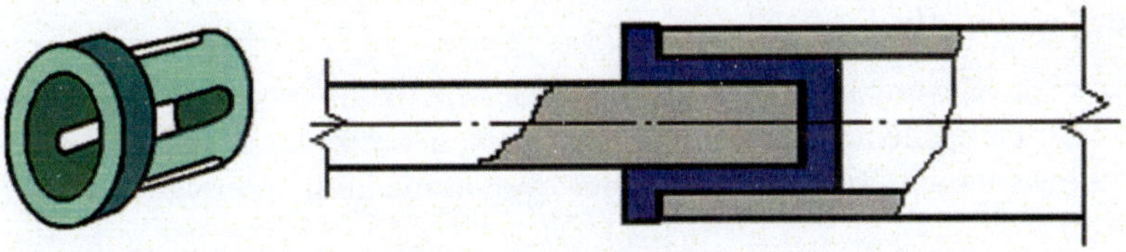

Fig. 13.4 Experimental setup (Lindholm and Yeakley 1968)

specimen limits the range of materials that can be tested using the described above scheme, which does not require any modification of traditional compressive SHPB to get a tensile experimental setup.

Nicholas proposed the scheme (Nicholas 1981), in which compressive pulse produced by the impact of the striker passes throw the specimen using a special hard surrounding ring (Fig. 13.5). The pulse turns into tensile loading pulse after reflection from a free end of the second bar. The main advantage of this scheme is the minimum modification of traditional compressive SHPB to get the tensile setup. The disadvantages are as follows:

- There is a small portion of compressive loading pulse reflected from interface «first bar - specimen» due to the differences in cross-section areas and material properties. This portion may interfere with useful signal that is used to calculate the force acting on specimen and make it more difficult to interpret registered signals

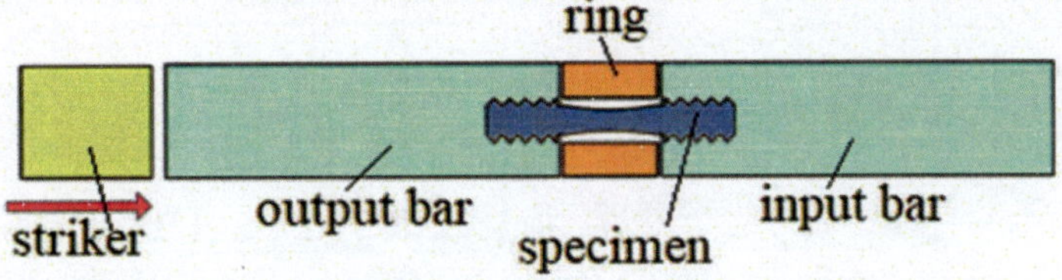

Fig. 13.5 Experimental setup (Nicholas 1981)

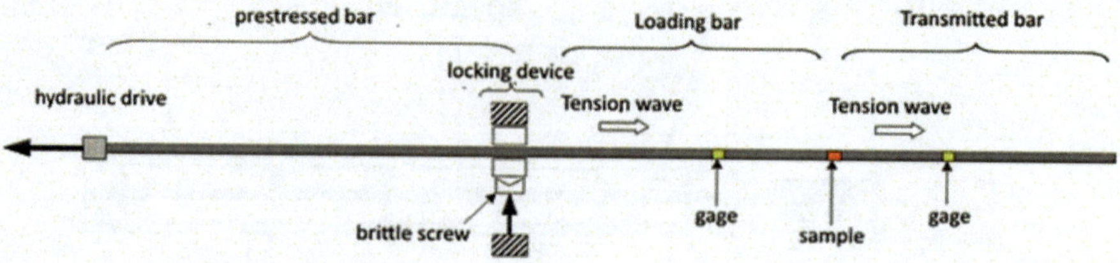

Fig. 13.6 Experimental setup (Caverzan et al. 2012)

- There can exist high frequency oscillations in the first measuring bar (used to measure the force acting on a specimen), remaining after compressive pulse due to dispersion effect.

The work (Caverzan et al. 2012) describes the setup based on prestressed bar (Fig. 13.6). This scheme makes it possible to obtain a very long tensile loading pulse which allows to achieve sufficient strains even at medium and relatively low strain rates.

A version of direct tensile impact loading using a rotating disk as a loading device (Fig. 13.7) is described in (Jiang and Zhang 2006).

The schemes with tube-strikers used for impact tensile loading are very popular of late. This class of setups is known as the Split Hopkinson Tension Bar (SHTB) technique. In (Harding 1992), such experimental device was used to investigate dynamic properties of composite materials. Similar loading schemes were employed in (Arthington et al. 2012; Chen et al. 2002; Gerlach 2008; Gerlach et al. 2012; Hasenpouth 2010; Huh et al. 2002; Noble and Harding 1994; Smerd et al. 2005; Taniguchi et al. 2007; Yokoyama 2003; Young 2015; Bragov et al. 2018).

Gerlach et al. (2012) proposed the novel construction of SHTB in which a tensile loading pulse is formed using U-shaped striker, accelerated by pneumatic cylinder (Fig. 13.8b). Disadvantages of tube-strikers are being discussed. It is evident that the use of tube-strikers gives no possibility for the measuring bar to be supported along striker displacement path (from gun to anvil). This leads to bending of the measuring bar and limits the length of the striker (loading pulse duration).

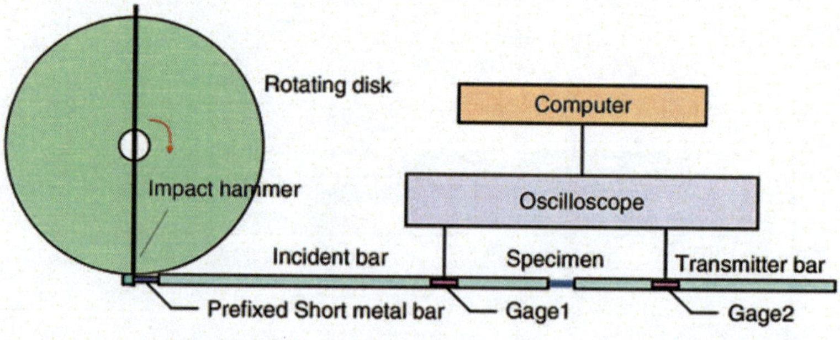

Fig. 13.7 Experimental setup (Jiang and Zhang 2006)

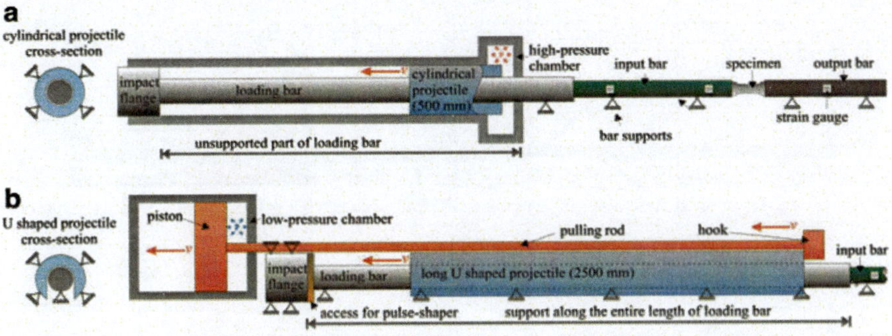

Fig. 13.8 Experimental setup (Gerlach et al. 2012)

Similar loading scheme with the tube striker sliding on rollers was used by Ganzenmuller et al. (2017). The striker was accelerated by pneumatic cylinder. The loading pulse duration amounted to 1.15 ms.

The disks made of plastic materials (copper, polymer) are placed between the tube striker and anvil to get different shapes of the loading pulse (Chen et al. 2002; Huang et al. 2010) as shown in Fig. 13.9A. Or sacrificial specimen is placed between loading and input bars (Guzman et al. 2011) as shown in Fig. 13.9B-1. The disadvantage of such approaches is the necessity for manufacturing specimens shapers for each test and the existence of the strain rate limit due to absorption of some portion of the loading pulse by pulse shapers. This complicates preparation for the experiment. Gerlach et al. (2011) proposed pulse shaping method at high rate tension with the help of special reusable insertion between loading bar and input bar as shown in Fig. 13.9B-2.

Dani (2017) analyzed the influence of a specimen size based on experimental results using the SHTB technique. It has been noted, that despite the wide use of this technique, there are no standards on specimen size. Numerical investigation of the effect of specimen sizes on obtained material characteristics was carried out using LS-DYNA. Firstly, at a fixed length of the specimen, its diameter was varied.

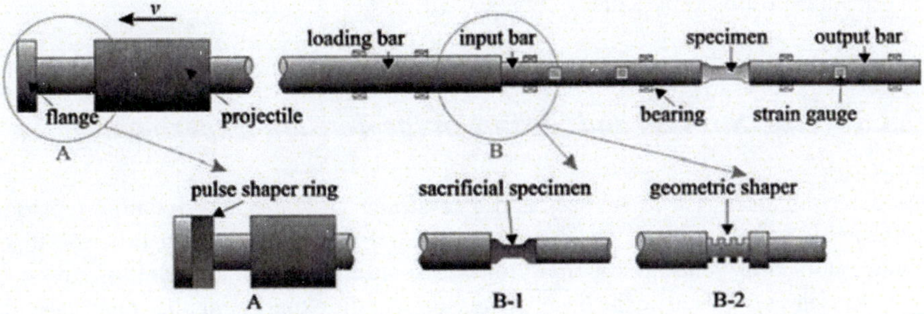

Fig. 13.9 Pulse shapers in direct high rate tension setups

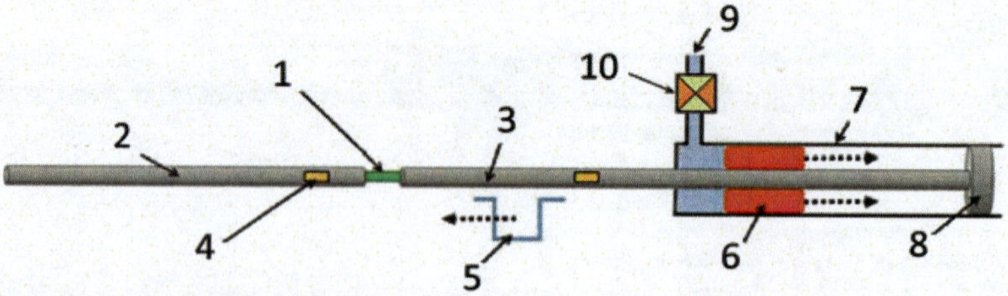

Fig. 13.10 Split Hopkinson tension bar

Secondly, at a fixed diameter of the specimen, its length was varied. The best result was obtained when L/D ratio was 0.75.

Nguyen et al. (2017) analyzed the influence of threated joint on distortion of signal in the Split Hopkinson Bar system during dynamic tension using numerical modeling. "Calibration sample" was used—the sample with the same material and diameter as the measuring bar. Ideally, loading pulse should pass through the specimen. In this study, thread diameter, thread form, flow stress of specimen material, and striker velocity were varied.

13.3.2 Experimental Setups

In the present work, the Split Hopkinson Tension Bar method (Bragov et al. 2018) was used to register the processes in the samples during high strain rate tension experiments. The sample *1* was fixed in measuring bars *2* and *3* using threaded joint. Recording of strain pulses in bars was carried out by strain gauges *4*. Loading tensile pulse *5* was formed by the impact of the tubular striker *6* on the anvil *8*. Acceleration of the striker *6* in the barrel *7* was due to the compressed air energy *9*, which was supplied when the pneumatic valve *10* was opened. The strain pulses in loading and output bars were used to obtain forces acting on the specimen during experiment and history of specimen interfaces displacements. These values were calculated using the formulas proposed by Kolsky (Fig. 13.10).

13.3.3 True Stresses and Strains in the Tension Experiments

The determination of true stresses and true strains in static and dynamic tension experiments is complicated by localization of deformation (necking process) that violates uniformity and homogeneity of stress–strain state. Calculation of stresses and strains in a specimen on the basis of measured in the experiment integral force and elongation become challenging under these conditions. In addition, strain localization affects strain rate history, which increases dramatically in the neck zone. These interpretation features are considered in works (Mirone 2013; Mirone et al. 2016). The appropriate correction of deformation diagram is required after the necking point (Bridgman 1955).

To evaluate effective true stresses, a number of methods have been proposed by (Alves and Jones 1999; Dietrich et al. 1970; Gromada et al. 2011; La Rosa et al. 2003; Ling 1996; Malinin and Rzysko 1981; Mirone 2004; Zhang et al. 1999; Bridgman 1955; Davidenkov and Spiridonova 1946).

The strain threshold of the initiation of localization can be determined by the following conditions:

$$\frac{d\sigma_0}{d\varepsilon} = 0 \text{ or } \frac{d\sigma}{d\varepsilon} = \frac{\sigma}{1+\varepsilon} \text{ or } \frac{d\sigma}{de} = \sigma \tag{13.1}$$

here σ, σ_0 are true and engineering stresses, respectively, e, ε are true (logarithmic) and engineering strains, respectively.

The Bridgman's assumption on uniformity of plastic strain in the neck section was experimentally confirmed by Aronofsky (1951). The assumption on similarity of longitudinal grid lines in the neck was confirmed by Davidenkov and Spiridonova (1946) The stress components distribution in the neck section proposed by Bridgman is in good agreement with Zhang's numerical analysis results (Zhang and Li 1994).

The main complexity of Bridgman's method (and similar approaches) is caused by the necessity of precise determination of minimum radius of the neck section and neck curvature. Another technique of deformation diagram correction, based on universal (material independent) polynomial, was proposed by Mirone (2004). The coefficients of the polynomial are derived by generalization of a large amount of experimental data. Mirone has showed (Mirone 2004, 2007; Mirone and Corallo 2010) that the proposed procedure is material independent and allows the proper determination of effective stress with account for strain localization. The effective true stress after necking prediction by Mirone's procedure exhibits an error of about 5% for quasi-static elasto-plasticity (Mirone 2013).

The high rate video registration was used in (Gilat et al. 2009; Kajberg and Wikman 2007; Li 2017; Noble 1997) to obtain the specimen minimum radius history.

In (Bazhenov et al. 2013; Ling 1996; Sasso et al. 2008; Markov 2018) numerical simulation was used to reconstruct the true deformation diagram based on tensile experimental data.

The procedure of true stress–strain calculation is considered in detail in the next section.

13.4 True Stress–Strain Determination from the Tension Experiments Data

13.4.1 *The Models for Estimation of Stress and Strains in Neck*

The integral forces acting on the specimen and the specimen gauge elongation are determined during the test. In the traditional method, an assumption of stress and strain fields uniformity is introduced to calculate stresses and strains based on forces and size changes. Therefore, it is only applicable till the moment of plastic strain localization (necking). This leads to losing a large part of stress–strain curve after the necking process.

The experimental and numerical analyses of specimen size effect on obtained stress–strain curves in impact tension experiments were carried out in (Konstantinov 2007). Figure 13.11 presents the comparison of the true curve (used in simulation to describe the properties of samples—black line) with diagrams constructed using Kolsky method for specimens with different gauge lengths. The virtual (numerical) experiments were used to obtain these data. The specimens with gauge lengths of 5, 10, and 15 mm and diameter of 5 mm were considered. It should be noted that the curve for 5 mm specimen differs from the true diagram even at the beginning (before necking) due to significant influence of specimen fixing parts.

Specimens with a longer working part allow more precisely to obtain true stress–strain curve up until localization process occurs. For the case under study, the threshold strain is about 15%. However, specimen fracture occurs at significantly

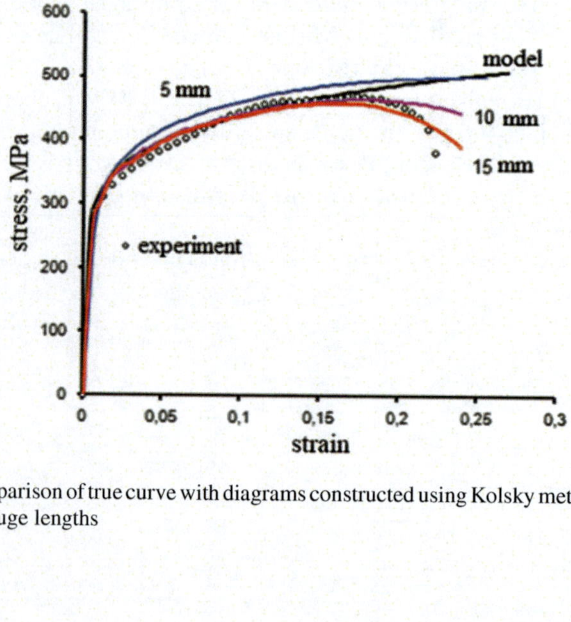

Fig. 13.11 Comparison of true curve with diagrams constructed using Kolsky method for specimens with different gauge lengths

greater plastic strain. The value of ultimate plastic strain equals to 38%. Therefore, the large portion of stress–strain curve (almost a half) is being lost.

The engineering strain and engineering stress can be calculated using strain pulses in measuring bars as follows:

$$\varepsilon_T(t) = \frac{1}{l_0}\int\limits_0^t \left[c_I\left(\varepsilon^I(\tau) - \varepsilon^R(\tau)\right) - c_T\varepsilon^T(\tau)\right]\mathrm{d}\tau \tag{13.2}$$

$$\sigma_T(t) = \frac{E_T \cdot S_T \cdot \varepsilon^T(t)}{\pi \cdot r_0^2} \tag{13.3}$$

here, l_0 and r_0 are initial gauge length and radius of working part of the specimen, ε^I, ε^R and ε^T are incident, reflected, and transmitted strain pulses registered in measuring bars, c_I and c_T are sound speeds of input and output bars, E_T and S_T are Young modulus and cross-section area of output bar.

The effective strain and stress prior strain localization can be found by formulas:

$$\sigma_{\mathrm{equ}}(t) = \sigma_T(t) \cdot (1 + \varepsilon_T(t)) \tag{13.4}$$

$$\varepsilon_{\mathrm{equ}}(t) = \ln(1 + \varepsilon_T(t)) \tag{13.5}$$

The deformation diagram calculated using formulas (13.4)–(13.5) does not reflect the real properties of material after necking (the fulfillment of conditions (13.1)). It is necessary to have the histories of neck radius $a(t)$ and neck curvature $R(t)$ to calculate the true strain and true stress after strain localization.

It has been shown (Konstantinov 2007) that maximum true strain in the specimen (which is located in neck) can be calculated by formula (13.6) for the whole tension process up until its failure.

$$\varepsilon_{\mathrm{true}} = \ln\left(\frac{S_0}{S}\right) = \ln\left(\frac{d_0^2}{d^2}\right) = 2 \cdot \ln\left(\frac{d_0}{d}\right) = 2 \cdot \ln\left(\frac{r_0}{a}\right) \tag{13.6}$$

The mean axial stress in neck section can be calculated as follows:

$$\sigma_{\mathrm{mean}} = \frac{F(t)}{\pi \cdot a^2(t)} \tag{13.7}$$

here, F is the integral force acting on a specimen.

It should be noted that there is a volumetric stress state in a neck. Thus, radial component of stress tensor exists. Therefore, formula (13.7) does not seem to be appropriate to estimate effective stress.

Bridgman's estimation of effective stress in neck with account for three-dimentionality of the stress is as follows (Bridgman 1955):

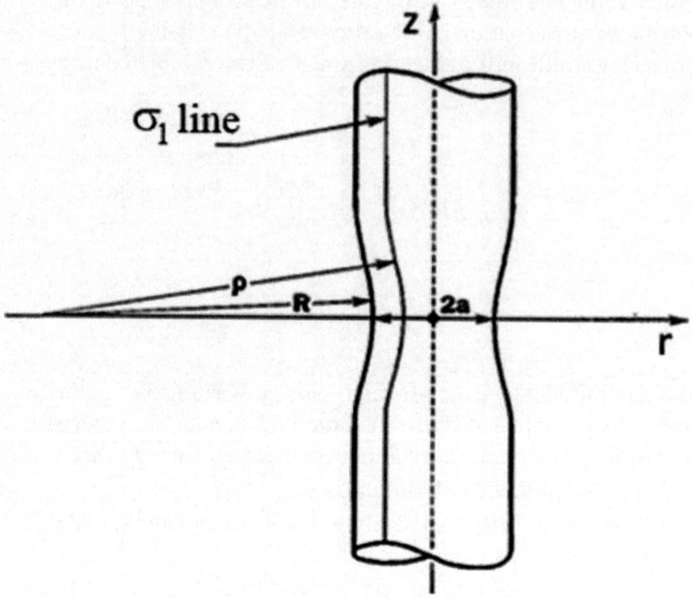

Fig. 13.12 Geometric characteristics of the neck

$$\sigma_{\text{real}}(t) = \frac{\sigma_{\text{mean}}(t)}{\left(1 + \frac{2R(t)}{a(t)}\right) \cdot \ln\left(1 + \frac{a(t)}{2R(t)}\right)} \tag{13.8}$$

here, R is the minimum cross-section curvature radius of a neck, a is the minimum cross-section radius of the deformed specimen (Fig. 13.12).

Davidenkov and Spiridonova have got the following estimation of effective true stress (Davidenkov and Spiridonova 1946):

$$\sigma_{\text{real}}(t) = \frac{\sigma_{\text{mean}}(t)}{\left(1 + \frac{a(t)}{4R(t)}\right)} \tag{13.9}$$

There are also other less known models.

Szczepinski (Dietrich et al. 1970) deduced the following equation:

$$\sigma_{\text{real}}(t) = \frac{\sigma_{\text{mean}}(t)}{\frac{2R(t)}{a(t)}\left[\exp\left(\frac{a(t)}{4R(t)}\right) - 1\right]} \tag{13.10}$$

Corrective formula of Malinin and Petrosian for stress calculation (Malinin and Rzysko 1981) is the following:

$$\sigma_{real}(t) = \frac{\sigma_{mean}(t)\cdot\left(1-\frac{a(t)}{8R(t)}+\frac{5}{384}\left(\frac{a(t)}{R(t)}\right)^2+\eta_1\left(\frac{a(t)}{R(t)}\right)^3+\xi_0^{(0)}\eta_2\left(\frac{a(t)}{R(t)}\right)^4\right)^2}{\left(1-\sqrt{2}\eta_2\left(\frac{a(t)}{R(t)}\right)^{\frac{9}{2}}\right)^2} \tag{13.11}$$

here, $\xi_0{}^{(0)} = 2.404825558$, $\eta_1 = -0.001684$, $\eta_2 = 0.00052196$.

Two empirical models are given in (Gromada et al. 2011):

$$\sigma_{real}(t) = \frac{\sigma_{mean}(t)}{\left(1+\frac{a(t)}{4R(t)}+\frac{19a(t)}{488R(t)}\right)} \tag{13.12}$$

$$\sigma_{real}(t) = \frac{\sigma_{mean}(t)}{\left(1+\frac{a(t)}{4R(t)}+\frac{a(t)}{56R(t)}\right)} \tag{13.13}$$

Mirone proposed a special polynomial to correct the stress after strain localization (Mirone 2004):

$$\mathrm{MLR}(\varepsilon_{AN}) = 1 - 0.6058\cdot\varepsilon_{AN}^2 + 0.6317\cdot\varepsilon_{AN}^3 - 0.2107\cdot\varepsilon_{AN}^4$$

here, $\varepsilon_{AN} = \varepsilon_{real} - \varepsilon_N$ is plastic strain after necking, ε_N is strain at which localization appears (conditions (13.1)).

The effective stress can be calculated by scaling the mean axial stress using above polynomial:

$$\sigma_{real} = \sigma_{mean}(\varepsilon_{real})\cdot\mathrm{MLR}(\varepsilon_{AN}) \tag{13.14}$$

Mirones's polynomial is material independent and its coefficients are obtained by generalization of a set of experimental and numerical investigations.

13.4.2 Numerical Analysis

The numerical simulation was carried out to analyze strain localization process under dynamic tension and numerical simulation was used to compare the accuracy of analytical models described above.

The geometry and boundary condition are shown in Fig. 13.13. Only the working part of specimen was considered. Axisymmetric finite element model was used. Simulation was carried out using open source code Calculix (https://www.calculix.de/). The explicit numerical scheme was used for time integration of equations. Zero velocities were applied to the bottom line. The radial velocities equal to zero

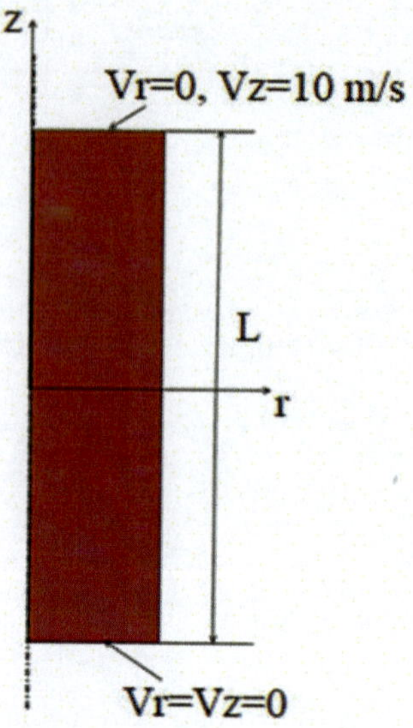

Fig. 13.13 Numerical simulation problem statement

on the top line. The existence of fixing parts of the specimen was replaced by this condition. The constant axial velocity $Vz = 10$ m/s was applied to top line. The gauge lengths of specimens L were 5 and 10 mm. The diameter of specimen was 5 mm.

The elastic–plastic models of plastic flow theory were used to describe the behavior of specimen's material. Two strain hardening models were considered:

$$\sigma = 400 + 1000 \cdot \varepsilon_p - \text{ model 1}$$

$$\sigma = 400 + 500 \cdot \varepsilon_p^{0.4} - \text{ model 2}$$

Model 1 presents a lager strain hardening, while model 2 shows softer material with nonlinear hardening Fig. 13.14.

Figure 13.15 illustrates the deformed specimens in times when maximum plastic strain reaches value 1. It should be noted that for softer material (model 2), strain localization is more pronounced and strain is located on fewer area. For harder material plastic strain is distributed over the larger zone.

It is clear from Fig. 13.16 that plastic strain 1 is reached:

- for hard 10 mm length specimen at total elongation 75%
- for soft 10 mm length specimen at total elongation 35%
- for hard 5 mm length specimen at total elongation 70%
- for soft 5 mm length specimen at total elongation 47%.

Thus, the effective plastic strain differs dramatically from averaged axial strain, especially for softer material.

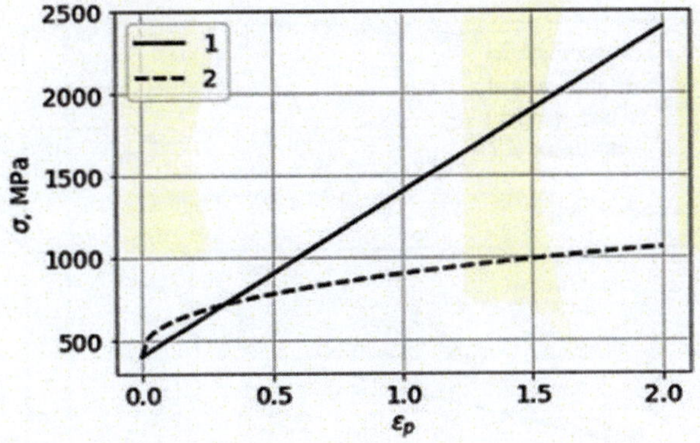

Fig. 13.14 Strain hardening models

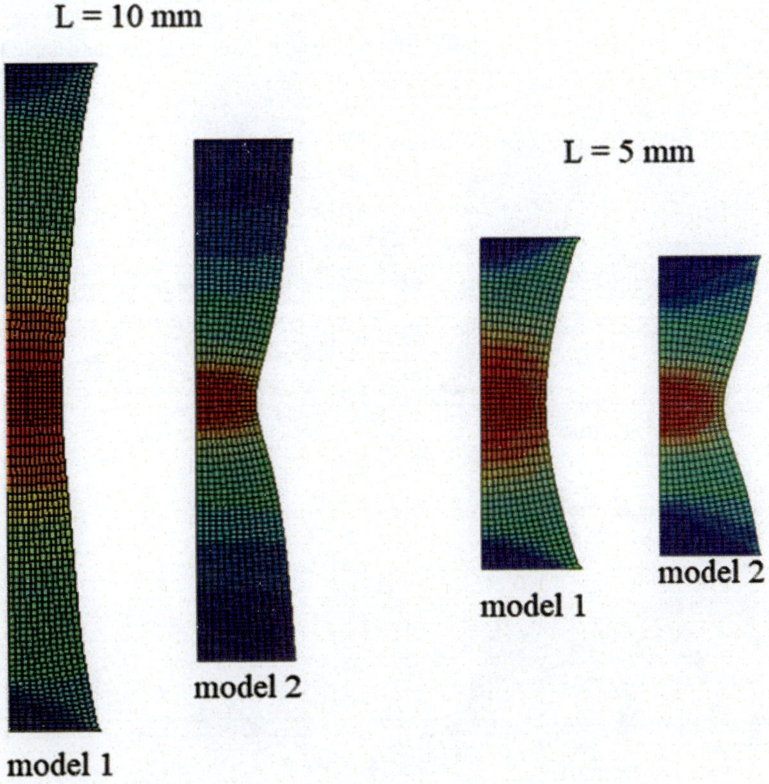

Fig. 13.15 Deformed specimens for considered cases

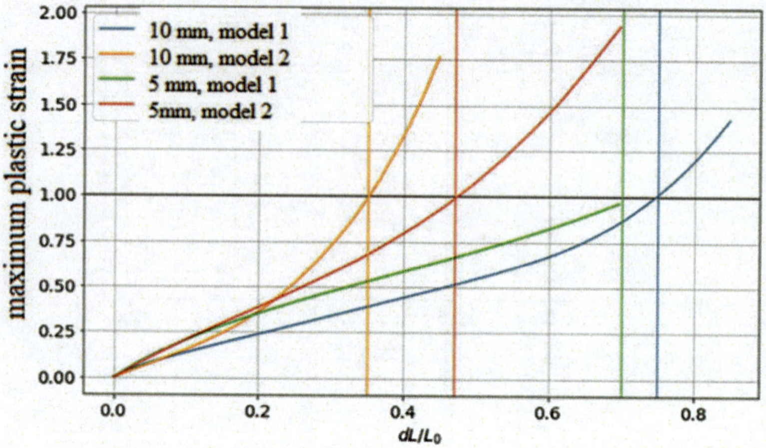

Fig. 13.16 Dependence of maximum plastic deformation in the working part of the sample against relative elongation

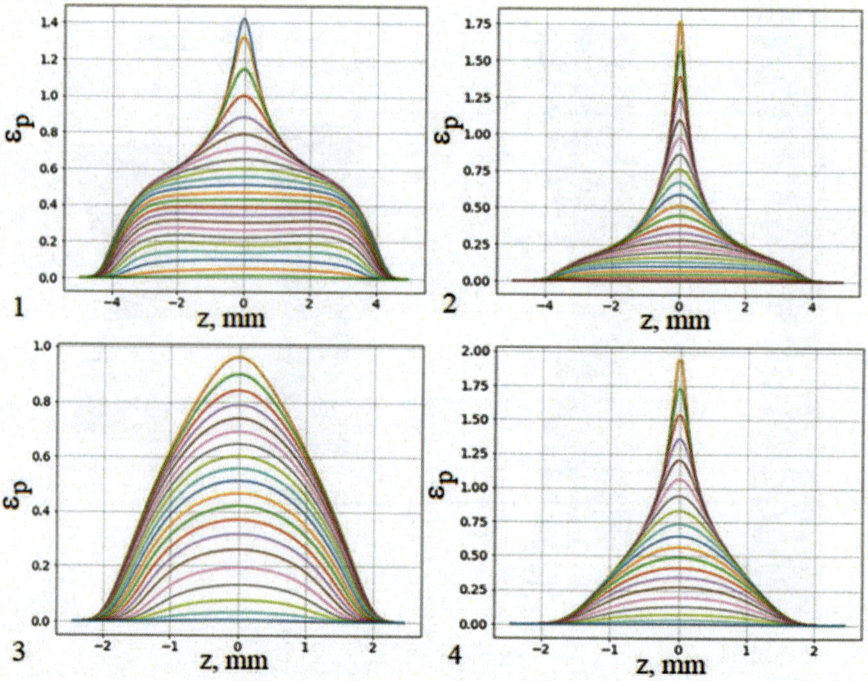

Fig. 13.17 Axial distribution of plastic strain at different times: 1—task 1, 2—task 2, 3—task 3, 4—task 4

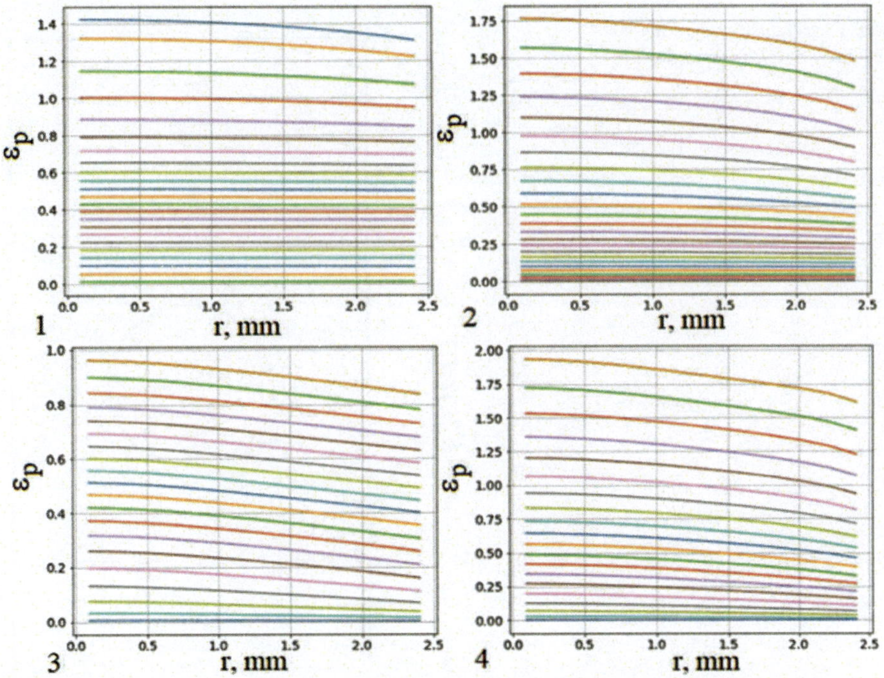

Fig. 13.18 Radial distribution in neck section of plastic strain at different times: 1—task 1, 2—task 2, 3—task 3, 4—task 4

Figure 13.17 shows plastic strain distribution along specimen axis at specific instants of time. It is to be noted that the plastic strain localization is more pronounced in soft materials and becomes more intensive at lower strains. Herewith, the uniform stage of tension ends earlier.

Figure 13.18 shows plastic strain distribution over the radius of minimum section in a neck versus time. The plastic strain is not constant. The plastic strain reaches its maximum at specimen axis. The uniformity of plastic strain in the minimum cross section is one of the basic assumptions in Bridgman's and Davidenkov's models.

Further, the comparison of different analytical models is given. The following data is used to calculate true stress–strain curves:

- $F(t)$ is the integral force acting on a specimen,
- $dL(t)$ is specimen elongation,
- $a(t)$ is time history of neck radius.
- This data was obtained from numerical simulation.
- For simplicity, the following definitions will be used:
- *Task 1*—10 mm length hard specimen (material model 1),
- *Task 2*—10 mm length soft specimen (material model 2),
- *Task 3*—5 mm length hard specimen (material model 1),
- *Task 4*—5 mm length soft specimen (material model 2).

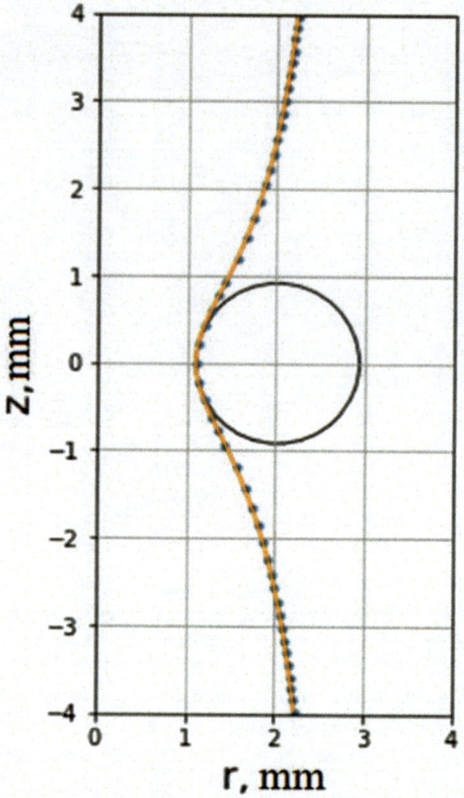

Fig. 13.19 Determination of the curvature *R* in a neck

The neck curvature radius is to be defined at each instant in all above models with exception of Mirone's model. For this, the points, representing external contour of specimen, are approximated by polynomial $r(z)$ (Fig. 13.19). The curvature radius in the point of the minimum section ($\min(r)$) is calculated by the formula:

$$R = \left. \frac{\left(1 + \left(\frac{\mathrm{d}r}{z}\right)^2\right)^{\frac{3}{2}}}{\left|\frac{\mathrm{d}^2 r(z)}{\mathrm{d}z^2}\right|} \right|_{z \text{ at } \min(r)}. \tag{13.15}$$

The model's nodes are shown as points in Fig. 13.19, solid lines represent the polynomial approximation; the black circle demonstrates contour curvature.

Figure 13.20 illustrates time histories of plastic strains in different points in a neck section (colored lines) and compares them with plastic strain estimation, calculated by formula (13.6). It is clear that plastic strain determined using current neck radius is in good agreement with the real mean value of plastic strain in a neck. The non-uniformity of plastic strain occurs at early stages of deformation for short 5 mm length specimen. The difference between maximum and minimum strain is significant.

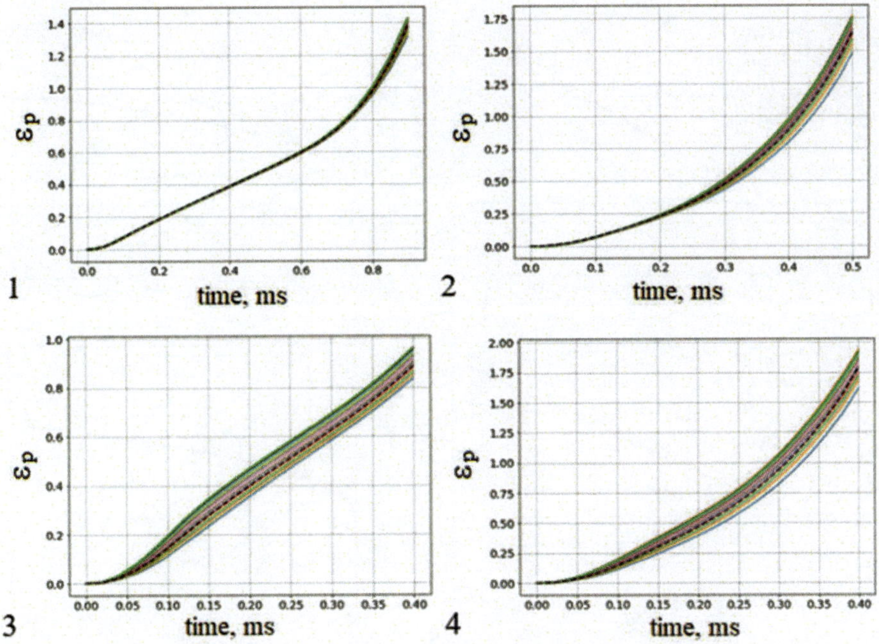

Fig. 13.20 Time histories of plastic strains in a neck section. Colored lines correspond to plastic strains in the different point in a neck section obtained in numerical simulation; black dashed line represents formula (13.6) estimation

Comparison of the deformation curves calculated by various analytical models with the true diagram (the model used in calculation) is given in Fig. 13.21. In the legend to Fig. 13.21, the numbers denote the following: the first number is the formula by which strain is calculated, the second one is the formula by which the stress is calculated. We can conclude that logarithmic diagram (13.4), (13.5) which does not account for localization of deformation allows us to describe material behavior only at a stage of uniform deformation of the working part of a specimen. In case of tension of the sample with 5 mm base, the obtained diagram slightly overestimates the true flow stress. The mean-stress diagram, calculated by formulas (13.6), (13.7) (average stress with account for real neck section) overestimates the value of flow stress that is especially noticeable in case of description of material behavior by model 2 (the model with less strain hardening). Bridgman and Davidenkov's models make it possible to calculate a true strain curve quite accurately, and the latter model gives the result as close as possible to the diagram embedded in the calculation. MLR adjustment underestimates the true flow stress at big strain levels. This is not surprising as polynomial parameters are received on the basis of generalization of experimental data to a strain level of about 1.2. Stress calculation out of the specified range is made in an extrapolation zone that does not guarantee reliable result. It should be noted that up to deformations of 1–1.2 level, all considered adjustment

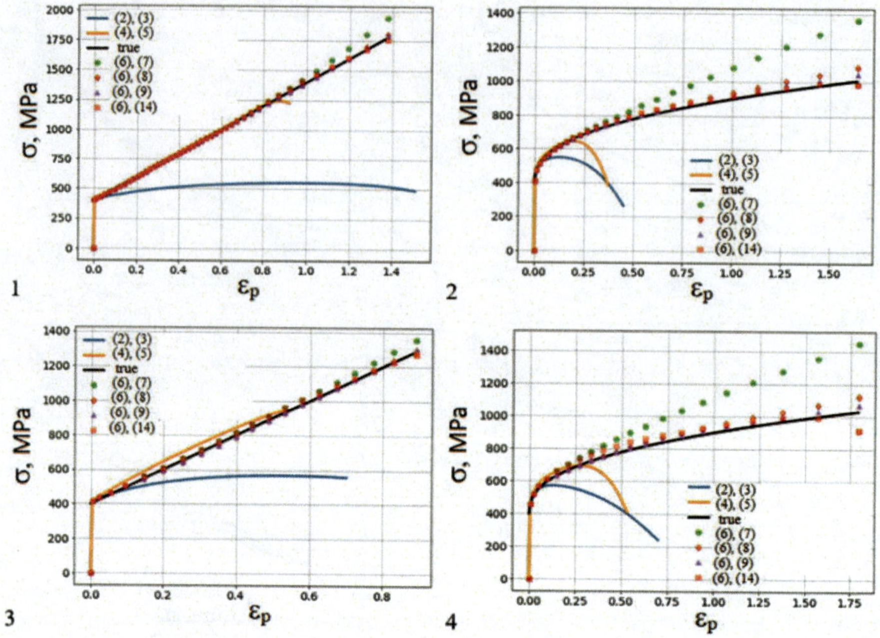

Fig. 13.21 Comparison stress–strain curves

schemes allow obtain a reliable result, which is sufficient for most materials since failure occurs at this deformation.

Bridgman and Davidenkov's adjustments are based on the geometrical formulas that connect curvature of principle stress in a specimen and on an external surface of the specimen in the zone of stress localization. Davidenkov assumed (Davidenkov and Spiridonova 1946):

$$\frac{1}{\rho} = \frac{r}{a \cdot R} \tag{13.16}$$

here, ρ is a curvature for a distance r from specimen axis, a is radius of a minimal section, R is a curvature of a minimal neck section (for $r = a$), as it is shown in Fig. 13.12.

Bridgman in his work used this formula:

$$\frac{1}{\rho} = \frac{r}{a \cdot R\left[1 + \frac{a}{2R}\left(1 - \left(\frac{r}{a}\right)^2\right)\right]} \tag{13.17}$$

To check formulas (13.16) and (13.17) at high-speed tension, their left and right parts for tasks 1–4 were calculated at the last instant along radius of a minimal

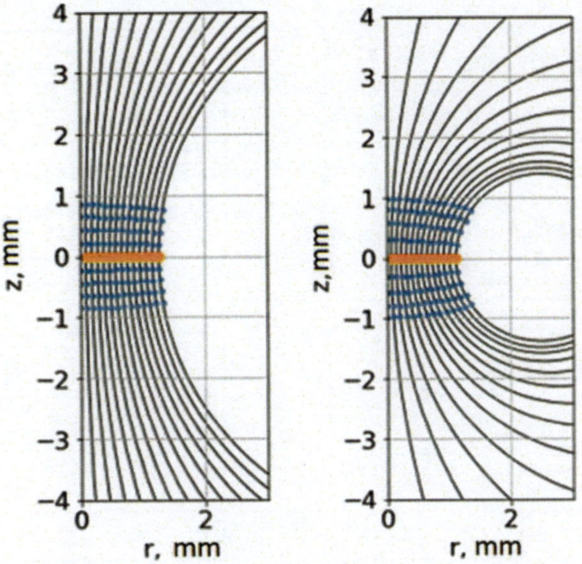

Fig. 13.22 Lines of the principal stress: left—for task 1, right—for task 2

section. Lines of the principal stress for tasks 1 and 2 are illustrated in Fig. 13.22. Comparison of the left and right parts of formulas (13.16) and (13.17) for different r is presented in Fig. 13.23.

Relation (13.16) works for all considered cases precisely enough however deviation of values is observed under condition (13.17) for the specimens with small strain hardening (strain localization). Apparently, it also leads to more exact prediction of a true stress–strain curve by Davidenkov's model.

Relative deviations in the diagrams, determined by different models and true curves for tasks 1–4, are given in Fig. 13.24. It is evident that for the soft material (task 2 and 4), Bridgman's adjustment exhibits the biggest error (8% in case of a specimen with 10 mm base and 10% for a specimen with 5 mm base). The best approximation of the true strain curve is given by the empirical formula (13.12).

Comparison of distributions of stress components for the minimum radial section of a specimen obtained using numerical modeling with analytical adjustments is of interest. In the Davidenkov model, the dependence of axial and radial stresses in the neck on the coordinate r is expressed by the formulas:

$$\sigma_z(r) = \sigma_{\text{real}} \cdot \left(1 + \frac{a}{2R} - \frac{r^2}{2Ra}\right),$$

$$\sigma_r(r) = \sigma_\theta(r) = \sigma_{\text{real}} \cdot \frac{a}{2R} \cdot \left(1 - \left(\frac{r}{a}\right)^2\right)$$

Bridgman's formulas:

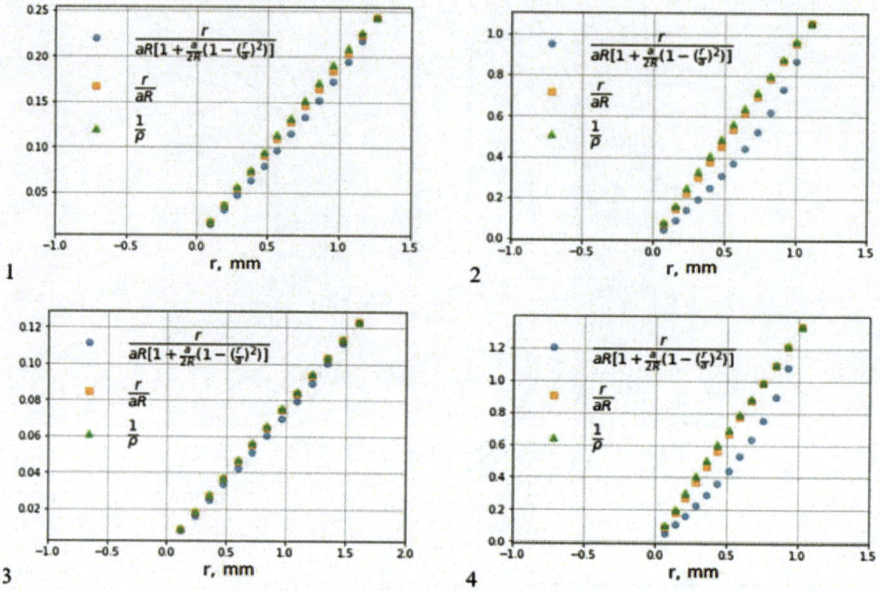

Fig. 13.23 Comparison of the left and right parts of formulas (13.16) and (13.17) for tasks 1–4

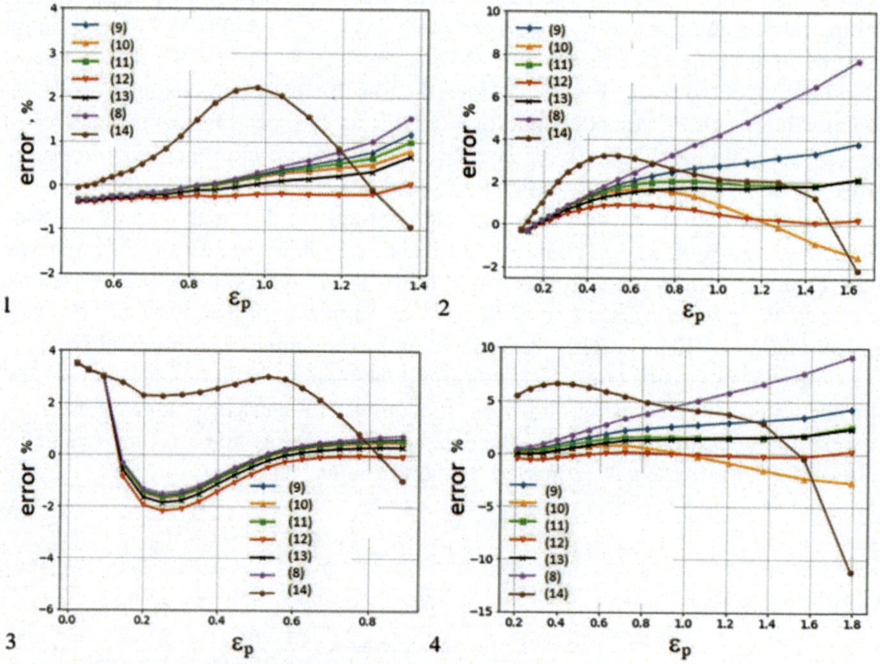

Fig. 13.24 Accuracy of analytical models' analysis

$$\sigma_z(r) = \sigma_{\text{real}} \cdot \left(1 + \ln\left(\frac{a^2 + 2aR - r^2}{2aR}\right)\right),$$

$$\sigma_r(r) = \sigma_\theta(r) = \sigma_{\text{real}} \cdot \ln\left(\frac{a^2 + 2aR - r^2}{2aR}\right).$$

Distribution of axial and radial stresses for the radius of the minimal neck section according to Shepinsky obeys the following laws:

$$\sigma_z(r) = \sigma_{\text{real}} \cdot \exp\left(\frac{a^2 - r^2}{2aR}\right),$$

$$\sigma_r(r) = \sigma_\theta(r) = \sigma_{real} \cdot \left[\exp\left(\frac{a^2 - r^2}{2aR}\right) - 1\right]$$

In Fig. 13.25, to the left, solid lines show the distributions of stress components over the radius of the minimum cross section at the final time, which were obtained in numerical simulation (problem 2). Markers correspond to the values calculated using Davidenkov's model, dashed lines—Bridgman's model, dash-dotted lines—Shepinsky's model. Estimation by Davidenkov's model is closer to numerical simulation.

The plastic strain across neck radius is shown to the right in Fig. 13.25. The difference between maximum (on specimen's axis) and minimum (at external surface) values of plastic strain is about 20%.

Time histories of neck radius and neck curvature during the whole tension process are to be used in the above analytical models, this being a challenging task for high rate experiments. The common approach for true stress–strain curve construction is an extension of the curve before necking using a fracture point. The plastic strain on fracture and effective true stress on fracture are determined using a and R, measured after testing the specimen. The main problem is the accurate determination of integral force acting on the specimen at the moment of its rupture. The shape of transmitted pulse, by which the force is measured, can be changed due to dispersion effect when using the Kolsky method (Bragov et al. 2019).

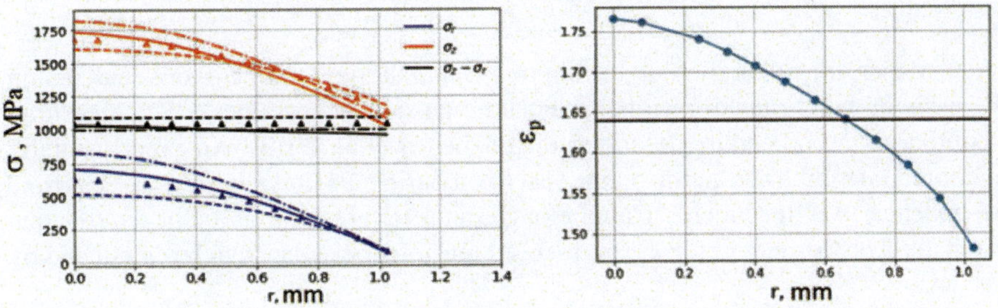

Fig. 13.25 Stress (to the left) and plastic strain (to the right) across neck section

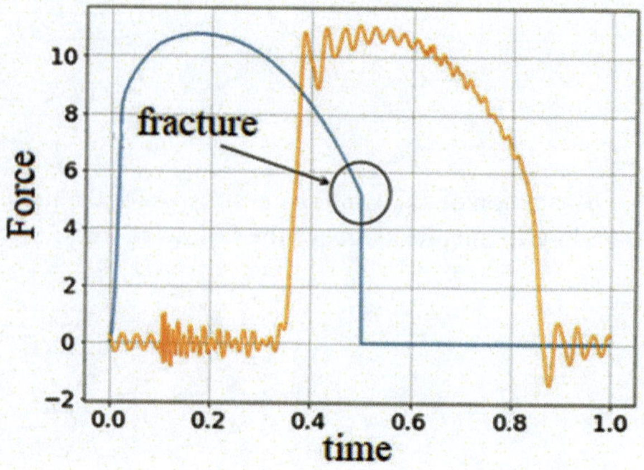

Fig. 13.26 Force blurring illustration

The sharp decrease in force occurs when specimen is fractured. Then propagating along measuring bar such signal is blurred due to dispersion effect as shown in Fig. 13.26. The accurate determination of fracture force becomes essential. Moreover, it is necessary to know conditions (strain rate, stress state, and temperature) under which the material characteristics are determined. The strain rate may increase significantly when deformation is localized. It is impossible to estimate the strain rate at fracture without the history of neck geometry.

The other method allowing true stress–strain curve construction including strain localization is the reverse identification method. It is based on numerical modeling of tension process. The main advantages of this approach are: there is no need for registration of specimen geometry during experiment and the method takes into account all features of the process and specimen (non-uniformities of stress and strain fields, complexity of geometry, inertia effects, and so on).

13.4.3 Experimental and Numerical Procedure of Construction a True Strain Curve According to Experiment on High-Speed Tension

In this paper, the restoration algorithm of a true material strain curve is realized on the basis of experimental results on dynamic tension of specimens. The algorithm is similar to the procedure described in (Bazhenov et al. 2013). An experimentally measured integral force acting on the specimen during tension $F_{\text{exp}}(t)$ as well as time dependence of a specimen's gauge elongation $V(t)$ are used as input data. When using the Kolsky method for the high-speed tension, these data can be calculated by formulas:

$$V(t) = c_I \cdot \left(\varepsilon^I(t) - \varepsilon^R(t)\right) - c_T \cdot \varepsilon^T(t) \quad (13.18)$$

$$F_{\exp}(t) = E_T \cdot S_T \cdot \varepsilon^T(t) \tag{13.19}$$

where c_I, c_T are bar sound speeds of the incident (subscript I) and the transmission (subscript T) bars, E_T and S_T are Young modulus and cross-section area of output bar, ε^I, ε^R and ε^T are incident, reflected, and transmitted pulses registered in measuring bars.

The following iterative procedure for determination of the true diagram is to be used:

1. The initial approximation of the material deformation diagram is selected (e.g., it can be a diagram obtained by extrapolation of a logarithmic curve, or an ideal-plastic model) as a table.
2. Process of dynamic tension of a specimen is simulated according to the scheme provided in Fig. 13.13, where boundary conditions at the top are $V_y(t) = V(t)$, $V_x(t) = 0$. Axial speed of the upper boundary $V(t)$ is calculated by formula (13.18).
3. The integrated reaction force $F_{\text{calc}}(t)$ on the fixed side is to be obtained.
4. Values $({\varepsilon_p}^i, {\sigma_m}^i)$ are determined for a discrete set of times t_i in a finite element in which the maximum effective plastic strain is realized (a finite element on a specimen's axis in a minimum section). Here, ${\varepsilon_p}^i$ is an effective plastic strain in the specified finite element, ${\sigma_m}^i$ is Mises stress in the specified finite element.
5. Tabular curve $\left(\left(\varepsilon_p^i, \sigma_m^i \cdot \frac{F_{\exp}(t_i)}{F_{\text{calc}}(t_i)}\right)\right)$ is accepted as an approximation of a true strain curve.
6. Steps 2–5 are repeated until the acceptable compliance of experimental and simulated forces is obtained or diagram at the next step stops changing significantly.

This iterative procedure is illustrated in Fig. 13.27. The experimental force and forces obtained on various iterations are shown on the left of the figure, while on the right, the history of diagram changing during consecutive adjustments is presented.

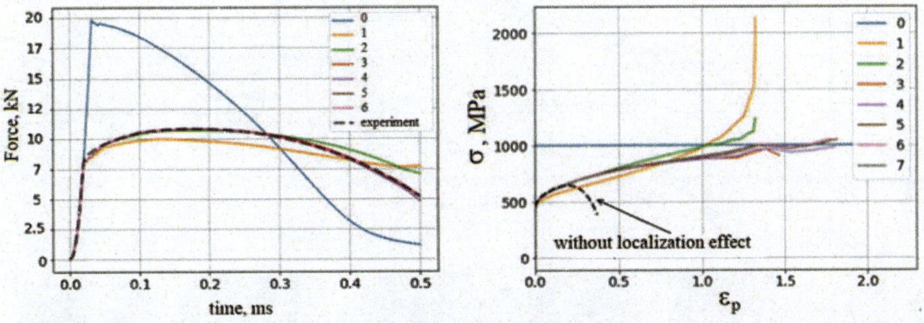

Fig. 13.27 Convergence of iterative procedure of definition of a true strain curve with the use of numerical modeling. On the left—comparison of forces, on the right—material diagrams for various iterations (number of iteration is shown in a legend)

It should be noted that the specified algorithm converges quickly enough, even with a not very successful choice of the initial approximation of the deformation curve (the model with a constant flow stress was used in the example).

13.4.4 Testing the Procedure on Experiments

Firstly, the procedure was tested on virtual experiments described in Sect. 13.4.1. Figure 13.28 compares true curves (used in virtual experiments models-black dashed lines) with the curves constructed using the implemented procedure (markers). The curves exhibit quite good agreement in a wide range of strains.

Secondly, the procedure was used to construct the true stress–strain curves for real materials. The data from SHTB experiments was used.

Experiments with the following features were selected (Fig. 13.29):

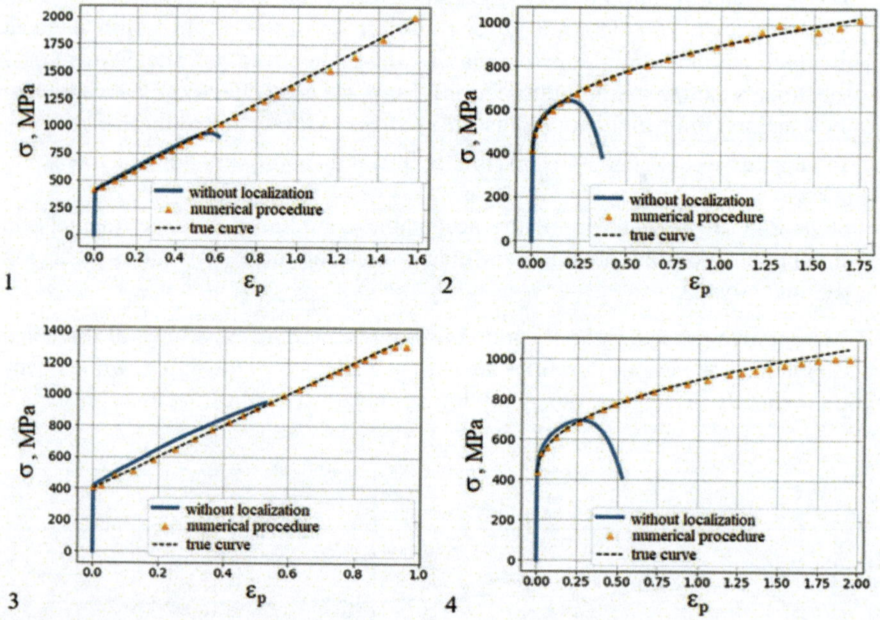

Fig. 13.28 Testing the procedure on virtual experiments

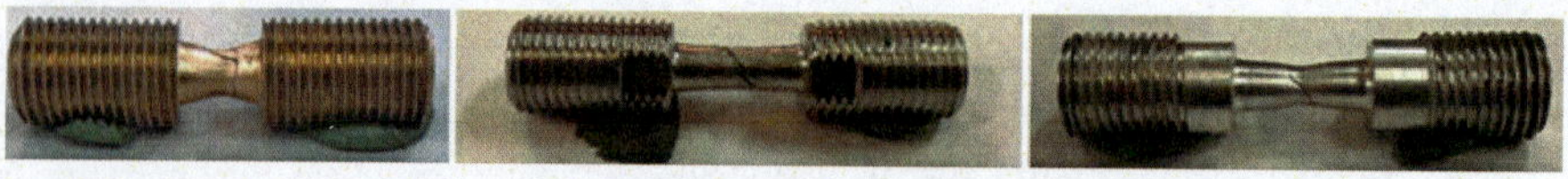

Fig. 13.29 Tested specimens: from left to right—cupper M1, EP-718 alloy, aluminum alloy 1575

- High rate tension tests of short (gauge length 5 mm, diameter 6 mm) specimen made of copper M1 were performed. Non-uniformity of stresses and strains in working part occurs due to closure of fixing parts.
- High rate tension of EP-718 alloy. The deformation of working part is mainly uniform until fracture occurs.
- High rate tension of aluminum alloy 1575 was carried out. Pronounced strain localization occurs.

The diagrams obtained for specimens described above are shown in Figs. 13.30, 13.31 and 13.32. They are compared with curves constructed using the traditional procedure which does not account for strain localization (solid blue lines). Dashed lines correspond to strain rate histories.

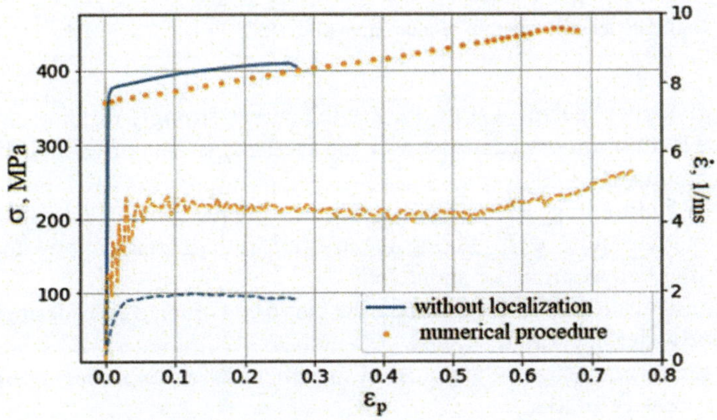

Fig. 13.30 High rate tension diagram of cupper M1

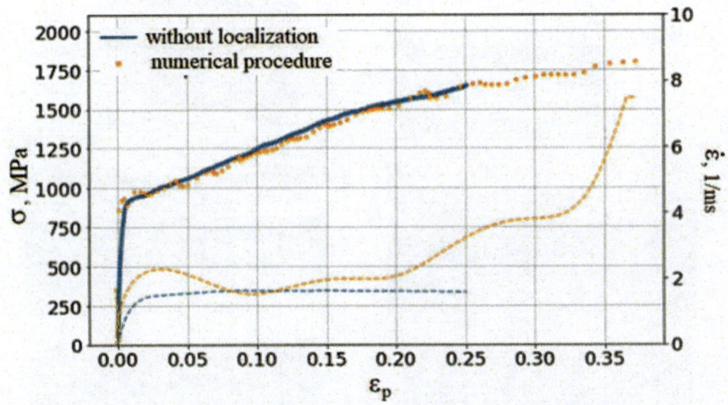

Fig. 13.31 High rate tension diagram of EP-718 alloy

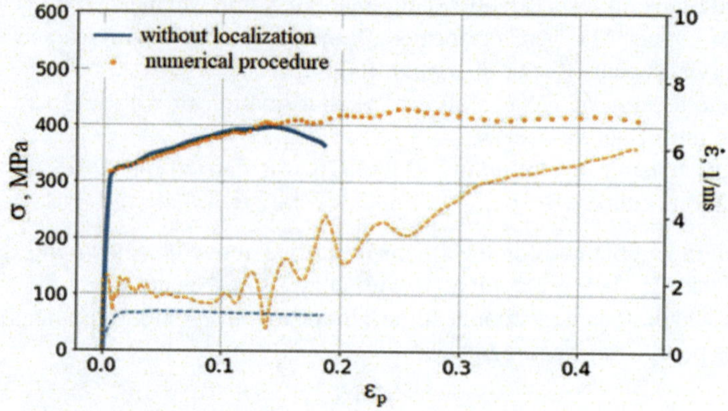

Fig. 13.32 High rate tension diagram of aluminum alloy 1575

For short copper specimen (Fig. 13.30), the curves strongly differ even at small strains. The local strain rate exceeds twice the averaged strain rate estimated without localization influence.

For EP-718 alloy (Fig. 13.31), the stress–strain curves constructed with different methods coincide quite well. Taking localization into account allows to slightly extend the diagram to strains of about 35%.

For aluminum alloy 1575 (Fig. 13.32), the curve constructed considering necking covers twice wider interval of strains.

Figure 13.33 demonstrates the shapes of specimens in real experiments and virtual tests. They are very similar.

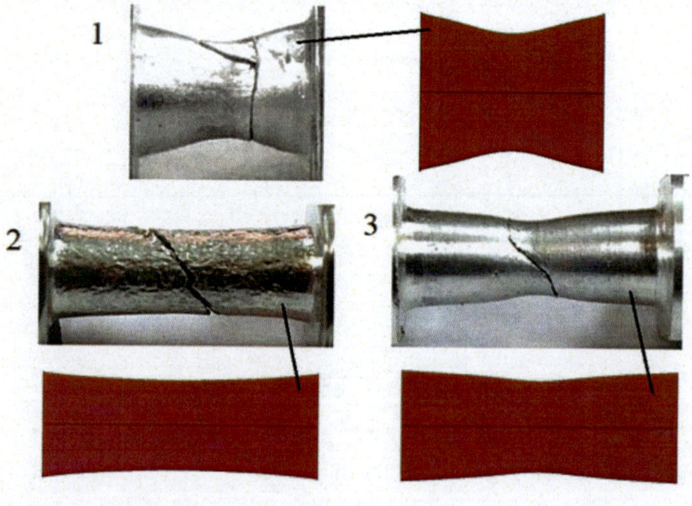

Fig. 13.33 Specimen shapes in real and virtual experiments: 1—M1, 2—Ep718, 3—1575

13.5 Conclusion

The review of the current state of high rate tension experiment shows that the most popular technique for testing materials at strain rates, ranging from 10^2 to 10^4 s^{-1} is the Kolsky method or the Split Hopkinson Pressure Bar method and its numerous modifications. However, there are a number of features intrinsic to high rate tension of visco-plastic specimens including stress and strain fields non-uniformity due to the closure of fixing parts of sample and strain localization (necking).

Numerical analysis of high rate tension process proves that the true stress–strain curve can be accurately constructed on the basis of the history of neck geometry during tension. If the neck geometry data is unavailable, one can apply numerical simulation and reverse analysis technique to obtain true dynamic diagram. Such procedure has been implemented and tested in the present work.

Acknowledgements The experimental study was supported by the Grant of the President of the Russian Federation for young scientists (MD-1221.2019.8). The theoretical investigations were supported by the grant of the Government of the Russian Federation (contract No. 14.Y26.31.0031).

References

Alibert, J. J., Seppecher, P., & dell'Isola, F. (2003). Truss modular beams with deformation energy depending on higher displacement gradients. *Mathematics and Mechanics of Solids, 8*(1).

Alves, M., & Jones, N. (1999). Influence of hydrostatic stress on failure of axisymmetric notched specimens. *Journal of the Mechanics and Physics of Solids, 47,* 643–667.

Aronofsky, J. (1951). Evaluation of stress distribution in the symmetrical neck of flat tensile bars. *Journal of Applied Mechanics,* 75–84.

Arthington, M. R., Siviour, C. R., & Petrinic, N. (2012). Improved materials characterisation through the application of geometry reconstruction to quasi-static and highstrain-rate tension tests. *International Journal of Impact Engineering, 46,* 86–96.

Auffray, N., dell'Isola, F., Eremeyev, V., Madeo A., & Rossi G. (2013). Analytical continuum mechanics à la Hamilton-Piola: least action principle for second gradient continua and capillary fluids. *Mathematics and Mechanics of Solids.*

Bacon, C., & Lataillade J.-L. (2001). Development of the Kolsky-Hopkinson techniques and applications for non-conventional testing. In W. K.Nowacki & J. R.Klepaczko (Eds.), *New experimental methods in material dynamics and impact. Trends in Mechanics of Materials* (pp. 1–58), Warsaw.

Barchiesi, E., Spagnuolo, M., & Placidi, L. Mechanical metamaterials: a state of the art. *Mathematics and Mechanics of Solids.*

Bazhenov, V. G., Lomunov, V. K., Osetrov S. L., & Pavlenkova, E. V. (2013). Experimental and computational method of studying large elastoplastic deformations of cylindrical shells in tension to rupture and constructing strain diagrams for an inhomogeneous stress-strain state. *Journal of Applied Mechanics and Technical Physics, 54*(1), 100–107

Bragov, A., Konstantinov, A., Kruszka, L., Lomunov, A., & Filippov, A. (2018). Dynamic properties of stainless steel under direct tension loading using a simple gas gun. *EPJ Web of Conferences, 183*(2035). https://doi.org/10.1051/epjconf/201818302035.

Bragov, A. M., Lomunov, A. K., Lamzin, D. A., et al. (2019). *Continuum Mechanics and Thermodynamics.* https://doi.org/10.1007/s00161-019-00776-0

Bridgman, P. W. (1955). *Studies in large plastic Flow and fracture (Russian translation)*. Moscow: Izd. Inostr. Lit.
Caverzan, A., Cadoni, E., & di Prisco, M. (2012). Tensile behaviour of high performance fibre-reinforced cementitious composites at high strain rates. *International Journal of Impact Engineering, 45*, 28–38.
Chen, W., Lu, F., & Cheng, M. (2002). Tension and compression tests of two polymers under quasistatic and dynamic loading. *Polymer Testing, 21*(2), 113–121.
Davidenkov, N. N., & Spiridonova, N. I. (1946). Analysis of the state of stress in the neck of a tensile test specimen. *Proceedings of ASTM, 46*, 1147–1158.
Davies, R. M. (1948). A critical study of the Hopkinson pressure bar. *Philosophical Transactions of the Royal Society of London. Series A, Mathematical and Physical Sciences, 240*, 375–457.
Del Vescovo, D., & Giorgio, I. (2014). Dynamic problems for metamaterials: Review of existing models and ideas for further research. *International Journal of Engineering Science, 80*, 153–172.
dell'Isola, F., Andreaus, U., & Placidi, L. (2015). At the origins and in the vanguard of peridynamics, non-local and higher-gradient continuum mechanics: An underestimated and still topical contribution of Gabrio Piola. *Mathematics and Mechanics of Solids, 20*(8).
dell'Isola, F., Della Corte, A., & Giorgio, I. (2016a). Higher-gradient continua: The legacy of Piola, Mindlin, Sedov and Toupin and some future research perspectives. *Mathematics and Mechanics of Solids*.
dell'Isola, F., Della Corte, A., Greco, L., & Luongo, A. (2016b). Plane bias extension test for a continuum with two inextensible families of fibers: a variational treatment with Lagrange multipliers and a perturbation solution. *International Journal of Solids and Structures*.
dell'Isola, F., Cuomo, M., Greco, L., & Della Corte, A. (2017). Bias extension test for pantographic sheets: numerical simulations based on second gradient shear energies. *Journal of Engineering Mathematics*.
dell'Isola, F., Giorgio, I., Pawlikowski, M., & Rizzi, N. (2016c). Large deformations of planar extensible beams and pantographic lattices: heuristic homogenization, experimental and numerical examples of equilibrium. *Proceedings of The Royal Society A, 472*(2185).
dell'Isola, F., Seppecher, P., & Alibert, J. J. (2019). Pantographic metamaterials: An example of mathematically driven design and of its technological challenges. *Continuum Mechanics and Thermodynamics, 31*(4), 851–884.
dell'Isola, F., Seppecher, P., & Madeo A. (2012). How contact interactions may depend on the shape of Cauchy cuts in Nth gradient continua: approach "à la D'Alembert". *Zeitschrift für angewandte Mathematik und Physik, 63*(6).
Dietrich, L., Miastkowski, J., & Szczepinski, W. (1970). *Limiting capacity of the construction elements*. Warsaw: PWN. ((in Polish)).
Eskandari, H., & Nemes J. A. (2000). Dynamic testing of composite laminates with a tensile split Hopkinson bar. *Journal of Composite Materials*.
Field J. E., Walley, S. M., Bourne, N. K., & Huntley, J. M. (1994). Experimental methods at high strain rate. *Journal de Physique, IV*(C3), 3–22.
Gama, B. A., Lopatnikov, S. L., & Gillespie, J. W., Jr. (2004). Hopkinson bar experimental technique: A critical review. *Applied Mechanics Reviews*. https://doi.org/10.1115/1.1704626
Ganzenmuller, G. C., Blaum, E., Mohrmann, D., Langhof, T., Plappert, D., Ledford, N., et al. (2017). A simplified design for a Split-Hopkinson Tension Bar with long pulse duration. *Procedia Engineering, 197*, 109–118.
Gerlach, R., Kettenbeil, C., & Petrinic, N. (2012). A new split Hopkinson tensile bar design. *International Journal of Impact Engineering, 50*, 63–67.
Gerlach, R., Sathianathan, S. K., Siviour, C., & Petrinic, N. (2011). A novel method for pulse shaping of Split Hopkinson tensile bar signals. *International Journal of Impact Engineering, 38*, 976–980.
Gilat, A., Schmid, T. E., & Walker, A. L. (2009). Full field strain measurement in compression and tensile split Hopkinson bar experiments. *Experimental Mechanics, 49*, 291–302.

Giorgio, I. (2016). Numerical identification procedure between a micro-Cauchy model and a macro-second gradient model for planar pantographic structures. *Zeitschrift für angewandte Mathematik und Physik*.

Gromada, M., Mishuris, G., & Öchsner A. (2011).Correction formulae for the stress distribution in round tensile specimens at neck presence. In *SpringerBriefs in applied sciences and technology computational mechanics* (pp. 1–89). https://doi.org/10.1007/978-3-642-22134-7.

Guzman, O., Frew, D. J., & Chen, W. (2011). A Kolsky tension bar technique using a hollow incident tube. *Measurement Science & Technology, 22*(4).

Harding, J., et al. (1992). Mechanical behaviour of composite materials under impact loading. In M. A. Meyers (Ed.), *Shock-wave and high strain rate phenomena in materials* (p. 21). New York: Dekker.

Hasenpouth, D. (2010). *Tensile high strain rate behavior of AZ31B magnesium alloy sheet*. Master's Thesis: University of Waterloo.

Hauser, F. E. (1966). Techniques for measuring stress-strain relations at high strain rates. *Experimental Mechanics, 6*(8), 395–402.

Huang, S., Chen, R., & Xia, K. W. (2010). Quantification of dynamic tensile parameters of rocks using a modified Kolsky tension bar apparatus . *Journal of Rock Mechanics and Geotechnical Engineering, 2*(2), 162–168.

Huh, H., Kang, W. J., & Han, S. S. (2002). A tension split Hopkinson bar for investigating the dynamic behavior of sheet metals. *Experimental Mechanics, 42*(1), 8–17.

Jiang, B., & Zhang, R. (2006). Tensile properties in the through-thickness direction for a carbon fiber woven reinforced composite at impact loading rate. *Journal de Physique IVFrance, 134*, 1071–1075.

Kajberg, J., & Wikman, B. (2007). Viscoplastic parameter estimation by high strain-rate experiments and inverse modelling—speckle measurements and high-speed photography. *International Journal of Solids and Structures, 44*, 145–164.

Kolsky, H. (1949). An investigation of the mechanical properties of materials at very high rates of loading. *Proceedings of the Physical Society. Section B, 62*, 676–700.

La Rosa, G., Mirone, G., & Risitano, A. (2003). Post-necking elastoplastic characterization: Degree of approximation in the Bridgman method and properties of the flow-stress/true-stress ratio. *Metallurgical and Materials Transactions A, 34A*(3), 615–624.

Lindholm, U. S., & Yeakley, L. M. (1968). High strain-rate testing: Tension and compression. *Experimental Mechanics, 8*(1), 1–9.

Ling, Y. (1996). Uniaxial true stress–strain after necking. *AMP Journal of Technology, 5*, 37–48.

Malinin, N. N., & Rzysko, J. (1981). *Mechanics of materials*. Warsaw: PWN. ((in Polish)).

Markov, I. P. (2018). Characterizing pantographic sheets by means of computational experiments. *Problems of Strength and Plasticity, 80*(3).

Mirone, G. (2013). The dynamic effect of necking in Hopkinson bar tension tests. *Mechanics of Materials, 58*, 84–96.

Mirone, G. (2004). A new model for the elastoplastic characterization and the stress–strain determination on the necking section of a tensile specimen. *International Journal of Solids and Structures, 41*, 3545–3564.

Mirone, G. (2007). Role of stress triaxiality in elastoplastic characterization and ductile failure prediction. *Engineering Fracture Mechanics, 74*(8), 1203–1221.

Mirone, G., & Corallo, D. (2010). A local viewpoint for evaluating the influence of stress triaxiality and Lode angle on ductile failure and hardening. *International Journal of Plasticity, 26*, 348–371.

Mirone, G., Corallo, D., & Barbagallo, R. (2016). Interaction of strain rate and necking on the stress-strain response of uniaxial tension tests by Hopkinson bar. *Procedia Structural Integrity, 2*, 974–985.

Nguyen, K., Kim, H. C., Shin, H., Yoo, Y. H., & Kim, J.-B. (2017). Numerical investigation into the stress wave transmitting characteristics of threads in the split Hopkinson tensile bar test. *International Journal of Impact Engineering, 109*, 253–263.

Nicholas, T. (1981). Tensile testing of materials at high rates of strain. *Experimental Materials, 21*(5), 177–185.

Noble, J. P., Goldthorpe, B. D., Church, P., & Harding, J. (1997). The use of the Hopkinson bar to validate constitutive relations at high rates of strain. *Journal of the Mechanics and Physics of Solids, 47,* 1187–1206.

Noble, J. P., & Harding, J. (1994). Temperature-measurement in the tensile Hopkinson bar test. *Measurement Science & Technology, 5*(9), 1163–1171.

Placidi, L., & Andreaus, U., & Giorgio I. (2017). Identification of two-dimensional pantographic structure via a linear D4 orthotropic second gradient elastic model. *Journal of Engineering Mathematics.*

Placidi, L., Barchiesi, E., Turco, E., Rizzi, N. L. (2016). A review on 2D models for the description of pantographic fabrics. *Zeitschrift für angewandte Mathematik und Physik, 67.*

Rahali, Y., Giorgio, I., Ganghoffer, J.-F., & dell'Isola, F. (2015). Homogenization à la Piola produces second gradient continuum models for linear pantographic lattices. *International Journal of Engineering Science, 97.*

Sasso, M., Newaz, G., & Amodio, D. (2008). Material characterization at high strain rate by Hopkinson bar tests and finite element optimization. *Materials Science and Engineering A, 487,* 289–300.

Sciarra, G., dell'Isola, F., & Coussy, O. (2007). Second gradient poromechanics. *International Journal of Solids and Structures, 44*(20).

Smerd, R., Winkler, S., Salisbury, C., Worswick, M., Lloyd, D., & Finn, M. (2005). High strain rate tensile testing of automotive aluminum alloy sheet. *International Journal of Impact Engineering, 32*(1–4), 541–560.

Taniguchi, N., Nishiwaki, T., & Kawada, H. (2007). Tensile strength of unidirectional CFRP laminate under high strain rate. *Advanced Composite Materials, 16*(2), 167–180.

Turco, E., Dell'Iola, F., Cazzani, A., & Rizzi, N. L. (2016). Hencky-type discrete model for pantographic structures: numerical comparison with second gradient continuum models. *Zeitschrift für angewandte Mathematik und Physik.*

Yokoyama, T. (2003). Impact tensile stressestrain characteristics of wrought magnesium alloys. *Strain, 39*(4), 167–175.

Young, K. (2015). *Development of a tensile split Hopkinson pressure bar testing facility.* PhD, University of Windsor, Ontario, Canada.

Volkov, I. A., Igumnov, L. A., Litvinchuk, SYu., & Vorobcov, I. V. (2018). Modeling dynamic deformation and failure of thin-walled structures under explosive loading. *EPJ Web of Conferences, 183,* 03016. https://doi.org/10.1051/epjconf/201818303016

Zhang, K. S., & Li, Z. H. (1994). Numerical analysis of the stress-strain curve and fracture initiation for ductile material. *Engineering Fracture Mechanics, 49,* 235–241.

Zhang, Z. L., Hauge, M., Odegard, J., & Thaulow, C. (1999). Determining material true stress–strain curve from tensile specimens with rectangular cross section. *International Journal of Solids and Structures, 36,* 2386–2405.

Chapter 14
Flocking Rules Governing Swarm Robot as Tool to Describe Continuum Deformation

Ramiro dell'Erba

Abstract In robotic swarm, the position of an element is often determined by the behaviour of its neighbours. Following this concept, we have realized a tool able to give a visually plausible simulation of continuum deformation. Without solving Newton's equation, we reproduced some behaviour of bidimensional deformable bodies both according to the standard Cauchy model and second gradient theory. Fracture can be easily managed. The tool has computational cost advantage, and it is very flexible to adapt for complex geometry samples.

Keywords Swarm robotics · Discrete mechanical systems

14.1 Introduction

The aim of this paper is not to compute physical processes but to generate visually plausible simulation results with low computational cost, sacrificing some accuracy, with respect to the solution of heavy differential equations, solved by finite element methods (FEM). The dynamic simulation of mechanics has its roots in computer graphics for videogames; classical methods are based on discretization (Lagrangian or Eulerian) of Newton's second law and formulate forces for each mechanical effect. Owing to the power of GPU, some new methods have become popular. The tool we propose is based on position-based dynamics (PBD), Bender et al. 2015) widely used it in computer animation due to its efficiency, robustness and simplicity. Like the robot swarm behaviour dell'Erba 2015; dell'Erba and Moriconi 2014), it does not determine forces or solve differential equations, but uses a position-based approach, where the new position of a particle is determined by its neighbour's positions, that approach being easily used to describe complex objects. So far, by using the flocking rules employed in underwater robotic swarms to compute the displacement of the elements in order to achieve an assigned swarm configuration (dell'Erba and Moriconi 2015; dell'Erba 2012; Moriconi and dell'Erba 2012), we adapted the swarm

R. dell'Erba (✉)
Robotics Laboratory, ENEA Technical Unit technologies for energy and industry, Rome, Italy
e-mail: ramiro.dellerba@enea.it

F. dell'Isola and L. Igumnov (eds.), *Dynamics, Strength of Materials and Durability in Multiscale Mechanics*, Advanced Structured Materials 137,
https://doi.org/10.1007/978-3-030-53755-5_14

control algorithm in PBD problems. Note that most PBD methods hide the dynamics inside their relationship; moreover, they ask for the knowledge of the velocity of the particles. We try to describe the deformation of a continuum medium by this tool useful for complex microstructures not easily analysed by Cauchy continuum. Classical Cauchy continua are not able to give accurate predictions in highly non-homogeneous microstructure: generalizations have to be introduced either considering additional degrees of freedom, to account for the kinematics at the level of the microstructure, or including in the deformation energy density higher gradients of the displacement than the first one. As part of generalized continua theory, second gradient models allow to study microstructured materials, by taking into account the gradient of strain instead of adding independent variables. That class of models is thus able to capture exotic behaviours. (Abali et al. 2015; Andreaus et al. 2018; Auffray et al. 2015; dell'Isola et al. 2019a; Eremeyev et al. 2018; Misra et al. 2018; Nejad-sadeghi et al. 2019; Placidi and El Dhaba 2017; Placidi et al. 2015, 2016, 2017; Turco et al. 2016b c, 2017). The derivation of such models has raised many mathematical challenges, such as that of homogenization methods. (Abdoul-Anziz and Seppecher 2018; Alibert et al. 2003). Such frameworks have permitted the development of generalized continua theory to go beyond Cauchy continua (Toupin 1964; Green 1965; Mindlin 1965; Maugin 2010). (This class of models has been used extensively in the recent years, for example to study microstructured materials. (Altenbach and Eremeyev 2009; Altenbach et al. 2013; Carcaterra et al. 2015; dell'Isola et al. 2015a, b, 2016d, 2017; Eremeyev 2018, 2019a, b; Eremeyev and Sharma 2019; Eremeyev et al. 2012; Misra and Poorsolhjouy 2015; Misra and Singh 2013; Misra and Singh 2015; Niiranen et al. 2016; Sharma and Eremeyev 2019; Spagnuolo and Andreaus 2019; Yildizdag et al. 2019). Many efforts have been made in the recent years towards the numerical modelling and simulation of those models (Balobanov et al. 2019; Barchiesi et al. 2019a, b; de Angelo et al. 2019; dell'Isola et al. 2016b, c, 2019b, c; Golaszewski et al. 2019; Khakalo and Niiranen 2017; Spagnuolo et al. 2017; Turco and Rizzi 2016; Turco et al. 2016a, 2018; Cazzani et al. 2016a, b; Spagnuolo et al. 2019; Turco et al. 2016b, c; Yang and Müller 2019; Yang et al. 2018). The latter is a particularly relevant topic if you consider the technological interest in developing exotic mechanical metamaterials able to perform targeted tasks; therefore, the investigation of new and efficient algorithms is of great interest at the moment. In our tool, according to our experience in control of robot swarm, the new position of a particle is determined by the spatial position of its neighbours. The system (dell'Erba 2018a, b; Battista et al. 2016) exhibits very different behaviours changing lattice type and its internal parameters. Therefore, the computation of new positions for a particle set can be considered as a constrained geometrical problem leading to a transformation operator between the matrices representing the particles configuration, C_t, for a discrete set of time steps $t1$, $t2$, … tn. In this paper, we summarize some results and try to examine the behaviour of a two-dimensional beam under bending stress, which still needs to be improved.

14.2 Tool Description

A complete description of the algorithm is reported in dell'Erba 2018a, c). Here, we briefly summarize that algorithm, breaking down its steps into "choices". The two-dimensional continuum is discretized into a finite number of particles occupying, in their initial configuration, the nodes of a lattice, chosen between the five plane Bravais lattices and the honeycomb lattice (Choice 1).

The lattice has four kinds of particles, but it is easy to introduce new kinds, to describe other properties, due to the modular structure of the algorithm.

1. The leaders; their motion is assigned and determine the displacements of the other particles.
2. The followers; their motion is determined by the interaction rule with other particles.
3. The frame; they are introduced so that any particle can have the same number of neighbours, to avoid edge effects. Their motion is determined by the frame rule.
4. The ghost; they are introduced to describe fracture mechanism.

Regarding the choice of the neighbours of any particles (Choice 2), typically *nc* particles, where *nc* is the coordination number of the lattice (first gradient): the position of those points will determine the displacement of the point under investigation. The chosen flocking rule (Choice 3) between the particles describes the position of a particle as function of its neighbour's positions. To avoid edge effects, it is necessary to build a frame surrounding the body, by defining an external shell of points, so that any follower interacts with the same number of elements. The displacement of the frame points follows the motion of an assigned follower. If the displacement of those assigned followers obeys a more complex rule (e.g. in a corner), then an average displacement, or a more generic rule (Choice 4), may be considered. Next, a threshold *df* (or a more complex rule) may be defined beyond which fracture appears (Choice 5). We assume that if the distance between points is larger than this threshold, they stop to influence each other so they are no longer taken into account in the calculation of the follower position and fracture is declared. To balance computation of the point's displacements, we introduce ghost points, to avoid collapsing. The possible positions of those ghost points (Choice 6) are chosen so that the original shape of the lattice may be recovered. Different choices lead to different results, and our intention will be to use different rules in order to approximate different constitutive equations. Now, by the displacements of the leaders, we can start an iterative work flow until an equilibrium state is reached. We would like to underline this point in each of the possible choices; as an example, if we take a larger coordination number *nc,* like the neighbours of the neighbours, we are in second gradient theory case. The algorithm is suitable to change particle role and rules dynamically at each time step. The most natural choice, concerning interaction rule, is the centre of gravity rule where the new x-coordinate of the particle j is

$$x_j(t) = \frac{\sum_{k=2}^{\text{all neighbours of } j} x_k(t)}{N} \tag{14.1}$$

Possible generalizations of Eq. 14.1 are geometric, power and weighted mean. Possible weight is the particles Euclidean distances dis(k, j) between the particles k and j. This can simulate Hooke's law, where the force is increased with increasing deformation. In Eq. 14.1, we note that the x and y coordinates are independent, so Poisson's effect cannot be obtained. A possible alternative is to use

$$y_j(t) = K * (x_j(t) - x_j(t_0)) * da + \frac{\sum_{k=2}^{\text{all neighbours of } j} y_k(t)}{N} \tag{14.2}$$

where da is a function of the distance from the central axis, K is a parameter determining the response force and $x(t_0)$ is the x-coordinate at time t_0. So far expansion of x-coordinate has effect on the y-coordinate. The Euclidean distance, dis(k, j), can be used as weight.

$$x_j(t) = \frac{\sum_{k=2}^{\text{all neighbours of } j} \text{dis}(k, j) x_k(t)}{\sum_{k=2}^{\text{all neighbours of } j} \text{dis}(k, j)} \tag{14.3}$$

We can also force the follower's movement to go beyond the barycentre equilibrium position, leading the lattice to oscillate.

$$x_j(t) = \frac{\sum_{k=2}^{\text{all neighbours of } j} w(k, j) x_k(t)}{\sum_{k=2}^{\text{all neighbours of } j} w(k, j)} + fd\left(\frac{\sum_{k=2}^{\text{all neighbours of } j} w(k, j) x_k(t)}{\sum_{k=2}^{\text{all neighbours of } j} w(k, j)} - MT(j, t_0)\right) \tag{14.4}$$

where $w(k, j)$ is the weight, fd is a feedback factor and $MT(i, t_0)$ is the x-coordinate of j point at t_0.

We now consider discussing the proposed model in a fully variational setting, which is by no means trivial but would provide clear methodological advantages (see Lanczos 2012) for an introduction and (Placidi et al. 2008; Dell'Isola and Placidi 2011; Dell'Isola et al. 2016a; Dell'Isola and Gavrilyuk 2012; Dell'Isola et al. 2014) for illustrative cases concerning continua with non-classical properties). More specifically, variational frameworks gives a systematic procedure allowing to build mathematically consistent models. (Abali et al. 2017; Steigmann and Faulkner 1993) To this aim, we introduce two formulations PE1 and PE2 concerning the pseudo-energy concept. The first is the sum, over the neighbours of a certain point, of squares of the differences between the distances of the point from its neighbours minus the distance in the initial configuration, i.e.

$$\text{PE1}(t, j) = \sum_{k=1}^{\text{allneighboursof} j} (\text{dis}(t, k, j) - \text{dis}(t_0, k, j))^2$$

where dis(t, k, j) is the Euclidean distance between points k and j at time t. This is the formula for the point j at time t. It is an attempt to simulate the potential energy of a material point obeying Hooke's law.

To compare time contiguous configuration C_t and C_{t-1},we define for each point j and each time t

$$\mathrm{PE2}(t, j) = ||C_t - C_{t-1}||$$

where || is the norm of the vector defined by the point j at time t and $t - 1$.

It must be underlined that this artifice has no direct connection with the usual energy definition (this is the reason we use the term pseudo-energy). Actually, it is only a graphic tool for a better understanding of the deformation. Extended use of them can be found in (dell'Erba 2018b).

14.3 Results from Previous Works

In this section, we briefly resume some results obtained by this tool in the preceding works (dell'Erba 2018a, c) where different strains of the leaders, with different choices, have been investigated. In preceding papers, we also have described the behaviour of some ASTM sample and the respect of Saint Venant principle; but now we want to focus our attention on other simple simulation to outline what we have to improve and what we need to understand better.

So far consider a case of simple strain and release in tensile test of a rectangular shape specimen. We are considering a square sample undergoing strain from one side (the other side is clamped) at constant velocity in the x-direction (speed 0.6 unit length/step time), with a square lattice of 10×10 units. At a certain time, the pull is released and the leaders return to original configuration (this means that the leaders have changed category and are now followers) attracted by the other points. The simple rule governing the followers' motion is that every point must be placed in the barycentre of its neighbours (Eq. 14.1); the neighbours are determined by the coordination number of the lattice; therefore, the leader's motion implies a displacement of the first layer that propagates in successive time steps to the other particles. This means the displacements, at each time step, involve a larger shell of points until all the lattice points are moving. In second gradient, (dell'Erba 2018c), we have considered also the neighbours of the first neighbours as second shell. In Fig. 14.1, we can see the configuration of the lattice over different time steps together with the PE1 contour plot. Red points are the leaders; blue are the followers; and orange are the frame. From the figure, we can outline that the x displacement of the points seems do not depend on the y-coordinate; however, looking at the PE1 picture we can note a light convexity that does mean this is not true.

A deeper examination of the points' displacements confirms that, close to the frame, the displacements along x are lower than the central points. This can be explained as an edge effect. In fact if we consider points on the same vertical lines

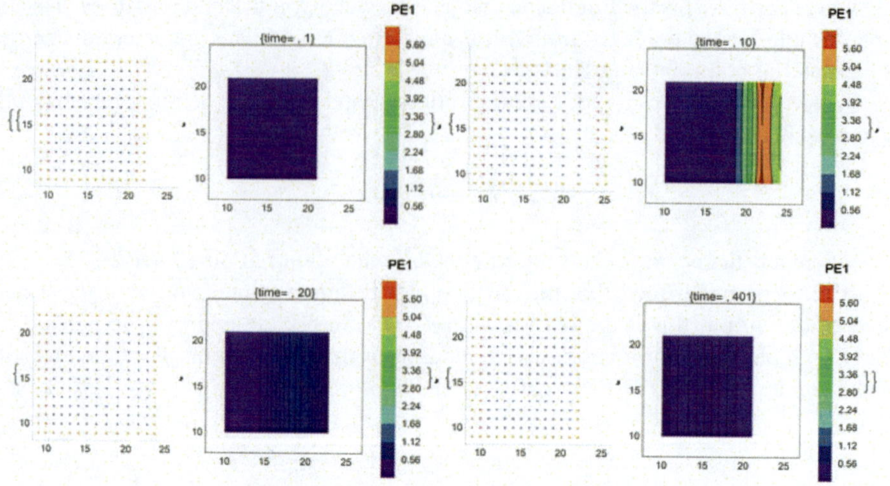

Fig. 14.1 Configuration of the lattice over different time (1, 10, 20 and 401) and PE1 contour plot

those that are close to the frame follow the neighbours with a little delay owing to the different rule determining the displacement of the frame and of the followers. Thus, they see a different situation with respect to, as an example, a central point. Note as that the maximum PE1 area (red color in the plot) is not on the leaders line but just one line on its left; this because, some points of the frame belong to the leader's neighbours and they always are close to them (they are subject to the frame rule) lowering the PE1 value. We can avoid this convexity effect using a different frame or mirroring the followers to obtain an infinite sample. Finally, it should be noted how, at $t = 401$, the lattice is not back completely to the reference configuration owing to the asymptotic process of relaxation. Consider now a central point $j = 67$ (sixth column, seventh row, points are numbered from left to right and from bottom to up) of the lattice. The value of the PE1 increases notably when points are pulled, after a delay owing to the propagation time as can be seen in Fig. 14.2; it decreases when the leaders become followers subjected only to the rules leading to equilibrium barycentre position. If we change point, the shape of the curve remains the same but can be less o more flared. In Figs. 14.3 and 14.4, the evolution with time of the coordinates of central point $j = 67$ and close to the leaders ($j = 115$) is shown. Also, in this picture, we can recognize the coordinate x increases linearly (velocity is constant), after a delay (less for $j = 115$), owing to the propagation time and later decreases to the original position.

A light modification can generate instabilities and oscillations; as an example, we can add to Eq. 1 a feedback term proportional to the difference between actual and initial position to overshoot the old equilibrium position. The result is showed in Figs. 14.5 and 14.6 (Same picture as 14.5 but different scale); note as also the reaction time is changed.

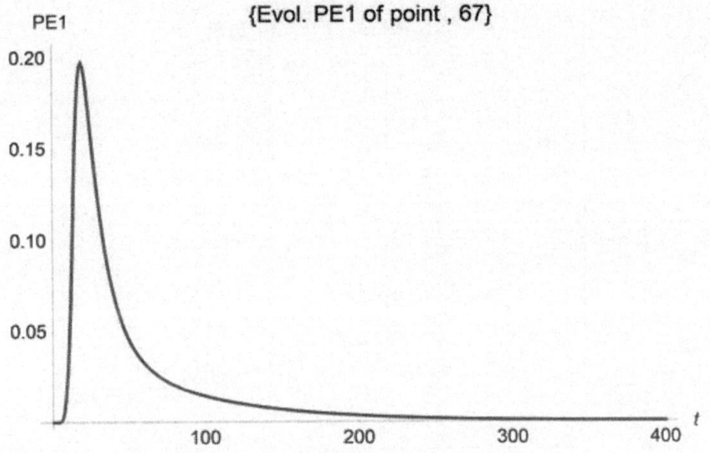

Fig. 14.2 PE1 of the central point's $j = 67$ versus time

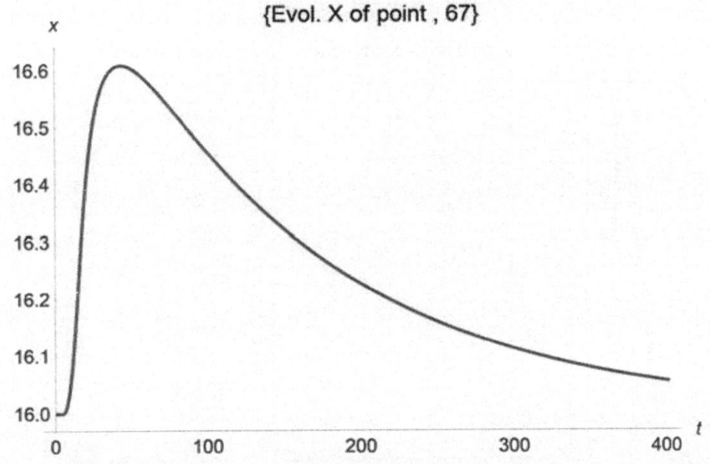

Fig. 14.3 X Evolution of the point $j = 67$ versus time

Consider now an ASTM specimen, for tensile test (see Fig. 14.8) during fracture. We are using a rectangular centred lattice with neighbour numbers reduced to five, to increase point's mobility. Distance fracture, df, is 11 units, and the speed was 0.6 step/unit time. The fracture occurs close to the top of the profile and not in the central area. Studies, still in progress, show how the fracture zone can be influenced by varying working conditions. We can render the fracture more or less brittle changing the model parameters, like neighbours' number, type of lattice, speed, etc. The PE1 plot in this case is less significant because it does not take into account the fictitious points; we are working on it to make this parameter meaningful also in fracture case.

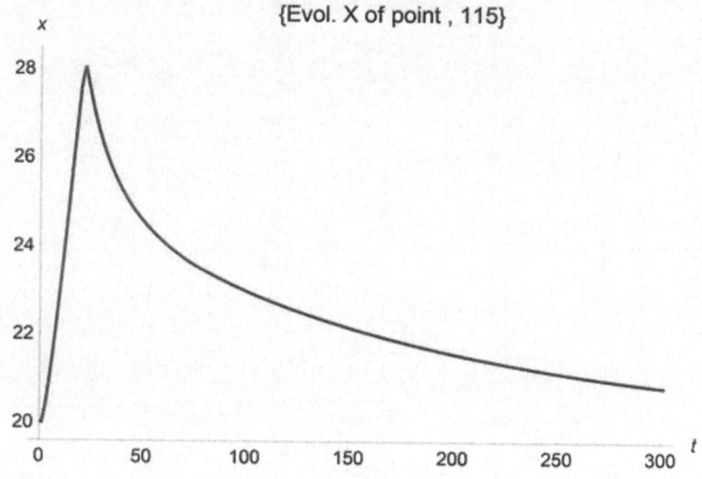

Fig. 14.4 X Evolution of the point $j = 115$ versus time

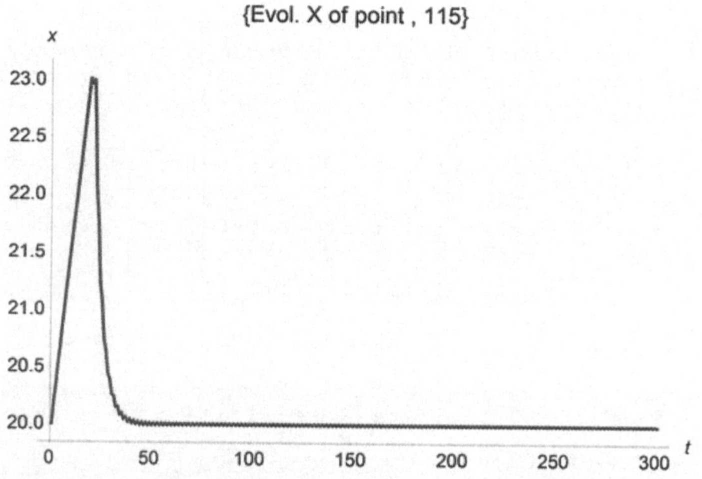

Fig. 14.5 X Evolution $j = 115$ versus time (modified rule with feedback) Full X range

14.4 Results for a Bending Beam

In this section, we want to investigate the behaviour of our tool applied to a bending beam loaded at one end and evaluate the differences with the numerical solutions of the differential equations. In the preceding papers, we got some success but we need to go deeply to a better understanding so we are looking for some cases that do not correspond to what we are expecting. We consider a bidimensional elastic short square beam (X and Y coordinate from 10 to 21 at the not deformed configuration)

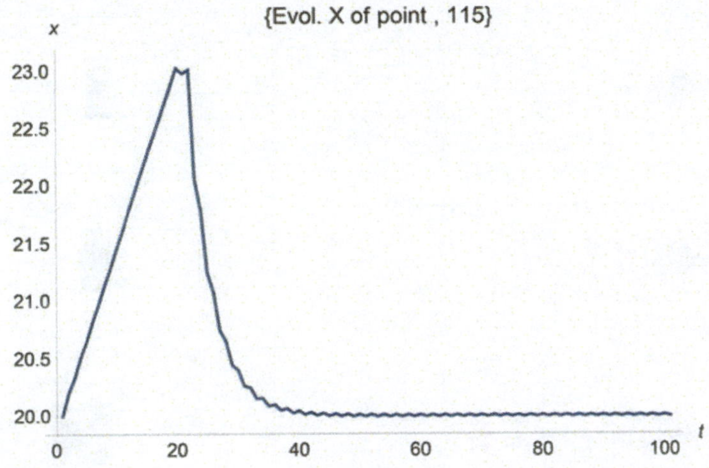

Fig. 14.6 *X* Evolution $j = 115$ versus time (modified rule with feedback) Reduced X range

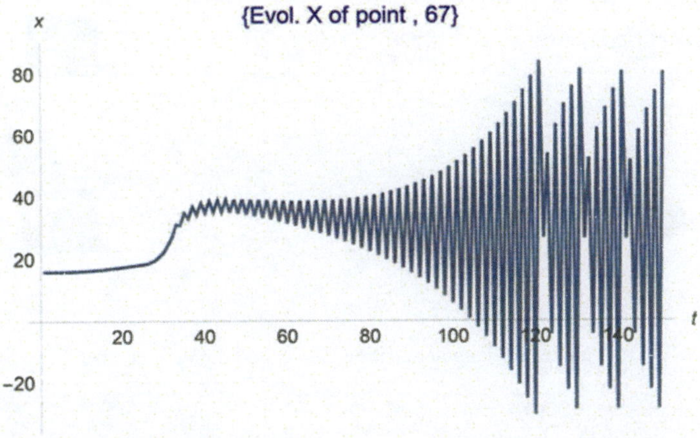

Fig. 14.7 *X* Evolution of the point $j = 67$ versus time (modified rule with feedback)

with materials parameters $Y = 1000$ N/(m^2) and $\nu = 0.33$, where Y is the Young's modulus, ν the Poisson coefficient; this value is close to that of a medium molecular weight polypropylene (Avella et al. 1993, 1995, 1996). Boundary load is on the right (Neumann condition) and no displacements on the left (Dirichlet conditions).

The equations to be solved are:

$$\frac{Y}{2(1+\nu)}\nabla^2\mathbf{u} + \frac{Y}{2(1-\nu)}\left(\frac{\partial}{\partial x}\mathbf{u} + \frac{\partial}{\partial y}\mathbf{v}\right) = 0$$

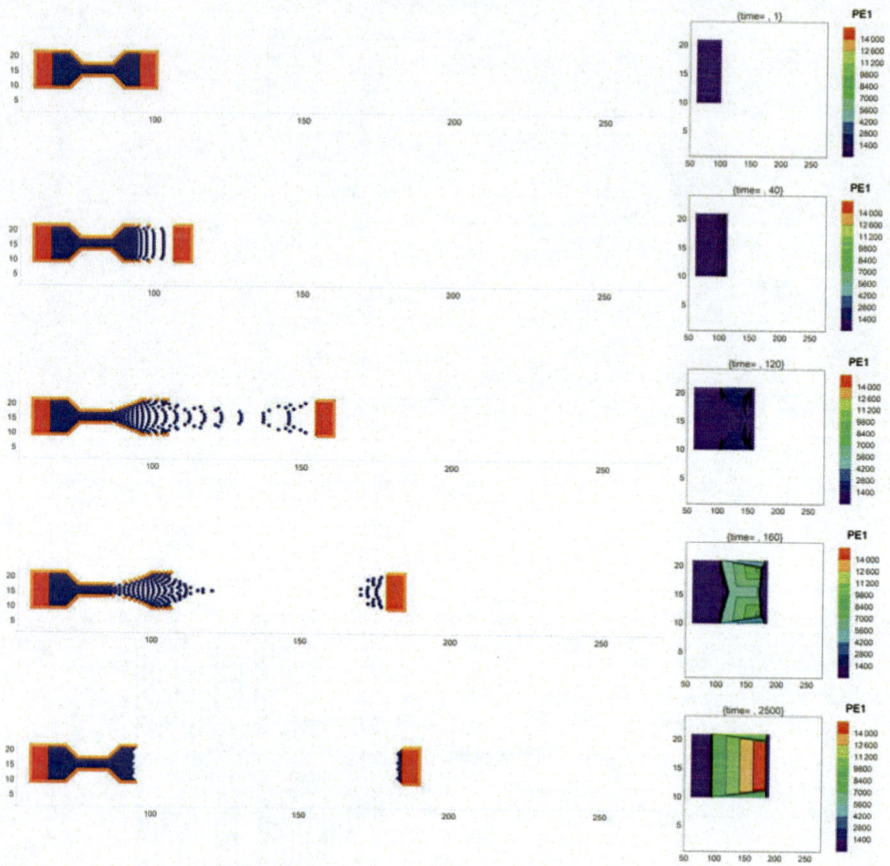

Fig. 14.8 ASTM fracture test rectangular centred lattice, coordination number 5

$$\frac{Y}{2(1+\nu)}\nabla^2 \mathbf{v} + \frac{Y}{2(1-\nu)}\left(\frac{\partial}{\partial x}\mathbf{u} + \frac{\partial}{\partial y}\mathbf{v}\right) = 0$$

$\mathbf{u}(x,y)$ and $\mathbf{v}(x,y)$ are the displacements function. We pose as boundary conditions 50 Pa as shear stress on the beam (Neumann condition for $x = 21$) and $\mathbf{u}(10,y) = \mathbf{v}(10,y) = 0$ as Dirichlet condition. These equations can be solved numerically; if we discretize our beam by a 10×10 square lattice, the solution is shown in Fig. 14.9 and the von Mises plot in Fig. 14.10; deformed mesh is plotted in red colour.

To apply our tool, we assign the displacements of the leader points, make some choice about the algorithm (lattice, interaction rules between the followers, etc.) and compute the strain when the followers readjust themselves, after a sufficiently long time. As leaders we have chosen the right and left side of the beam, so we assign the displacements of these points as obtained from the FEM equations and investigate the arrangement of the other points. The important thing to point out is that we still

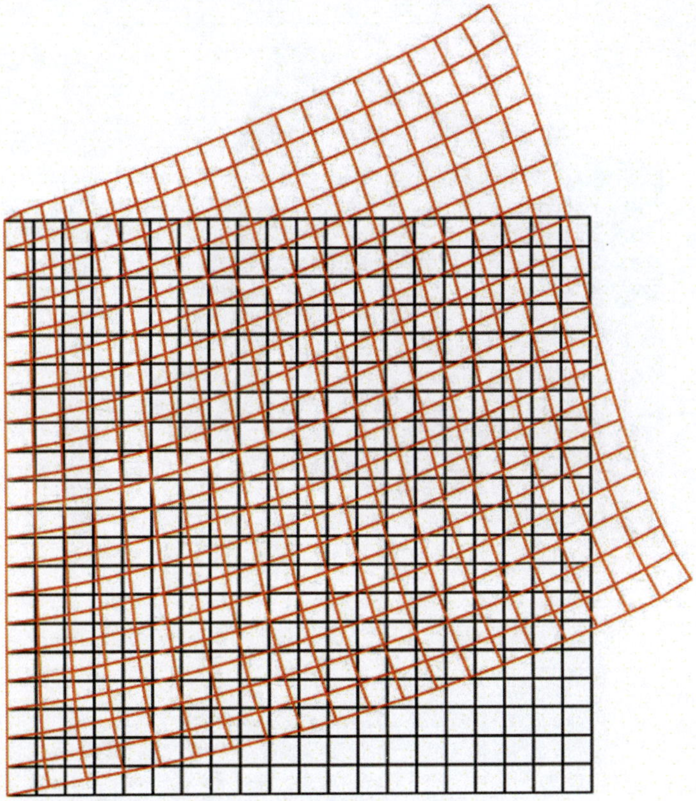

Fig. 14.9 FEM solutions of bidimensional square

have no criteria about the choice of lattice, interaction law between followers, etc. So, as a first attempt, we use a square lattice and no weight in the computation of the followers coordinates. In Fig. 14.11, the obtained configuration together with the FEM solution (red points) are shown; in Fig. 14.12, the corresponding von Mises plot is shown. The points on the left and on the right of the beam are overlapped because they are the leaders and we have imposed their displacement as the FEM solution beam deformation. We may outline that the external configuration of the beam is quite the same, but the internal displacement of the points, i.e. the strain, is different. This can be highlighted if we look the von Mises plot. Changes in the tool's parameters lead to different configurations, corresponding to different strains of the beam; almost none of them is satisfactory.

In Figs. 14.13 and 14.14, a second gradient model was used; no differences can be appreciated but a quantitative measure of the discrepancies with the FEM solutions shows a light worsening, visible in von Mises plot.

We can try many combinations of the parameters tool to fit the deflection of our beam but it is meaningless; we need to connect the constitutive parameters with the

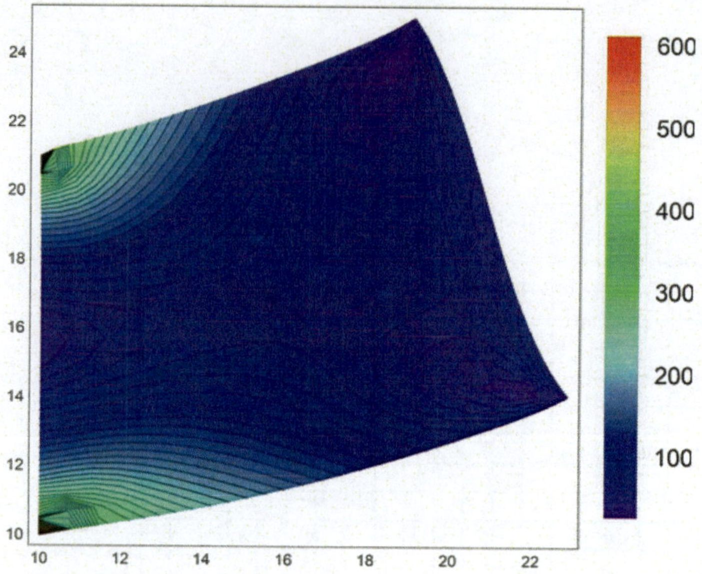

Fig. 14.10 Von Mises plot of FEM solutions beam under shear stress

Fig. 14.11 Beam deformation

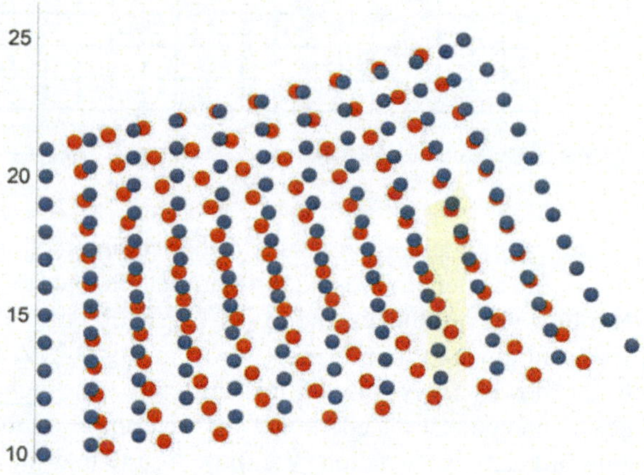

tool. If we increase the ratio length/width of our beam from 1 to 5, the results are still different (see Figs. 14.15, 14.16, 14.17 and 14.18). The results do not behave as we expected and show a wide range of possibilities in which to choose. Similar results were obtained if we increase the ratio length/width up to 10 or more. Although the external shape of the beam can sometimes be acceptable, we can see that the displacement of the internal particles is quite different from the FEM solutions; it is clear that the differences are sometimes noticeable. This can be evidenced by von Mises' graphs. In several attempts, almost none emerges as the best. With these

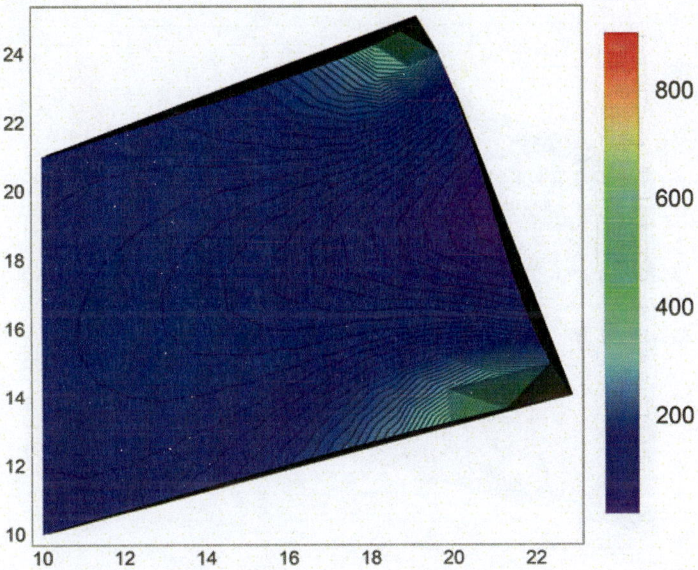

Fig. 14.12 Corresponding von Mises plot

Fig. 14.13 Beam deformation

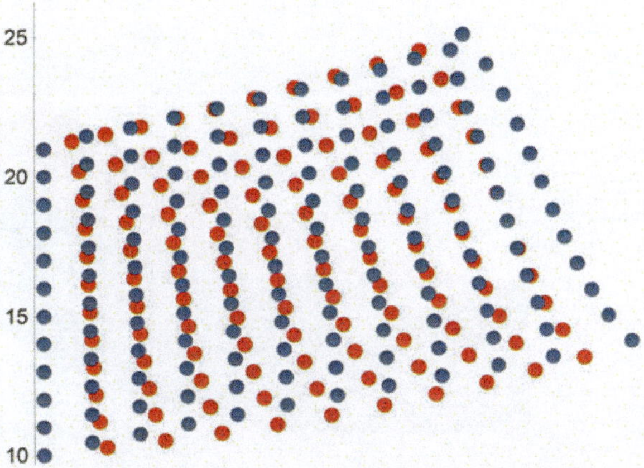

results, we can conclude that the deformation of the beam can sometimes be very similar to the FEM solution but the stress diagram of von Mises is quite always unsatisfactory. We now stop trial and errors to think about the needed connection between the tool's parameters and the constitutive features of the material. So far we have to think about the needs to a better understanding of the physic behind the tool and what should be driving our choices in the tool to describe a material continuum. We have to remember that the material parameters Y and ν do not appear explicitly in our algorithm but they are hidden in the interaction relationship between

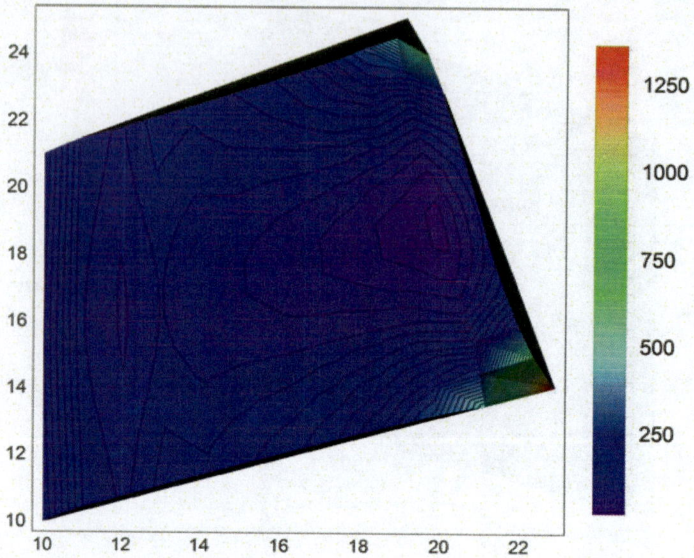

Fig. 14.14 Corresponding von Mises plot

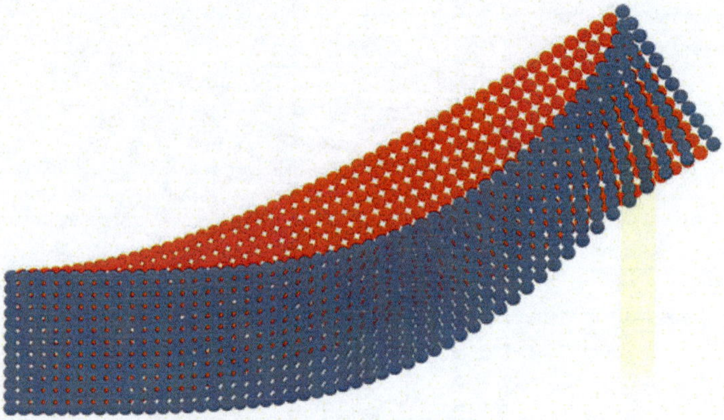

Fig. 14.15 Square lattice

the followers, the neighbours and the choice of the lattice. So far we have no idea on how to select our choices to match the problems. This will be the object of the next paper.

Fig. 14.16 Square lattice reduced neighbours numbers

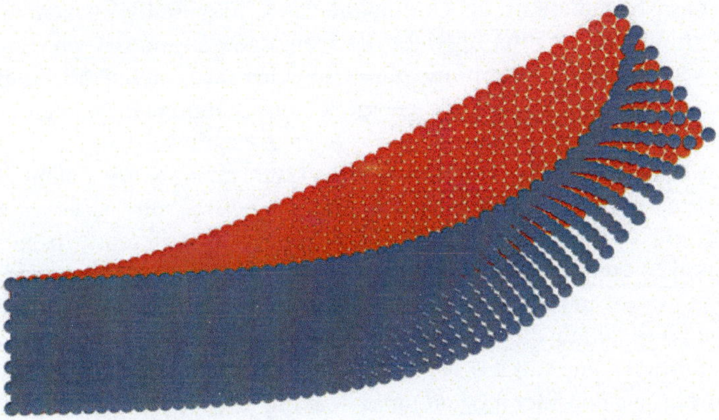

Fig. 14.17 Hexagonal lattice

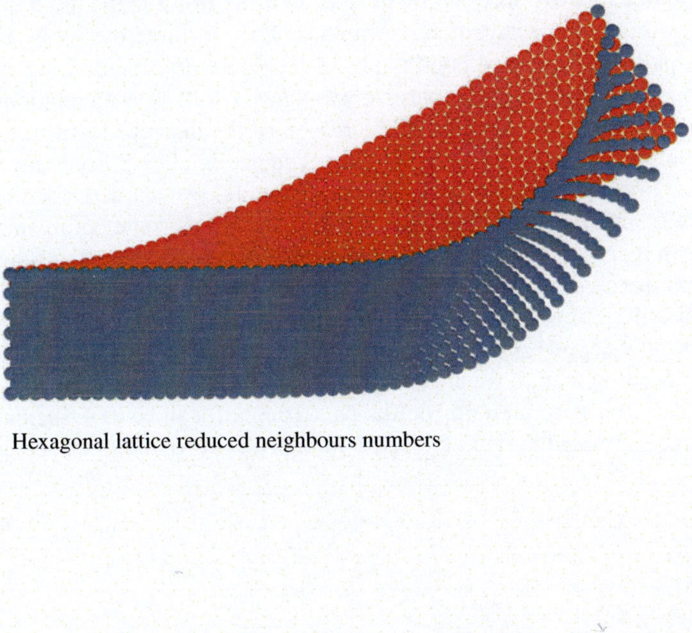

Fig. 14.18 Hexagonal lattice reduced neighbours numbers

14.5 Future Work and Conclusion

Flocking rule, used in robot swarm, can be used to describe the deformation of a bidimensional continuum by a simple algorithm highly customizable and able to adapt to take into account complex physical effects in a plausible way. The strain is imposed on leader particles whose motion is assigned and the "follower" particles move according to "position rules". These displacements are determined, like in a bird swarm, by the relative position of their neighbours. Therefore, the deformed configuration is calculated not by Newton's law but only by the relative positions between the particles of the system saving machine time for computing; edge effects are taken into account by a frame, and fracture mechanism is described by a threshold effect. Changes of some parameters, like lattice, interaction rules, fracture distance and numbers of neighbours lead to different behaviours. In previous works, we showed that the results of this tool have, in some cases, good similarity with the predictions of standard FEM simulations, also in fracture cases. Essentially we have to compute the action of a transformation operator between matrices and the job can be parallelized between the GPU cores of the powerful video card, saving the computational cost needed to solve FEMs. Pseudo-energetic considerations have been introduced to describe different deformation regimes, such as elastic and plastic, but only in a preliminary form that still has to be cleared; the future aim is to find potential descriptive interactions depending on the relative distance between the particles, which are able to reproduce the well-known physical behaviour. Differently from the PBD methods used in computer graphics, we still do not ask for the knowledge of the velocity and do not introduce any kind of forces to take into account mechanical effects. The presented results are interesting but they still are at a preliminary stage. We have collected some success showing plausible deformations in different conditions, but when we consider a beam under loading the need to connect constitutive equations with the parameters of our tool emerges powerfully. Thus, the most important topic is to understand how material parameters are related with the choices we make in our tool, to make a connection with the usual methods of continuum mechanics. Actually, we have no criteria about how to direct our tools' choices to describe a particular material continuum. In the beam deformation, we have used Young's modulus and Poisson's coefficient assigned but there is no relationship with the parameters of our model. This is the reason of the discrepancy in the resulting deformation. However, the tool has demonstrated enough flexibility to give chances that, once connected with the constitutive parameters, we can describe many other behaviours. Generalization in 3D would be easy but needs some optimization in the code to keep the computation time in the order of seconds, by using a normal PC Desktop. We are relaxing the hypothesis of Lagrangian neighbours to describe liquid and gas; this forces us to add a calculation step. Indeed, it is necessary to compute the neighbours, now defined as the particles inside a specified volume, at each time step. Another interesting feature we are introducing is to add constraints on the particle's motion to describe structured object like pantograph. It can be described as a set of beams with constrained point in the pivot; in the Hencky vision, it can be conceived as a set of point interconnected by springs. Further developments are concerning different fracture mechanism, different frame to avoid edge effects, other interaction rules and adaptive lattices. Cellular automata seems to be a good candidate to enhance our work; cellular automata is a simple computational mechanism that, for example, changes the colour of each cell on a lattice based on the colour of neighbours cells according to a transformation rule. Some attempts to use them in mechanics have been done, but one of their principal limits is that they do not evolve sufficiently, so they quickly reach a limited asymptote in their order of complexity. An evolutionary process involving conflict and competition is needed, like in biology systems. Moreover, there is no way to predict the outcome of a cellular process without actually running the process. So even though our decisions are determined, there is no way to predetermine what these decisions will be. But the system has succeeded, especially in fluid dynamics to describe complex behaviour. The question of whether one can work on patterns of information, rather than matter and energy, as the more fundamental building blocks of reality is still open, and we would like to make a connection with our tool.

Funding The authors received no financial support for the research, authorship, and/or publication of this article.

References

Abali, B. E., Müller, W. H., & Eremeyev, V. A. (2015). Strain gradient elasticity with geometric nonlinearities and its computational evaluation. *Mechanics of Advanced Materials and Modern Processes, 1*(1), 4.

Abali, B. E., Müller, W. H., & Dell'Isola, F. (2017). Theory and computation of higher gradient elasticity theories based on action principles. *Archive of Applied Mechanics, 87*(9), 1495–1510.

Abdoul-Anziz, H., & Seppecher, P. (2018). Strain gradient and generalized continua obtained by homogenizing frame lattices. *Mathematics and Mechanics of Complex Systems, 6*(3), 213–250.

Alibert, J. J., Seppecher, P., & Dell'Isola, F. (2003). Truss modular beams with deformation energy depending on higher displacement gradients. *Mathematics and Mechanics of Solids, 8*(1), 51–73.

Altenbach, H., & Eremeyev, V. A. (2009). On the linear theory of micropolar plates. *ZAMM-Journal of Applied Mathematics and Mechanics/Zeitschrift Für Angewandte Mathematik Und Mechanik, 89*(4), 242–256.

Altenbach, H., Bîrsan, M., & Eremeyev, V. A. (2013). Cosserat-type rods. In *Generalized Continua from the Theory to Engineering Applications* (pp. 179–248). Springer, Vienna.

Andreaus, U., Spagnuolo, M., Lekszycki, T., & Eugster, S. R. (2018). A Ritz approach for the static analysis of planar pantographic structures modeled with nonlinear Euler-Bernoulli beams. *Continuum Mechanics and Thermodynamics, 30*(5), 1103–1123.

Auffray, N., Dirrenberger, J., & Rosi, G. (2015). A complete description of bi-dimensional anisotropic strain-gradient elasticity. *International Journal of Solids and Structures, 69*, 195–206.

Avella, M., dell'Erba, R., Martuscelli, E., & Ragosta, G. (1993). Influence of molecular mass, thermal treatment and nucleating agent on structure and fracture toughness of isotactic polypropylene. *Polymer, 34*(14), 2951–2960.

Avella, M., dell'Erba, R., D'Orazio, L., & Martuscelli, E. (1995). Influence of molecular weight and molecular weight distribution on crystallization and thermal behavior of isotactic polypropylene. *Polym. Netw. Blends, 5*(1), 47–54.

Avella, M. delL'Erba, R., Martuscelli E. «Fiber reinforced polypropylene: Influence of iPP molecular weight on morphology, crystallization, and thermal and mechanical properties», *Polym. Compos.*, vol. 17, n. 2, pagg. 288–299, 1996.

Balobanov, V., Kiendl, J., Khakalo, S., & Niiranen, J. (2019). Kirchhoff-Love shells within strain gradient elasticity: Weak and strong formulations and an H3-conforming isogeometric implementation. *Computer Methods in Applied Mechanics and Engineering, 344,* 837–857.

Barchiesi, E., Ganzosch, G., Liebold, C., Placidi, L., Grygoruk, R., & Müller, W. H. (2019a). Out-of-plane buckling of pantographic fabrics in displacement-controlled shear tests: Experimental results and model validation. *Continuum Mechanics and Thermodynamics, 31*(1), 33–45.

Barchiesi, E., Spagnuolo, M., & Placidi, L. (2019b). Mechanical metamaterials: A state of the art. *Mathematics and Mechanics of Solids, 24*(1), 212–234.

Battista, A., Rosa, L., dell'Erba, R., & Greco, L. (2016). Numerical investigation of a particle system compared with first and second gradient continua: Deformation and fracture phenomena. *Mathematics and Mechanics of Solids*, 1081286516657889.

Bender, J., Müller, M., & Macklin, M. (2015). Position-based simulation methods in computer graphics. *Eurographics* (*Tutorials*). Available at: https://www.researchgate.net/profile/Jan_Bender/publication/274940214_Position-Based_Simulation_Methods_in_Computer_Graphics/links/552cc4a40cf29b22c9c466df/Position-Based-Simulation-Methods-in-Computer-Graphics.pdf. [Consultato: 06-set-2017].

Carcaterra, A., dell'Isola, F., Esposito, R., & Pulvirenti, M. (2015). Macroscopic description of microscopically strongly inhomogenous systems: A mathematical basis for the synthesis of higher gradients metamaterials. *Archive for Rational Mechanics and Analysis, 218*(3), 1239–1262.

Cazzani, A., Malagù, M., & Turco, E. (2016a). Isogeometric analysis: A powerful numerical tool for the elastic analysis of historical masonry arches. *Continuum Mechanics and Thermodynamics, 28*(1–2), 139–156.

Cazzani, A., Malagù, M., Turco, E., & Stochino, F. (2016b). Constitutive models for strongly curved beams in the frame of isogeometric analysis. *Mathematics and Mechanics of Solids, 21*(2), 182–209.

De Angelo, M., Spagnuolo, M., D'annibale, F., Pfaff, A., Hoschke, K., Misra, A., & Pawlikowski, M. (2019). The macroscopic behavior of pantographic sheets depends mainly on their microstructure: Experimental evidence and qualitative analysis of damage in metallic specimens. *Continuum Mechanics and Thermodynamics*, 1–23.

dell'Erba, R. (2012). *The localization problem for an underwater swarm.* ENEA: Technical Report.

dell'Erba, R., & Moriconi, C. (2014). Bio-inspired robotics—it. Available at: https://www.enea.it/it/produzione-scientifica/edizioni-enea/2014/bio-inspirede-robotics-proceedings. [Consultato: 15-dic-2014].

dell'Erba, R. (2015) Determination of spatial configuration of an underwater swarm with minimum data. *International Journal of Advanced Robotic Systems, 12*(7), 97.

dell'Erba, R. (2018a). Swarm robotics and complex behaviour of continuum material. *Continuum Mechanics and Thermodynamics*. Available at:https://doi.org/10.1007/s00161-018-0675-1.

dell'Erba, R. (2018b). Swarm robotics and complex behaviour of continuum material. *Continuum Mechanics and Thermodynamics*, 1–26.

dell'Erba, R. (2018c). Position-based dynamic of a particle system: a configurable algorithm to describe complex behaviour of continuum material starting from swarm robotics. *Continuum Mechanics and Thermodynamics, 30*(5), 1069–1090.

dell'Erba, R. (2018d). Position-based dynamic of a particle system: A configurable algorithm to describe complex behaviour of continuum material starting from swarm robotics. *Continuum Mechanics and Thermodynamics*, 1–22.

dell'Erba, R., & Moriconi, C. (2015). High power leds in localization of underwater robotics swarms. *IFAC-Paper, 48*(10), 117–122.

dell'Isola, F., & Placidi, L. (2011). Variational principles are a powerful tool also for formulating field theories. In *Variational models and methods in solid and fluid mechanics* (pp. 1–15). Springer, Vienna.

dell'Isola, F, Auffray, N., Eremeyev, V. A., Madeo, A., Placidi, L., Rosi, G. (2014) Least action principle for second gradient continua and capillary fluids: A lagrangian approach following piola's point of view. In *The complete works of Gabrio Piola: Volume I*, Springer, pp. 606–694.

Dell'Isola, F., Gavrilyuk, S. (2012). *Variational models and methods in solid and fluid mechanics*, vol. 535. Springer Science & Business Media.

dell'Isola, F., Seppecher, P., & Corte, A. D. (2015a). The postulations á la D'Alembert and á la Cauchy for higher gradient continuum theories are equivalent: A review of existing results. *Proceedings of the Royal Society A: Mathematical, Physical and Engineering Sciences, 471*(2183), 20150415.

dell'Isola, F., Lekszycki, T., Pawlikowski, M., Grygoruk, R., & Greco, L. (2015b). Designing a light fabric metamaterial being highly macroscopically tough under directional extension: First experimental evidence. *Zeitschrift Für Angewandte Mathematik Und Physik, 66*(6), 3473–3498.

dell'Isola, F., Della Corte, A., Greco, L., & Luongo, A. (2016a). Plane bias extension test for a continuum with two inextensible families of fibers: A variational treatment with lagrange multipliers and a perturbation solution. *International Journal of Solids and Structures, 81*, 1–12

dell'Isola, F., d'Agostino, M. V., Madeo, A., Boisse, P., & Steigmann, D. (2016b). Minimization of shear energy in two dimensional continua with two orthogonal families of inextensible fibers: The case of standard bias extension test. *Journal of Elasticity, 122*(2), 131–155.

dell'Isola, F., Della Corte, A., Greco, L., & Luongo, A. (2016c). Plane bias extension test for a continuum with two inextensible families of fibers: A variational treatment with Lagrange multipliers and a perturbation solution. *International Journal of Solids and Structures, 81*, 1–12.

dell'Isola, F., Madeo, A., & Seppecher, P. (2016d). Cauchy tetrahedron argument applied to higher contact interactions. *Archive for Rational Mechanics and Analysis, 219*(3), 1305–1341.

dell'Isola, F., Corte, A. D., & Giorgio, I. (2017). Higher-gradient continua: The legacy of Piola, Mindlin, Sedov and Toupin and some future research perspectives. *Mathematics and Mechanics of Solids, 22*(4), 852–872.

dell'Isola, F., Seppecher, P., Spagnuolo, M., Barchiesi, E., Hild, F., Lekszycki, T., Eugster, S. R., et al. (2019a). Advances in pantographic structures: Design, manufacturing, models, experiments and image analyses. *Continuum Mechanics and Thermodynamics*, 1–52.

Dell'Isola, F., Seppecher, P., Alibert, J. J., Lekszycki, T., Grygoruk, R., Pawlikowski, M., & Gołaszewski, M., et al. (2019b). Pantographic metamaterials: An example of mathematically driven design and of its technological challenges. *Continuum Mechanics and Thermodynamics, 31*(4), 851–884.

dell'Isola, F., Turco, E., Misra, A., Vangelatos, Z., Grigoropoulos, C., Melissinaki, V., & Farsari, M. (2019c). Force–displacement relationship in micro-metric pantographs: experiments and numerical simulations. *Comptes Rendus Mécanique, 347*(5), 397–405.

Dell'Erba, R. (2001). Rheo-mechanical and rheo-optical characterisation of ultra high molecular mass poly (methylmethacrylate) in solution. *Polymer, 42*(6), 2655–2663.

Eremeyev, V. A. (2018). A nonlinear model of a mesh shell. *Mechanics of Solids, 53*(4), 464–469.

Eremeyev, V. A. (2019a). Strongly anisotropic surface elasticity and antiplane surface waves. *Philosophical Transactions of the Royal Society A, 378*(2162), 20190100.

Eremeyev, V. A. (2019b). Two-and three-dimensional elastic networks with rigid junctions: Modeling within the theory of micropolar shells and solids. *Acta Mechanica, 230*(11), 3875–3887.

Eremeyev, V. A., & Sharma, B. L. (2019). Anti-plane surface waves in media with surface structure: discrete vs. continuum model. *International Journal of Engineering Science, 143*, 33–38.

Eremeyev, V. A., Lebedev, L. P., & Altenbach, H. (2012). *Foundations of micropolar mechanics*. Springer Science & Business Media.

Eremeyev, V. A., Dell'Isola, F., Boutin, C., & Steigmann, D. (2018). Linear pantographic sheets: Existence and uniqueness of weak solutions. *Journal of Elasticity, 132*(2), 175–196.

Golaszewski, M., Grygoruk, R., Giorgio, I., Laudato, M., & Di Cosmo, F. (2019). Metamaterials with relative displacements in their microstructure: Technological challenges in 3D printing, experiments and numerical predictions. *Continuum Mechanics and Thermodynamics, 31*(4), 1015–1034.

Green, A. E. (1965). Micro-materials and multipolar continuum mechanics. *International Journal of Engineering Science, 3*(5), 533–537.

Khakalo, S., & Niiranen, J. (2017). Isogeometric analysis of higher-order gradient elasticity by user elements of a commercial finite element software. *Computer-Aided Design, 82,* 154–169.

Lanczos, C. (2012). *The variational principles of mechanics*. Courier Corporation.

Maugin, G. A. (2010). Generalized continuum mechanics: what do we mean by that?. In *Mechanics of Generalized Continua* (pp. 3–13). Springer, New York.

Mindlin, R. D. (1965). Second gradient of strain and surface-tension in linear elasticity. *International Journal of Solids and Structures, 1*(4), 417–438.

Misra, A., & Poorsolhjouy, P. (2015). Granular micromechanics model for damage and plasticity of cementitious materials based upon thermomechanics. *Mathematics and Mechanics of Solids*, 1081286515576821.

Misra, A., & Singh, V. (2013). Micromechanical model for viscoelastic materials undergoing damage. *Continuum Mechanics and Thermodynamics, 25*(2–4), 343–358.

Misra, A., & Singh, V. (2015). Thermomechanics-based nonlinear rate-dependent coupled damage-plasticity granular micromechanics model. *Continuum Mechanics and Thermodynamics, 27*(4–5), 787–817.

Misra, A., Lekszycki, T., Giorgio, I., Ganzosch, G., Müller, W. H., & Dell'Isola, F. (2018). Pantographic metamaterials show atypical pointing effect reversal. *Mechanics Research Communications, 89,* 6–10.

Moriconi, C., & dell'Erba, R. (2012). The localization problem for harness: A multipurpose robotic swarm. In *SENSORCOMM 2012, The Sixth International Conference on Sensor Technologies and Applications*, pp. 327–333. Available at: https://www.thinkmind.org/index.php?view=article&articleid=sensorcomm_2012_14_20_10138. [Consultato: 04-apr-2014].

Nejadsadeghi, N., De Angelo, M., Drobnicki, R., Lekszycki, T., dell'Isola, F., & Misra, A. (2019). Parametric experimentation on pantographic unit cells reveals local extremum configuration. *Experimental Mechanics*, 1–13.

Niiranen, J., Khakalo, S., Balobanov, V., & Niemi, A. H. (2016). Variational formulation and isogeometric analysis for fourth-order boundary value problems of gradient-elastic bar and plane strain/stress problems. *Computer Methods in Applied Mechanics and Engineering, 308,* 182–211.

Placidi, L., & El Dhaba, A. R. (2017). Semi-inverse method à la Saint-Venant for two-dimensional linear isotropic homogeneous second-gradient elasticity. *Mathematics and Mechanics of Solids, 22*(5), 919–937.

Placidi, L., Dell'Isola, F., Ianiro, N., & Sciarra, G. (2008) Variational formulation of pre-stressed solid-fluid mixture theory, with an application to wave phenomena. *European Journal of Mechanics A Solids, 27*(4), 582–606.

Placidi, L., Andreaus, U., Della Corte, A., & Lekszycki, T. (2015). Gedanken experiments for the determination of two-dimensional linear second gradient elasticity coefficients. *Zeitschrift Für Angewandte Mathematik Und Physik, 66*(6), 3699–3725.

Placidi, L., Greco, L., Bucci, S., Turco, E., & Rizzi, N. L. (2016). A second gradient formulation for a 2D fabric sheet with inextensible fibres. *Zeitschrift Für Angewandte Mathematik Und Physik, 67*(5), 114.

Placidi, L., Andreaus, U., & Giorgio, I. (2017). Identification of two-dimensional pantographic structure via a linear D4 orthotropic second gradient elastic model. *Journal of Engineering Mathematics, 103*(1), 1–21.

Sharma, B. L., & Eremeyev, V. A. (2019). Wave transmission across surface interfaces in lattice structures. *International Journal of Engineering Science, 145,* 103173.

Spagnuolo, M., & Andreaus, U. (2019). A targeted review on large deformations of planar elastic beams: Extensibility, distributed loads, buckling and post-buckling. *Mathematics and Mechanics of Solids, 24*(1), 258–280.

Spagnuolo, M., Barcz, K., Pfaff, A., Dell'Isola, F., & Franciosi, P. (2017). Qualitative pivot damage analysis in aluminium printed pantographic sheets: Numerics and experiments. *Mechanics Research Communications, 83,* 47–52.

Spagnuolo, M., Peyre, P., & Dupuy, C. (2019). Phenomenological aspects of quasi-perfect pivots in metallic pantographic structures. *Mechanics Research Communications, 101,* 103415.

Steigmann, D. J., & Faulkner, M. G. (1993). Variational theory for spatial rods. *Journal of Elasticity, 33*(1), 1–26.

Toupin, R. A. (1964). Theories of elasticity with couple-stress.

Turco, E., & Rizzi, N. L. (2016). Pantographic structures presenting statistically distributed defects: Numerical investigations of the effects on deformation fields. *Mechanics Research Communications, 77,* 65–69.

Turco, E., Barcz, K., Pawlikowski, M., & Rizzi, N. L. (2016a). Non-standard coupled extensional and bending bias tests for planar pantographic lattices. Part I: Numerical simulations. *Zeitschrift für angewandte Mathematik und Physik, 67*(5), 122.

Turco, E., Golaszewski, M., Cazzani, A., & Rizzi, N. L. (2016b). Large deformations induced in planar pantographic sheets by loads applied on fibers: Experimental validation of a discrete Lagrangian model. *Mechanics Research Communications, 76,* 51–56.

Turco, E., Dell'Isola, F., Rizzi, N. L., Grygoruk, R., Müller, W. H., & Liebold, C. (2016c). Fiber rupture in sheared planar pantographic sheets: Numerical and experimental evidence. *Mechanics Research Communications, 76,* 86–90.

Turco, E., Giorgio, I., Misra, A., & Dell'Isola, F. (2017). King post truss as a motif for internal structure of (meta) material with controlled elastic properties. *Royal Society Open Science, 4*(10), 171153.

Turco, E., Misra, A., Pawlikowski, M., Dell'Isola, F., & Hild, F. (2018). Enhanced Piola-Hencky discrete models for pantographic sheets with pivots without deformation energy: Numerics and experiments. *International Journal of Solids and Structures, 147,* 94–109.

Yang, H., & Müller, W. H. (2019). Computation and experimental comparison of the deformation behavior of pantographic structures with different micro-geometry under shear and torsion. *Journal of Theoretical and Applied Mechanics, 57.*

Yang, H., Ganzosch, G., Giorgio, I., & Abali, B. E. (2018). Material characterization and computations of a polymeric metamaterial with a pantographic substructure. *Zeitschrift Für Angewandte Mathematik Und Physik, 69*(4), 105.

Yildizdag, M. E., Tran, C. A., Barchiesi, E., Spagnuolo, M., dell'Isola, F., & Hild, F. (2019). A multi-disciplinary approach for mechanical metamaterial synthesis: A hierarchical modular multiscale cellular structure paradigm. In *State of the Art and Future Trends in Material Modeling* (pp. 485–505). Springer, Cham.

Chapter 15
Homogenization-Based Mechanical Behavior Modeling of Composites Using Mean Green Operators for Infinite Inclusion Patterns or Networks Possibly Co-continuous with a Matrix

Patrick Franciosi and Mario Spagnuolo

Abstract The Fourier–Green homogenization method for estimating the behavior of composites was first developed for aggregates and inhomogeneity-reinforced (-weakened) matrices, based on Eshelby (Proc R Soc Lond, A 421:379–396, 1957, Proc R Soc Lond, A 252:561–569, 1959) solution of the isolated inclusion problem. The need to address increasingly complex structures opened fruitful development routes, firstly solving the inhomogeneity pair interaction problem and the one of heterogeneous (double or multilayered) inhomogeneities, in order to account for inclusion dense concentrations and patterns. This work reports recent developments from the authors and co-workers which examined in that framework possibly infinite inclusion patterns, possibly arranged into an infinite network possibly co-continuous with the embedding matrix and possibly evolving under strain. The proposed modeling amounts to determining the representative mean Green operator (mGO) for the infinite pattern or network in its current (strain evolving) state. Once the method foundations being summarized, previously solved "elementary" cases are recalled, concerning infinite coaxial alignments of spheres, spheroids or finite cylinders and planar alignments of infinite parallel cylinders or rectangular beams. It is next shown how other complex patterns or networks could be represented in combining such elementary ones. The mGO solution for a new family of inhomogeneous axial inclusion alignments is reported to support the discussion. Potential other application fields are evoked.

Keywords Composites · Inclusion patterns · Phase co-continuity · Green operators · Homogenization

P. Franciosi (✉)
CNRS, UPR3407 LSPM, Université Sorbonne Paris-Nord, 93430 Villetaneuse, France
e-mail: patrick.franciosi@univ-paris13.fr

M. Spagnuolo
International Research Center M&MoCS, Universitàdegli Studi dell'Aquila, L'Aquila, Italy

F. dell'Isola and L. Igumnov (eds.), *Dynamics, Strength of Materials and Durability in Multiscale Mechanics*, Advanced Structured Materials 137,
https://doi.org/10.1007/978-3-030-53755-5_15

15.1 Introduction

The modeling of the overall behavior of heterogeneous macro-homogeneous materials and the estimate of their effective properties have experienced considerable developments from the Eshelby solution of the stress and strain fields in and out a finite ellipsoidal inhomogeneity in an infinite matrix subjected to stress or strain loading conditions at infinity (Eshelby 1957, 1959). Having first established that interior fields in an ellipsoid resulting from uniform loading at infinity were also uniform, a "boulevard" was open for applications to, on the one hand composites of the reinforced-matrix type, say a uniform matrix embedding particles of one or several other phases, yielding to Hashin and Shtrikman (1963) variational estimate type of effective properties (Walpole 1981; Willis 1981), and on the other hand granular materials, including polycrystals, where each grain type could in turn be considered as embedded into some equivalent uniform matrix to the aggregate, with the self-consistent scheme being the leading one for these structures (Hill 1952, 1965; Kröner 1958, 1990; Zeller and Dederich 1973).

A considerable and still rapidly expanding, more or less all "Eshelby-based", literature followed, exploding with the progress in computational means meanwhile, impossible to summarize here where only a few ones will be cited, with several key steps being taken in the following decades and up to now: from linear elasticity to other (as dielectric or thermal) properties and to nonlinear (including plastic) mechanical behavior (Levin 1967; Kröner 1990; Ponte-Castaneda and Suquet 1998; Suquet 1997; Clyne et al. 2005; Buryachenko and Brun 2012), from isolated inclusions to groups, either in statistical terms (Ponte-Castaneda and Willis 1995; Bornert et al. 1996) or in deterministic manner from the solution of the (ellipsoidal) inhomogeneity pair interaction problem (Berveiller et al. 1987), also from homogeneous to heterogeneous, multi-phased inclusions, from either the proposal of composite inclusion models (Christensen and Lo 1979) or the solution of the double inclusion problem, an inhomogeneity inside an inhomogeneity (Hori and Nemat-Nasser 1993; Hu and Weng 2000) and for particular (spherical, cylindrical) multilayered structures (Hervé and Zaoui 1993, 1995) to only mention major milestones.

With regard to this state of the art from the last century, it is noteworthy that, within the so-called Fourier–Green homogenization framework applications (as recalled next on, the Eshelby tensor is intimately related to a Green operator with "better" properties), a missing composite structure was the one where possibly several to all phases are co-continuous, with still several phases possibly being discontinuous (say embedded), although several attempts for two-phase co-continuity without follow-up were presented in (Postma 1955; Boucher 1974; Christensen 1979a). A new start was proposed in (Franciosi and El Omri 2011) in terms of "laminate system schemes" for two-phase bi-continuous composites first and then for any p phase number among n in (Franciosi and Charles 2016a; Franciosi 2020a). Although the bi-continuous two-phase composites are the most frequent co-continuous structures with a variety that goes from sponge-like or foam-like structures (Roberts and Garboczi 2002; Gong et al. 2005; Bender et al. 2008) to dual phase metallic materials (Delannay et al.

2006; McCue et al. 2016) or ceramics (Pavese et al. 2007) in passing by woven textile composites (Udhayaraman and Mulay 2017), bones (Kinney et al. 2005), metal matrix (Peng et al. 2001), compacted powders (Poquillon et al. 2002) or metamaterials (Yang et al. 2018), existing three-phase structures with phase concentration domains exhibiting tri-continuity (Han et al. 2009; Ryoo 2009) as well as assemblages where phases duplicate in partly continuous and partly discontinuous (Veenstra et al. 2000; Limodin et al. 2007; Torres et al. 2012) made this extension to multiple co-continuity of phases worthy of interest as applications show in (Franciosi 2020a).

In this enormous flux of developments for analysing the overall behavior of increasingly varied composite structures several issues are still scarcely investigated: non-ellipsoidal inhomogeneity shapes (Walpole 1967; Kinoshita and Mura 1971; Chiu 1977; Rodin 1996; Franciosi 2005; Traxl and Lackner 2018), which are still so far abusively approximated by ellipsoids [a sphere would then represent both a cube and an octahedron which are opposite polyhedrons (Franciosi and Lormand 2004) and a finite cylinder is badly represented by a spheroid of same aspect ratio (Franciosi et al. 2016)]; aside of inclusions distributions (Ponte-Castaneda and Willis 1995; Buryachenko 2001; Franciosi and Lebail 2004), many specific inclusion patterns with potentially significant interactions when the group density becomes high and for any pattern size, possibly up to infinite elements, still need improved descriptions. Limit cases of interest are when the elements in the group become at contact, what may possibly yield to connectivity inversions (Franciosi and Gaertner 1998) and to the transition toward interconnected pattern elements. This last problem is the one to solve for describing an inclusion pattern turning from an embedded (discontinuous) phase into a continuous (networked) one or a multi-continuous state change in a n-phase composite.

The authors and co-workers have addressed several of these questions in recent works (Franciosi 2010; Franciosi et al. 2015, 2019; Spagnuolo et al. 2020; Franciosi 2020b), offering some results which may serve further to estimate effective properties for several composite structures comprising a uniform matrix and a second phase (or more): A second phase can be described by several infinite arrangements (as linear or planar arrays) of elements which can then in turn be spatially organized into a single pattern having one- or multi-directional (mD) co-continuity with a reference matrix phase (the volume fraction of which possibly going down to zero, that is infinitesimal). Such arrangements can thus realize a mD-networked pattern of second phase ($m = 1, 2, 3, \ldots$, omni), the element interconnections being either physically present or mathematically described, as will be commented and exemplified. We here summarize, with a minimum of equations to be complemented by the referenced papers, these previous results with this particular viewpoint in mind, and we aim at projecting them toward potential diversified applications on complex composite structures and architectures at different (from micro to macro) scales. A new type of inhomogeneous axial alignment of inclusions, built from a mixed alignment of several inclusion shapes, is presented and solved (not in the details) for spheroids and finite cylinders axial associations, serving as support to the discussion.

Section 15.2 briefly recalls the foundations of the homogenization-based model we make use, namely the (Ponte-Castaneda and Willis 1995) framework, to be denoted as "PCW", it points the essential role in it of the (rapidly reintroduced) GOs, which are an advantageous equivalent to the Eshelby tensors, and it presents the calculation of these operators from the Radon and inverse (*RT-IRT*) transform method which was used in all the cited works by the authors team on infinite inclusion patterns. Section 15.3 reports and discusses some of the obtained results, in terms of their representative mean GOs (mGOs) which determine the effective properties of the composites embedding them, as well as the (mean) stress and strain fields in the pattern in comparison with those applied to the composite: from the cases of axial infinite alignments of spheres and spheroids (Franciosi 2010) and of finite, flat or long, cylinders (Franciosi et al. 2015) and then from the cases of planar alignments of infinite parallel cylinders (Franciosi et al. 2019) and rectangular and square beams (Franciosi 2018; Spagnuolo et al. 2020). Based on the results of these works on "elementary" alignments, it is next in Sect. 15.4 discussed and exemplified how mD networks can be built to be then embedded in a matrix and possibly be evolving under straining of the composite structure (the new case of inhomogeneous axial alignment of inclusions is introduced there), as was successfully attempted in building 1D fiber bundles and 3D fiber networks from planar fiber arrays. The last example of the building of a 3D fiber network from assembling planar arrays of beams, reported from (Spagnuolo et al. 2020) with the array interconnections being inspired by those of pantographic 2D fiber architectures (Alibert et al. 2003; Andreaus et al. 2018; Barchiesi and Placidi 2017; Barchiesi et al. 2019; Boutin et al. 2017; Cuomo et al. 2017; dell'Isola et al. 2015, 2016, 2017; Eremeyev et al. 2016, 2019; Eremeyev 2018; Placidi et al. 2016, 2017, 2018, 2019), points that mathematical constraints describing the physical interconnection mechanisms can substitute them efficiently if hard to account for in the pattern description. Section 15.5 evokes different potential other application fields of the presented results beyond the one of linear static elastic-like (and dielectric-like) ones, in pointing the remaining difficulties to overcome for progressing further. Section 15.6 concludes.

15.2 Effective (Piece-Wise) Linear Elastic-like Properties of Composites of the Matrix-Reinforced Type

We here select the two-point statistic framework of (Ponte-Castaneda and Willis 1995), which is an extension of the well-defined variational approach from Hashin and Shtrikman (1963), to be referred to as "PCW" and "HS", respectively, for a uniform medium with (stiffness) properties $\boldsymbol{C}^M$ in which are embedded elements of one or n phases (Vj), with properties $\boldsymbol{C}^{Vj}$ and volume fractions f_{Vj}, with an a priori single request of not overpassing space filling at $\sum_{j=1}^{n} f_{Vj} = 1$ (that is a null matrix phase fraction $f_M = 0$). For the following, it is enough to recall that it takes the generic form:

$$\boldsymbol{C}^{Vj/\mathrm{SDist}}_{\mathrm{effPCW}} = \boldsymbol{C}^M - \left(\left(\sum_{j=1}^{n}\left(f_{Vj}\left(\left(\boldsymbol{C}^M - \boldsymbol{C}^{Vj}\right)^{-1} - \overline{\boldsymbol{t}^{Vj}_{\boldsymbol{C}^M}}\right)\right)^{-1}\right)^{-1} + \boldsymbol{t}^{\mathrm{SDist}}_{\boldsymbol{C}^M}\right)^{-1} \tag{15.1a}$$

where $\overline{\boldsymbol{t}^{Vj}_{\boldsymbol{C}^M}}$ is formally the mGO of some representative domain Vj for phase j and $\boldsymbol{t}^{\mathrm{SDist}}_{\boldsymbol{C}^M}$ is the uniform GO representing some spatial distribution of ellipsoidal symmetry for all these Vj domains. These GOs and mGOs attached to some domain V to be denoted $\boldsymbol{t}^{V}_{\boldsymbol{C}^M}(\boldsymbol{r})$ and $\overline{\boldsymbol{t}^{V}_{\boldsymbol{C}^M}}$, respectively, (and $\boldsymbol{t}^{V}_{\boldsymbol{C}^M}$ for uniform GOs of ellipsoids) are the second spatial derivative of the strain Green elastic (rank-2) tensor $\boldsymbol{G}(\boldsymbol{r} - \boldsymbol{r}')$ which is defined from the relation $\boldsymbol{u}(\boldsymbol{r}) = \boldsymbol{G}(\boldsymbol{r} - \boldsymbol{r}')\boldsymbol{f}(\boldsymbol{r}')$ between the displacement $\boldsymbol{u}(\boldsymbol{r})$ at any point $\boldsymbol{r}$ and the force $\boldsymbol{f}(\boldsymbol{r}')$ at another point $\boldsymbol{r}'$ in the material. Although, as a result of a variational approach, Eq. (15.1a) can be obtained as a result of a variational approach (see the cited PCW and HS references) in minimizing a strain energy functional for some selected reference medium.[1] It can also be explicated from solving the local stress equilibrium condition $\mathrm{Div}\boldsymbol{\sigma}(\boldsymbol{r}) = 0$ in a heterogeneous macro-homogeneous infinite material of elastic property tensor field $\boldsymbol{C}(\boldsymbol{r}) = \boldsymbol{C}^M + \Delta\boldsymbol{C}(\boldsymbol{r})$. They are related to the local and mean (or uniform) Eshelby tensors for V, say $\boldsymbol{E}^{V}_{\boldsymbol{C}^M}(\boldsymbol{r})$, $\overline{\boldsymbol{E}^{V}_{\boldsymbol{C}^M}}$, $\boldsymbol{E}^{V}_{\boldsymbol{C}^M}$, through the elastic properties of the reference matrix as $\boldsymbol{t}^{V}_{\boldsymbol{C}^M}(\boldsymbol{r}) = \boldsymbol{E}^{V}_{\boldsymbol{C}^M}(\boldsymbol{r}) : \boldsymbol{C}^{M-1}$, $\overline{\boldsymbol{t}^{V}_{\boldsymbol{C}^M}} = \overline{\boldsymbol{E}^{V}_{\boldsymbol{C}^M}} : \boldsymbol{C}^{M-1}$ and without overbar when uniform (ellipsoid V), where $\boldsymbol{C}^{M-1}$ is the compliance (inverse stiffness) tensor of the $n+1$th reference matrix phase. In addition to minor symmetries (i, j), (k, l) GOs and mGOs for elastic properties which are consequently rank-4 operators also have major (ij, kl) symmetry, and they are positive definite, characteristics that Eshelby tensors to not benefit from.

The main improvement of the PCW estimate with regard to the HS one is to specifically account for that (homogeneous) spatial distribution of all the embedded inhomogeneities, a remaining restriction being that this distribution must obey an ellipsoidal symmetry for the PCW estimate to rigorously hold as formulated (accounting for real inclusion distributions which in general are hardly ellipsoidal and mono-modal (Franciosi and Lebail 2004), remains an open question). Yet, this major improvement has clarified why the HS estimate failed, especially at high inclusion concentrations, to yield relevant effective properties when all the assembled inhomogeneities in the matrix were not all congruent to a same shape, whether them be of a same (j) phase or of different ($i, j, k,\ldots$) ones. The HS framework proved to implicitly assume a spatial distribution of the inclusions arbitrarily "defined" by some average of the assembled inclusions shapes, possibly yielding inconsistent results (as effective properties with incorrect symmetries) and not guarantee to be valid up to total (compact) space filling, with the consequence of violating established optimal bounds for effective

[1] In contrast the also widely used MT (Mori and Tanaka 1973) estimate which, although identifying with the HS one in the simple case when the latter coincides with the PCW solution, does not and fails to satisfy the operator requested supersymmetries (and the estimate bounds) in most of the other cases (see Benveniste 1987).

properties when trespassing concentrations limits. Hence, the PCW estimate form which is recalled in Eq. (15.1a) involves one more GO than the HS estimate, independently on the mGOs for the inclusion shapes but with a request of consistency with these shapes and their relative concentrations. In contrast with the inclusion mGOs, owing to the request on the spatial distribution symmetry to be ellipsoidal, this GO is uniform for it is the one of an ellipsoidal shape. If all the inclusions have a same shape, they all have a same mGO (or GO if ellipsoids), and if this shape is ellipsoidal and is represented by the same ellipsoidal GO as the one characterizing the spatial distribution, total space filling by the inclusions is permitted and only then. The same reasoning can be transposed from isolated inclusions to patterns of them, with a spatial distribution GO applying to the patterns, with the smaller the "lost" matrix volume fraction at the pattern boundary, when the larger the pattern is. Infinite inclusion patterns become then a limit case for which the spatial distribution GO calls for a different interpretation. The cases of multi-phased pattern are a more complex issue for which Eq. (15.1a) needs be modified (Fassi-Fehri et al. 1989; Franciosi and Charles 2016b) not to be considered here, yet evoked in the last section. The PCW estimate for a single second phase consequently simplifies into

$$\boldsymbol{C}^{V/\mathrm{SDist}}_{\mathrm{effPCW}} = \boldsymbol{C}^M - f_V\left(\left(\left(\boldsymbol{C}^M - \boldsymbol{C}^V\right)^{-1} - \overline{\boldsymbol{t}^V_{\boldsymbol{C}^M}}\right) + f_V \boldsymbol{t}^{\mathrm{SDist}}_{\boldsymbol{C}^M}\right)^{-1}. \qquad (15.1\mathrm{b})$$

In Eq. (15.1b), the single included phase has a representative inclusion pattern V with $\overline{\boldsymbol{t}^V_{\boldsymbol{C}^M}}$ being its mGO, a volume fraction $f_V = 1 - f_M$ and elastic properties $\boldsymbol{C}^V$, and these patterns are spatially distributed according to some (ellipsoidal) symmetry represented by the $\boldsymbol{t}^{\mathrm{SDist}}_{\boldsymbol{C}^M}$ uniform GO. When such a composite is deformed, both the characteristic pattern of the embedded inclusions and their spatial distribution evolve. In the particular case of infinite inclusion patterns, it was shown in (Franciosi et al. 2019) that when the pattern evolution is well accounted for, its distribution symmetry can be taken determined by the influence zone symmetry around each (here fiber) element.

Now, explicit GOs or mGOs related to an infinite medium serving as a reference matrix are only at easy hand when this matrix has highly symmetric properties, the simplest case corresponding to isotropic properties and to less extents, transversally isotropic ones, or cubic ones. The higher is the matrix symmetry, the larger is the number of case for which a GO (for ellipsoidal shapes, including symmetry ones) or a mGO (for other shapes) can be calculated in manageable analytical form (otherwise, more or less costly computational calculations are always available). In most analytically treated cases, the reference matrix is taken as isotropic. In this limit, several mGOs corresponding to a variety of non-ellipsoidal shapes have been obtained in analytical (exact or closely approximate) forms by the authors and co-workers, with the help of using the (*RT-IRT*) Radon transform method and its inversion formula (Gel'fand et al. 1966; Natterer 1986; Helgason 1980; Ramm and Katsevich 1996). When in the literature mostly mathematical interests of the *RT-IRT* are called for, full benefit was taken that it allows to multiplicatively dissociate the (m)GOs into a purely geometrical part (through a "shape function" for a single inclusion or for patterns of it, or for an (ellipsoidal) distribution symmetry) and a purely matrix-property-dependent part, say the property type (elastic-like, dielectric-like or else) and the property anisotropy (through a set of elementary uniform GOs, one for each direction around the unit sphere Ω in R3 space). According to this *RT-IRT* form, the (m)GO operator of a *V* domain or pattern reads (skipping the dependency on $\boldsymbol{C}^M$), respectively, at each interior point ($\boldsymbol{r}$) of *V* and in terms of mean value over *V*:

$$t^V_{pqjn}(\boldsymbol{r}) = \int\limits_{\Omega} t^e_{pqjn}(\boldsymbol{\omega})\psi_V(\boldsymbol{\omega}, \boldsymbol{r})\mathrm{d}\boldsymbol{\omega}, \tag{15.2a}$$

$$\overline{t^V_{pqjn}} = \int\limits_{\Omega} t^e_{pqjn}(\boldsymbol{\omega})\overline{\psi_V(\boldsymbol{\omega})}\mathrm{d}\boldsymbol{\omega}. \tag{15.2b}$$

The way the elementary GOs $\boldsymbol{t}^e(\boldsymbol{\omega})$ showing up in Eqs. (15.2a, b) are obtained is summarized in most of the cited authors and co-worker references, starting with (Franciosi and Lormand 2004). Their form will be recalled in Eq. (15.12) and commented there. They identify with the (uniform) GOs of infinitely flat (oblate) spheroids, sometimes called platelets, or of laminate layers. Their obtaining is recalled in Appendix in the case of isotropic elastic (or elastic-like, that is rank-four) tensorial properties, represented by two constant $A = -B/2(1-\nu), B = 1/\mu$, and they are recalled in Table 15.1 [this elastic form contains the dielectric-like, rank-2, operator form in skipping the terms linked to the constant *A* while $B = 1/D$, with *D* the dielectric constant, as explained in the cited references, starting from (Franciosi 2005)]. These elementary GOs related to a transversally isotropic matrix can be found in (Franciosi 2013) with extension to so-called generalized elastic-like

Table 15.1 Nonzero terms of the elementary operator, with "$c\theta$" and "$s\theta$" for "cosθ" and "sinθ" (resp. ϕ)

	1	22	33	23	31	12
11	$As^4\theta c^4\phi$ $Bs^2\theta c^2\phi$	$As^4\theta c^2\phi s^2\phi$ 0	$As^2\theta c^2\theta c^2\phi$ 0			
22	$As^4\theta c^2\phi s^2\phi$ 0	$As^4\theta s^4\phi$ $Bs^2\theta s^2\phi$	$As^2\theta c^2\theta s^2\phi$ 0			
33	$As^2\theta c^2\theta c^2\phi$ 0	$As^2\theta c^2\theta s^2\phi$ 0	$As^4\theta$ $Bc^2\theta$			
23				$As^2\theta c^2\theta s^2\phi$ $B(s^2\theta s^2\phi +$ $c^2\theta)/4$		
31					$As^2\theta c^2\theta c^2\phi$ $B(s^2\theta c^2\phi +$ $c^2\theta)/4$	
12						$As^4\theta c^2\phi s^2\phi$ $Bs^2\theta/4$

properties, of the coupled magneto-electro-elastic type. These GO sets remain valid for any shape function to be considered according to Eq. (15.2a).

Conversely, the shape function in Eq. (15.2a) is only determined from the geometrical characteristics of the V " domain" (single inclusion or pattern) under concern. Once it is determined for some V, it can be used with any set of elementary GOs. Of course, the difficulty remains in analytically solving the integrals in Eq. (15.2a) for each independent term of the GO or mGO to be obtained.

Equations (15.3) recall how the shape functions are to be derived (in general) from the geometrical shape characteristics of some V domain, using the *RT-IRT* approach. Details are to be found in the cited references which each explicates the (sometimes tedious) calculation for one or several specific shape or pattern of inclusions. With $\mathrm{d}\boldsymbol{r}' = \mathrm{d}s_V(z',\boldsymbol{\omega})\mathrm{d}z'$ and $\boldsymbol{\omega} = (\sin\theta\cos\varphi, \sin\theta\sin\varphi, \cos\theta)$ running over the unit sphere, one has

$$\psi_V(\boldsymbol{\omega},\boldsymbol{r}) = -\int_V \frac{\delta''(\boldsymbol{\omega}.(\boldsymbol{r}-\boldsymbol{r}'))}{8\,\pi^2}\mathrm{d}\boldsymbol{r}'$$

$$= -\int_{z'=D_V^-(\boldsymbol{\omega})}^{z'=D_V^+(\boldsymbol{\omega})}\left(\int_{s_V(z',\omega)} \mathrm{d}s_V(z',\boldsymbol{\omega})\right)\frac{\delta''(z-z',\boldsymbol{\omega})}{8\,\pi^2}\mathrm{d}z' = -\frac{\tilde{s}''_V(z,\boldsymbol{\omega})}{8\,\pi^2}, \tag{15.3a}$$

where $\int_{s_V(z',\omega)} \mathrm{d}s_V(z',\boldsymbol{\omega}) = \mathrm{s}_V(z,\boldsymbol{\omega})$ is the planar section area of V when cut by the plane of equation $z = \boldsymbol{\omega}.\boldsymbol{r}$ that passes through the point $\boldsymbol{r}$ and is normal to the direction $\boldsymbol{\omega}$. Also, $\tilde{s}''_V(z,\boldsymbol{\omega})$ is the second z-derivative of $s_V(z,\boldsymbol{\omega})$ in the sense of distributions (Gel'fand et al. 1966; Franciosi 2010). With v the volume of V, the mean shape function can finally be written

$$\overline{\psi_V(\boldsymbol{\omega})} = -\frac{1}{8\,\pi^2\mathrm{v}}\int_{D_V^-(\boldsymbol{\omega})}^{D_V^+(\boldsymbol{\omega})} \tilde{s}''_V(z,\boldsymbol{\omega})\mathrm{s}_V(z,\boldsymbol{\omega})\mathrm{d}z = \frac{1}{8\,\pi^2\mathrm{v}}\int_{D_V^-(\boldsymbol{\omega})}^{D_V^+(\boldsymbol{\omega})} \left(\tilde{s}'_V(z,\boldsymbol{\omega})\right)^2\mathrm{d}z \tag{15.3b}$$

The interval $\left[D_V^-(\boldsymbol{\omega}), D_V^+(\boldsymbol{\omega})\right] = 2D_V(\boldsymbol{\omega})$ corresponds to the breadth of V in the $\boldsymbol{\omega}$ direction (i.e., the distance between the two opposite tangent planes to V, of $\boldsymbol{\omega}$-normal). It is immediate from this *RT-IRT* form that for ellipsoids whose planar sections are ellipses (quadrics in z), their second z-derivatives are constant for each $\boldsymbol{\omega}$ direction, and the shape function (hence the GO) is uniform inside V. When V is a pattern, $V = \bigcup_{i=1}^{n} V_i$ of n (possibly infinite) number of non-overlapping V_i inclusion, at any interior point $\boldsymbol{r}$ of V, that is of one of the V_i elements, the shape function and the related GO comprise the interior term from that V_i and all the exterior contributions from the other Vj domains ($j \neq i$). The mean shape function for an (up to infinite) number of inclusions reads from Eq. (15.3b):

$$\overline{\psi_V(\boldsymbol{\omega})} = \frac{1}{8\,\pi^2 v} \int\limits_{D_V^-(\boldsymbol{\omega})}^{D_V^+(\boldsymbol{\omega})} \left(\sum_{i=1}^{n} \tilde{s}'_{V_i}(z,\boldsymbol{\omega}) \right) \left(\sum_{j=1}^{n} \tilde{s}'_{Vj}(z,\boldsymbol{\omega}) \right) \mathrm{d}z, \quad \text{with } v = \sum_{i=1}^{n} v_i, \tag{15.4a}$$

$$\overline{\psi_V(\omega)} = \sum_{i=1}^{n} \frac{v_i}{v} \overline{\psi_{V_i}(\omega)} + \sum_{i=1}^{n} \sum_{j=i+1}^{n} \left(\frac{v_i + v_j}{v} \right) \overline{\psi_{V_i,V_j}(\omega)}. \tag{15.4b}$$

It assembles the average of the interior mean shape functions of the elements and the average of all the pair interactions between them. From comparing Eq. (15.4a) with Eq. (15.3a), it is also immediate to write

$$\overline{\psi_{V_i}(\omega)} = \int\limits_{D_{V_i}^-(\omega)}^{D_{V_i}^+(\omega)} \frac{\left(\tilde{s}'_{V_i}(z,\omega)\right)^2}{8\,\pi^2 v_i} \mathrm{d}z; \quad \overline{\psi_{V_i,V_j}(\omega)} = \int\limits_{\sup\left(D_{V_i}^-(\omega), D_{V_j}^-(\omega)\right)}^{\inf\left(D_{V_i}^+(\omega), D_{V_j}^+(\omega)\right)} \frac{\tilde{s}'_{V_i}(z,\omega)\tilde{s}'_{V_j}(z,\omega)}{8\,\pi^2 (v_i + v_j)} \mathrm{d}z. \tag{15.4c}$$

The pattern mGO follows with a same partition into mean interior and pair interaction parts as:

$$\overline{t^V} = \sum_{i=1}^{n} \frac{v_i}{v} \overline{t^{V_i}} + \sum_{i=1}^{n} \sum_{j=i+1}^{n} \left(\frac{v_i + v_j}{v} \right) \overline{t^{V_i,V_j}}, \quad \text{with } v = \sum_{i=1}^{n} v_i. \tag{15.4d}$$

The main terms only depend on the element shapes, and the interaction part depends on the detailed element organization in the pattern. In Eqs. (15.4a, b), the mean shape function terms have been written in using the right hand side of Eq. (15.3b) for in many cases, the square of the first derivative of the section areas is of easier use. Yet, the first form given in Eq. (15.3b) (as products of the section areas by their second derivative) can be preferable, as will appear in a (new) case treated in the following. Also useful next, if alignments alternate two different element shapes E and F, the Eqs. (15.4b, d) become

$$\begin{aligned}
\overline{\psi_V(\omega)} &= \sum_{i=1}^{nE} \frac{v_i^E}{v} \overline{\psi_{V_i}^E(\omega)} + \sum_{i=1}^{nF} \frac{v_i^F}{v} \overline{\psi_{V_i}^F(\omega)} \\
&+ \sum_{i=1}^{nE} \sum_{j=i+1}^{nE} \left(\frac{v_i^E + v_j^E}{v} \right) \overline{\psi_{V_i,V_j}^{E,E}(\omega)} + \sum_{i=1}^{nF} \sum_{j=i+1}^{nF} \left(\frac{v_i^F + v_j^F}{v} \right) \overline{\psi_{V_i,V_j}^{F,F}(\omega)} \\
&+ 2 \sum_{i=1}^{nE} \sum_{j=1}^{nF} \left(\frac{v_i^E + v_j^F}{v} \right) \overline{\psi_{V_i,V_j}^{E,F}(\omega)}.
\end{aligned} \tag{15.5a}$$

$$\begin{aligned}\overline{t^V} = &\sum_{i=1}^{nE} \frac{v_i^E}{v} \overline{t_E^{V_i}} + \sum_{i=1}^{nF} \frac{v_i^F}{v} \overline{t_F^{V_i}} + \sum_{i=1}^{nE} \sum_{j=i+1}^{nE} \left(\frac{v_i^E + v_j^E}{v}\right) \overline{t_{E,E}^{V_i,V_j}} \\ &+ \sum_{i=1}^{nF} \sum_{j=i+1}^{nF} \left(\frac{v_i^F + v_j^F}{v}\right) \overline{t_{E,E}^{V_i,V_j}} + 2 \sum_{i=1}^{nE} \sum_{j=1}^{nF} \left(\frac{v_i^E + v_j^F}{v}\right) \overline{t_{E,F}^{V_i,V_j}}\end{aligned}, \quad \text{with } v = \sum_{i=1}^{nE} v_i^E + \sum_{i=1}^{nF} v_i^F. \tag{15.5b}$$

In Eq. (15.5a), one retrieves the main and interaction parts for each E and F sets of identical inclusions plus a complementary mixed interaction part from all the mixed (E, F) element pairs. That is combining two patterns (E) and (F) amounts to determine each homogeneous pattern mGO and the representative mGO for one(E–F) mixed pair interaction which provides access to the global mixed interaction additional term. An example will be given.

An essential property of the shape function at interior points of any V domain as well as of its mean value over V is to have an ω-integral over the unit sphere Ω necessarily equal to unity, for this integral is the interior characteristic (indicator) function of V, equal to one inside V or in average over V and equal to zero at any exterior point. The shape function as expressed in Eqs. (15.3, 15.4, 15.5) is more precisely the *IRT* form of this characteristic function. When V is a pattern, only the main part of the shape function contributes to this integral over Ω equals to one, the integral of the interaction part is always zero, with all the integrals for the individual pair interactions being zero separately.

In Sect. 15.3 are first recalled the obtained mean shape functions and mGO forms for (i) axial n-alignments of equally sized spheres, and for (ii) planar n-alignments of parallel infinite cylinders (with equally sized circular cross sections), with n possibly going to infinity in both cases, as presented, respectively, in (Franciosi 2010) and (Franciosi et al. 2019). Relatively to case (i), will also be commented axial alignments of spheroids (Franciosi 2010) or of finite—flat to long—cylinders (Franciosi et al. 2015) for they have interesting limit connections with the infinite cylinder when the alignment becomes compact. These "elementary" patterns with quite simple forms of mean shape functions and GOs (for isotropic matrices) are shown usable for the building of more complicated patterns made of these elementary ones. Possible axial interconnections between the elements of such axial alignments, mimicking sorts of "skewers" to make these alignments continuous in their axial direction, were considered in (Franciosi 2010) using an arbitrary average between the alignment mGO and the GO of an infinite cylinder. Here, a better founded (although approximated) solution will be originally given for such sphere or spheroid skewers, treating them as compact alternated axial alignments of finite cylinders and spheroids (or spheres). This new result also serves as a practical example of the *RT-IRT* calculation method for a pattern mGO, applying the recalled formulae in Sect. 15.2. Relatively to case (ii), will also be mentioned infinite planar alignments of parallel rectangular beams treated in (Franciosi 2018) and in (Spagnuolo et al. 2020) for the square section case, their limit connection being also interestingly with laminate layers when the beams are at contact. The second section part recalls how a 1D fiber bundle and a 3D fiber (cylinders or beams) network have been built from planar arrays and submitted to

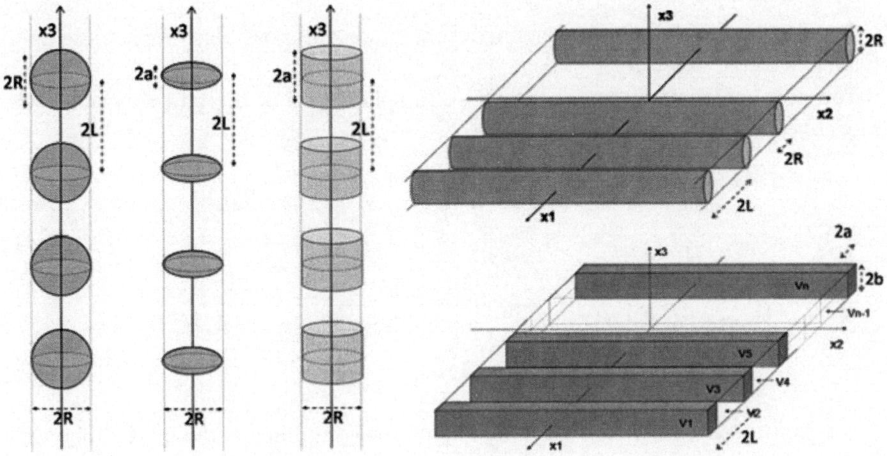

Fig. 15.1 Examples of elementary axial or planar alignments of inclusions and fibers

deformation once embedded in a matrix. In the 3D network case, fiber layer interconnections were mathematically accounted for, inspired from the behavior of 2D pantographic structures (Alibert et al. 2003; Barchiesi and Placidi 2017; Eremeyev et al. 2018, 2019; Eremeyev and Turco 2020). At last, some new examples of structures built from elementary alignments are commented. Since the spheroids coincide with laminate layers when infinitely flat, the new examined case of "spheroid skewers" is shown to bring at hand approximate descriptions for other structure types as the case of laminate structures with pillars in between layers.

15.3 Elementary Axial and Planar Alignments of Axially Symmetric or Fiber-like Elements

The brief equation recalls in Sect. 15.2 amount to pointing that in estimating effective properties of composite structures, the essential need, aside of phase properties and relative volume fractions is, as it is now quite well known, the description of their morphology, in the sense of shapes and spatial organization of the constitutive elements. In that respect, determining a mGO for representing the substructure in the matrix is essential, equivalently to determining a mean Eshelby tensor for it but more easily thanks to more helpful specific properties.

Figure 15.1 presents different *regular*[2] "elementary" alignments that have been working on, in the recent years, either in terms of axial patterns (left) or planar ones

[2]These regular cases of quite simple extension to infinite alignments are taken from general less regular ones.

(right), up to possibly infinite element number and at all possible compactness up to contact. We pose $\rho_0 = R/L$ or a/L.

The nonzero components of the GO of a sphere and of a circular cylinder are, respectively:

$$t_{\mathrm{iiii}}^{\mathrm{Sph}} = (3A+5B)/15;\ t_{\mathrm{(i,j),(i,j)}}^{\mathrm{Sph}} = (2A+5B)/30;$$
$$t_{\mathrm{ii,jj}}^{\mathrm{Sph}} = t_{\mathrm{iiii}}^{\mathrm{Sph}} - 2t_{\mathrm{(i,j),(i,j)}}^{\mathrm{Sph}},\quad i, j = 1, 2, 3, \tag{15.6a}$$

$$t_{iiii}^{\mathrm{Cyl}} = (3A+4B)/8;\ t_{(i,3),(i,3)}^{\mathrm{Cyl}} = \mathrm{B}/8;\ t_{(1,2),(1,2)}^{\mathrm{Cyl}} = (A+2B)/8,$$
$$t_{11,22}^{\mathrm{Cyl}} = t_{iiii}^{\mathrm{Cyl}} - 2t_{(1,2),(1,2)}^{\mathrm{Cyl}},\ i = 1, 2. \tag{15.6b}$$

For their infinite (respectively axial and planar) alignments, the mGOs take very comparable forms:

$$\overline{\boldsymbol{t}_{\rho_0}^{\infty\mathrm{Sph}}} = \boldsymbol{t}^{\mathrm{Sph}} + \left(\boldsymbol{v}_0^{(\infty)\mathrm{Sph}}\rho_0^3 + \boldsymbol{w}_0^{(\infty)\mathrm{Sph}}\rho_0^5\right) = \boldsymbol{t}^{Sph} + 2\left(Z(3)\boldsymbol{v}_0^{2\mathrm{Sph}}\rho_0^3 + Z(5)\boldsymbol{w}_0^{2\mathrm{Sph}}\rho_0^5\right), \tag{15.7a}$$

$$\overline{\boldsymbol{t}_{\rho_0}^{\infty\mathrm{Cyl}}} = \boldsymbol{t}^{\mathrm{Cyl}} + \left(\boldsymbol{v}_0^{(\infty)\mathrm{Cyl}}\rho_0^2 + \boldsymbol{w}_0^{(\infty)\mathrm{Cyl}}\rho_0^4\right) = \boldsymbol{t}^{\mathrm{cyl}} + 2\left(Z(2)\boldsymbol{v}_0^{2\mathrm{Cyl}}\rho_0^2 + Z(4)\boldsymbol{w}_0^{2\mathrm{Cyl}}\rho_0^4\right), \tag{15.7b}$$

where $Z(k)$ is the Zeta function for $k = (2, 3, 4, 5)$, $\rho_0 = R/L$ and the respective tensor pairs $(\boldsymbol{v}_0, \boldsymbol{w}_0)$ are defined from the pair interaction mGOs, all to be found in (Franciosi et al. 2019).

Among the various obtained results concerning these possibly infinite patterns, Fig. 15.2 compares the variation with the alignment compactness of these *exact* mGOs obtained for the sphere axial infinite alignment case and the cylinder planar one,[3] borrowed from (Franciosi 2010) and (Franciosi et al. 2019). Although the plots are in inverted and not identical presentations (respectively versus $R/2L = \rho_0/2$ plots for aligned spheres and $L/R = 1/\rho_0$ plots for cylinders, both of radius R and $2L$ axis to axis inter-distance, the contact limit being $\rho_0 = 1$ in both cases at $2L = 2R$), the two major insights are obvious in both cases: The first one is that the interaction effects in an infinite row are cumulative (in the sphere case, the plot also reports in grey the sphere pair mGO), and the second is that this cumulating is bounded by above since beyond a critical distance the pair interactions vanish.

Also, the influence zone extension is clearly less for the sphere axial alignment (typically five times the sphere size) than for the planar alignment of cylinders (between ten to twenty times the cylinder diameter). This example corresponds to an incompressible isotropic matrix, but the corresponding cases with a Poisson ratio of 0.3, to be found in the sources, give the same dimension of "influence zone." The

[3] The plots for the sphere pair mGO terms also appear in gray on Fig. 15.2 left, not for the cylinder pair at right.

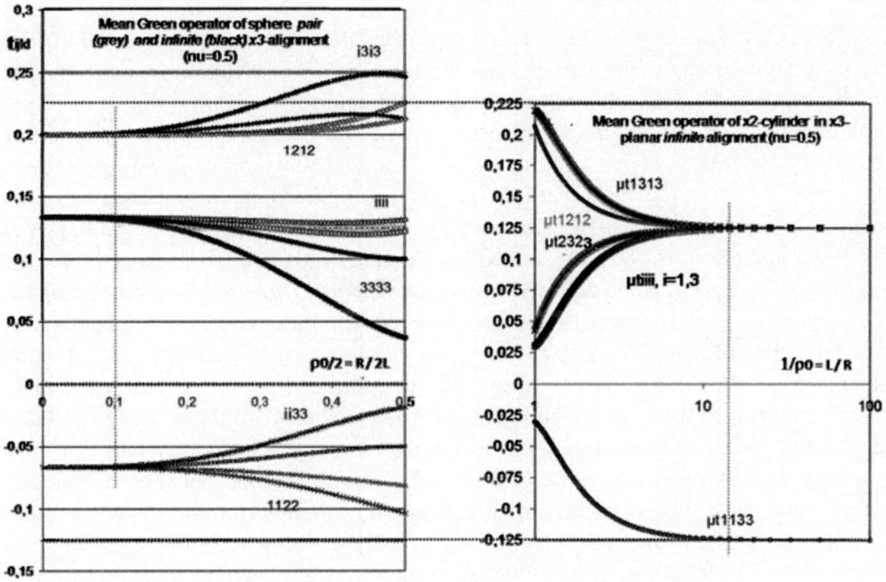

Fig. 15.2 Evolution of the mGO terms for sphere (left) and cylinder (right) alignments versus inter-distance

one for planar cylinder alignments is confirmed for similar arrays of square beams in (Franciosi 2018; Spagnuolo et al. 2020).

A net consequence of this influence zone finiteness is on the second GO showing up in the PCW estimate (Eq. 15.1b) which represents some spatial distribution of the inclusions or patterns of them: If the considered pattern is infinite, considering distributions of infinite patterns does not make sense anymore, and this distribution operator more likely corresponds to a finite "window" on the infinite matrix substructure related to a representative zone in terms of interactions. It then well corresponds to that influence zone out of which the rest of the pattern does not contribute to the overall properties of the composite. Under this viewpoint, the zone is more characterized by the (possibly evolving) matrix properties and by the details of the infinite pattern structure. This assumption has been successfully tested in (Franciosi et al. 2019) for transversally compressed 1D bundles of parallel fibers embedded in an isotropic matrix: Estimates from the PCW modeling (Eq. 15.1b) where the distribution GO was kept invariant (cylindrical symmetry) while the fiber bundle mGO was evolving (such as to change the number and position of the fibers in the influence zone) gave consistent results (in terms of effective elastic modulus evolutions) with numerically calculated ones for comparisons. In contrast, a distribution symmetry taken to evolve homothetically to the fiber arrangement in a finite pattern did not yield relevant estimates.

In comparison with the simple forms of Eqs. (15.7a, b) for spheres and infinite cylinders, the mGOs obtained for coaxial finite cylinders and for rectangular or square beams (Franciosi et al. 2015; Franciosi 2018; Spagnuolo et al. 2020) did not have so nice ones. Yet, these cases are among those having an interaction part nicely expressed from a linear combination of interior mGOs of the same type, noticing that intermediates empty spaces between elements are also finite cylinders or rectangular beams, say have same shape as the aligned elements. This was formalized as a specific "decomposition method" in (Franciosi 2018), a consequence from the geometric nature of the *RT-IRT* method. The interior mGOs for finite cylinders and for rectangular beams were explicated in (Franciosi 2014; 2018), respectively. An approximate, more manageable form was provided in (Spagnuolo et al. 2020) for square beams. In particular, it is easily retrieved that the mGO of a compact set of coaxial finite cylinders of same radius (resp. of aligned rectangular beams of same height) equal the (uniform) GO of an infinite cylinder (respectively of a laminate layer), confirming the known "blindness to connectedness" of the mGOs, as related they are to the covariance function, between parts in contact in any partition of a compact domain (Santalo 1976; Serra 1982). The difference between a compact set of elements and a connected identical one is in their different behaviors under loading: While the former will lose its "integrity" and dissociates, the latter will not. These are examples where missing physical interconnections have to be represented by mathematical (here mechanical) conditions.

Regarding, for the axial alignments of spheres (and spheroids), possibilities of describing more general interconnections between such aligned elements (either when at contact or not), an arbitrary modification of the alignment mGO was proposed (Franciosi 2010) in just averaging it with the GO of a uniform infinite cylinder of same orientation. This was obviously a simple way to stiffen the alignment in the axial direction "as if" an axial connection was present between them but without much support for so doing. A more relevant description would correspond to Fig. 15.3 **left**, where the links between spheroids are ensured by finite cylinders (the rest of the cylinder passing inside the spheroids indistinguishably of them) in a compact

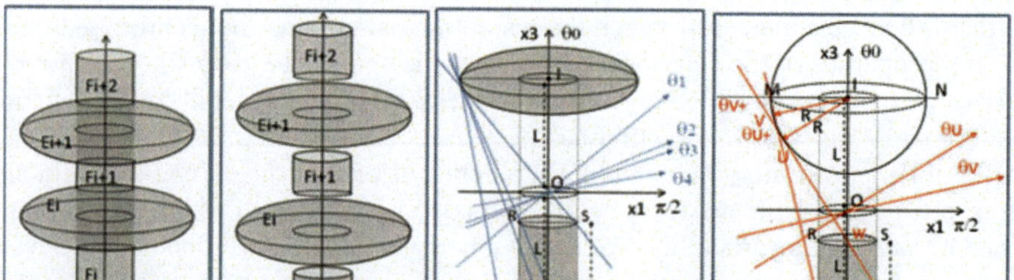

Fig. 15.3 From left to right, compact and noncompact mixed spheroid–cylinder alignments, the mixed pair interaction to be solved exactly and approximately with the simplest sphere case

axial array. This corresponds to the particular case of Fig. 15.3 **middle left** with not contacting alternated spheroids and finite cylinder elements (at the contact limit, the spheroid large section must be assumed larger enough than the cylinder one to consider the surface contact to be flat). For such mixed alignments, a solution is at hand from calculating the interaction between a mixed (finite cylinder with spheroid or sphere) element pair, according to Fig. 15.3 **middle right and right**. Resolving this pair interaction problem provides the mGO that remains needed according to Eq. (15.5a) in the case of two (E, F) element shapes present in a same alignment, for which the respective alignment mGOs (main plus interaction parts) are known. Note that in terms of potential applications, the structures in Fig. 15.3 left correspond, among other possibilities, and when n is a finite number, to dendritic-like ones in metallic materials.

Determining a pair interaction mGO from the *RT-IRT* method can be exemplified on this new case: One calculates as recalled the section areas and derivative by planes (of θ normal here since the problem is axially symmetric) passing *simultaneously* through both elements. One element belonging to the ellipsoid family is a typical case where the first form of the mean interaction weight function in Eq. (15.3b) is of easier use, writing, with $v = v_E + v_F$ the volume of the element pair:

$$\overline{\psi_{V_E,V_F}(\boldsymbol{\omega})} = -\frac{1}{8\,\pi^2 \mathrm{v}} \int_{D_V^-(\boldsymbol{\omega})}^{D_V^+(\boldsymbol{\omega})} \tilde{s}''_{V_E}(z,\boldsymbol{\omega}) \mathrm{s}_{V_F}(z,\boldsymbol{\omega})\mathrm{d}z = -\frac{\tilde{s}''_{V_E}(z,\theta)}{4\pi v} \int_{D_V^-(\theta)}^{D_V^+(\theta)} \mathrm{s}_{V_F}(z,\theta)\mathrm{d}z. \tag{15.8a}$$

The area second derivative for the ellipsoidal (or spherical) element (E) being constant, it simplifies the remaining integral as being all or a part of the volume of the second element (F) and the ϕ integral around the alignment $(x3)$ axis contributes the 2π factor that appeared in Eq. (15.8a).

As shown in Fig. 15.3 **middle right**, there are five angular θ sectors—defined by the orientations of the tangent planes to the spheroid passing through the cylinder vertical section "corners" P, Q, R, S to separately "treat": The first one $(0\text{–}\theta 1)$ corresponds to no plane simultaneously cutting both elements, and the last one $(\theta 4\text{–}\pi/2)$ corresponds to directions for which the entire cylinder volume is swept out by the simultaneously cutting planes (the supplementary $(\pi/2\text{–}\pi)$ sector is equivalent by symmetry). In the three intermediate sectors $(\theta 1\text{–}\theta 2, \theta 2\text{–}\theta 3, \theta 3\text{–}\theta 4)$, only a θ-varying part of the cylinder volume is swept out by crossing planes. In the central θ-angular part limited by the cylinder "corners" R and Q, this cylinder volume fraction equal to $\int_{D_V^-(\theta)}^{D_V^+(\theta)} \mathrm{s}_{V_F}(z,\theta)\mathrm{d}z$ reads from simple geometry $\pi r^2(2h - 2r \sin\theta)$, with $2h$ the cylinder height and with r the cylinder radius taken smaller enough than the sphere radius R, for the assumption of a flat contact section to hold up to the limit contact case. The integrals to solve to fully determine the interaction shape function are at quite easy hand provided manipulations of little interest here. For room saving and not developing out of purpose calculations, a simplified solution is sketched in simplifying the five angular domains to be treated into the remaining three in

Fig. 15.3 right (small sectors 2 and 4 are shared between the first, central and last ones) and also considering spheres, from which the spheroid case follows provided some more manipulations not of much interest either here and the differences being in a different expression for the sector limiting θ_U and θ_V orientations (and of a well-known θ-dependent section area derivative for spheroids). These approximate median limit orientations are taken to, respectively, correspond to the tangent planes to the sphere that pass through the middle points w and y of the RS and PQ diameters of the cylinder, respectively. We thus have the mean interaction shape function for the (E, F) pair as:

$$\overline{\psi_{V_E,V_F}(\theta)} = -\frac{\tilde{s}''_{V_E}(z,\theta)}{4\pi(v_E + v_F)} \times \begin{cases} 0 & \text{for } \theta \in (0 - \theta_U) \\ \pi r^2(2h - 2r\sin\theta) & \text{for } \theta \in (\theta_U - \theta_V) \\ v_F & \text{for } \theta \in (\theta_V - \pi/2), \end{cases} \tag{15.8b}$$

with $\tilde{s}''_{V_E}(z,\theta) = -3\ v/2\ D_{\text{ell}}(\theta)^3$ for spheroids with breadth (thickness) $2D_{\text{ell}}(\theta)$ in each θ direction which becomes uniformly $-4\pi R^3/2R^3 = -2\pi$ for a sphere of radius R (equal to D). The θ-integral of $\overline{\psi_{V_E,V_F}(\theta)}$ reads, with $C = 0.5(v_E + v_F)^{-1} = 0.5(4\pi R^3/3 + 2\pi r^2 h)^{-1}$:

$$\begin{aligned} 2\int_0^{\pi/2} \overline{\psi_{V_E,V_F}(\theta)} \sin\theta \mathrm{d}\theta &= 2C\left(2\pi r^2 \int_{\theta_U}^{\theta_V} (h - r\sin\theta)\sin\theta \mathrm{d}\theta + 2\pi r^2 h \int_{\theta_V}^{\pi/2} \sin\theta \mathrm{d}\theta\right) \\ &= 4\pi r^2 C\left(h\cos\theta_U - \frac{r}{2}\left((\theta_V - \theta_U) - \frac{\sin 2\theta_V - \sin 2\theta_U}{2}\right)\right). \end{aligned} \tag{15.9}$$

The point here is not to calculate it but to simply show that the necessary integrals to obtain the related mixed pair interaction mGO (in an isotropic elastic matrix) are at hand, say the integrals for $p = 1, 2$:

$$\begin{aligned} I^{2p} &= 2\int_0^{\pi/2} \overline{\psi_{V_E,V_F}(\theta)} \cos^{2p}\theta \sin\theta \mathrm{d}\theta \\ &= 4\pi r^2 C\left(h\int_{\theta_U}^{\pi/2} \cos^{2p}\theta \sin\theta \mathrm{d}\theta - r\int_{\theta_U}^{\theta_V} \cos^{2p}\theta \sin^2\theta \mathrm{d}\theta\right). \end{aligned} \tag{15.10}$$

This is the case for these two integral pairs have simple primitives which are not to be explicated further here (the shape function in Eq. (15.9) corresponds to the integral of Eq. (15.10) when $p = 0$). These results allow to specifying the mixed series showing up in Eqs. (15.5a, b) for sphere/cylinder pairs and to obtaining a mGO for the mixed n (possibly infinite) axial alignment. For the spheroidal shapes, the section derivatives $\tilde{s}''_{V_E}(z,\theta) = -3\ v/2\ D_{\text{ell}}(\theta)^3$ read $-2\pi\zeta\left(1 + (\zeta^2 - 1)\cos^2\theta\right)^{-3/2}$ from

$D_{\text{ell}}(\theta) = R\sqrt{1 + \left(\zeta^2 - 1\right)\cos^2\theta}$ and $\zeta = a/R < 1$ with regard to the spheroid direction along the $x3$ alignment axis such that $\tilde{s}''_{V_E}(z, \theta)$ contributes to the θ integrals. This is likely at hand for the integrals related to spheroid shapes have been solved already, but the total resolution is let to further works, together with finalizing the sphere case.

An important point here is, as stressed already, that such an alignment with elements at contact is not equivalent to connected elements although they have the same (undifferentiated) mGOs: Depending on the case, it would be necessary to additionally assign the elements in the alignments to remain at contact or not. What means substituting mathematical interconnection (connectedness) conditions to supply for missing physical ones. If the exemplified mixed alignment is kept undissociated by appropriately chosen conditions, the axial stiffness for the spheroid array will be enhanced by the added cylinders in relevant manner with regard to the structure. In comparison, averaging the spheroid alignment mGO with the GO of an infinite cylinder to similarly give axial stiffness to such alignments as if the array was connected is an efficient assumption that saves more needs of interface conditions, but it is arbitrary. Being necessary to finalize the mGO calculation for such mixed alignments in order to comparing and discussing further, this is let to forthcoming papers.

15.4 From Elementary Alignments to Bundles or Networks

The second pointed fact in this work is that determining mGOs for various patterned of networked substructures is simplified when fully using the geometrical nature of the *RT-IRT* method. This section therefore focuses on the mGOs obtained that way in the case of substructures that can be represented by axial and planar alignments of similar (possibly being composite) elements, possibly up to infinite numbers, from which more complicated patterns or networked could be represented, as is exemplified. This section addresses the more complex structures one can represent/describe from the treated elementary ones recalled in Sect. 15.3.

Using axial alignments of spheres, spheroids, finite cylinders or mixtures of them allows to building more complex patterns or networks as 1D bundles of such alignments, or multi-directional (mD) ones which can also be omni-directional as an isotropic structure for example would be. Thanks to the knowledge of the mGOs for these axial (axially symmetric) arrays, this is as easily at hand than creating bundles of fibers, with only the knowledge of the rotation tensor to apply on the 1D mGO and averaging over directional mGOs is a relevant approximate for the mGO of such a, pretty much networked already, pattern. It is noteworthy that depending on how the averaging is weighted, it is possible to build isotropic or anisotropic patterns, the use of ellipsoidal spatial distributions being still an efficient weighting method. In similar way, planar arrays of parallel fibers can be used to build 1D fiber bundles, with a representative mGO obtained in averaging the mGOs of the planar fiber arrays around a common direction of the constitutive fibers. Multi-directional (mD) bundles can then also be obtained in averaging several 1D bundles. Depending on the chosen method to reach a same structure objective, the account for the interactions between the elements will be different. Figure 15.4, which was used in (Franciosi et al. 2019) to describe different possible assumptions on the element spatial distribution evolution in transverse compression of a 1D fiber bundle (oriented normally to the figure), also holds for a sphere 3D structure. For dilute concentrations (no element interactions), the spatial distribution symmetry of individual elements varies like the matrix deformation (left). The element arrangement may not remain dilute in all directions, but new interactions cannot be accounted for. At larger concentrations which need to consider patterns of elements with interactions, if the pattern is still assumed to deform as the matrix (middle), significant interactions will also be lost in directions where the element density increases. Considering infinite patterns and all interactions in it makes the strain-independent influence zone shape (right) more relevant as a representative symmetry, with new elements entering in directions of increasing density and elements getting out where density lowers.

The particular "simple" case of 1D bundles of cylindrical fibers built from several planar arrays of common fiber direction as examined in (Franciosi et al. 2019) served to validate that a simple averaging of the array mGOs around a common fiber direction, with appropriate relative weights to represent the bundle transverse anisotropy details, was relevant to easily obtain an estimate of the elastic response of a matrix reinforced by such a 1D network when transversally compressed. Using the PCW estimate form of Eq. (15.1b), it was shown that a pretty good match with numerical simulations was obtained with taking a distribution symmetry GO defined by a cylindrical interaction (influence) zone shape around the fiber bundle. Figure 15.5 recalls a part of the results from the last cited reference where, among the various distribution options represented in Fig. 15.4, only the reference to the influence zone symmetry as representative spatial distribution of the fibers succeeded in capturing the evolutions of the effective moduli of a composite comprising an isotropic matrix

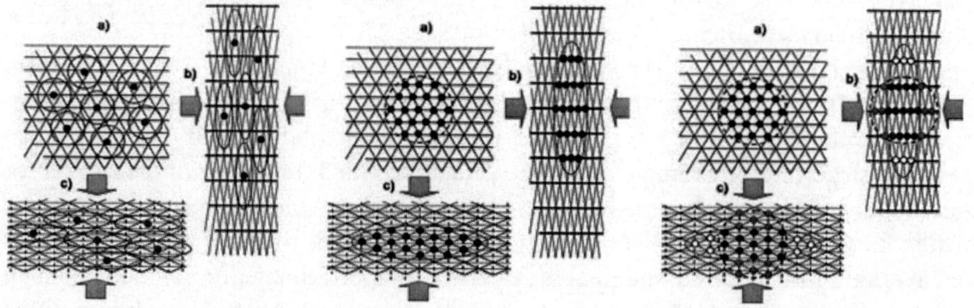

Fig. 15.4 Possible evolution description of element spatial distribution under axial or lateral compression: left and middle, matrix domains for individual and finite patterns of elements evolve as strained; right, a finite part of an infinite pattern of elements evolves with strain inside a strain-invariant influence zone

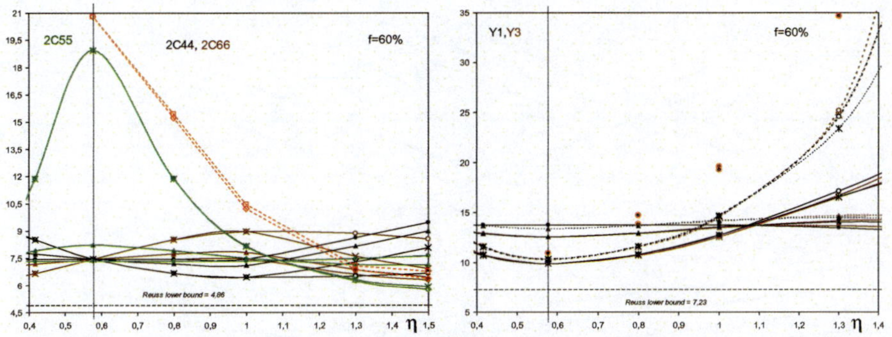

Fig. 15.5 Several estimated evolutions of effective elastic stiffness moduli in a 60% 1D fiber bundle-reinforced matrix under lateral compression compared with numerical simulations. The largest variations are captured by referring the element spatial distribution to the influence zone shape as defined from interactions

and a large (60%) volume fraction of such 1D fiber bundle. The shape of this influence zone, taken invariant cylindrical in the bundle direction there, is fixed by the anisotropy properties of the matrix when not isotropic, making not all the directions equivalent then, and also by the shape of the assembled elements (square beams will not interact over same distances whether aligned normally or diagonally with respect to their sides). It is expected to evolve with them if they do, in indirect straining effect.

From these first results on matrix-embedded 1D fiber bundles, a 3D fiber networked structure, to be as well embedded in a compliant matrix as a two-phase 3D bi-continuous composite, was built from piling parallel arrays of parallel square beams (those exemplified in Fig. 15.1, bottom right), alternately misoriented by a $\pm\theta$ angle with regard to an exterior (loading) frame. This 3D structure is exemplified in Fig. 15.6 **left** where physical interconnections exist between the fiber layers to make the network 3D interconnected. Fig. 15.6 **middle and right** shows the considered piling of planar array prior and after alternated misorientations around the z (left) or $x3$ (middle, right) axis to represent that networked structure.

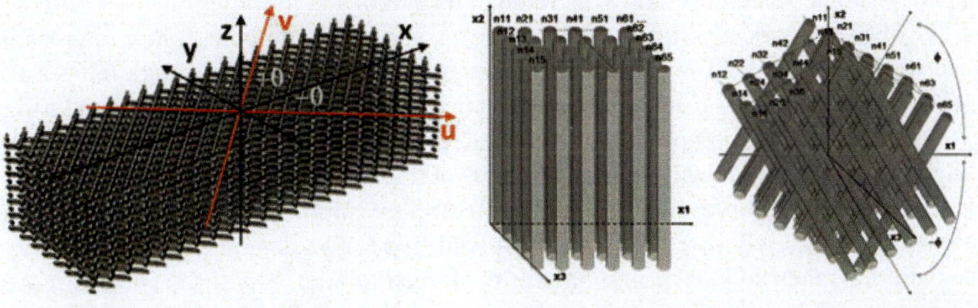

Fig. 15.6 Left, a 3D "pantographic-inspired" fiber network; middle, piled identically $x2$-oriented fiber planar arrays; right, planar fiber arrays alternately misoriented of a $\pm\varphi$ angle with regard to $x1$ axis

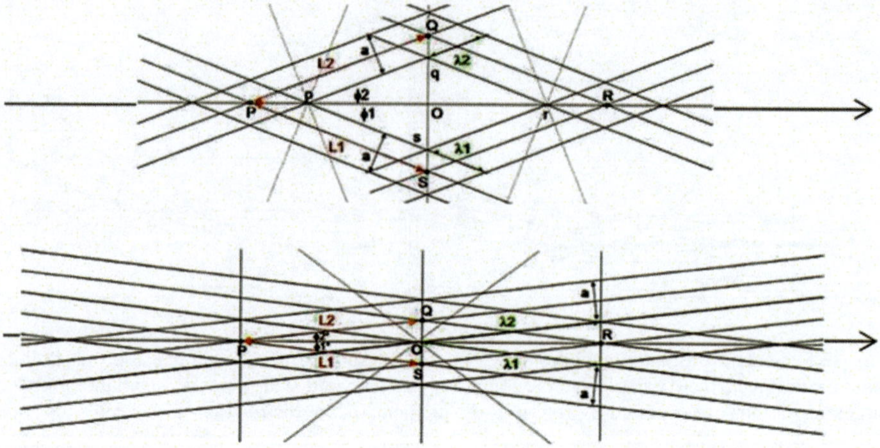

Fig. 15.7 Current (top) and final (bottom) extension description of a structural element for a 2D pantographic-like bilayered fiber arrangement, assuming the layer interconnecting pivots (at *P*, *Q*, *R*, *S* points) to not bend

Although no physical interconnections between the layers are represented in this alternated piling of arrays (Fig. 15.6 right), the deformation under in-plane extension was correctly ruled from mathematically describing so-called geometrical descriptors that constrain the layers to remaining aligned and behave *as if* physical connectors were linking them alike in Fig. 15.3 left. Typically, these geometrical descriptors connect, as described in the minimal representative "cell" of the structure in Fig. 15.7, the network extension to the change of misorientation angles between the layers and to the inter-distance between the fibers or beams in each layer.

The representative mGO for this layered structure being obtained from averaging the two oppositely rotated planar arrays, with considering the current fiber inter-distance in the layers and the layer misorientations to change as in the reference pantographic bilayer description in Fig. 15.7, the current mGO is "pantographic-like evolving" with the applied extension. This mechanical behavior being typical of pantographic structures, allowing large deformations at low energy cost, that type of 3D fiber network as built as well as the two-phase composite comprising such a network and a matrix were called "pantographic-inspired" in (Spagnuolo et al. 2020). Some estimates, from the cited reference, of stiffness moduli evolution during extension of such a composite are recalled in Fig. 15.8a, b, showing pretty well satisfying comparisons with numerical extension simulations.

An opening example on a new type of complex infinite networked structure is inspired by the newly discussed (partly solved) type of inclusion pair problem, say the cylinder/spheroid axial combinations examined in Sect. 3.1: If the spheroids are taken to be increasingly flat and large, the structures exemplified in Fig. 15.3 left turn into the one drawn in Fig. 15.9 **left** where finite cylinders between large and laminate layer like spheroids act as pillar-like physical interconnections between layers of a laminate structure. The so obtained mGO form, which is included in the cases solved

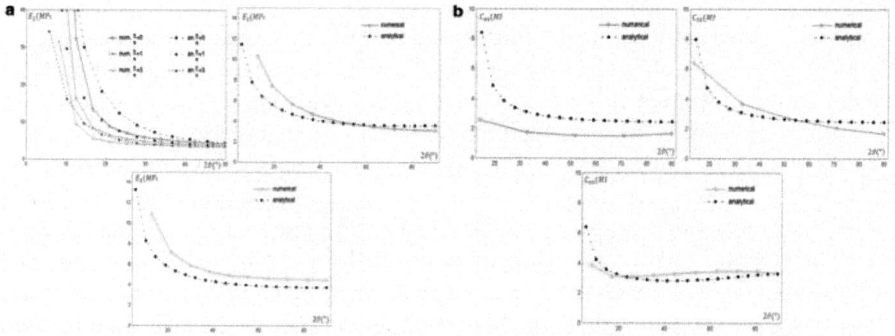

Fig. 15.8 **a** Estimated and numerically calculated effective Young moduli evolutions under $x2$ extension of a pantographic-inspired composite ($h = 0$ means alternated fiber layers into contact). **b** Estimated and numerically calculated effective shear moduli evolutions under $x2$ extension of a pantographic-inspired composite ($h = 0$ means alternated fiber layers into contact)

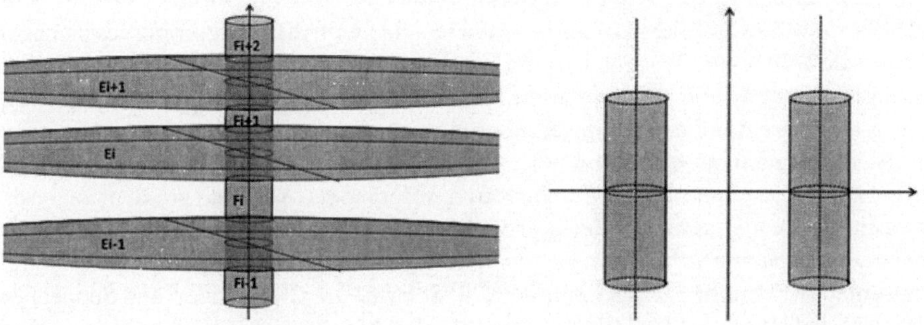

Fig. 15.9 Isolated pillar-connected laminate layers (left), a case included in the mixed spheroid-cylinder alignment examined in Sect. 3.1; the pillar interaction problem (right) of examination interest

from Eqs. (15.8, 15.9, 15.10), is relevant as far as the pillar density is dilute enough for each being considered as isolated. The resolution for patterns with dense finite pillars to be placed between large, laminate-like, spheroids remains to be solved in terms of mGO. It is likely at hand provided solving the pair interaction problem between two *finite* cylinders in lateral positions, as drawn in Fig. 15.9 **right**. The known solution for the interaction between infinite parallel cylinders can also provide an approximate solution. These are examples among relatively simple problems to solve with practical, sometimes surprising, applications as, at quite the other scale end than micro-devices, chips and circuits concerned with physico-mechanical coupled properties (Liu 2011), when fibers are pipes, tunnels or even tanks, embedded in soils submitted to seismic loadings (Manolis et al. 2013; Chen et al. 2018).

Prior to finish this sort or overview on specific inclusion patterns and networks in an elastic deformation context, it is worth to briefly evoke the more essential cases of multi-phased patterns, said in introduction to be out of the present scope. Indeed, it remains not permitted to treat multi-phased inclusion patterns or fiber networks in the PCW framework of Eqs. (15.1), for a single mGO for elements of different phases (in the sense of different properties) cannot manage with phase to phase interactions in a mixed network. The route to follow in such important cases of multi-phased inclusion patterns is indicated by Eq. (15.5a) which consider two different (E, F) inclusion types: whether the type differences be shapes or properties, the global mGO shares into three parts, the mGOs for the sub-patterns E and F and the third mixed interaction mGO, as here partly explicated for a mixture of spheroids and of finite cylinder coaxial elements. As far as the E, F elements have same physical (elastic or else) properties and are only shape-different, a single mGO can be used as indicated. If the cylinders and the spheroids do not have same properties, then three distinct mGOs need be separately accounted for, say shortly speaking the E and the F uniform ones and the E–F interaction one, using the n-site homogenization method from (Fassi-Fehri et al. 1989) specialized to two elastic phases in (Franciosi and Charles 2016b). The same three parts would be needed if aligning all shape-identical inclusions but with two different property types. Equations (15.5a, b) here clearly show how infinite patterns mixing elements with different properties can also be treated following the same here proposed method, with this specific variant that a single equation for a single phase in terms of properties becomes a n-set equation when the phase number with different properties in the pattern is n.

At last, it cannot be ignored here, as far as possibly large elastic deformations may be concerned, the as well tremendous developments of second gradient analyses especially devoted to highly deformable fiber networks also mostly inspired by the pantograph specific structure type, owing to torsion and bending components of the deformation (Mindlin 1965; Germain 1973; Sciarra et al. 2007; Pideri and Seppecher 1997; Rahali et al. 2015; dell'Isola et al. 2019) which are often hard to disregard as here done. However, explicating GOs for second gradient deformation behavior to extend the here discussed type of homogenization schemes is a highly challenging domain. Hence, if the here built structures are "pantographic-inspired", they can only deform moderately as described. If this constitutes a widely opened field for further research on innovative composites, some other ones are briefly discussed in the next section for which, to various extents, applications of the presented results for describing complicated structures from a global mGO are more at hand. Second gradient approaches of classical Eshelby/Green problems with more or less implicit connections to the RT-IRT method can be found in (Drugan and Willis 1996; Gao and Ma 2009; Wu et al. 2015), with references cited there in.

15.5 Other Potential Application Extension Directions

Not much far after the Eshelby seminal publications in the fifties, a lot of research started to explore further the potential application domains of the Eshelby results on embedded ellipsoidal inclusions at various, from nano to mega, scales and "generalized" Eshelby tensors for inclusions have been calculated in various other contexts as, in addition to already cited thermal or dielectric properties in introduction, to acoustic or dynamic problems, plasticity and nonlinear properties and behavior, generalized media,..., showing first that the tensor uniformity property in the linear, elastic-like or dielectric-like, contexts were in general not preserved, reinforcing the interest for mean values.

As was synthesized in the abstract, the present work aimed at presenting recent developments from the authors and collaborators which extended the seminal Eshelby isolated inclusion problem to large and possibly infinite inclusion patterns with all pair interactions between elements being accounted for, to patterns possibly turning into an infinite network arrangement possibly then being co-continuous with the embedding matrix and with the possibility to account for that arrangement evolving under strain in the specific field of mechanical behavior. The proposed method which is based on using the *RT-IRT* to determine one representative global mGO for the infinite pattern or network in its current state (such that only having a single inclusion to solve for obtaining an equivalent homogeneous material say) presents the helpful advantage for calculations of separating, for linear and static problems, the morphological and the physical features in the mGO expression. Beyond the formally simple validity extension to other linear static physical (dielectric, magnetic, thermal,...) application fields as mentioned in introduction, it is expectable that all the performed work on the morphological mGO parts remains valid and can be used more broadly, for instance when the behavior of concern is not linear, when properties are coupled or when dynamical effects cannot be ignored. Yet, even in the simplest of these problems, say when the reference matrix properties are "not too much anisotropic", the full calculation of the mGOs is not necessarily made so simple with using the *RT-IRT*, especially when aiming at analytical well enough approximate solutions of easy use in larger modeling developments. Computational help will frequently remain necessary to attain a solution. For the three mentioned directions (each of which is a research field), simple extension possibilities of the here presented work are briefly exemplified, pointing some already realized and easily realizable openings and locking points to solve.

15.5.1 From Linear to Nonlinear Behavior

The extension to nonlinear (static) effective behavior of composites calls for an appropriate development of the homogenization framework to use. Remaining in the examples of mechanical behavior, the incremental homogenization method of (Hill 1965) using a piece-wise linearization of the phase constitutive behavior to linearly increment the composite response has been used for years to estimating the plastic response of aggregates and polycrystals. The step-wise reference to comparison materials of affine thermo-elastic (i.e., linear elastic with eigenstrain) behavior type for the constitutive phases provided a rigorous improvement which permits in quite simple manner the extension of a linear problem to nonlinear ones. That is, using constitutive laws of the form, at each point $\mathbf{r}$ in the medium with behavior $\boldsymbol{\sigma}(\mathbf{r}) =$

$f(\boldsymbol{\varepsilon}(\mathbf{r}))$:

$$\boldsymbol{\sigma}(\mathbf{r}) = \boldsymbol{\sigma}_o(\mathbf{r}) + \boldsymbol{l}_{\boldsymbol{\varepsilon}_o(\mathbf{r})}(\mathbf{r}) : (\boldsymbol{\varepsilon}(\mathbf{r}) - \boldsymbol{\varepsilon}_o(\mathbf{r})) = \boldsymbol{l}_{\boldsymbol{\varepsilon}_o(\mathbf{r})}(\mathbf{r}) : \left(\boldsymbol{\varepsilon}(\mathbf{r}) - \boldsymbol{\varepsilon}_o^*(\mathbf{r})\right), \quad (15.11)$$

with $\boldsymbol{\varepsilon}_o^*(\mathbf{r}) = \boldsymbol{\varepsilon}_o(\mathbf{r}) - \boldsymbol{l}^{-1}_{\boldsymbol{\varepsilon}_o(\mathbf{r})}(\mathbf{r}) : \boldsymbol{\sigma}_o(\mathbf{r})$ and $\boldsymbol{l}_{\boldsymbol{\varepsilon}_o(\mathbf{r})}(\mathbf{r}) = \partial f(\boldsymbol{\varepsilon}(\mathbf{r}))/\partial\boldsymbol{\varepsilon}|_{\boldsymbol{\varepsilon}=\boldsymbol{\varepsilon}_o}$ the tangent moduli at previous iteration "o", to step-wise determining the thermo-elastic (affine) effective composite of comparison (see Masson et al. 2000 and given references there). Using the Levin results (1967), the thermal properties entered quite easily into the picture of elastic behavior modeling, hence the affine, thermo-elastic-like, model success in practice. Also, with progress in second moment theory, it was shown that the inner phase strain and stress fields were better represented by their second moment than by simple averages (first order moments) owing to their nonuniformity from the phase nonlinear behavior (Suquet 1995, 1997; Ponte-Castaneda and Suquet 1998; Ponte-Castaneda and Willis 1999; Buryachenko 2001: Brenner et al. 2001; Lahellec and Suquet 2004). These two directions have allowed substantial modeling improvements, although still complicated to deal with unless in elementary cases. This so-called affine procedure could have been applied to the follow-up of the here examined evolving microstructures, at least in its classical first order form, although even the modified, second order, form seems quite at hand too,[4]. This framework could likely be selected to develop a pretty well rigorous modeling of the here examined structure types, what we did not need to do so far, for it remains quite complex to fully implement. Simpler alternatives are often preferred, like the use of secant moduli as linear comparison materials or even the true elastic behavior of the phases in the so-called NTFA[5] modeling (Michel and Suquet 2004). Such an attempt has been reported in (Franciosi and Berbenni 2007, 2008) for polycrystal plasticity, with a particular use of the *RT-IRT*. Available analytical forms for the mGOs representative of microstructures in concern (as those exemplified in this lecture) can avoid the computations to become too huge machinaries. The same route can be applied to other pairs of conjugate variables generalizing stress and strain tensors.

15.5.2 From Elasticity-Type to Coupled Piezo-Type Properties

These coupled problems (Li and Dunn 1998) which assemble the sub-combinations from the thermo-electro-magneto-elastic (TMEE) type share into several more or less difficult problems according to which properties are carried by the matrix phase: namely, either the matrix has uncoupled properties (they can be isotropic then) or it

[4]It consists in substituting the phase ϕ mean strains $\langle\varepsilon^{\phi}\rangle$ with the square root of their mean second moments $\langle\varepsilon^{\phi}\otimes\varepsilon^{\phi}\rangle^{1/2}$ of relatively easy calculations from the derivatives of the effective elastic energy by the moduli of the phase (Brenner et al. 2001) compared with other difficulties.

[5]Nonuniform Transformation Field Analysis.

has one property type among the TMEE sub-cases what needs at most transversal isotropy (TI) symmetry. Owing to the vast literature in these cases, an important classification to be done is whether the works address the specific determination of GOs in a uniform infinite medium of most general anisotropy, or directly face estimating overall properties of composite structures. In the former group, a very instructive work of that type is the one of (Buroni and Saez 2010) which partly uses the *RT-IRT* method plus the Stroh formalism to explicit all the necessary integral roots in the general anisotropic elasticity case. The obtained complicated analytical form can be used in further numerical calculations. In the second group, when the matrix has isotropic, hence uncoupled, properties (Poizat and Sester 1999; Clyne et al. 2005; Liu 2011), it can be considered that the problem of an infinite inclusion pattern or fiber network having any TMEE properties of any anisotropic symmetry can be treated as a generalized thermo-elastic problem with an isotropic matrix and an embedded phase of the here discussed morphology, possibly with TMEE properties (remind that as pointed in the previous section problems of a multi-phased network in a matrix cannot be treated by a single equation as for a single-phase pattern or network). These isotropic matrix cases already represent quite a lot of structures of interest.

When not isotropic, the matrix phase properties are hardly found in the literature to be less than transversally isotropic, what is already so challenging that numerical calculations are the most often used. And even so, most results remain restricted to axially symmetric problems that is reinforcing phase properties and shapes axially symmetric with the matrix TI symmetry. See (Huang et al. 1998; Mikata 2001; Hou and Leung 2004; Chen et al. 2004; Lee et al. 2005) to cite a few. For example, to the authors knowledge, there are not even Eshelby-based approaches proposing fully explicit analytical solutions for off-symmetric structures in a TI matrix, although the formalism for them is frequently presented. In the already cited application of the present *RT-IRT* method in (Franciosi 2013) for MEE axially symmetric (spheroids, laminates and fibers) inclusions in a MEE matrix, a newly obtained result toward the follow-up of evolving patterns was the analytical mGO calculations for oblique orientations of the inclusion axis with regard to the symmetry axis of the matrix, which to the authors knowledge has no anterior proposal. The resulting effective properties do not have TI symmetry anymore, yet the result for spheroids is obtained from implementing analytical explicit integrals (oblique infinite fibers and laminates in a TI matrix can be treated in full analytical manner). Not all published results on the GO variations with orientations for spheroids were presented in (Barboura and Franciosi 2016) in the TI elastic case. In addition to the few presented graphs in the cited short reference, Fig. 15.10 illustrates the case of stiffer axial than transverse properties, the (not shown) converse being indeed very different and not just reversed. Applications to piezo-elastic problems are directly at hand, just needing numerical implementation. Studying this inclusion orientation effect in not isotropic (here TI) matrices shows that "tuning" a property does not necessarily correspond to in-axes orientation choices but can be (case per case) in between.

When an inclusion pattern has no specific orientation parameter as a 1D fiber bundle, the orientation changes with some matrix property anisotropy becomes a

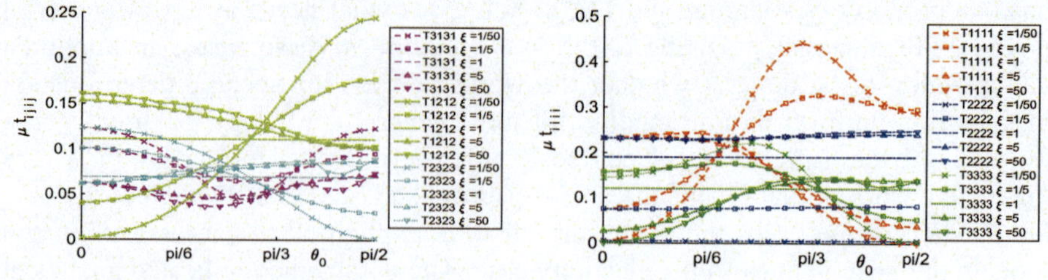

Fig. 15.10 Off-symmetry variations of effective elastic GO terms for a spheroid from flat (1/50) to long (50) aspect ratios, in a TI symmetric matrix stiffer in the symmetry axis direction $x3$ (Barboura and Franciosi 2016)

key information to follow, and in that respect, the results shown in Fig. 15.10 are important. In order to go further from oblique spheroids toward obliquely embedded patterns in a TI matrix with any properties of the TMEE type, a still missing essential analytical information is the calculation of pair interaction operators between two *oblique* same inclusions with general same orientation, as the fiber pair of Fig. 15.9 right. Since these calculations do not look to be at easy hand analytically, it is a case where approximating these interactions with the forms obtained in an isotropic matrix could be a "better than nothing" option (these isotropic forms are those exemplified in Fig. 15.2 and reported in Eqs. 15.7a, b).

15.5.3 From Static to Dynamical Problems

Dynamic problems in mechanics and in other fields of physics also are per se at the origin of a since long huge literature concerned with propagation/dispersion of various wave types in increasingly many composite structures and metamaterials (Kennett et al. 1978; Grossa et al. 2007; Willis 2019). It is consequently difficult to here do more than pointing some features in relation with our studies of global mGOs for large to infinite patterns and networks of inclusions and giving useful references. It is worth to first point that regarding the use of the PCW (1995) framework in the presented studies, a result of (Weng 2010) "validated" this framework dedicated to static elasticity of composites as consistent with a dynamical foundation, a plus to that author owing to other micromechanics models which do not have a dynamic counterpart.[6] Next considering the foundations of the *RT-IRT* method which also are at the basis of the presented results, those look a priori pretty well adapted to elastodynamic-like problems: On the one hand, the elementary operators involved in the GOs (Eqs. 15.2a, b) which read (for an isotropic elastic-like matrix of moduli $\boldsymbol{C}$, using the spherical coordinates $\omega 1 = \cos\phi\ \sin\theta$, $\omega 2 = \sin\phi\ \sin\theta$,

[6]The Mori-Tanaka (MT) model is also validated as having a dynamical foundation, yet it is in fact only in its restricted validation domain of all congruent ellipsoids in homothetic ellipsoidal distribution, which coincides with both the HS estimate and the PCW one in this case.

$\omega 3 = \cos\theta$, with $\theta = 0$ (resp. $\phi = 0$) in the $x3$ (resp. $x1$) direction of a $(x1, x2, x3)$ frame):

$$t^{e}_{pqjn}(\omega) = \left((M^{-1})_{pj}(\omega)\omega_q\ \omega_n\right)_{(pq),(jn)}, \tag{15.12}$$

are built on the acoustic 3 × 3 tensor $\boldsymbol{M}$ ($M_{mp}(\boldsymbol{\omega}) = C_{mnpq}\omega_n\omega_q$) which defines the Fourier transform of the (static) strain GO as $k^2\widehat{G}_{pj}(\boldsymbol{k}) = M^{-1}_{pj}(\boldsymbol{\omega})$ (see recalls in Spagnuolo et al. 2020); on the other hand, the shape function form in the mGOs involves explicitly the plane wave decomposition (Gel'fand and Shilov 1964) of the delta function $\delta(\mathbf{r} - \mathbf{r}') = (-1/8\ \pi^2)\int_\Omega \delta''(z - z',\omega)\mathrm{d}\omega$, in comparing with Eq. (15.3) where $s''_V(z,\omega) = \int_{-\infty}^{\infty} s_V(z',\omega)\delta''(z - z',\omega)\mathrm{d}z'$ in $\int_\Omega \psi_V(\omega,\mathbf{r})\mathrm{d}\omega = 1 = (-1/8\ \pi^2)\int_\Omega s''_V(z,\omega)\mathrm{d}\omega = (-1/8\ \pi^2)\ \int_\Omega s''_{(V)}(z,\omega)\mathrm{d}\omega = X_V(\mathbf{r}) = \int_V \delta(\mathbf{r} - \mathbf{r}')\mathrm{d}\mathbf{r}'$, and $X_V(\mathbf{r})$ being the indicator function at point $\mathbf{r}$ in V. Thanks to this latter property, the specific use of the *RT-IRT* method in dynamic elasticity of homogeneous media appears in many works having proposed a dynamic analogous to a static approach, with also extensions to various coupled property types entering the same formalism as elasticity (Burridge 1967; Norris 1994; Khutorianski and Sosa 1995; Chen et al. 2007; Zielinski 2010).

The method has been defended as especially efficient for numerical calculations in cases of anisotropic symmetries (Wang and Achenbach 1993), and recent applications can be found in (Tavaf et al. 2018; Shrestha and Banerjee 2018) for example. It was already applied analytically in (Willis 1980a, b), using results from (Korringa 1973), for studying waves through a matrix reinforced with isolated or aligned spheroids. These Willis' works proposed a GO formulation where "static and dynamic terms uncouple completely, to allow a very direct low-frequency perturbation solution, starting from the solution of an associated static problem."

From the presently used framework, in solving the motion equation $\mathrm{Div}\boldsymbol{\sigma}(\boldsymbol{r}, t) = \rho(\boldsymbol{r})\partial^2\boldsymbol{u}(\boldsymbol{r}, t)/\partial t^2$ (with as for $\boldsymbol{C}(\boldsymbol{r})$ writing $\rho(\boldsymbol{r}) = \rho^M + \Delta\rho(\boldsymbol{r})$ for the density field), instead of the stress equilibrium $\mathrm{Div}\boldsymbol{\sigma}(\boldsymbol{r}) = 0$ and with a dynamic Green function of the form $u_p(\boldsymbol{r}, t) = G_{pj}(r - r', t - t')f_j(r', t')$ at place of the static one, one easily verifies (following the static calculations in Spagnuolo et al. 2020) that the dynamical effects on the GOs under the form of Eqs. (15.2a, b) do not alter, as expected, the static shape function characteristic of a given inclusion pattern and that they essentially modify the elementary static GO operators, through a frequency-dependent contribution. The latter quite simply enters $\boldsymbol{M}$ in Eq. (15.12) for the isotropic elastic cases with also extensions to thermo- and poro-elastic domains which remain of isotropic nature, while piezoelectric and magnetic coupling do not for they require at most a TI symmetry (Norris 1994; Chen et al. 2007).

It is then permitted to expect that in long wavelength ranges where a composite can be considered to dynamically behave as a homogeneous equivalent (effective) medium, although calculations may remain far from simple, applications of wave propagation analyses to the here considered structures (also allowing to account for some dispersive features as is pointed in (Christensen 1979b) are possible. Yet, from these different cited works among many, it appears that even if the dynamic problems would in general be solved by a substantial amount when the effective elastic static behavior for the medium, in terms of effective properties, is known first, overall calculations to estimate specific wave features, as propagation or dispersion (scattering cross section) resonance, remain difficult, even for simple cases as aligned spheroids in an isotropic matrix, as shown in details in (Willis 1980b). Prior to safely extend from static to dynamic problems the here presented results on mGOs for "pattern complexity direction," a major issue to examine is to what extent considering a global pattern mGO in a single equation does not lose too much of local effects in wave characteristic concerns, compared with the coupled equation system that usually describes an inclusion arrangement. Difficulties are expectable too for the bi- and multi-continuous composite structures which are still not that much examined as here done in static property contexts. To the authors knowledge, more efficient frame in terms of microstructure simplifying is still to be proposed.

15.6 Conclusion

The presented results establish that considering infinite patterns of inclusions embedded in a matrix provides relevant estimates of static linear properties within a classical first gradient homogenization framework for such composite materials, using a mean Green operator (mGO) for these patterns at the place of single inclusions or of finite groups of them as is the most frequently done, with a single equation to solve instead of a system (the substitution is not permitted if the inclusions do not all have same properties). One shown consequence to referring to infinite patterns is to modify the notion of spatial distribution of elements, whether them be single elements or groups. It has been proposed and checked efficient that the relevant representative domain for evolving infinite patterns was the shape of the influence (interaction) zone for its elements, the variation of which was firstly related to the embedding medium spatial anisotropy symmetry and the shape anisotropy characteristics of the elements, both possibly modified but in second row, by an applied strain. These mGOs are of interest only if they account correctly for the pair interaction between the elements and are not only a mean of the element mGOs. This type of mGO is unfortunately not at easy analytical hand in all situations of interest and especially when the matrix properties are far from being isotropic. Even when at hand analytically, its major interest is to possibly evolve easily in following the current deformed state of the composite it represents. In that respect, it has been shown through the several results from the authors and co-workers here synthesized that the geometrical nature of the *RT-IRT* method, what has additionally provided the quite simple so-called decomposition method for determining mGOs of various inclusion patterns, is a powerful route toward describing complex arrangements of infinite and possibly networked patterns as this work has hopefully shown. Applications toward coupled properties, linearizable ones and dynamic ones are discussed.

Appendix

The form of the elementary operators t^e for isotropic elasticity. (from Franciosi and Lormand 2004; Franciosi 2005), to be retrieved in (Spagnuolo et al. 2020). Each $\boldsymbol{t}^{\mathrm{e}}(\boldsymbol{\omega}) = \boldsymbol{t}^{\mathrm{e}}(\theta,\varphi)$ elementary operator in the GO $\boldsymbol{t}^V(\boldsymbol{r})$ or mGO $\overline{\boldsymbol{t}^V}$ of V is an axisymmetric operator, defined in reference to the parallel planes of $\boldsymbol{\omega}$ normal direction in some reference medium frame. For general elasticity anisotropy, considering the $(0, 0)$-oriented $\boldsymbol{t}^e(0, 0)$ elementary operator for $(\theta,\varphi) = (0, 0)$, the only nonzero terms correspond to $\boldsymbol{\omega} = (0, 0, 1)$ what only involves the C_{m3p3} elastic moduli of the infinite medium, for $\boldsymbol{C}$ expressed in the operator axes frame as $\boldsymbol{C}(0, 0)$. In this frame, the nonzero terms of the $\boldsymbol{t}^e(0, 0)$ operator are the $t^e_{p3j3}(0,0)$ terms which make a symmetric 3×3 matrix, $\boldsymbol{\Delta t}$ say, such that $t^e_{p3j3} = \Delta t_{pj}$. For elastic isotropy, since in all frames $C_{3333} = \lambda + 2\,\mu = 2\,\mu\,(1 - \nu)/(1 - 2\,\nu)$ and $C_{1313} = C_{2323} = \mu$, the $\boldsymbol{C}$ frame identification is made useless. It remains $\Delta t_{11} = 1/C_{1313}$, $\Delta t_{22} = 1/C_{2323}$, $\Delta t_{33} = 1/C_{3333}$ and $t_{((2,3),(2,3))} = t_{((3,1),(3,1))} = 1/4\ \mu = B/4$, $t_{3333} = (1/\mu)(1 - 0.5/(1 - \nu)) = B + A$, $\forall(\theta,\varphi) \equiv \boldsymbol{\omega}$. One arrives at $t^e_{pqjn}(\boldsymbol{\omega}) = A\,\tau^A_{pqjn}(\omega) + B\,\tau^B_{pqjn}(\boldsymbol{\omega})$ with $\tau^{\mathrm{A}}_{\mathrm{pqjn}}(\boldsymbol{\omega}) = \omega_{\mathrm{j}}\omega_{\mathrm{p}}\omega_{\mathrm{n}}\omega_{\mathrm{q}}$ and $\tau^B_{pqjn}(\boldsymbol{\omega}) = \left(\delta_{jp}\ \omega_n\omega_q\right)\big|_{(p,q),(j,n)}$ (Table 15.2).

These terms are defined by even trigonometric functions $\cos^{2r}\varphi\ \sin^{2s}\varphi$, not vanishing upon integration over the unit sphere when multiplied by a positive and even function in (θ,φ) as the mean shape function $\overline{\psi_V(\theta,\varphi)}$ of V. Owing to the 5 $((1, 0), (0, 1))$ and $((2, 0), (1, 1), (0, 2))$ possible values taken by both the (l, m) and (r, s) exponent pairs, and owing to the dependency relations between the trigonometric functions, all terms can be expressed in using only two of the five θ-functions, one in each exponent pair set within brackets, and similarly two of the five ϕ-ones, such as for example $(l, m) = (1, 0), (2, 0)$ and $(r, s) = (1, 0), (2, 0)$ which, respectively, correspond to the two functions $\cos^2(\iota)$ and $\cos^4(\iota)$ for each angle $\iota = \theta,\varphi$. Thus, the mGO terms over some domain V read $\overline{t^V_{pqjn}} = \int_\Omega t^e_{pqjn}(\boldsymbol{\omega})\overline{\psi_V(\boldsymbol{\omega})}\mathrm{d}\boldsymbol{\omega}$ and explicating the elementary (isotropic elastic) operator part in it, the integrals which

Table 15.2 *iijj* (top) and *ijij* (bottom) terms of $\boldsymbol{t}^{\mathrm{e}}(\theta,\varphi)$, with "$c\theta$", "$s\theta$" for "$\cos\theta$", "$\sin\theta$" (resp. φ)

	11	22	33
11	$As^4\theta c^4\phi$ $Bs^2\theta c^2\phi$	$As^4\theta c^2\phi s^2\phi$ 0	$As^2\theta c^2\theta c^2\phi$ 0
22	$As^4\theta c^2\phi s^2\phi$ 0	$As^4\theta s^4\phi$ $Bs^2\theta s^2\phi$	$As^2\theta c^2\theta s^2\phi$ 0
33	$As^2\theta c^2\theta c^2\phi$ 0	$As^2\theta c^2\theta s^2\phi$ 0	$Ac^4\theta$ $Bc^2\theta$

2323	3131	1212
$As^2\theta c^2\theta s^2\phi$ $B(s^2\theta s^2\phi + c^2\theta)/4$	$As^2\theta c^2\theta c^2\phi$ $B(s^2\theta c^2\phi + c^2\theta)/4$	$As^4\theta c^2\phi s^2\phi$ $Bs^2\theta/4$

then need to be calculated are two pairs selected among those of the form:

$$\overline{t_{pqjn}^{V}} = \int_{\theta=0}^{\pi} \int_{\varphi=0}^{2\pi} \overline{\psi_V(\theta, \varphi)} \left(\cos^{2l}\theta \, \sin^{2m}\theta \, \cos^{2r}\varphi \, \sin^{2s}\varphi\right) \sin\theta \mathrm{d}\theta \mathrm{d}\varphi.$$

References

Alibert, J. J., Seppecher, P., & dell'Isola, F. (2003). Truss modular beams with deformation energy depending on higher displacement gradients. *Mathematics and Mechanics of Solids, 8*(1), 51–73.

Andreaus, U., Spagnuolo, M., Lekszycki, T., & Eugster, S. R. (2018). A Ritz approach for the static analysis of planar pantographic structures modeled with nonlinear Euler-Bernoulli beams. *Continuum Mechanics and Thermodynamics, 30,* 1103–1123.

Barboura, S., & Franciosi, P. (2016). Analytical inclusion Green operators in transversally isotropic media. In *Proceeding of the 24th ICTAM conference, Montreal, Canada* (vol. 3, pp. 2350–2351).

Barchiesi, E., & Placidi, L. (2017). A review on models for the 3D statics and 2D dynamics of pantographic fabrics. In *Wave dynamics and composite mechanics for microstructured materials and meta-materials* (pp. 239–258). Singapore: Springer.

Barchiesi, E., Ganzosch, G., Liebold, C., Placidi, L., Grygoruk, R., & Müller, W. H. (2019). Out-of-plane buckling of pantographic fabrics in displacement-controlled shear tests: Experimental results and model validation. *Continuum Mechanics and Thermodynamics, 31*(1), 33–45.

Bender, J., Jarvoll, P., Nydén, M., & Engström, S. (2008). Structure and dynamics of a sponge phase in the methyl δ; aminolevulinate/monoolein/water/propylene glycol system. *Journal of Colloid and Interface Science, 317,* 577–584.

Benveniste, Y. (1987). A new approach to the application of Mori–Tanaka's theory in composite materials. *Mechanics of Materials, 6,* 147–157.

Berveiller, M., Fassi-Fehri, O., & Hihi, A. (1987). The problem of two plastic and heterogeneous inclusions in an anisotropic medium. *International Journal of Engineering Sciences, 25*(6), 691–709.

Bornert, M., Stolz, C., & Zaoui, A. (1996). Morphologically representative pattern-based bounding in elasticity. *Journal of the Mechanics and Physics of Solids, 44*(3), 307–331.

Boucher, S. (1974). On the effective moduli of isotropic two-phase elastic composites. *Journal of Composite Materials, 8*(1), 82–89.

Brenner, R., Castelnau, O., & Gilormini, P. (2001). *A modified affine theory for the overall properties of nonlinear composites.* Comptes Rendus de l'Académie des Sciences, Paris, Mechanics series, 329(9), 649–654.

Burridge, R. (1967). The singularities on the plane lids of the wave surface of elastic media with cubic symmetry. *The Quarterly Journal of Mechanics and Applied Mathematics, 20*(1), 41–56.

Buroni, C., & Saez, A. (2010). Three-dimensional Green functions and its derivatives for materials with general anisotropic magneto-electro-elastic coupling. *Proceedings of the Royal Society A, 466,* 515–537.

Buryachenko, V. A. (2001). Multiparticle effective field and related methods in micromechanics of composite materials. *Applied Mechanics Review, 54,* 1–47.

Buryachenko, V. A., & Brun, M. (2012). Thermo elastic effective properties and stress concentrator factors of composites reinforced by heterogeneities of non canonical shape. *Mechanics of Materials, 53,* 91–110.

Boutin, C., Giorgio, I., & Placidi, L. (2017). Linear pantographic sheets: Asymptotic micro-macro models identification. *Mathematics and Mechanics of Complex Systems, 5*(2), 127–162.

Chen, J., Pan, E., & Chen, H. (2007). Wave propagation in magneto-electro-elastic multilayered plates. *International Journal of Solids and Structures, 44,* 1073–1085.
Chen, W. Q., Lee, K. Y., & Ding, H. J. (2004). General solution for transversely isotropic magneto-electro-thermo-elasticity and the potential theory method. *International Journal of Engineering Science, 42,* 1361–1379.
Chen, Z., Jeffrey, R. G., & Pandurangan, V. (2018). The far-field deformation caused by a hydraulic fracture in an inhomogeneous elastic half-space. *International Journal of Solids and Structures, 130,* 220–231.
Chiu, Y. P. (1977). On the stress field due to initial strains in a cuboid surrounded by an infinite elastic space. *Journal of Applied Mechanics, 12,* 587–590.
Christensen, R. M. (1979a). Isotropic properties of platelet reinforced media. *Journal of Engineering Materials and Technology, 101*(3), 299–303.
Christensen, R. M. (1979b). *Mechanics of composite materials*. New York: Wiley.
Christensen, R. M., & Lo, K. H. (1979). Solutions for effective shear properties in 3 phase sphere and cylinder models. *Journal of the Mechanics and Physics of Solids, 27*(4), 315–330.
Clyne, T. W., Markaki, A. E., & Tan, J. C. (2005). Mechanical and magnetic properties of metal fibre networks with and without a polymeric matrix. *Composites Science and Technology, 65,* 2492–2499.
Cuomo, M., dell'Isola, F., Greco, L., & Rizzi, N. L. (2017). First versus second gradient energies for planar sheets with two families of inextensible fibres: Investigation on deformation boundary layers, discontinuities and geometrical instabilities. *Composites Part B Engineering, 115,* 423–448.
Delannay, L., Lani, F., Pardoen, T., & Delannay, F. (2006). Mean field modeling of the plastic behaviour of co-continuous dual-phase alloys with strong morphological anisotropy. *International Journal of Plasticity, 22,* 2327–2345.
dell'Isola, F., Lekszycki, T., Pawlikowski, M., Grygoruk, R., & Greco, L. (2015). Designing a light fabric metamaterial being highly macroscopically tough under directional extension: First experimental evidence. *Zeitschrift für Angewandte Mathematik und Physik, 66*(6), 3473–3498.
dell'Isola, F., Giorgio, I., Pawlikowski, M., & Rizzi, N. L. (2016). Large deformations of planar extensible beams and pantographic lattices: Heuristic homogenization, experimental and numerical examples of equilibrium. *Proceedings of the Royal Society London A, 472*(2185), 20150790.
dell'Isola, F., Cuomo, M., Greco, L., & Della, Corte A. (2017). Bias extension test for pantographic sheets: Numerical simulations based on second gradient shear energies. *Journal of Engineering Mathematics, 103*(1), 127–157.
dell'Isola, F., Seppecher, P., Spagnuolo, M., et al. (2019). Advances in pantographic structures: Design, manufacturing, models, experiments and image analyses. *Continuum Mechanics and Thermodynamics, 31,* 1231–1282.
Drugan, W. J., & Willis, J. R. (1996). A micro-mechanics-based non local constitutive equation and estimate of representative volume element size for elastic composites. *Journal of the Mechanics and Physics of Solids, 44*(4), 497–524.
Eremeyev, V. A., dell'Isola, F., Boutin, C., & Steigmann, D. (2018). Linear pantographic sheets: Existence and uniqueness of weak solutions. *Journal of Elasticity, 132*(2), 175–196.
Eremeyev, V. A., Alzahrani, F. S., Cazzani, A., dell'Isola, F., Hayat, T., Turco, E., et al. (2019). On existence and uniqueness of weak solutions for linear pantographic beam lattices models. *Continuum Mechanics and Thermodynamics, 31*(6), 1843–1861.
Eremeyev, V. A., & Turco, E. (2020). Enriched buckling for beam-lattice metamaterials. *Mechanics Research Communications, 103,* 103458.
Eremeyev, V. A., Skrzat, A., & Stachowicz, F. (2016). On finite element computations of contact problems in micropolar elasticity. *Advances in Materials Science and Engineering.* https://doi.org/10.1155/2016/9675604.
Eremeyev, V. A. (2018). On the material symmetry group for micromorphic media with applications to granular materials. *Mechanics Research Communications, 94,* 8–12.

Eshelby, J. D. (1957). The determination of the elastic field of an ellipsoidal inclusion and related problems. *Proceedings of the Royal Society London, A, 421,* 379–396.

Eshelby, J. D. (1959). The elastic field outside an ellipsoidal inclusion. *Proceedings of the Royal Society London, A, 252,* 561–569.

Fassi-Fehri, O., Hihi, A., & Berveiller, M. (1989). Multiple site self consistent scheme. *International Journal of Engineering Sciences, 27*(5), 495–502.

Franciosi, P. (2005). On the modified Green operator integral for polygonal, polyhedral and other non-ellipsoidal inclusions. *International Journal of Solids and Structures, 42*(11/12), 3509–3531.

Franciosi, P. (2010). The boundary-due terms in the Green operator of inclusion patterns from distant to contact and to connected situations using Radon transforms: Illustration for spheroid alignments in isotropic media. *International Journal of Solids and Structures, 47*(2), 304–319.

Franciosi, P. (2013). Transversally isotropic magneto-electro-elastic composites with co-(dis)continuous phases. *International Journal of Solids Structure, 50,* 1013–1031.

Franciosi, P. (2014). Mean and axial Green and Eshelby tensors for an inclusion with finite cylindrical shape. *Mechanics Research Communications, 59,* 26–36.

Franciosi, P. (2018). A decomposition method for obtaining mean Green interaction operators between inclusions in patterns. Application to beam arrays in media with 2D or 3D isotropic properties. *International Journal of Solids and Structures, 147,* 1–19.

Franciosi, P. (2020a). Multiple continuity of phases in composite materials: Overall property estimates from a laminate system scheme. *International Journal of Solids and Structures, 184,* 40–62.

Franciosi, P. (2020b). Uniformity of the Green operator and Eshelby tensor for hyperboloidal domains in infinite media. *Mathematics and Mechanics of Solids, 25*(8), 1610–1642.

Franciosi, P., Barboura, S., & Charles, Y. (2015). Analytical mean Green operators/Eshelby tensors for patterns of coaxial finite long or flat cylinders in isotropic matrices. *International Journal of Solids and Structures, 66*(1), 1–19.

Franciosi, P., & Berbenni, S. (2007). Heterogeneous crystal and poly-crystal plasticity modeling from a transformation field analysis within a regularized Schmid law. *Journal of the Mechanics and Physics of Solids, 55,* 2265–2299.

Franciosi, P., & Berbenni, S. (2008). Multi-Laminate plastic strain organization for non-uniform TFA modelling of poly-crystal regularized plastic flow. *International Journal of Plasticity, 24*(9), 1549–1580.

Franciosi, P., & Charles, Y. (2016a). Effective properties of n-phase composites with from all to none continuous phases. *International Journal of Solids and Structures, 96,* 110–125.

Franciosi, P., & Charles, Y. (2016b). Mean Green operators and Eshelby tensors for hemispherical-inclusions and hemisphere interactions in spheres. Application to bi-material spherical inclusions in isotropic spaces. *Mechanics Research Communications, 75,* 57–66.

Franciosi, P., & El Omri, A. (2011). Effective properties of fiber and platelet systems and related phase arrangements in n-phase heterogeneous media. *Mechanics Research Communications, 38,* 38–44.

Franciosi, P., & Gaertner, R. (1998). On phase connectivity descriptions in modeling viscoelasticity of fiber or sphere reinforced composites. *Polymer Composites, 19*(1), 81–96.

Franciosi, P., & Lebail, H. (2004). Anisotropy features of phase and particle spatial pair distributions, in various matrix/inclusions structures. *Acta Materialia, 52*(10), 3161–3172.

Franciosi, P., & Lormand, G. (2004). Using the Radon transform to solve inclusion problems in elasticity. *International Journal of Solids and Structures, 41*(3/4), 585–606.

Franciosi, P., Spagnuolo, M., & Salman, O. U. (2019). Mean Green operators of deformable fiber networks embedded in a compliant matrix and property estimates. *Continuum Mechanics and Thermodynamics, 31*(1), 101–132.

Gao, X.-L., & Ma, H. M. (2009). Green's function and Eshelby's tensor based on a simplified strain gradient elasticity theory. *Acta Mechanica, 207,* 163–181.

Gel'fand, I. M., Graev, M. I., & Vilenkin, N. Y. (1966). *Generalized functions, v5: Integral geometry and representation theory.* New York: Academic Press.

Gel'fand, I. M., & Shilov, G. E. (1964). *Generalized functions, v1: Properties and operations*. New York: Academic Press.

Germain, P. (1973). La méthode des puissances virtuelles en mécanique des milieux continus, première partie: théorie du second gradient. *Journal de mécanique, 12*(2), 235–274.

Gong, L., Kyriakides, S., & Jang, W.-Y. (2005). Compressive response of open-cell foams. Part I: Morphology and elastic properties. *International Journal of Solids and Structures, 42,* 1355–1379.

Grossa, D., Dineva, P., & Rangelov, T. (2007). BIEM solution of piezoelectric cracked finite solids under time-harmonic loading. *Engineering Analysis with Boundary Elements, 31,* 152–162.

Han, Y., Zhang, D., Chung, L. L., Sun, J., Zhao, L., Zou, X., et al. (2009). A tri-continuous mesoporous material with a silica pore wall following a hexagonal minimal surface. *Nature Chemistry, 1,* 123–127.

Hashin, Z., & Shtrikman, S. (1963). A variational approach to the theory of elastic behavior of multi-phase materials. *Journal of the Mechanics and Physics of Solids, 11,* 127–140.

Helgason, S. (1980). *The Radon transform*. Progress in Mathematics, (vol. 5). Boston: Birkhausser.

Hervé, E., & Zaoui, A. (1993). N-layered inclusion-based micromechanical modeling. *International Journal of Engineering Science, 31*(1), 1–10.

Hervé, E., & Zaoui, A. (1995). Elastic behaviour of multiply coated fibre reinforced composites. *International Journal of Engineering Science, 33*(10), 1419–1433.

Hill, R. (1952). The elastic behavior of a crystalline aggregate. *Proceedings of the Physical Society, A, 65*(5), 349–354.

Hill, R. (1965). Continuum micro-mechanics of elastoplastic polycrystals. *Journal of the Mechanics and Physics of Solids, 13*(2), 89–101.

Hori, M., & Nemat-Nasser, S. (1993). Double-inclusion model and overall moduli of multi-phase composites. *Mechanics of Materials, 14*(3), 189–206.

Hou, P. F., & Leung, A. Y. T. (2004). A spheroidal inclusion in an infinite magneto-electroelastic material. *International Journal of Engineering Sciences, 42,* 1255–1273.

Hu, G. K., & Weng, G. J. (2000). The connections between the double inclusion model and the Ponte Castaneda-Willis, Mori Tanaka, and Kuster-Toksoz models. *Mechanics of Materials, 32,* 495–503.

Huang, J. H., Chiu, Y. H., & Liu, H. K. (1998). Magneto-electro-elastic Eshelby tensors for a piezoelectric–piezomagnetic composite reinforced by ellipsoid inclusions. *Journal of Applied Physics, 83*(10), 5364–5370.

Kennett, B. L. N., Kerry, N. J., & Woodhouse, J. H. (1978). Symmetries in the reflection and transmission of elastic waves. *Geophysical Journal, Royal Astronomical Society, 52*(2), 215–229.

Khutorianski, N., & Sosa, H. (1995). Dynamic representation formulas and fundamental solutions for piezoelectricity. *International Journal of Solids and Structures, 32*(22), 3307–3325.

Kinney, J. H., Stölken, J. S., Smith, T. S., Ryaby, J. T., & Lane, N. E. (2005). An orientation distribution function for trabecular bone. *Bone, 36*(2), 193–201.

Kinoshita, N., & Mura, T. (1971). Elastic fields of inclusions in anisotropic media. *Physica Status Solidi A, 5,* 759–768.

Korringa, J. (1973). Theory of elastic constants of heterogeneous media. *Journal of Mathematical Physics, 14,* 509–513.

Kröner, E., (1958). Berechnung der elastischen constanten des vielkristalls aus den konstanten des einkristalls. Zeitschrift fur Physik, 151–504.

Kröner, E. (1990). Modified Green functions in the theory of heterogeneous and/or anisotropic linearly elastic media. In Weng, G. J., Taya, M., & Abe, H. (Eds.), *Micromechanics and inhomogeneity* (pp. 197–211). New York: Springer.

Lahellec, N. L., & Suquet, P. (2004). Nonlinear composites: A linearization procedure, exact to second-order in contrast and for which the strain-energy and affine formulations coincide. *Comptes Rendus de l'Académie des Sciences, Paris, Mechanics series, 332*(9), 693–700.

Lee, J., Boyd, J. G., & Lagoudas, D. C. (2005). Effective properties of three-phase electromagneto-elastic composites. *International Journal of Engineering Science, 43,* 790–825.

Levin V. M. (1967). Thermal expansion coefficients for heterogeneous materials. *Mekhanika Tverdogo Tela, 2*, 88–94 (English translation: *Mechanics of Solids, 11*, 58–61).

Li, J. Y., & Dunn, M. L. (1998). Micromechanics of magnetoelectroelastic composite materials: Average fields and effective behaviour. *Journal of Intelligent Material Systems and Structures, 9*, 404–416.

Limodin, N., Salvo, L., Suery, M., & Di Michiel, M. (2007). In situ investigation by X-ray tomography of the overall and local microstructural changes occurring during partial remelting of an Al–15.8 wt.% Cu alloy. *Acta Materialia, 55*, 3177–3191.

Liu, M. F. (2011). An exact deformation analysis for the magneto-electro-elastic fiber-reinforced thin plate. *Applied Mathematical Modeling, 35*, 2443–2461.

Manolis, G. D., Makra, K., Dineva, P. S., & Rangelov, T. V. (2013). Seismic motions in a non-homogeneous soil deposit with tunnels by a hybrid computational technique. *Earthquakes and Structures, 5*(2), 161–205.

Masson, R., Bornert, M., Suquet, P., & Zaoui, A. (2000). An affine formulation for the prediction of the effective properties of non linear composites and poly-crystals. *Journal of the Mechanics and Physics of Solids, 48*(6/7), 1203–1227.

McCue, I., Ryan, S., Hemker, K., Xu, X., Li, N., Chen, M., et al. (2016). Size effects in the mechanical properties of bulk bicontinuous ta/cu nanocomposites made by liquid metal dealloying. *Advances in Engineering Materials, 18*(1), 46–50.

Michel, J. C., & Suquet, P. (2004). Computational analysis of nonlinear composite structures using the non uniform transformation field analysis. *Computer Methods in Applied Mechanics and Engineering, 193*, 5477–5502.

Mikata, Y. (2001). Explicit determination of piezo-electric Eshelby tensor for a spheroidal inclusion. *International Journal of Solids and Structures, 38*, 7045–7063.

Mindlin, R. D. (1965). Second gradient of strain and surface-tension in linear elasticity. *International Journal of Solids and Structures, 1*(4), 417–438.

Mori, T., & Tanaka, K. (1973). Average stressing the matrix and average elastic energy of materials with misfitting inclusions. *Acta Metallurgica, 21*, 571–574.

Natterer, F. (1986). *The mathematics of computerized tomography*. Stuttgart: Teubner.

Norris, A. N. (1994). Dynamic Green's functions in anisotropic piezoelectric, thermoelastic and poroelastic solids. *Proceedings of the Royal Society London A, 447*, 175–188.

Pavese, M., Valle, M., & Badini, C. (2007). Effect of porosity of cordierite preforms on microstructure and mechanical strength of co-continuous ceramic composites. *Journal of European Ceramic Society, 27*, 131–141.

Peng, H. X., Fan, Z., & Evans, J. R. G. (2001). Bi-continuous metal matrix composites. *Materials Science and Engineering A, 303*, 37–45.

Pideri, C., & Seppecher, P. (1997). A second gradient material resulting from the homogenization of an heterogeneous linear elastic medium. *Continuum Mechanics and Thermodynamics, 9*(5), 241–257.

Placidi, L., Barchiesi, E., Turco, E., & Rizzi, N. L. (2016). A review on 2D models for the description of pantographic fabrics. *Zeitschrift für Angewandte Mathematik und Physik, 67*(5), 121.

Placidi, L., Andreaus, U., & Giorgio, I. (2017). Identification of two-dimensional pantographic structure via a linear D4 orthotropic second gradient elastic model. *Journal of Engineering Mathematics, 103*(1), 1–21.

Placidi, L., Barchiesi, E., & Misra, A. (2018). A strain gradient variational approach to damage: A comparison with damage gradient models and numerical results. *Mathematics and Mechanics of Complex Systems, 6*(2), 77–100.

Placidi, L., Rosi, G., & Barchiesi, E. (2019). Analytical solutions of 2-dimensional second gradient linear elasticity for Continua with cubic-D4 microstructure. In *New Achievements in Continuum Mechanics and Thermodynamics* (pp. 383–401). Cham: Springer.

Poizat, C., & Sester, M. (1999). Effective properties of composites with embedded piezoelectric fibres. *Computational Materials Science, 16*(1–4), 89–97.

Ponte-Castaneda, P., & Suquet, P. (1998). Non linear composites. *Advances in Applied Mechanics, 34,* 172–302.

Ponte-Castaneda, P., & Willis, J. R. (1995). The effect of spatial distribution on the effective behavior of composite materials and cracked media. *Journal of the Mechanics and Physics of Solids, 43*(12), 1919–1951.

Ponte-Castaneda, P., & Willis, J. R. (1999). Variational second-order estimates for nonlinear composites. *Proceedings of the Royal Society London A, 455*(1985), 1799–1811.

Poquillon, D., Lemaitre, J., Baco-Carles, V., Tailhades, Ph, & Lacaze, J. (2002). Cold compaction of iron powders—Relations between powder morphology and mechanical properties. *Powder Technology, 126,* 65–74.

Postma, G. W. (1955). Wave propagation in a stratified medium. *Geophysics, 20*(4), 780–806.

Rahali, Y., Giorgio, I., Ganghoffer, J. F., & dell'Isola, F. (2015). Homogenization à la Piola produces second gradient continuum models for linear pantographic lattices. *International Journal of Engineering Science, 97,* 148–172.

Ramm, A. G., & Katsevich, A. I. (1996). *The radon transform and local tomography*. Boca Raton, FL: CRC Press.

Roberts, A. P., & Garboczi, E. J. (2002). Elastic properties of model random three-dimensional open-cell solids. *Journal of the Mechanics and Physics of Solids, 50,* 33–55.

Rodin, G. J. (1996). Eshelby's inclusion problem for polygons and polyhedra. *Journal of the Mechanics and Physics of Solids, 44,* 1977–1995.

Ryoo, R. (2009). Porous materials: A tricontinuous mesoporous system. *Nature Chemistry, 1,* 105–106.

Santalo, A. (1976). Integral geometry and geometric probability. In *Encyclopedia of mathematics and its applications*. Reading, MA: Addison-Wesley Pub. Co.

Shrestha, S., & Banerjee, S. (2018). Virtual nondestructive evaluation for anisotropic plates using symmetry informed sequential mapping of anisotropic Green's function. *Ultrasonics, 88,* 51–63.

Sciarra, G., dell'Isola, F., & Coussy, O. (2007). Second gradient poromechanics. *International Journal of Solids and Structures, 44*(20), 6607–6629.

Serra, J. (1982). *Image analysis and mathematical morphology*. London Academic Press.

Spagnuolo, M., Franciosi, P., & dell'Isola, F. (2020). A Green operator-based elastic modeling for two-phase pantographic-inspired bi-continuous materials. *International Journal of Solids and Structures, 188,* 282–308.

Suquet, P., (1995). *Overall properties of nonlinear composites: A modified secant moduli theory and its link with Ponte Castañeda's nonlinear variational procedure*. Comptes Rendus de l'Académie des Sciences, Paris, Mechanics Series, 320(11), 563–571.

Suquet, P. (1997). Effective behavior of non linear composites. In P. Suquet (Ed.), *Continuum micromechanics, CISM course and lecture* (Vol. 377, pp. 220–259). Berlin: Springer.

Tavaf, V., Saadatzi, M., Shrestha, S., & Banerjee, S. (2018). Quantification of material degradation and its behavior of elastodynamic Greens function for computational wave field modeling in composites. *Materials Today Communications, 17,* 402–412.

Torres, Y., Pavon, J. J., & Rodrıguez, J. A. (2012). Processing and characterization of porous titanium for implants by using NaCl as space holder. *Journal of Materials Processing Technology, 212,* 1061–1069.

Traxl, R., & Lackner, R. (2018). Consideration of arbitrary inclusion shapes in the framework of isotropic continuum micromechanics: The replacement Eshelby tensor approach. *Mechanics of Materials, 126,* 126–139.

Udhayaraman, R., & Mulay, S. S. (2017). Multi-scale approach based constitutive modeling of plain woven textile composites. *Mechanics of Materials, 112,* 172–192.

Veenstra, H., Verkooijen, P. C. J., Van Lent, B. J. J., Van Dam, J., De Boer, A. P., & Nijhof, A. P. H. J. (2000). On the mechanical properties of co-continuous polymer blends: Experimental and modeling. *Polymer, 41,* 1817–1826.

Walpole, L. J. (1967). The elastic field of an inclusion in an anisotropic medium. *Proceedings of the Royal Society London A, 300,* 270–289.

Walpole, L. J. (1981). Elastic behavior of composite materials: Theoretical foundations. *Advances in Applied Mechanics, 21,* 169–242.

Wang, G. Y., & Achenbach, J. D. (1993). A new method to obtain 3-D Green's functions for anisotropic solids. *Wave Motion, 18,* 273–289.

Weng, G. J. (2010). A dynamical theory for the Mori-Tanaka and Ponte Castañeda-Willis estimates. *Mechanics of Materials, 42*(9), 886–893.

Willis, J. R. (1980a). A polarization approach to the scattering of elastic waves-I. Scattering by a single inclusion. *Journal of the Mechanics and Physics of Solids, 28,* 287–305.

Willis, J. R. (1980b). A polarization approach to the scattering of elastic waves-II. Multiple scattering by inclusions. *Journal of the Mechanics and Physics of Solids, 28,* 307–327.

Willis, J. R. (1981). Variational and related methods for the overall properties of composites. *Advances in Applied Mechanics, 21,* 1–78.

Willis, J. R. (2019). From statics of composites to acoustic metamaterials. *Philosophical Transactions of the Royal Society A, 377,* 20190099.

Wu, T.-H., Chen, T., & Weng, C.-N. (2015). Green's functions and Eshelby tensors for an ellipsoidal inclusion in a non-centrosymmetric and anisotropic micropolar medium. *International Journal of Solids and Structures, 64*(65), 1–8.

Yang, H., Ganzosch, G., Giorgio, I., & Abali, B. E. (2018). Material characterization and computations of a polymeric metamaterial with a pantographic substructure. *Zeitschrift für Angewandte Mathematik und Physik, 69*(4), 105.

Zeller, R., & Dederich, P. H. (1973). Elastic constants of polycrystals. *Physic State Solids B, 55,* 831–842.

Zielinski, T. G. (2010). Fundamentals of multiphysics modelling of piezo-poro-elastic structures. *Archives of Mechanics, 62*(5), 343–378.

Chapter 16
Strain Gradient Models for Growing Solid Bodies

Zineeddine Louna, Ibrahim Goda, and Jean-François Ganghoffer

Abstract A unifying mechanical constitutive framework for growing solid bodies exhibiting scale effects is presented, relying on the principle of virtual power, and energy, and entropy principles. We focus in this chapter on strain gradient constitutive models and expose different variants of higher gradient theories, adopting a phenomenological viewpoint. Incorporation of strain gradient terms in the constitutive models developed for biological tissues is motivated by the occurrence of pronounced microscopic strain gradients within their internal architecture showing scale hierarchy and strong contrasts of mechanical properties between different phases. A strain gradient model for bone remodeling is developed following a micromechanical approach in order to highlight how such models can be constructed starting from the microstructural level.

16.1 Introduction

The focus of this contribution is the setting up of a modeling framework based on strain gradient materials for growing solid bodies experiencing a change of mass due to mass production and an irreversible mass flux across their boundary. Recent contributions of the authors in the field of bone remodeling account for both first- and second-order deformation gradients, relying on the thermodynamics of surfaces and configurational forces for the simulation of the evolution of the external bone surface induced by mechanical stimulations (Louna et al. 2018, 2019). It has been shown

Z. Louna
LMFTA, Faculté de Physique, USTHB, BP 32 El Alia, 16111 Bab Ezzouar, Algiers, Algeria

I. Goda
Arts et Métiers ParisTech, I2M Bordeaux, UMR CNRS 5295, 33400 Talence, France

J.-F. Ganghoffer (✉)
LEM3—Université de Lorraine, CNRS, 7 Rue Félix Savart, 57073 Metz, France
e-mail: jean-francois.ganghoffer@univ-lorraine.fr

F. dell'Isola and L. Igumnov (eds.), *Dynamics, Strength of Materials and Durability in Multiscale Mechanics*, Advanced Structured Materials 137,
https://doi.org/10.1007/978-3-030-53755-5_16

in particular that strain gradient models allow for the description of microstructure-related size effects which are known to be important in hierarchically heterogeneous materials like trabecular bones, but more generally in many biological tissues, soft or hard, which exhibit the same features motivating the recourse to generalized continuum mechanics.

In terms of mass change (growth), a clear distinction can be made between volumetric and surface growth, as emphasized in (Skalak et al. 1997; Epstein and Maugin 2000; Ganghoffer et al. 2014; Goda et al. 2016). Volumetric growth refers to the processes taking place in the bulk of the material, while surface growth involves deposition of mass at a surface, mostly occurring in hard tissues. The structural hierarchy of living tissues with microstructures, either soft or hard, such as trabecular bones (in the hard category) or tendons (a soft tissue) plays an important role in determining their macroscopic mechanical behavior as well as the stress and strain distribution at the macroscopic scale. Such microstructural effects become especially pronounced near the bone–implant interfaces and more generally in zones with high strain gradients, like in any heterogeneous material. This issue can be investigated using generalized continuum mechanics theories.

To incorporate the scale of the microstructure of a heterogeneous material within the continuum framework, a number of phenomenological remedies to the lack of microstructural features of classical continuum mechanics (first gradient) have been proposed in the past decades requiring to abandon the local action hypothesis of classical continuum mechanics. Such enhanced continuum models aim to incorporate information on the microstructure and can be categorized into three main classes: (i) non-local integral models (Kröner 1976; Eringen and Edelen 1972) (ii) higher-order gradient models (Alibert et al. 2003; dell'Isola et al. 2015a, b; Madeo et al. 2011, 2012; Alibert and Della Corte 2015; Goda and Ganghoffer 2016; Berkache et al. 2017; Giorgio et al. 2017; Reda et al. 2017; dell'Isola et al. 2019; Barchiesi et al. 2019) and (iii) Cosserat theories and variants of it (Cosserat and Cosserat 1909; Altenbach and Eremeyev 2009; Goda et al. 2012; Goda and Ganghoffer 2015a; Eremeyev 2019; Giorgio et al. 2019).

Evolution laws for growing solid bodies incorporating strain gradients are developed along the line of the phenomenology in the next section.

16.2 Evolution Laws for Growing Solid Bodies with Strain Gradient Effects

Solid bodies experience growth due to new mass being produced in their bulk or mass is being exchanged by a convective flux across their external boundary. In order to set the stage, we recall the balance laws of mass and momentum written in (Ganghoffer 2010). The general form of the mass balance **in Eulerian format** for an evolving domain Ω_t is given in terms of the actual density ρ as

$$\frac{\mathrm{D}}{\mathrm{D}t}\int_{\Omega_t}\rho\mathrm{d}x=\int_{\Omega_t}\left(\frac{\mathrm{D}\rho}{\mathrm{D}t}+\rho\nabla.\mathbf{v}\right)\mathrm{d}x=\int_{\Omega_t}\pi\,\mathrm{d}x+\int_{\partial\Omega_t}m\mathrm{d}s\equiv\int_{\Omega_t}\Gamma\rho\mathrm{d}x$$

with $\rho(\mathbf{x},t)$ the actual density, π the physical source of mass due to chemical reactions producing new species, and $m := \mathbf{m}.\mathbf{n}$ the irreversible scalar physical mass flux across the domain boundary, projection of the flux (vector) $\mathbf{m}$ onto the unit exterior normal $\mathbf{n}$. Localization of previous integral equation gives the local mass balance

$$\frac{\mathrm{D}\rho}{\mathrm{D}t}=\pi+\nabla.\mathbf{m}-\rho\nabla.\mathbf{v}$$

with $\mathbf{v}(\mathbf{x},t) := \left(\frac{\partial\mathbf{x}}{\partial t}\right)_X$ the Eulerian velocity. This balance law is consistent with (and equivalent to) the more physical writing (Ganghoffer and Haussy 2005)

$$\dot{\rho}+\rho\mathrm{div}(\mathbf{v})=\Phi_\rho+\sigma_\rho$$

with $\Phi_\rho\equiv\nabla.\mathbf{m}$ the total flux of conduction and $\sigma_\rho\equiv\pi$ the volumetric source of mass. Expressing the total mass of the growing domain Ω_g as $m(\Omega_\mathrm{g})=\int_{\Omega_\mathrm{g}}\rho(x)\mathrm{d}x$, the mass variation due to the transport phenomena is written as the following integral accounting for source terms

$$\left(\frac{\mathrm{d}m}{\mathrm{d}t}\right)_{\text{source}}:=\int_{\Omega}\pi\,\mathrm{d}x=\int_{\Omega}\mathrm{d}x\Rightarrow\pi=\Gamma\rho$$

with Γ the rate of mass variation due to growth, a quantity having the dimension of the inverse of time. All balance laws for growing solid bodies experiencing a variation of mass due to mass production or/and mass exchanges across the boundary can be obtained based on the material derivative of integrals of specific quantities (defined per unit mass) $a=a(\mathbf{x},t)$:

$$\frac{\mathrm{D}}{\mathrm{D}t}\int_{\Omega_t}\rho a\mathrm{d}x=\int_{\Omega_t}\left\{\rho\frac{\mathrm{D}a}{\mathrm{D}t}+a(\pi+\nabla.\mathbf{m})\right\}\mathrm{d}x$$

Using the mass balance, the **Eulerian version** of the balance of momentum writes (Epstein and Maugin 2000)

$$\frac{\mathrm{D}}{\mathrm{D}t}\int_{\Omega_t}\rho\mathbf{v}\mathrm{d}x=\int_{\Omega_t}\mathbf{f}\mathrm{d}x+\int_{\partial\Omega_t}\mathbf{n}.\boldsymbol{\sigma}\mathrm{d}\sigma_t+\int_{\Omega_t}\pi\mathbf{v}\mathrm{d}x+\int_{\partial\Omega_t}\mathbf{n}.(\mathbf{m}\otimes\mathbf{v})\mathrm{d}\sigma_t$$

with $\boldsymbol{\sigma}$ therein the Cauchy stress and $\mathbf{f}$ the body forces per unit physical volume. The right-hand side of previous equality represents the power of external forces; localizing previous balance law and using the mass balance gives the balance of linear momentum

$$\rho\frac{\mathrm{D}\mathbf{v}}{\mathrm{D}t} = \mathbf{f} + \mathrm{div}\boldsymbol{\sigma} + (\mathbf{m}.\nabla)\mathbf{v} \equiv p_{\mathrm{i}}$$

The density of the internal power of internal forces, the scalar p_{i}, is accordingly defined as the right-hand side of previous relation. The material derivative of the total energy can be expanded as (Ganghoffer 2010):

$$\frac{\mathrm{D}E}{\mathrm{D}t} = Q + \int_{\Omega_t}\left(\Pi_e + \mathrm{div}(e\mathbf{m}) + \boldsymbol{\sigma} : (\mathrm{grad}\,\mathbf{v})^{\mathrm{S}}\right)\mathrm{d}x = -P_{\mathrm{i}} + Q + \Pi_E$$

$$\Pi_E = \int_{\Omega_t}(\Pi_e + \mathrm{div}(e\mathbf{m}))\mathrm{d}x = \int_{\Omega_t}\Pi_e\mathrm{d}x + \int_{\partial\Omega_t} e\mathbf{m}.\mathbf{n}\mathrm{d}s$$

introducing therein the quantity of heat, the scalar $Q := -\int_{\Omega_t}(\mathrm{div}\mathbf{q})\mathrm{d}x$, and where the second equality defines the energy source term Π_E. The second principle of thermodynamics in presence of source terms due to growth writes

$$\int_{\Omega_t}\rho\theta\frac{\mathrm{D}s}{\mathrm{D}t}\mathrm{d}x \geq Q + \int_{\Omega_t}\mathbf{q}.\frac{\mathrm{grad}\,\theta}{\theta}\mathrm{d}x - \int_{\Omega_t}\theta s(\mathrm{div}\mathbf{m})\mathrm{d}x$$

Let rewrite

$$\int_{\Omega_t}\rho\theta\frac{\mathrm{D}s}{\mathrm{D}t}\mathrm{d}x = \int_{\Omega_t}\rho\frac{\mathrm{D}e}{\mathrm{D}t}\mathrm{d}x - \int_{\Omega_t}\rho s\frac{\mathrm{D}\theta}{\mathrm{D}t}\mathrm{d}x - \int_{\Omega_t}\rho\frac{\mathrm{D}\psi}{\mathrm{D}t}\mathrm{d}x$$
$$\equiv \frac{\mathrm{D}E}{\mathrm{D}t} - \int_{\Omega_t}(\pi e + e\mathrm{div}\mathbf{m})\mathrm{d}x - \int_{\Omega_t}\rho s\frac{\mathrm{D}\theta}{\mathrm{D}t}\mathrm{d}x - \int_{\Omega_t}\rho\frac{\mathrm{D}\psi}{\mathrm{D}t}\mathrm{d}x$$

involving the free energy density $\psi = e - \theta s$. Previous writings deliver the Clausius–Duhem inequality

$$-P_{\mathrm{i}} - \int_{\Omega_t}\rho s\frac{\mathrm{D}\theta}{\mathrm{D}t}\mathrm{d}x - \int_{\Omega_t}\rho\frac{\mathrm{D}\psi}{\mathrm{D}t}\mathrm{d}x + \int_{\Omega_t}\mathbf{m}.(\mathrm{grad}\,e)\mathrm{d}x \geq \int_{\Omega_t}\mathbf{q}.\frac{\mathrm{grad}\,\theta}{\theta}\mathrm{d}x$$
$$-\int_{\Omega_t}\theta s(\mathrm{div}\mathbf{m})\mathrm{d}x$$

Note that all writings hold in both the small and large strains regimes. Selecting a free energy density depending upon the absolute temperature and the elastic part of the transformation gradient delivers the state laws and the residual dissipation involving the growth strain rate, the second-order tensor $\mathbf{D}_{\mathrm{g}}$:

$$\psi = \psi(\mathbf{F}_{\mathrm{e}}, \theta) \to \boldsymbol{\sigma} = \rho\partial_{F_{\mathrm{e}}}\psi; \quad s = -\partial_\theta\psi(\mathbf{F}_{\mathrm{e}}, \theta)$$

$$\boldsymbol{\sigma} : \mathbf{D}_g + (s\mathbf{m} - \mathbf{q}/\theta).\text{grad}\,\theta + \theta\mathbf{m}.\text{grad}\,s + \mathbf{m}.\text{grad}\,\psi - \rho s\dot{\theta} \geq 0$$

Isolating the thermal dissipation gives the following two independent inequalities:

$$\boldsymbol{\sigma} : \mathbf{D}_g + \mathbf{m}.\text{grad}\,\psi \geq 0$$
$$(s\mathbf{m} - \mathbf{q}/\theta).\text{grad}\,\theta + \theta\text{div}(\mathbf{m}s) - \rho s\dot{\theta} \geq 0$$

The quantity $\theta\text{div}(\mathbf{m}s)$ in previous inequality represents an irreversible entropy source due to the (irreversible) mass flux, and the term $\mathbf{m}.\text{grad}\,\psi$ is an irreversible entropy flux accounting for the mass flux, which gives rise to second gradient effects. We accordingly reach the conclusion that modeling solid body growth in the framework of a continuum theory requires to incorporate strain gradient effects. It becomes indeed clear after expanding the gradient of the free energy density that the second term in previous inequality leads to a strain gradient model. This important fact was mentioned in Epstein and Maugin (2000) and receives here an independent proof. It is accordingly the aim of this contribution to formulate different types of enriched constitutive models for growing solid bodies in the framework of the mechanics of continuous media. The point of view adopted in deriving previous inequalities is that of the phenomenology, so no recourse is made to any underlying evolutive microstructure prone to growth or remodeling.

The mass flux and growth rate obey an evolution law which shall satisfy the previous Clausius–Duhem inequality; since the mass flux (a vector) and the growth rate (a second-order tensor) are of a different tensor order, they are according to Curie principle uncoupled when writing the kinetic laws satisfied by the internal variables.

The next section is devoted to the illustration of such strain gradient models to bone remodeling, whereby an effective strain gradient evolutive bone model is constructed based on the homogenization of the trabecular microstructure accounting for its evolution due to remodeling.

16.3 Micromechanical Second Gradient Models for Bone Growth in the Framework of Thermodynamics of Irreversible Processes

In bone biomechanics, important size effects are known at different scales of the hierarchical bone microstructure: It concerns the elastic behavior of single osteons (Lakes 1995), cortical bone (Frasca et al. 1981; Yang and Lakes 1982; Park and Lakes 1986; Buechner and Lakes 2003), and trabecular bone (Harrigan et al. 1988; Ramézani et al. 2012; Goda et al. 2012, 2013, 2014; Goda and Ganghoffer 2015b; Giorgio et al. 2016; Andreaus et al. 2014; Lekszycki and dell'Isola 2012). For a single osteon, the size effects are attributed to the compliance of the interfaces separating the laminae. As for trabecular bone, experimental evidence shows that the cement lines considered as compliant interfaces account for most of the stiffness difference between osteons and the entire bone. Despite the development of many continuum models of trabecular bone over the last two decades under the umbrella of classical elasticity (e.g., Taylor et al. 2002; Bowman et al. 1998), those models ignore microstructure-related scale effects on the macroscopic mechanical properties. They accordingly do not provide a satisfactory description of the bone behavior when the microstructural size of bone is comparable to the macroscopic length scale. It is accordingly the aim of the developed strain gradient continuum theory of bone to remedy these deficiencies and to properly account for microstructural effects when modeling the evolutive bone microstructure.

Strain gradient models for bone remodeling rely on micromechanical analyses performed at the scale of a representative volume element of trabecular bone structure (Louna et al. 2018). Both static and evolutive homogenized properties of a periodic network of bone trabeculae are evaluated by combining a methodology for the evaluation of the average kinematic and static variables over a trabecular cell and numerical simulations with controlled imposed first and second strain rates. In fact, the use of a strain gradient model accounts explicitly for the precise microstructure of the porous trabecular network, i.e., for the geometric distribution of porosity and the size of the pores inside the representative volume element. The strain gradient type constitutive model has been identified within the umbrella of the thermodynamics of irreversible processes, adopting a split of the kinematic and static tensors into their deviator and hydrostatic contributions. The elaborated effective growth constitutive law at the scale of the homogenized set of trabeculae consequently relates the average first and second growth strain rates to the homogenized Cauchy stress and hyperstress tensors, weighted by a nonlinear function of the (evolving) apparent density.

We herewith summarize the constitutive model developed in (Louna et al. 2019). The bone microstructures are based on real 3D images of trabecular bone, from which a 2D section is obtained (Fig. 16.1).

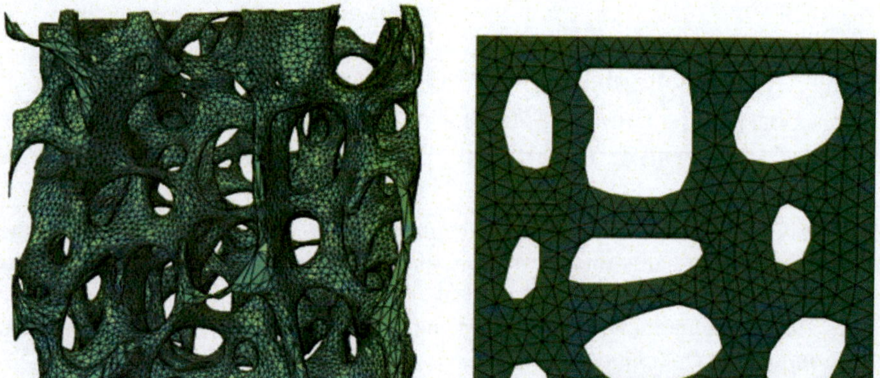

Fig. 16.1 3D trabecular bone sample (left) and 2D section (right)

The Cauchy stress and hyperstress tensors can be constructed based on the extension of Hill–Mandel equivalence principle, viz

$$\langle \boldsymbol{\sigma} : \dot{\boldsymbol{\varepsilon}} \rangle = \langle \boldsymbol{\sigma} \rangle : \langle \dot{\boldsymbol{\varepsilon}} \rangle + \langle \boldsymbol{\sigma} \otimes \mathbf{x} \rangle \therefore \dot{\mathbf{K}} \rightarrow \boldsymbol{\Sigma} := \langle \boldsymbol{\sigma} \rangle, \, \boldsymbol{\Sigma}^{\mathrm{S}} := \langle \boldsymbol{\sigma} \otimes \mathbf{x} \rangle$$

defining the effective Cauchy stress and hyperstress tensors, the second- and third-order tensors $\boldsymbol{\Sigma}$, $\boldsymbol{\Sigma}^{\mathrm{S}}$, respectively. This writing also provides the average kinematics in terms of the average displacement, the linearized displacement gradient, and the strain gradient tensor, successively given by

$$\mathbf{U}(\mathbf{X}) := \langle \mathbf{u}(\mathbf{x}) \rangle_{V(\mathbf{X})},$$

$$\mathbf{E}(\mathbf{X}) := \mathbf{U}(\mathbf{X}) \otimes \nabla_X := \langle \mathbf{u}(\mathbf{x}) \otimes \nabla_x \rangle_{V(\mathbf{X})} \rightarrow \boldsymbol{\varepsilon}(\mathbf{X}) \equiv \frac{1}{2}\left(\mathbf{U}(\mathbf{X}) \otimes \nabla_X + \mathbf{U}(\mathbf{X}) \otimes \nabla_X^{\mathrm{T}}\right)$$

$$\mathbf{K}(\mathbf{X}) := \boldsymbol{\varepsilon}(\mathbf{X}) \otimes \nabla_X \rightarrow K_{ijk} = \varepsilon_{ij,k} = \varepsilon_{ji,k} = K_{jik}$$

The averaging of the microscopic fields therein indicated by the bracket notation $\langle . \rangle_{V(\mathbf{X})}$ is done over a representative volume element $V(\mathbf{X})$ centered around the mesoscopic point $\mathbf{X}$. The Cauchy stress and hyperstress tensors satisfy the following static equilibrium equation (inertia terms can be neglected considering the very long time scales of the bone remodeling process).

$$\left(\boldsymbol{\Sigma} - \boldsymbol{\Sigma}^{s}.\nabla_X\right).\nabla_X + \mathbf{f} = 0$$

with $\mathbf{f}$ the body weight. Since we restrict to a small strain rate framework in the present context of bone remodeling, the average kinematic first and second gradient tensors $\mathbf{E}(\mathbf{X})$ and $\mathbf{K}(\mathbf{X})$ introduced previously can be assimilated to their small strain versions; thus, one is entitled to recourse to the following symmetrized tensors

$$\boldsymbol{\varepsilon}(\mathbf{X}) \equiv \frac{1}{2}\left(\mathbf{U}(\mathbf{X}) \otimes \nabla_X + \mathbf{U}(\mathbf{X}) \otimes \nabla_X^{\mathrm{T}}\right) \rightarrow \varepsilon_{ij} = \varepsilon_{ji}$$

$$\mathbf{K}(\mathbf{X}) := \boldsymbol{\varepsilon}(\mathbf{X}) \otimes \nabla_X \rightarrow K_{ijk} = \varepsilon_{ij,k} = \varepsilon_{ji,k} = K_{jik}$$

Note especially that we use here the symmetrized form (with respect to the first and second indices) of the second gradient of the displacement field $\mathbf{K}$. Since bone growth at the mesoscopic level is a slow process occurring at a typical time scale of a few weeks, the average strain rate scan be linearized, using their small strains rate counterparts. It accordingly holds the following approximations of the total strain rate and its growth and elastic parts (the index 2 in any tensor therein indicates a third-order tensor representative of second gradient effects):

$$\overline{\mathbf{D}}_1 \cong \dot{\boldsymbol{\varepsilon}}, \; \overline{\mathbf{D}}_{1\mathrm{g}} \cong \dot{\boldsymbol{\varepsilon}}_{\mathrm{g}}, \; \overline{\mathbf{D}}_{1\mathrm{e}} \cong \dot{\overline{\boldsymbol{\varepsilon}}}_{1\mathrm{e}} = \dot{\boldsymbol{\varepsilon}}_{\mathrm{e}} \rightarrow \dot{\boldsymbol{\varepsilon}} = \dot{\boldsymbol{\varepsilon}}_{\mathrm{e}} + \dot{\boldsymbol{\varepsilon}}_{\mathrm{g}}$$

$$\overline{\mathbf{D}}_2 \cong \dot{\overline{\boldsymbol{\varepsilon}}} \otimes \nabla = \dot{\overline{\boldsymbol{\varepsilon}}}_2, \; \overline{\mathbf{D}}_{2\mathrm{g}} \cong \dot{\overline{\boldsymbol{\varepsilon}}}_{\mathrm{g}} \otimes \nabla = \dot{\overline{\boldsymbol{\varepsilon}}}_{2\mathrm{g}} = \dot{\mathbf{K}}_{\mathrm{g}},$$

$$\overline{\mathbf{D}}_{2\mathrm{e}} \cong \dot{\overline{\boldsymbol{\varepsilon}}}_{\mathrm{e}} \otimes \nabla = \dot{\overline{\boldsymbol{\varepsilon}}}_{2\mathrm{e}} = \dot{\mathbf{K}}_{\mathrm{e}} \rightarrow \dot{\mathbf{K}} = \dot{\mathbf{K}}_{\mathrm{e}} + \dot{\mathbf{K}}_{\mathrm{g}}$$

Following the standard Coleman–Noll procedure, the homogenized constitutive law is formulated based on an elastic potential:

$$\boldsymbol{\Sigma} := \frac{\partial \psi_e(\mathbf{E}_e, \mathbf{K}_e)}{\mathbf{E}_e} = \mathbf{C} : \mathbf{E}_e, \quad \boldsymbol{\Sigma}^{\mathbf{S}} := \frac{\partial \psi_e(\mathbf{E}_e, \mathbf{K}_e)}{\mathbf{K}_e} = \mathbf{D} \,\vdots\, \mathbf{K}_e$$

The general writing of the present growth model in tensor format shall a priori incorporate a combination of isotropic and kinematic contributions. Let define the non-negative scalar $\overline{p}_{1g} := \int_0^t \left(\frac{2}{3}\overline{\mathbf{D}}_{1g} : \overline{\mathbf{D}}_{1g}\right)^{1/2} \mathrm{d}t,\ \overline{p}_{g2} = \int_0^t \left(\frac{1}{2}\overline{\mathbf{D}}_{2g} \,\vdots\, \overline{\mathbf{D}}_{2g}\right)^{1/2} \mathrm{d}t$ as the cumulative growth strain. The growth part of the free energy density is set as $\psi_g = \psi_g\left(\mathbf{E}_g := \overline{\boldsymbol{\varepsilon}}_{1g}, \mathbf{k}_g := \overline{\boldsymbol{\varepsilon}}_{2g}, r_g\right)$, in which the scalar variable r_g is the isotropic growth hardening or softening variable, which shall have the ability to account for a possible growth recovery, given versus the effective plastic strain rate as $\dot{r}_g = \frac{\partial \Omega}{\partial R_g} = \dot{\overline{p}}_g$. The thermodynamic variables conjugated to the introduced internal variables $\left(\mathbf{E}_g := \overline{\boldsymbol{\varepsilon}}_{1g}, \mathbf{k}_g = \mathbf{E}_{2g} := \overline{\boldsymbol{\varepsilon}}_{2g}, r_g\right)$ are the radius R_g representing the size of the dissipation equipotential and the center of the growth domain (conjugated to r_g) and the second- and third-order internal stress tensors $\mathbf{X}_{1g}, \mathbf{X}_{2g}$, elaborated as the following partial derivatives:

$$R_g = \frac{\partial \psi_g\left(\mathbf{E}_g, \mathbf{E}_{2g}, r_g\right)}{\partial r_g}, \mathbf{X}_{1g} := \frac{\partial \psi_g\left(\mathbf{E}_g, \mathbf{E}_{2g}, r_g\right)}{\partial \mathbf{E}_g}, \mathbf{X}_{2g} := \frac{\partial \psi_g\left(\mathbf{E}_g, \mathbf{E}_{2g}, r_g\right)}{\partial \mathbf{E}_{2g}}$$

These variables successively represent the size and position of the growth domain. The local reduced dissipation incorporating these driving forces is then identified following the standard procedure (Lemaitre and Chaboche 2009) as

$$\Phi = \left(\boldsymbol{\Sigma} - \mathbf{X}_{1g}\right) : \dot{\mathbf{E}}_g + \left(\boldsymbol{\Sigma}^{\mathbf{S}} - \mathbf{X}_{2g}\right) \,\vdots\, \dot{\mathbf{E}}_{2g} - R_g \dot{r}_g \geq 0$$

The dissipation potential is next formally introduced in stress space as

$$\begin{aligned} \varphi^* &= \Omega\left(\Sigma_{eq}\left(\boldsymbol{\Sigma} - \mathbf{X}_{1g}, \boldsymbol{\Sigma}^{\mathbf{S}} - \mathbf{X}_{2g}\right) - R_g - \sigma_g\right) \\ &\equiv \Sigma_{eq}\left(\boldsymbol{\Sigma} - \mathbf{X}_{1g}, \boldsymbol{\Sigma}^{\mathbf{S}} - \mathbf{X}_{2g}\right) - R_g - \Sigma_g \end{aligned}$$

The scalar quantity Σ_g therein is the growth threshold corresponding to the minimal effective stress below which no remodeling occurs, associated to the lazy zone. The contributions $\Sigma_{eq}\left(\boldsymbol{\Sigma} - \mathbf{X}_{1g}\right) - R_g$ and $\mathbf{X}_{1g}$ account successively for isotropic and kinematic growth hardening. The superscript 'D' in previous and subsequent relations denotes the deviator part of the corresponding tensor, elaborated in the 2D context from the harmonic decomposition of any second-order tensor (Olive and Auffray 2014). The first- and second-order tensors $\mathbf{X}_{1g}, \mathbf{X}_{2g}$ are the center of the actual equipotential surface in stress and hyperstress spaces, respectively, accounting for a possible kinematic growth hardening.

We select the following linear combination of the two invariants to define the effective (equivalent) stress

$$\Sigma_{eq} := \left\{\alpha J_1(\boldsymbol{\Sigma} - \mathbf{X}_{1g}) + \beta J_2(\boldsymbol{\Sigma} - \mathbf{X}_{1g}) + \gamma J_1(\boldsymbol{\Sigma}^{S} - \mathbf{X}_{2g}) + \delta J_2(\boldsymbol{\Sigma}^{S} - \mathbf{X}_{2g})\right\}^{1/2}$$

The equivalent stress incorporates both the stress and hyperstress tensors through the first and second invariants of the differences $(\boldsymbol{\Sigma} - \mathbf{X}_{1g})$, $(\boldsymbol{\Sigma}^{S} - \mathbf{X}_{2g})$, respectively, the scalar quantities

$$\begin{aligned}
J_1(\boldsymbol{\Sigma} - \mathbf{X}_{1g}) &= \mathrm{Tr}(\boldsymbol{\Sigma} - \mathbf{X}_{1g}),\\
J_2(\boldsymbol{\Sigma} - \mathbf{X}_{1g}) &= \left(\frac{3}{2}(\boldsymbol{\Sigma} - \mathbf{X}_{1g})^{D} : (\boldsymbol{\Sigma} - \mathbf{X}_{1g})^{D}\right)^{1/2}\\
J_1(\boldsymbol{\Sigma}^{S} - \mathbf{X}_{2g}) &= \big(V_i^{\nabla\,\mathrm{str}}(\boldsymbol{\Sigma}^{S} - \mathbf{X}_{2g}).V_i^{\nabla\,\mathrm{str}}(\boldsymbol{\Sigma}^{S} - \mathbf{X}_{2g})\\
&\quad + V_i^{\mathrm{rot}}(\boldsymbol{\Sigma}^{S} - \mathbf{X}_{2g}).V_i^{\mathrm{rot}}(\boldsymbol{\Sigma}^{S} - \mathbf{X}_{2g})\big)^{1/2}\\
J_2(\boldsymbol{\Sigma}^{S} - \mathbf{X}_{2g}) &= \left(\frac{1}{2}(\boldsymbol{\Sigma}^{S} - \mathbf{X}_{2g})^{D} \vdots (\boldsymbol{\Sigma}^{S} - \mathbf{X}_{2g})^{D}\right)^{1/2}
\end{aligned}$$

The first invariant of a third-order tensor is constructed from its non-deviator parts, so that $V_i^{\nabla\,\mathrm{str}}$, V_i^{rot} denote here the vector parts of the harmonic decomposition of the third-order tensor $(\boldsymbol{\Sigma}^{S} - \mathbf{X}_{2g})$, see Louna et al. (2018). The growth model is then elaborated from a growth potential $\Omega_g(f)$, selected as a scalar valued function of the driving force of von Mises type depending on the two parameters K, n, leading to the dissipation potential φ^*:

$$\begin{aligned}
\Omega_g(f) &:= \frac{K}{n+1}\left(\frac{f}{K}\right)^{n+1} \to \varphi^* = \Omega_g\big(\Sigma_{eq}(\boldsymbol{\Sigma} - \mathbf{X}_{1g}, \boldsymbol{\Sigma}^{S} - \mathbf{X}_{2g}) - R_g - \Sigma_g\big)\\
&:= \frac{K}{n+1}\left(\frac{\Sigma_{eq} - R_g - \Sigma_g}{K}\right)^{n+1}
\end{aligned}$$

The first and second gradient average growth rate tensors are obtained from the viscoplastic type dissipation potential φ^* based on the normality rule as follows:

$$\begin{aligned}
\overline{\mathbf{D}}_{1g} &= \frac{\partial \varphi^*(\boldsymbol{\Sigma}, \mathbf{X}_{1g}, R_{1g}, \boldsymbol{\Sigma}^{S}, \mathbf{X}_{2g}, R_{2g})}{\partial \boldsymbol{\Sigma}} = \frac{\partial \varphi^*(\boldsymbol{\Sigma}, \mathbf{X}_{1g}, R_{1g}, \boldsymbol{\Sigma}^{S}, \mathbf{X}_{2g}, R_{2g})}{\partial \Sigma_{eq}}\frac{\partial \Sigma_{eq}}{\partial \boldsymbol{\Sigma}},\\
\frac{\partial \Sigma_{eq}}{\partial \boldsymbol{\Sigma}} &= \frac{\partial \Sigma_{eq}}{\partial J_1(\boldsymbol{\Sigma} - \mathbf{X}_{1g})}\frac{\partial J_1(\boldsymbol{\Sigma} - \mathbf{X}_{1g})}{\partial \boldsymbol{\Sigma}} + \frac{\partial \Sigma_{eq}}{\partial J_2(\boldsymbol{\Sigma} - \mathbf{X}_{1g})}\frac{\partial J_2(\boldsymbol{\Sigma} - \mathbf{X}_{1g})}{\partial \boldsymbol{\Sigma}}\\
&= \left\{\alpha \mathbf{I} + \frac{3}{2}\beta\frac{(\boldsymbol{\Sigma} - \mathbf{X}_{1g})^{D}}{J_2(\boldsymbol{\Sigma} - \mathbf{X}_{1g})}\right\}/\Sigma_{eq} \Rightarrow \overline{\mathbf{D}}_{1g}\\
&= \bar{\dot{p}}_{1g}\left\{\alpha J_1(\boldsymbol{\Sigma} - \mathbf{X}_{1g})\mathbf{I} + \frac{3}{2}\beta\frac{(\boldsymbol{\Sigma} - \mathbf{X}_{1g})^{D}}{J_2(\boldsymbol{\Sigma} - \mathbf{X}_{1g})}\right\}/\Sigma_{eq},
\end{aligned}$$

$$\overline{\mathbf{D}}_{2g} = \frac{\partial \varphi^*(\boldsymbol{\Sigma}, \mathbf{X}_{1g}, R_g, \boldsymbol{\Sigma}^S, \mathbf{X}_{2g}, R_{2g})}{\partial \boldsymbol{\Sigma}^S} = \frac{\partial \varphi^*(\boldsymbol{\Sigma}, \mathbf{X}_{1g}, R_{1g}, \boldsymbol{\Sigma}^S, \mathbf{X}_{2g}, R_{2g})}{\partial \Sigma_{eq}} \frac{\partial \Sigma_{eq}}{\partial \boldsymbol{\Sigma}^S}$$
$$= \bar{\bar{p}}_{2g} \left\{ \gamma \mathbf{I} + \frac{1}{2} \delta \frac{(\boldsymbol{\Sigma}^S - \mathbf{X}_{2g})^D}{J_2(\boldsymbol{\Sigma}^S - \mathbf{X}_{2g})} \right\} / \Sigma_{eq}.$$

The internal variables $(\boldsymbol{\alpha}_g, r_g)$ associated to the driving forces $\mathbf{X}_g$, R_g are then given by the following normality rule:

$$\dot{\boldsymbol{\alpha}}_{1g} = -\frac{\partial \Omega(\boldsymbol{\Sigma}, \mathbf{X}_{1g}, \boldsymbol{\Sigma}^s, \mathbf{X}_{2g}, R_g)}{\partial \mathbf{X}_{1g}} = \frac{\partial \Omega(\boldsymbol{\Sigma}, \mathbf{X}_{1g}, \boldsymbol{\Sigma}^s, \mathbf{X}_{2g}, R_g)}{\partial \boldsymbol{\Sigma}},$$
$$\dot{\boldsymbol{\alpha}}_{2g} = -\frac{\partial \Omega(\boldsymbol{\Sigma}, \mathbf{X}_{1g}, \boldsymbol{\Sigma}^s, \mathbf{X}_{2g}, R_g)}{\partial \mathbf{X}_{2g}} = \frac{\partial \Omega(\boldsymbol{\Sigma}, \mathbf{X}_{1g}, \boldsymbol{\Sigma}^s, \mathbf{X}_{2g}, R_g)}{\partial \boldsymbol{\Sigma}^s},$$
$$\dot{r}_g = -\frac{\partial \Omega(\boldsymbol{\Sigma}, \mathbf{X}_{1g}, \boldsymbol{\Sigma}^s, \mathbf{X}_{2g}, R_g)}{\partial R_g} = \frac{\partial \Omega(\boldsymbol{\Sigma}, \mathbf{X}_{1g}, \boldsymbol{\Sigma}^s, \mathbf{X}_{2g}, R_g)}{\partial \Sigma_{eq}}$$

The growth model has been simplified to a perfect viscoplastic model without isotropic hardening. Time is indeed not a physical parameter influencing directly growth, but instead the applied stress (or displacement) over the RUC dictates the growth rate, and no time hardening is present. Isotropic hardening can accordingly be neglected (like viscoplastic models neglecting primary creep, Lemaitre and Chaboche 2009), since growth develops at constant applied stress in the microscopic external remodeling law, provided that the effective stress lies outside the lazy zone described by the scalar parameter Σ_g. The average growth model is thus a pure viscoplastic model with no growth hardening.

The effective first and second gradient bone moduli evolve with bone remodeling and shall accordingly be evaluated for a frozen state of growth; we provide in Fig. 16.2 the strain distribution within the 2D trabecular bone samples resulting from some of the loading cases used for the identification of the corresponding strain gradient moduli.

The evolution of the effective first- and second-order rigidity components versus the remodeling time step is given in Fig. 16.3; throughout the growth process, the effective first gradient moduli (tensile and shear coefficients) increase faster than the second gradient coefficients D_{ijk}, showing that second gradient moduli are less influenced by growth in comparison with the first gradient effective moduli.

The characteristic length indicates the nature and strength of non-classical phenomena described by a strain gradient effective continuum in the response of a medium with microstructure like trabecular bone. In the present 2D context, three internal lengths associated to the independent classical moduli are defined by the relations involving the homogenized first and second gradient stiffness moduli:

$$l_{11} = \left(\frac{D_{111} + D_{112}}{C_{11}} \right)^{1/2}, \quad l_{22} = \left(\frac{D_{212} + D_{222}}{C_{22}} \right)^{1/2}, \quad l_{12} = \left(\frac{D_{112} + D_{122}}{C_{33}} \right)^{1/2}$$

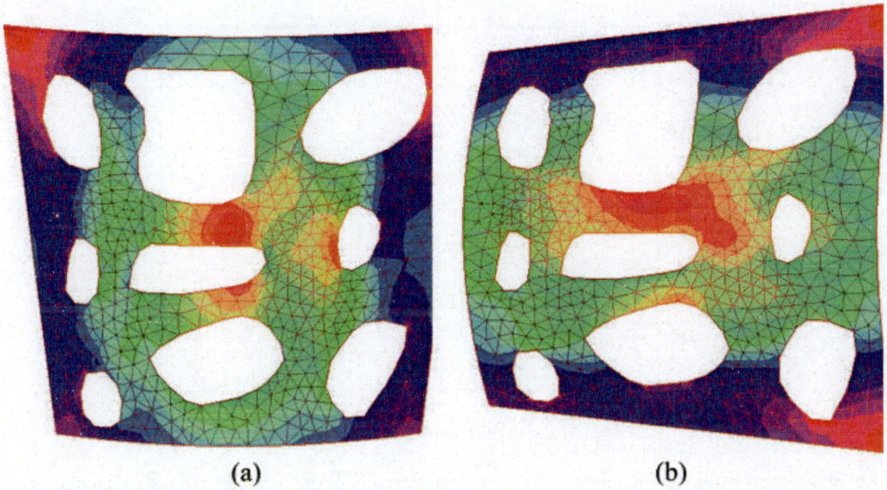

Fig. 16.2 Strain distributions in the RUC of trabecular bone due to the load cases corresponding to the strain gradient. **a** $k_{yxx} = \varepsilon_{xy,x}$, and **b** $k_{xyy} = \varepsilon_{xy,y}$

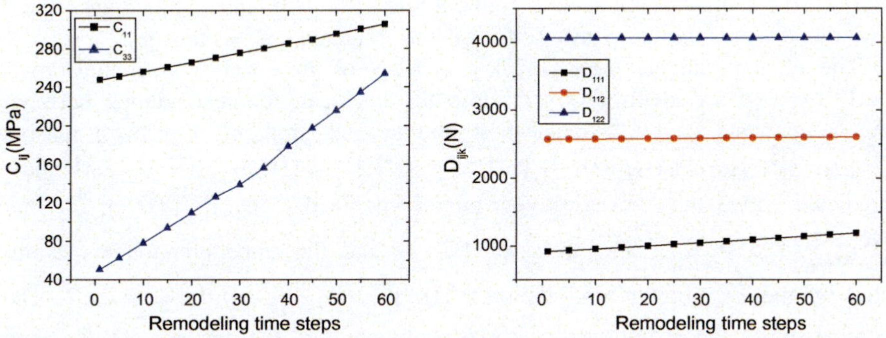

Fig. 16.3 First and second gradient elasticity stiffness tensors C_{ij} (left) and D_{ijk} (right) versus the remodeling time step

The ratio of the characteristic internal lengths to the characteristic unit cell size plotted in Fig. 16.4 versus the remodeling time has an order of magnitude close to unity, indicating that strain gradient effects significantly impact the continuum behavior.

A nearly linear evolution of two components of the deviator of the average second gradient rate of growth tensor is obtained, as shown in Fig. 16.5. Instead of considering the evolution of individual components of the second gradient rate of growth tensor, we analyze and record the condensed information provided by its first and second invariants; a nearly linear influence of the hydrostatic and deviatoric part of the hyperstress versus (traduced by the first and second invariants, respectively) versus

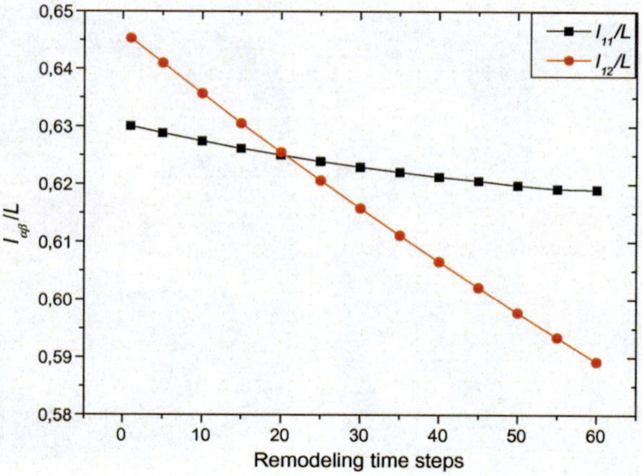

Fig. 16.4 Ratio of the characteristic strain gradient lengths to the unit cell size of trabecular bone versus the remodeling time step

the corresponding kinematic invariants is recorded (Fig. 16.6). The isotropic and deviatoric parts of the average second gradient rate of growth tensor show quasi-linear evolutions versus the same invariants of the average hyperstress tensor.

The six material parameters $\alpha, \beta, \gamma, \delta, K, N$ of the constitutive model previously exposed are identified by minimizing the mean square deviation between the components of the average first and second gradient of growth tensors $(\overline{\mathbf{D}}_{1g})_{11}, (\overline{\mathbf{D}}_{1g})_{22}, (\overline{\mathbf{D}}_{1g})_{12}, (\overline{\mathbf{D}}_{2g})_{111}, (\overline{\mathbf{D}}_{2g})_{222}, (\overline{\mathbf{D}}_{2g})_{112}, (\overline{\mathbf{D}}_{2g})_{212}, (\overline{\mathbf{D}}_{2g})_{211}, (\overline{\mathbf{D}}_{2g})_{122}$ predicted by the constitutive model—denoted $(\overline{\mathbf{D}}_{1g})_{ij}^{\mathrm{CM}}, (\overline{\mathbf{D}}_{2g})_{ijk}^{\mathrm{CM}} = (\overline{\mathbf{D}}_{2g})_{ikj}^{\mathrm{CM}}(\alpha, \beta, \gamma, \delta, K, N), i, j, k = 1 \ldots 2$—and the same components evaluated by the FE simulations—denoted $(\overline{\mathbf{D}}_{1g})_{ij}^{\mathrm{FE}}, (\overline{\mathbf{D}}_{2g})_{ijk}^{\mathrm{FE}}, i, j, k = 1 \ldots 2$. The comparison of the evolution of the two components of the deviator $(\overline{\mathbf{D}}_{2g})_{112}$ and $(\overline{\mathbf{D}}_{2g})_{122}$ versus the same driving hyperstress deviator components predicted by the constitutive model and by direct FE simulations for a strain gradient loading (applied to the representative unit cell) combining $(\boldsymbol{\Sigma}^{\mathrm{S}} - \mathbf{X}_{2g})_{122}^{\mathrm{D}}$ and $(\boldsymbol{\Sigma}^{\mathrm{S}} - \mathbf{X}_{2g})_{112}^{\mathrm{D}}$ exhibits a very good agreement (Fig. 16.7). This highlights the capability of the identified second gradient growth model to predict the response of trabecular bone microstructures for general loadings at the scale of the representative trabecular bone unit cell.

The formulated mesoscopic growth model shall prove useful for simulating bone sample microstructural evolutions at the macrolevel of entire bone structures, with a good compromise between numerical efficiency and accuracy.

We shall in the next section formulate different classes of strain gradient growth models, following the classification proposed in Forest and Sievert (2003).

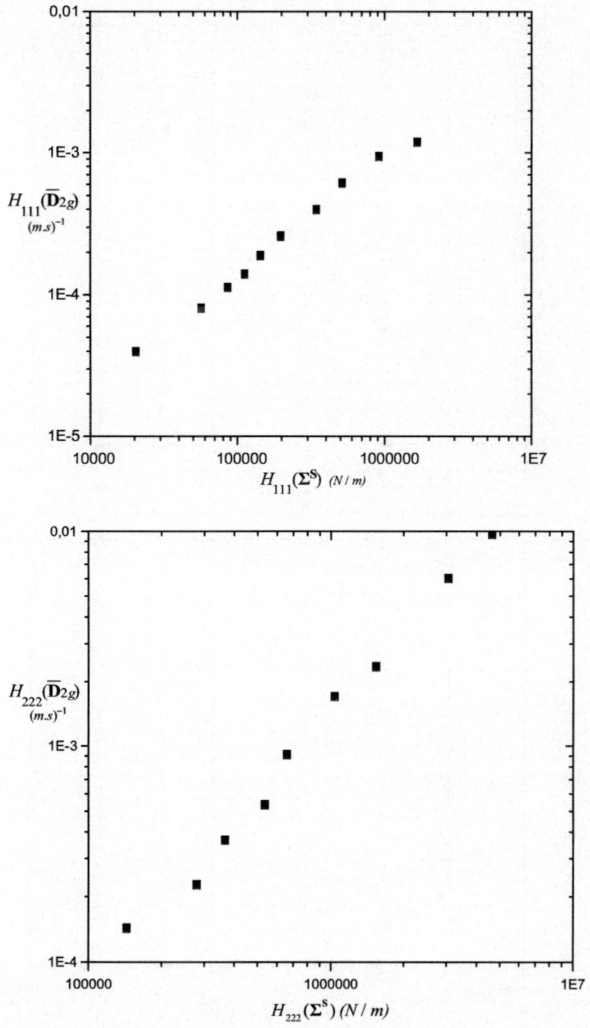

Fig. 16.5 Components of the third-order deviator tensor H_{ijk} of $\overline{\mathbf{D}}_{2g}$ versus the corresponding components of the hyperstress deviator

16.4 Formulation of Different Classes of Enhanced Growth Models: Standard Strain Gradient Growth, Strain Gradient Materials with Growth, Gradient of Internal Variable Approach of Growth

The framework of small strain rates is conveniently adopted in the sequel, considering that growth phenomena occur at time scales that are much larger than the typical time

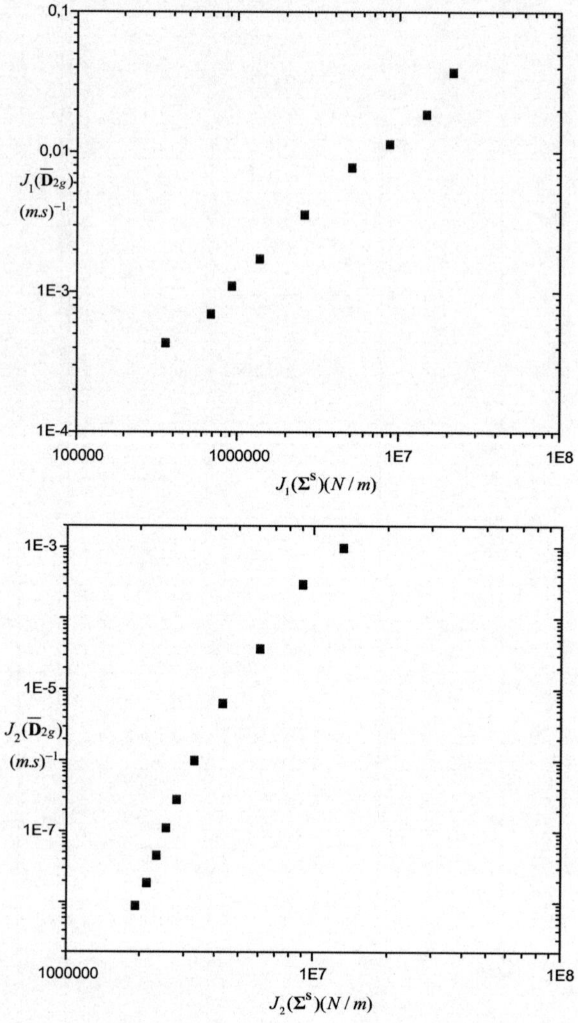

Fig. 16.6 First and second invariant of $\overline{\mathbf{D}}_{2g}$ versus the first and second invariants of the hyperstress tensor

scale of return to equilibrium of the material at any point. For each type of continuum, the balance equations, the state laws, and the evolution laws of the internal variables (especially the growth strain rate) are derived using the principle of virtual power, the balance of energy, and the entropy principle. The method of virtual power is a powerful tool in continuum mechanics with internal variables to derive the field equations satisfied by the unknown fields and their associated boundary conditions. It relies on the definition of the set of virtual motions and the set of variables chosen in the model to enter the virtual power of internal and contact forces (Maugin 1980). In

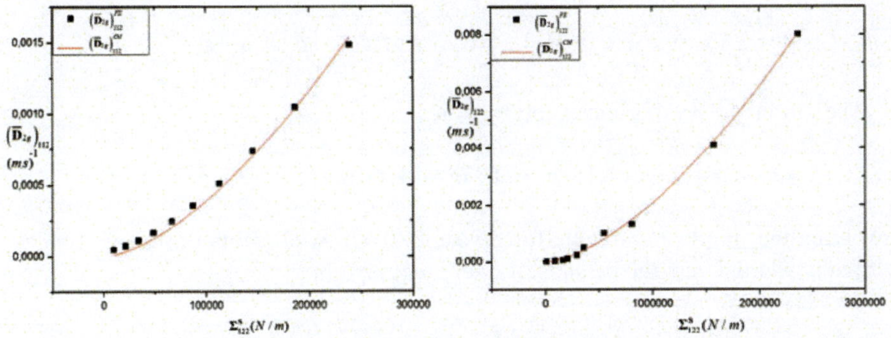

Fig. 16.7 Evolution of two components of the deviator $\left(\overline{\mathbf{D}}_{2g}\right)^{\mathrm{D}}$ versus their counterpart for the driving hyperstress deviator $\left(\boldsymbol{\Sigma}^{\mathrm{S}} - \mathbf{X}_{2g}\right)^{\mathrm{D}}$ predicted by the model and by direct FE simulations

a static situation, the virtual power principle states that the virtual power of external forces is equal to the virtual power of internal forces. Those powers are integral over the domain and its boundary, with volumetric densities denoted, respectively, p_{i}, p_{c}, chosen as linear forms of the generalized virtual motions. The requirement of objectivity allows to restrict the form of the virtual power of internal forces, so that for instance it will not be affected by rigid body motions. The combination of the energy and entropy principles leads to the Clausius–Duhem inequality (Forest and Sievert 2003):

$$-\rho\left(\dot{\psi} + s\dot{\theta}\right) + p_{\mathrm{i}} - \mathbf{q}.\frac{\operatorname{grad}\theta}{\theta} \geq 0$$

with ψ the Helmholtz free energy density, s the entropy density, and θ the absolute temperature therein.

16.4.1 Standard Strain Gradient Growth Model

Strain gradient materials with the growth strain rate $\dot{\boldsymbol{\varepsilon}}_{\mathrm{g}}$ as a DOF defines a first category of strain gradient models for growing solid bodies. The set of DOF's entering the virtual power of internal and contact forces are

$$V = \{\dot{\mathbf{u}}, \dot{\mathbf{u}} \otimes \nabla, \dot{\mathbf{u}} \otimes \nabla \otimes \nabla\}, \quad V^{\mathrm{c}} = \{\dot{\mathbf{u}}, D_{\mathrm{n}}\dot{\mathbf{u}}\}$$

where D_{n} is the normal gradient operator. The virtual power of internal forces involves the Cauchy stress and hyperstress tensors as the conjugate driving forces of the strain and strain gradient tensors (respectively, second- and third-order tensors), thereby defining the virtual power of internal forces:

$$\boldsymbol{\varepsilon} := \frac{1}{2}(\mathbf{u} \otimes \nabla + \mathbf{u} \otimes \nabla^{\mathrm{T}}), \quad \mathbf{K} := \boldsymbol{\varepsilon} \otimes \nabla \to p_{\mathrm{i}} = \boldsymbol{\sigma} : \dot{\boldsymbol{\varepsilon}} + \mathbf{S} \therefore \dot{\mathbf{K}}$$

The virtual power of contact forces writes

$$p_{\mathrm{c}} = \mathbf{t} : \dot{\mathbf{u}} + \mathbf{M}.D_{\mathrm{n}}\dot{\mathbf{u}}$$

involving the simple and double force vectors **t**, **M**. Application of the principle of virtual power delivers the balance of linear momentum

$$\operatorname{div}(\sigma - \operatorname{div}\mathbf{S}) = 0$$

together with boundary conditions involving the surface tractions and double tractions, which will, however, not be detailed.

Regarding the kinematics of the growing continuum, the total strain and its gradient decompose additively into growth and elastic parts as

$$\boldsymbol{\varepsilon} = \boldsymbol{\varepsilon}_{\mathrm{e}} + \boldsymbol{\varepsilon}_{\mathrm{g}}, \quad \mathbf{K} = \mathbf{K}_{\mathrm{e}} + \mathbf{K}_{\mathrm{g}}$$

The free energy density is selected as a function of the elastic strain and strain gradient, together with an internal variable $\boldsymbol{\gamma}$ (of scalar or tensor nature) describing hardening (isotropic hardening is sufficient in the case of bone, as evidenced in Sect. 16.3), viz $\psi(\boldsymbol{\varepsilon}_{\mathrm{e}}, \mathbf{K}_{\mathrm{e}}, \boldsymbol{\gamma})$. The intrinsic dissipation writes accordingly

$$D = \boldsymbol{\sigma} : \dot{\boldsymbol{\varepsilon}} + \mathbf{S} \therefore \dot{\mathbf{K}} - \rho\dot{\psi} = \left(\boldsymbol{\sigma} - \rho\frac{\partial\psi}{\partial\boldsymbol{\varepsilon}_{\mathrm{e}}}\right) : \dot{\boldsymbol{\varepsilon}}_{\mathrm{e}}$$
$$+ \left(\mathbf{S} - \rho\frac{\partial\psi}{\partial\mathbf{K}_{\mathrm{e}}}\right) : \dot{\mathbf{K}}_{\mathrm{e}} + \boldsymbol{\sigma} : \dot{\boldsymbol{\varepsilon}}_{\mathrm{g}} + \mathbf{S} \therefore \dot{\mathbf{K}}_{\mathrm{g}} - \rho\frac{\partial\psi}{\partial\boldsymbol{\gamma}}\dot{\boldsymbol{\gamma}} \geq 0$$

This leads, following the standard Coleman–Noll procedure, to the state laws

$$\boldsymbol{\sigma} = \rho\frac{\partial\psi}{\partial\boldsymbol{\varepsilon}_{\mathrm{e}}}, \quad \mathbf{S} = \rho\frac{\partial\psi}{\partial\mathbf{K}_{\mathrm{e}}}, \quad R = \rho\frac{\partial\psi}{\partial\boldsymbol{\gamma}}$$

defining successively the stress, hyperstress tensor, and the driving force for the internal irreversible processes, that is growth or remodeling. This entails the residual dissipation

$$D = \boldsymbol{\sigma} : \dot{\boldsymbol{\varepsilon}}_{\mathrm{g}} + \mathbf{S} \therefore \dot{\mathbf{K}}_{\mathrm{g}} - \rho\frac{\partial\psi}{\partial\boldsymbol{\gamma}}\dot{\boldsymbol{\gamma}} \geq 0$$

As illustrated in Sect. 16.3 for bone, the classical theory of standard materials is extended to second gradient materials by choosing a viscoplastic dissipation potential $\Omega(\sigma, \mathbf{S}, R)$ such that the evolution laws of the internal variables write:

$$\dot{\boldsymbol{\varepsilon}}_g = \frac{\partial \Omega(\boldsymbol{\sigma}, \mathbf{S}, R)}{\partial \boldsymbol{\sigma}}, \quad \dot{\mathbf{K}}_g = \frac{\partial \Omega(\boldsymbol{\sigma}, \mathbf{S}, R)}{\partial \mathbf{S}}, \quad \dot{\gamma} = -\frac{\partial \Omega(\boldsymbol{\sigma}, \mathbf{S}, R)}{\partial R}$$

The convexity of the dissipation potential is sufficient to guarantee the positivity of the intrinsic dissipation. Note that the dissipation potential is a function of third-rank tensors, so that the notion of harmonic decomposition proves useful to split these tensors into their deviator and isotropic parts, as detailed in Louna et al. (2018). Observe that this model requires to set a specific and independent flow rule for the growth part of the strain gradient tensor, since the growth part of the strain gradient is not the gradient of the growth part of the total strain. A more simple model as to dissipative aspects can be constructed assuming that the elastic and growth strain gradient terms are the spatial gradients of the elastic and growth 'strains,' respectively; this theory has been put forward in the case of strain gradient plasticity in Forest and Sievert (2003). Such a theory shall deserve the name strain gradient growth model, as exposed in the next section.

16.4.2 *Strain Gradient Growth Model with Growth Strain as an Additional DOF*

It holds in this model the following definitions of the elastic and growth parts of the strain gradient:

$$\mathbf{K}_e := \boldsymbol{\varepsilon}_e \otimes \nabla, \quad \mathbf{K}_g := \boldsymbol{\varepsilon}_g \otimes \nabla$$

Following the argumentation given in Forest and Sievert (2003), one can show that the free energy density has to incorporate the gradient of the elastic 'strain' as a state variable; this leads to the enlarged sets of virtual motions and contact velocities:

$$V = \left\{\dot{\mathbf{u}}, \dot{\mathbf{u}} \otimes \nabla, \dot{\mathbf{u}} \otimes \nabla \otimes \nabla, \dot{\boldsymbol{\varepsilon}}_g, \dot{\boldsymbol{\varepsilon}}_g \otimes \nabla\right\}, \quad V^c = \left\{\dot{\mathbf{u}}, D_n \dot{\mathbf{u}}, \dot{\boldsymbol{\varepsilon}}_g\right\}$$

The densities of power of internal and contact forces are taken as linear forms of the elements of previous sets, as follows

$$\begin{aligned} p_i &= \boldsymbol{\sigma} : \dot{\boldsymbol{\varepsilon}} + \mathbf{S} \,\therefore\, \dot{\mathbf{K}} + \mathbf{A}_g : \dot{\boldsymbol{\varepsilon}}_g + \left(\mathbf{B}_g - \mathbf{S}\right) \therefore \left(\dot{\boldsymbol{\varepsilon}}_g \otimes \nabla\right) \\ p_c &= \mathbf{t} : \dot{\mathbf{u}} + \mathbf{M}.D_n \dot{\mathbf{u}} + \mathbf{A}_g^c : \dot{\boldsymbol{\varepsilon}}_g \end{aligned}$$

with $\mathbf{A}_g$, $\mathbf{B}_g$ internal driving forces for growth, respectively, a second- and third-order tensor; the surface generalized force tensor $\mathbf{A}_g^c$ has been introduced into the density of power of contact forces. The principle of virtual power leads to the balance equation for the effective stress as in previous section; the additional contribution $\left(\mathbf{B}_g - \mathbf{S}\right)$ stems from the fact mentioned in Forest and Sievert (2003) in the case of plasticity that these two tensors develop work on two different tensors, namely $\mathbf{K}_e$, $\mathbf{K}_g$, respectively. The intrinsic dissipation is based on the free energy density

$\psi(\varepsilon_e, \varepsilon_e \otimes \nabla, \varepsilon_g \otimes \nabla, \gamma)$; it leads in a straightforward manner to the state laws

$$\sigma = \rho \frac{\partial \psi}{\partial \varepsilon_e}, \quad \mathbf{S} = \rho \frac{\partial \psi}{\partial \mathbf{K}_e}, \quad \mathbf{B}_g = \rho \frac{\partial \psi}{\partial \mathbf{K}_g}, \quad R = \rho \frac{\partial \psi}{\partial \gamma}$$

The residual mechanical dissipation is finally obtained as

$$D = (\sigma + \mathrm{div}(\mathbf{B}_g - \mathbf{S})) : \dot{\varepsilon}_g - R\dot{\gamma} \geq 0$$

highlighting the driving force for growth as the second order tensor $\tau_g^{\mathrm{eff}} := \sigma + \mathrm{div}(\mathbf{B}_g - \mathbf{S})$. The evolution laws for the growth variables are then given by a convex potential of dissipation $\Omega\left(\tau_g^{\mathrm{eff}}, R\right)$ such that

$$\dot{\varepsilon}_g = \frac{\partial \Omega\left(\tau_g^{\mathrm{eff}}, R\right)}{\partial \tau_g^{\mathrm{eff}}}, \quad \dot{\gamma} = -\frac{\partial \Omega\left(\tau_g^{\mathrm{eff}}, R\right)}{\partial R}$$

16.4.3 Gradient of Growth Model

This strategy is in line with the so-called gradient of internal variable approach, whereby the internal variable for growth includes an internal variable γ_g as an additional DOF in both sets of virtual motions and contact velocities:

$$V = \{\dot{\mathbf{u}}, \dot{\mathbf{u}} \otimes \nabla, \dot{\gamma}_g, \dot{\gamma}_g \otimes \nabla\}, \quad V^c = \{\dot{\mathbf{u}}, \dot{\gamma}_g\}$$

The densities of power of internal and contact forces are taken as linear forms of the elements of previous sets, so that it holds

$$p_i = \sigma : \dot{\varepsilon} + A_g : \dot{\gamma}_g + \mathbf{B}_g \therefore (\dot{\gamma}_g \otimes \nabla)$$
$$p_c = \mathbf{t} : \dot{\mathbf{u}} + A_g^c : \dot{\gamma}_g$$

Applying the principle of virtual power leads to the following balance laws and boundary conditions:

$$\mathrm{div}\sigma = \mathbf{0}, \quad A_g = \mathrm{div}(\mathbf{B}_g)$$
$$\mathbf{t} = \sigma.\mathbf{n}, \quad \mathbf{A}_g^c = \mathbf{B}_g.\mathbf{n}$$

Considering a Helmholtz free energy density of the form $\psi(\varepsilon_e, \dot{\gamma}_g \otimes \nabla, q)$, straightforward computations lead to the state laws

$$\boldsymbol{\sigma} = \rho \frac{\partial \psi}{\partial \boldsymbol{\varepsilon}_{\mathrm{e}}}, \quad \mathbf{B}_{\mathrm{g}} = \rho \frac{\partial \psi}{\partial \boldsymbol{\gamma}_{\mathrm{g}}}, \quad R = \rho \frac{\partial \psi}{\partial q}$$

The residual intrinsic dissipation writes:

$$D = \boldsymbol{\sigma} : \dot{\boldsymbol{\varepsilon}}_{\mathrm{g}} + A_{\mathrm{g}} \dot{\boldsymbol{\gamma}}_{\mathrm{g}} - R\dot{q} \geq 0$$

Note that the driving force $\mathbf{A}_{\mathrm{g}}$ appears as a back stress due to growth or internal remodeling phenomena.

The evolution equations may be formulated based on the choice of a dissipation pseudo-potential $\Omega(\boldsymbol{\sigma}, A_{\mathrm{g}}, R)$, so that it holds

$$\dot{\boldsymbol{\varepsilon}}_{\mathrm{g}} = \frac{\partial \Omega(\boldsymbol{\sigma}, \mathbf{A}_{\mathrm{g}}, R)}{\partial \boldsymbol{\sigma}}, \quad \dot{\boldsymbol{\gamma}}_{\mathrm{g}} = \frac{\partial \Omega(\boldsymbol{\sigma}, \mathbf{A}_{\mathrm{g}}, R)}{\partial A_{\mathrm{g}}}, \quad \dot{q} = -\frac{\partial \Omega(\boldsymbol{\sigma}, \mathbf{A}_{\mathrm{g}}, R)}{\partial R}$$

If the internal variable is identified to the growth 'strain,' one recovers the strain gradient growth model described in Sect. 16.4.2; the residual intrinsic dissipation becomes

$$D = \boldsymbol{\tau}_{\mathrm{g}}^{\mathrm{eff}} : \dot{\boldsymbol{\varepsilon}}_{\mathrm{g}}, \quad \boldsymbol{\tau}_{\mathrm{g}}^{\mathrm{eff}} = \boldsymbol{\sigma} + A_{\mathrm{g}}$$

Instead of three evolution equations written in Sect. 16.4.2, one obtains two evolution equations involving the effective stress

$$\dot{\boldsymbol{\varepsilon}}_{\mathrm{g}} = \frac{\partial \Omega\left(\boldsymbol{\tau}_{\mathrm{g}}^{\mathrm{eff}}, R\right)}{\partial \boldsymbol{\tau}_{\mathrm{g}}^{\mathrm{eff}}} = \frac{\partial \Omega\left(\boldsymbol{\tau}_{\mathrm{g}}^{\mathrm{eff}}, R\right)}{\partial \boldsymbol{\sigma}} = \frac{\partial \Omega\left(\boldsymbol{\tau}_{\mathrm{g}}^{\mathrm{eff}}, R\right)}{\partial A_{\mathrm{g}}}, \quad q = -\frac{\partial \Omega\left(\boldsymbol{\tau}_{\mathrm{g}}^{\mathrm{eff}}, R\right)}{\partial R}$$

It is worth emphasizing that since growth is not necessarily associated to irreversibility, none of the proposed strategies may be satisfactory, since they all lead to the growth modeled as an irreversible process. It shall nevertheless be emphasized that the introduced irreversible growth 'strain' is the net effect of microstructural phenomena (evolution of internal density, mechanical properties, shape changes) leading to an overall 'strain' representative of modeling, remodeling or growth phenomena.

A reversible growth model can be constructed, letting the Helmholtz free energy density which depends on the growth strain, viz $\psi(\boldsymbol{\varepsilon}_{\mathrm{e}}, \boldsymbol{\varepsilon}_{\mathrm{g}})$, without involving any internal variable. The stress can then be additively decomposed into an elastic and a growth contribution, so that application of the principle of virtual power, energy, and entropy principles leads to a non-dissipative growing continuum with state laws

$$\boldsymbol{\sigma} = \boldsymbol{\sigma}_{\mathrm{e}} + \boldsymbol{\sigma}_{\mathrm{g}} \rightarrow \boldsymbol{\sigma}_{\mathrm{e}} = \rho \frac{\partial \psi(\boldsymbol{\varepsilon}_{\mathrm{e}}, \boldsymbol{\varepsilon}_{\mathrm{g}})}{\partial \boldsymbol{\varepsilon}_{\mathrm{e}}}, \quad \boldsymbol{\sigma}_{\mathrm{g}} = \rho \frac{\partial \psi(\boldsymbol{\varepsilon}_{\mathrm{e}}, \boldsymbol{\varepsilon}_{\mathrm{g}})}{\partial \boldsymbol{\varepsilon}_{\mathrm{g}}}$$

The physical meaning of the introduced growth stress is that of an internal stress, the counterpart of the back stress in plasticity theories.

16.5 Conclusion

A unifying mechanical constitutive framework for growing solid bodies exhibiting scale effects is developed, based on the virtual power, energy, and entropy principles. Different variants of higher gradient theories are exposed, adopting the viewpoint of the phenomenology. The consideration of strain gradient terms in the constitutive models developed for biological tissues like bone is motivated by the occurrence of pronounced microscopic strain gradients within their internal architecture showing a scale hierarchy and strong contrasts of mechanical properties between the different phases. A strain gradient model for bone remodeling has been developed based on the homogenization of the trabecular architecture in order to exemplify the construction of such strain gradient models starting from the microstructural level.

The consideration of surface growth effects in constitutive models based on the mechanics of generalized continua shall deserve future contributions. Moreover, since the present theories are applicable to hard tissues like bone, they restrict to a small strains situation; it thus remains to extend them to large strains situations as exhibited by soft biological tissues.

References

Alibert, J.-J., & Della Corte, A. (2015). Second-gradient continua as homogenized limit of pantographic microstructured plates: A rigorous proof. *Zeitschrift fur Angewandte Mathematik und Physik, 66*(5), 2855–2870.

Alibert, J. J., Seppecher, P., & dell'Isola, F. (2003). Truss modular beams with deformation energy depending on higher displacement gradients. *Mathematics and Mechanics of Solids, 8*(1), 51–73.

Altenbach, H., & Eremeyev, V. A. (2009). On the linear theory of micropolar plates. *Zeitschrift für Angewandte Mathematik und Mechanik, 89*(4), 242–256.

Andreaus, U., Giorgio, I., & Lekszycki, T. (2014). A 2-D continuum model of a mixture of bone tissue and bio-resorbable material for simulating mass density redistribution under load slowly variable in time. *ZAMM-Journal of Applied Mathematics and Mechanics/Zeitschrift für Angewandte Mathematik und Mechanik, 94*(12), 978–1000.

Barchiesi, E., Eugster, S. R., Placidi, L., & dell'Isola, F. (2019). Pantographic beam: A complete second gradient 1D-continuum in plane. *Zeitschrift für Angewandte Mathematik und Physik, 70*(5), 135.

Berkache, K., Deogekar, S., Goda, I., Picu, R. C., & Ganghoffer, J.-F. (2017). Construction of second gradient continuum models for random fibrous networks and analysis of size effects. *Composite Structures, 181*, 347–357.

Bowman, S. M., et al. (1998). Creep contributes to the fatigue behavior of bovine trabecular bone. *Journal of Biomechanical Engineering, 120*, 647–654.

Buechner, P. M., & Lakes, R. S. (2003). Size effects in the elasticity and viscoelasticity of bone. *Biomechanics and Modeling in Mechanobiology, 1*(4), 295–301.

Cosserat, E., & Cosserat, F. (1909). *Théorie des Corps Déformables*. Paris: Librairie Scientifique A. Hermann et Fils.

dell'Isola, F., Andreaus, U., & Placidi, L. (2015a). At the origins and in the vanguard of peridynamics, non-local and higher-gradient continuum mechanics: An underestimated and still topical contribution of Gabrio Piola. *Mathematics and Mechanics of Solids, 20*(8), 887–928.

dell'Isola, F., Seppecher, P., Alibert, J. J., Lekszycki, T., Grygoruk, R., Pawlikowski, M., ... Hild, F. (2019). Pantographic metamaterials: An example of mathematically driven design and of its technological challenges. *Continuum Mechanics and Thermodynamics, 31*(4), 851–884.

dell'Isola, F., Seppecher, P., & Della Corte, A. (2015b). The postulations á la D'Alembert and á la Cauchy for higher gradient continuum theories are equivalent: A review of existing results. *Proceedings of the Royal Society of London, Series A: Mathematical, Physical and Engineering Sciences, 471*, 2183.

Epstein, M., & Maugin, G. A. (2000). Thermomechanics of volumetric growth in uniform bodies. *International Journal of Plasticity, 16*, 951–978.

Eremeyev, V. A. (2019). On non-holonomic boundary conditions within the nonlinear Cosserat continuum. In *New achievements in continuum mechanics and thermodynamics* (pp. 93–104). Cham: Springer.

Eringen, A. C., & Edelen, D. G. B. (1972). On nonlocal elasticity. *International Journal of Engineering Science, 10*(3), 233–248.

Forest, S., & Sievert, R. (2003). Elastoviscoplastic constitutive frameworks for generalized continua. *Acta Mechanica, 160*, 71–111.

Frasca, P., Harper, R., & Katz, J. L. (1981). Strain and frequency dependence of shear storage modulus for human single osteons and cortical bone micro samples—Size and hydration effects. *Journal of Biomechanics, 14*(10), 679–690.

Ganghoffer, J. F. (2010). Mechanical modeling of growth considering domain variation—Part II: Volumetric and surface growth involving Eshelby tensors. *Journal of the Mechanics and Physics of Solids, 58*(9), 1434–1459.

Ganghoffer, J. F., & Haussy, B. (2005). Mechanical modeling of growth considering domain variation. Part I: Constitutive framework. *International Journal of Solids and Structures, 42*(15), 4311–4337.

Ganghoffer, J. F., Plotnikov, P. I., & Sokołowski, J. (2014). Mathematical modeling of volumetric material growth. *Archive of Applied Mechanics, 84*(9–11), 1357–1371.

Giorgio, I., Andreaus, U., dell'Isola, I., & Lekszycki, T. (2017). Viscous second gradient porous materials for bones reconstructed with bio-resorbable grafts. *Extreme Mechanics Letters, 13*, 141–147.

Giorgio, I., Andreaus, U., & Madeo, A. (2016). The influence of different loads on the remodeling process of a bone and bioresorbable material mixture with voids. *Continuum Mechanics and Thermodynamics, 28*(1–2), 21–40.

Giorgio, I., De Angelo, M., Turco, E., & Misra, A. (2019). A Biot–Cosserat two-dimensional elastic nonlinear model for a micromorphic medium. *Continuum Mechanics and Thermodynamics*, 1–13.

Goda, I., Assidi, M., Belouettar, S., & Ganghoffer, J.-F. (2012). A micropolar anisotropic constitutive model of cancellous bone from discrete homogenization. *Journal of the Mechanical Behavior of Biomedical Materials, 16*, 87–108.

Goda, I., Assidi, M., & Ganghoffer, J.-F. (2014). A 3D elastic micropolar model of vertebral trabecular bone from lattice homogenization of the bone microstructure. *Biomechanics and Modeling in Mechanobiology, 13*, 53–83.

Goda, I., & Ganghoffer, J.-F. (2015a). 3D plastic collapse and brittle fracture surface models of trabecular bone from asymptotic homogenization method. *International Journal of Engineering Science, 87*, 58–82.

Goda, I., & Ganghoffer, J.-F. (2015b). Identification of couple-stress moduli of vertebral trabecular bone based on the 3D internal architectures. *Journal of the Mechanical Behavior of Biomedical Materials, 51*, 99–118.

Goda, I., & Ganghoffer, J.-F. (2016). Construction of first and second order grade anisotropic continuum media for 3D porous and textile composite structures. *Composite Structures, 141*, 292–327.

Goda, I., Ganghoffer, J. F., & Maurice, G. (2016). Combined bone internal and external remodeling based on Eshelby stress. *International Journal of Solids and Structures, 94–95*, 138–157.

Goda, I., Rahouadj, R., & Ganghoffer, J.-F. (2013). Size dependent static and dynamic behavior of trabecular bone based on micromechanical models of the trabecular. *International Journal of Engineering Science, 72*, 53–77.

Harrigan, T. P., Jasty, M. J., Mann, R. W., & Harris, W. H. (1988). Limitations of the continuum assumption in cancellous bone. *Journal of Biomechanics, 21*, 269–275.

Kröner, E. (1976). Elasticity theory of materials with long range cohesive forces. *International Journal of Solids and Structures, 3*(5), 731–742.

Lakes, R. (1995). On the torsional properties of single osteons. *Journal of Biomechanics, 28*(1409–1410), 1.

Lekszycki, T., & dell'Isola, F. (2012). A mixture model with evolving mass densities for describing synthesis and resorption phenomena in bones reconstructed with bio-resorbable materials. *ZAMM-Journal of Applied Mathematics and Mechanics/Zeitschrift für Angewandte Mathematik und Mechanik, 92*(6), 426–444.

Lemaitre, J., & Chaboche, J. L. (2009). *Mécanique des matériaux solides*. Paris: Dunod.

Louna, Z., Goda, I., & Ganghoffer, J. F. (2018). Identification of a constitutive law for trabecular bone samples under remodeling in the framework of irreversible thermodynamics. *Continuum Mechanics and Thermodynamics, 30*(3), 529–551.

Louna, Z., Goda, I., & Ganghoffer, J. F. (2019). Homogenized strain gradient remodeling model for trabecular bone microstructures. *Continuum Mechanics and Thermodynamics, 31*(5), 1339–1367.

Madeo, A., George, D., Lekszycki, T., Nierenberger, M., & Rémond, Y. (2012). A second gradient continuum model accounting for some effects of micro-structure on reconstructed bone remodeling. *Comptes Rendus Mécanique, 340*(8), 575–589.

Madeo, A., Lekszycki, T., & dell'Isola, F. (2011). Continuum model for the bio-mechanical interactions between living tissue and bio-resorbable graft after bone reconstructive surgery. *Comptes Rendus Mécanique, 339*(10), 625–682.

Maugin, G. A. (1980). The method of virtual power in continuum mechanics: Application to coupled fields. *Acta Mechanica, 35*, 1–70.

Olive, M., & Auffray, N. (2014). Isotropic invariants of a completely symmetric third-order tensor. *Journal of Mathematical Physics, American Institute of Physics (AIP), 55*(9), 1.4895466.

Park, H. C., & Lakes, R. S. (1986). Cosserat micromechanics of human bone: Strain redistribution by a hydration sensitive constituent. *Journal of Biomechanics, 19*(5), 385–397.

Ramézani, H., El-Hraiech, A., Jeong, J., & Benhamou, C.-L. (2012). Size effect method application for modeling of human cancellous bone using geometrically exact Cosserat elasticity. *Computer Methods in Applied Mechanics and Engineering, 237*, 227–243.

Reda, H., Goda, I., Ganghoffer, J. F., L'Hostis, G., & Lakiss, H. (2017). Dynamical analysis of homogenized second gradient anisotropic media for textile composite structures and analysis of size effects. *Composite Structures, 161*, 540–551.

Skalak, R., Farrow, D. A., & Hoger, A. (1997). Kinematics of surface growth. *Journal of Mathematical Biology, 35*, 869–907.

Taylor, M., Cotton, J., & Zioupos, P. (2002). Finite element simulation of the fatigue behaviour of cancellous bone. *Meccanica, 37*, 419–429.

Yang, J. F. C., & Lakes, R. S. (1982). Experimental study of micropolar and couple stress elasticity in compact bone in bending. *Journal of Biomechanics, 15*(2), 91–98.

Chapter 17
Microplane Modeling for Inelastic Responses of Shape Memory Alloys

Mohammad Reza Karamooz-Ravari, Mahmoud Kadkhodaei, and Mohammad Elahinia

Abstract Shape memory alloys (SMAs) are a class of smart materials. In these alloys, an inelastically deformed configuration can recover to their original shape upon heating to a specific temperature. One of the main challenges in modeling these materials under multiaxial loadings is that the so-called normality rule does not necessarily hold true as the direction of inelastic strain rate vector does not coincide with the deviatoric stress vector for nonproportional loadings. Therefore, any generalization of 1-D constitutive equations to 3-D cases based on J_2 or J_2-J_3 plasticity is valid only for proportional loadings. Microplane modeling approach is a promising candidate for overcoming this challenge since 1-D constitutive models in this method are generalized to 3-D through a particular homogenization technique. All the material parameters can be obtained using uniaxial tension–compression tests. These features make microplane theory an efficient approach in constitutive modeling of shape memory alloys. In this chapter, first, the basic concepts of microplane theory are reviewed. Then, a microplane model for SMAs along with an efficient technique in numerical implementation of the constitutive equations is presented. Introduction of tension–compression asymmetry is further discussed and verified. Finally, modeling of plastic and cyclic response is explained, and the theoretical results are validated against experimental findings.

Keywords Microplane theory · Shap memory alloy · SMA · Tension-compression asymmetry · Residual strain · Cyclic loading · Strain accumulation · Numerical implementation

M. R. Karamooz-Ravari
Faculty of Mechanical and Materials Engineering, Graduate University of Advanced Technology, 76318-18356 Kerman, Iran

M. Kadkhodaei (✉)
Department of Mechanical Engineering, Isfahan University of Technology, Isfahan 84156-83111, Iran
e-mail: kadkhodaei@iut.ac.ir

M. Elahinia
Department of Mechanical, Industrial, and Manufacturing Engineering, The University of Toledo, Toledo, OH 43606, USA

F. dell'Isola and L. Igumnov (eds.), *Dynamics, Strength of Materials and Durability in Multiscale Mechanics*, Advanced Structured Materials 137,
https://doi.org/10.1007/978-3-030-53755-5_17

17.1 An Introduction to the Basics of Microplane Modeling

The so-called microplane modeling approach originated from the work of Taylor (1938) who achieved constitutive equations of polycrystalline metals by developing relations between stress and strain vectors on generic planes of arbitrary orientations in a material point so that the macroscopic stress or strain tensors were determined as a resultant of all these vectors. This concept was later modified by others and was commonly known as "slip theory of plasticity". As slip is not the source of inelastic response for all types of materials, Bažant (1984) introduced the neutral term "microplane theory" which can be used for general inelastic behaviors. In this approach, 1-D constitutive laws for each stress vector and its associated strain vector are sufficient to generate a macroscopic 3-D model by considering either one of two main formulations in microplane theory named "static constraint" and "kinematic constraint". In static constraint formulation, it is assumed that the stress vector acting on each microplane is the projection of the macroscopic stress tensor. In kinematic constraint formulation, the strain vector on any microplane is considered as the projection of the macroscopic strain tensor. Moreover, there are some particular cases, called "double constraint" formulation, where both static and kinematic constraints co-exist.

Microplane theory with static constraint inspired by the works of Bažant et al. (1996) and Carol and Bažant (1997) is first briefly reviewed in this section. Referring to Fig. 17.1, if a microplane is considered at a material point, the stress vector $\vec{\sigma}_N$ on the plane is the projection of macroscopic stress tensor, i.e.,

$$\sigma_N = N_{ij}\sigma_{ij} \tag{17.1}$$

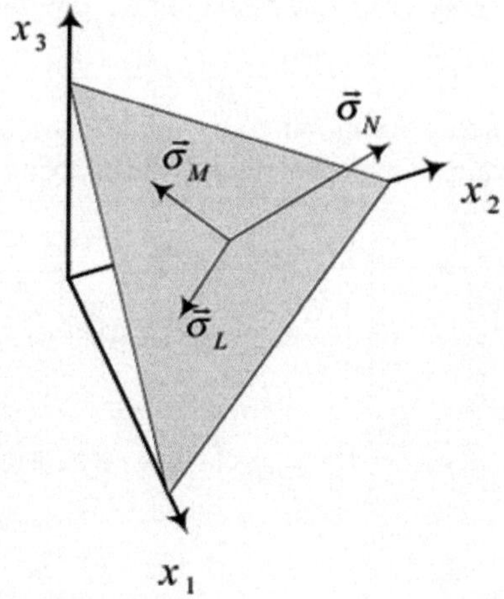

Fig. 17.1 Stress components on a microplane

where $N_{ij} = n_i n_j$ and n_i represent the Cartesian components of the unit normal vector $\boldsymbol{n}$ to a microplane.

Shear stress on each microplane is characterized by its components along two perpendicular directions M and L on the plane. The corresponding unit vectors are denoted by $\boldsymbol{m}$ and $\boldsymbol{l}$ with components m_i and l_i. These shear stresses can be expressed as,

$$\sigma_M = M_{ij}\sigma_{ij}, \sigma_L = L_{ij}\sigma_{ij} \tag{17.2}$$

in which $M_{ij} = \left(n_i m_j + n_j m_i\right)/2$ and $L_{ij} = \left(n_i l_j + n_j l_i\right)/2$ (Bažant and Prat 1988a, b). The principle of complementary virtual work yields:

$$\frac{4\pi}{3}\varepsilon : \delta\sigma = 2\int_{\Omega} (\varepsilon_N \delta\sigma_N + \varepsilon_M \delta\sigma_M + \varepsilon_L \delta\sigma_L)\mathrm{d}\Omega \tag{17.3}$$

where Ω is the surface of a unit hemisphere representing all possible orientations at a point. This equation states that the virtual work done by macroscopic stress and strain tensors over the volume of a unit hemisphere is equal to the virtual work done by the microplane stresses and strains over its surface (Caner et al. 2019). More details about the basics of Eq. (17.3) is provided in Bažant et al. (1996).

Substituting Eqs. (17.1) and (17.2) into (17.3), by taking into account the independence of individual components of virtual stress tensor, yields:

$$\varepsilon_{ij} = \frac{3}{2\pi}\int_{\Omega} \left(\varepsilon_N N_{ij} + \varepsilon_M M_{ij} + \varepsilon_L L_{ij}\right)\mathrm{d}\Omega \tag{17.4}$$

To use microplane formulation with static constraint, stress components on each plane passing through a material point are first calculated by the projection rule stated in Eqs. (17.1) ad (17.1). Then, 1-D microplane laws are required to determine normal as well as shear strains. Equation (17.4), which is in fact homogenization of all microplane strains, is finally utilized to obtain macroscopic strains. Compared to conventional modeling approaches, where strain tensor is directly either an explicit or an implicit function of stress tensor, macroscopic strains in microplane modeling with static constraint are indirectly obtained from the macroscopic stress components. Dual formulation of static constraint is kinematic constraint, in which projection rule is applied for strain as:

$$\varepsilon_N = N_{ij}\varepsilon_{ij}, \varepsilon_M = M_{ij}\varepsilon_{ij}, \varepsilon_L = L_{ij}\varepsilon_{ij} \tag{17.5}$$

Similar to Eqs. (17.3) and (17.4), the principle of virtual work is applied to obtain the following homogenization for determination of macroscopic stress tensor:

$$\sigma_{ij} = \frac{3}{2\pi}\int\limits_{\Omega}(\sigma_N N_{ij} + \sigma_M M_{ij} + \sigma_L L_{ij})\mathrm{d}\Omega \tag{17.6}$$

As a matter of fact, projection rules for stress and strain are simultaneously taken to be valid in conventional modeling approaches, but only one of them holds true in each microplane formulation. It should be noted that projection rule for stress and the principle of virtual work are both interpretations of equilibrium and that projection rule for strain (for small deformations) and the principle of complementary virtual work are both interpretations of geometric consistency in deformations of a body. Consequently, equilibrium in statically constrained microplane formulation is considered by projection rule for stress, while geometric consistency is taken into account by the principle of complementary virtual work. Similarly, in kinematically constrained formulation, equilibrium and geometric consistency are, respectively, satisfied by the principle of virtual work and projection rule for strain. These cause microplane formulations to be generalized versions of conventional modeling methods with the capability of modeling behaviors which cannot be directly taken into account in physical foundations of constitutive equations for various types of materials and responses (Mehrabi and Kadkhodaei 2013; Badnava et al. 2016). Note that, if Eq. (17.5) is satisfied, Eq. (17.4) will be valid too, but Eq. (17.4) does not necessarily lead to Eq. (17.5). This relation is the same between Eqs. (17.6) and (17.1) and (17.2). These all emphasize that microplane models are extensions of conventional ones. In other words, ordinary modeling approaches, in which projection rules for stress and strain simultaneously hold true, are a special case and may be referred to as "microplane formulations with double constraint".

There are almost no strict rules according to which one can select either static or kinematic constraint is more appropriate and efficient for specific modeling purposes, and comparison of theoretical predictions with experimental observations may be the most reliable criterion to select a formulation type. However, it is believed that when shear is the only source of inelastic response on microplanes, the kinematic constraint poses some limitations to the material compliance (Brocca and Bažant 2000).

To reveal more details about the microplane theory, linear elastic behavior is modeled based on the static constraint formulation. The relevant microplane laws are assumed as:

$$\varepsilon_N = \frac{\sigma_N}{E_N^0}, \varepsilon_{M,L} = \frac{\sigma_{M,L}}{E_T^0} \tag{17.7}$$

where E_N^0 and E_T^0 are deemed as local components of the linear elastic stiffness tensor, i.e., local elastic moduli of the material. By substituting these relationships into Eq. (17.4), evaluating the integral, and comparing the result with the constitutive equations of linear elasticity:

$$\varepsilon_{ij} = \frac{1+\nu}{E}\sigma_{ij} - \frac{\nu}{E}\sigma_{kk}\delta_{ij} \tag{17.8}$$

The following relations between the local and the macrolevel moduli are obtained:

$$E_N^0 = \frac{E}{1-2\nu}, E_T^0 = \frac{E}{1+3\nu} \tag{17.9}$$

in which E and ν are the Young's modulus and Poisson's ratio. These local moduli do not make sense and are not predictable before comparison with the known constitutive equations. Therefore, this may be a drawback when modeling unknown responses for which no governing equations have been already proposed. If kinematic formulation is applied, the following relationships are achieved (Carol and Bazant 1997; Caner et al. 2019):

$$E_N^0 = \frac{E}{1-2\nu}, E_T^0 = \frac{1-4\nu}{(1+\nu)(1-2\nu)}E \tag{17.10}$$

where ν has to be less than 0.25. This is also unreasonable since the theory of elasticity is not limited to such ranges for material parameters. To overcome these drawbacks, it is recommended (Kadkhodaei et al. 2007a, b) to derive formulations based on the volumetric–deviatoric split of normal stress and strain on a microplane:

$$\sigma_V = \frac{\delta_{ij}}{3}\sigma_{ij}, \sigma_D = \sigma_N - \sigma_V = \left(N_{ij} - \frac{\delta_{ij}}{3}\right)\sigma_{ij} \tag{17.11}$$

$$\varepsilon_N = \varepsilon_V + \varepsilon_D = \frac{\delta_{ij}}{3}\varepsilon_{ij} + \left(N_{ij} - \frac{\delta_{ij}}{3}\right)\varepsilon_{ij} \tag{17.12}$$

Using these relations, for statically constrained formulation, Eq. (17.4) takes the following form:

$$\varepsilon_{ij} = \varepsilon_V\delta_{ij} + \frac{3}{2\pi}\int_{\Omega}\left(\varepsilon_D N_{ij} + \varepsilon_M M_{ij} + \varepsilon_L L_{ij}\right)\mathrm{d}\Omega \tag{17.13}$$

If microplane laws are now considered as

$$\varepsilon_V = \frac{\sigma_V}{E_V^0}, \varepsilon_D = \frac{\sigma_D}{E_D^0}, \varepsilon_{M,L} = \frac{\sigma_{M,L}}{E_T^0} \tag{17.14}$$

the following relationships are obtained:

$$E_V^0 = \frac{E}{1-2\nu}, E_D^0 = E_T^0 = \frac{E}{1+\nu} \tag{17.15}$$

In fact, the local moduli are equal to the macroscopic ones. Moreover, the same local moduli are achieved by using kinematic constraint. Consequently, not only the aforementioned deficiencies are resolved but also a formulation with double constraint is derived since both static and kinematic constraints yield the same constitutive equations. As a result, for microplane modeling of any responses, the volumetric–deviatoric split is recommended to be applied (Kadkhodaei et al. 2007a) so that macroscopic elastic moduli are applicable for microplane laws and that double constraint formulation is obtained for elastic responses.

By developing proper microplane laws based on 1-D macroscopic constitutive equations, microplane models for several kinds of materials and responses (Badnava et al. 2016; Brocca and Bažant 2000; Kadkhodaei et al. 2007a, b; Ožbolt et al. 2001; Bažant and Zi 2003; Bažant et al. 2000; Zreid and Kaliske 2016; Prat and Bažant 1991; Salviato et al. 2016; Carol et al. 1991; Bažant and Prat 1987; Kuhl 2001; Bažant and Di Luzio 2004; Caner et al. 2007; Chang and Sture 2006; Caner and Carol 2006; Li et al. 2017; Kirane et al. 2015; Chatti et al. 2019; Steinke et al. 2019; Etse et al. 2003; Jin 2016) have been so far presented. This theory has been extended for large deformations as well (Bažant et al. 2000; Carol et al. 2004; Indriyantho et al. 2019). The main advantage of microplane theory over conventional modeling approaches is that only 1-D constitutive equations are required to derive a 3-D constitutive model; however, appropriate adjustments of 1-D macroscopic models to obtain microplane laws are challenging aspects of this technique (Kadkhodaei et al. 2007a).

Prior to proposition of a microplane model for shape memory alloys (SMAs), one more improvement to conventional microplane formulations is required to be developed. It can be shown (Kadkhodaei et al. 2007a, b) that expressing shear stress within each microplane by two resolved components on the plane causes inaccurate results and even may lead to prediction of shear strain during pure axial loading under certain conditions or axial strain during pure shear loading of isotropic materials. The main reason for this discrepancy is believed to be the so-called directional bias due to the same nonlinear stress–strain laws for the shear directions $\boldsymbol{m}$ and $\boldsymbol{l}$. To avoid this, shear stress on each microplane can be described by using the resultant shear stress vector on that plane, instead of resolving into two components on arbitrary directions $\boldsymbol{m}$ and $\boldsymbol{l}$ perpendicular to each other. For instance, when static constraint is used, one shear stress on each microplane is considered as is shown in Fig. 17.2.

Shear stress within each microplane is characterized by its resultant value on the plane, and it can be expressed in the form of:

$$\sigma_T = T_{ij}\sigma_{ij} \tag{17.16}$$

where $T_{ij} = \left(n_i t_j + n_j t_i\right)/2$ in which,

$$t_i = \frac{\sigma_{ik}n_k - \sigma_N n_i}{\sqrt{\sigma_{jr}\sigma_{js}n_r n_s - \sigma_N^2}} \tag{17.17}$$

represents Cartesian components of the unit vector $\boldsymbol{t}$ along the direction of resultant shear stress on the plane. Accordingly, the principle of complementary virtual work yields the following relation for strain tensor:

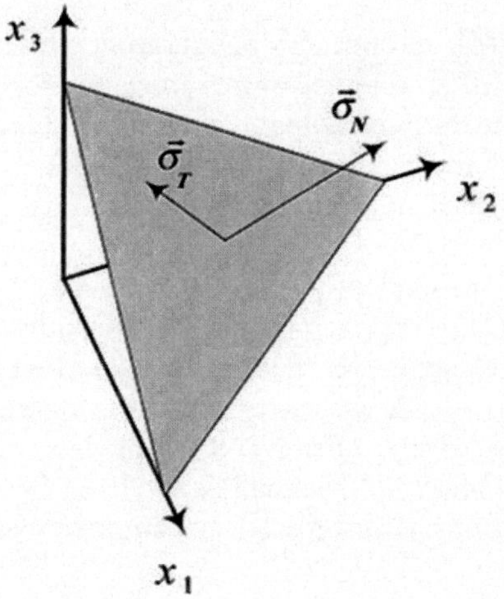

Fig. 17.2 Stress components on a microplane by considering one shear direction

$$\varepsilon_{ij} = \varepsilon_V \delta_{ij} + \frac{3}{2\pi} \int_{\Omega} \left(\varepsilon_D N_{ij} + \varepsilon_T T_{ij} \right) \mathrm{d}\Omega \tag{17.18}$$

Following the idea of one resultant shear component on each microplane, other similar approaches have been proposed (Kuhl and Ramm 1998; Leukart and Ramm 2003; Leukart and Ramm 2006). Moreover, thermodynamically consistent approaches have been developed to obtain microplane formulations in a continuum framework (Mehrabi et al. 2014a; Vrech et al. 2016; Dean et al. 2016). The readers are referred to these works for more details, but Eq. (17.18) will be utilized in the rest of this chapter.

17.2 Microplane Modeling of Shape Memory Alloys

Shape memory alloys usually exhibit two stable phases of austenite and martensite. Under certain circumstances, however, there may be an additional intermediate phase as well, but such details are not discussed here. In a stress-free state, when an SMA is heated, transformation from martensite to austenite starts at the temperature of A_s and ends at A_f. Cooling an austenitic SMA causes it to begin transforming back to martensite at M_s, and this transformation finishes at M_f. When stress is applied to temperature-induced martensite, transition to stress-induced martensite starts at a critical amount of stress and finishes at another critical value. These temperatures and critical stresses are important characteristics of an SMA. The formation of stress-induced martensite causes inelastic strains, which remain after complete unloading. However, this residual strain is recoverable upon heating to above A_f so that the material regains its original, unreformed configuration. This behavior is referred to as shape memory effect (SME). At temperatures above A_f, austenite also transforms to stress-induced martensite during loading; but, elastic and inelastic deformations are altogether spontaneously recovered upon unloading with no need of further heating so that the material reaches its initial configuration once the load is completely removed. This property is called superelasticity (SE) or pseudoelasticity (PE).

To develop a microplane model for shape memory alloys, appropriate microplate laws should be derived based on macroscopic 1-D constitutive equations. One of the simplest, though very efficient, 1-D models was proposed by Brisnon (1993) and is employed here to provide the basic concepts of a microplane model for shape memory alloys. Needless to say, other 1-D SMA models can also be utilized in the microplane modeling of an SMA (Mehrabi et al. 2012a). Neglecting thermal strain compared to inelastic strains, Brinson's 1-D constitutive model can be written as,

$$\varepsilon = \varepsilon^e + \varepsilon^t = \frac{\sigma}{E} + \varepsilon^* \xi_s \tag{17.19}$$

where ε^t is transformation-induced strain, ε^* is the maximum recoverable strain, and ξ_s is the stress-induced martensite volume fraction whose summation with the temperature-induced martensite volume fraction yields total martensite volume fraction. These fractions are usually stated as functions of instantaneous stress and temperature as well as history of the alloy in conjunction with a so-called stress–temperature phase diagram of an SMA. Moreover, the Young's modulus depends on total martensitic volume fraction, so it is a function of stress and temperature.

Microplane modeling of SMAs was first performed by Brocca et al. (2002), and this approach was further extended by Mehrabi and Kadkhodaei (2013), Kadkhodaei et al. (2007a, b), Mehrabi et al. (2012a, b, 2014a, b) from various aspects. As martensitic transformations are driven by shear (displacive) deformations, the total inelastic strain is only due to shear strains on microplanes. A linear elastic stress–strain relation is thus considered for the normal direction, and a 1-D SMA constitutive law is used for the shear direction on each microplane. Accordingly, with the use of Eq. (17.8), the strain at each point is decomposed as:

$$\varepsilon_{ij} = \varepsilon_{ij}^e + \varepsilon_{ij}^t \tag{17.20}$$

where

$$\varepsilon_{ij}^e = -\frac{\nu}{E}\sigma_{ss}\delta_{ij} + \frac{1+\nu}{E}\sigma_{rt} \cdot \frac{3}{2\pi}\int\limits_{\Omega}\left(N_{rs}N_{ij} + T_{rs}T_{ij}\right)\mathrm{d}\Omega \tag{17.21}$$

$$\varepsilon_{ij}^t = \frac{3}{2\pi}\varepsilon^*\xi_s\int\limits_{\Omega}T_{ij}\mathrm{d}\Omega \tag{17.22}$$

Although transformation-induced strain is the only part of inelastic response in the utilized 1-D microplane laws, it can be shown that the resultant 3-D constitutive equations predicts inelastic strain due to martensite variants reorientation as well (Mehrabi and Kadkhodaei 2013). Reorientation of martensite variants is an important phenomenon under nonproportional loadings, which causes deviation from normality rule in constitutive modeling of SMAs, and the presented formulation takes it into account although no expression is explicitly attributed to this kind of strain. However, one can introduce separate terms for reorientation-induced strains in microplane modeling of shape memory alloys (Zhou et al. 2019).

For numerical implementation of the derived model, according to the Voigt notation, the following stress and strain vectors are considered:

$$\boldsymbol{E} = \left\{ \epsilon_{11}\ \epsilon_{22}\ \epsilon_{33}\ \gamma_{12}\ \gamma_{13}\ \gamma_{23} \right\}^T = \left\{ E_1\ E_2\ E_3\ E_4\ E_5\ E_6 \right\}^T \tag{17.23}$$

$$\boldsymbol{\Sigma} = \left\{ \sigma_{11}\ \sigma_{22}\ \sigma_{33}\ \sigma_{12}\ \sigma_{13}\ \sigma_{23} \right\}^T = \left\{ \Sigma_1\ \Sigma_2\ \Sigma_3\ \Sigma_4\ \Sigma_5\ \Sigma_6 \right\}^T \tag{17.24}$$

Consequently, the constitutive relation might be reformulated as (Karamooz-Ravari and Shahriari 2017):

$$E_i = \frac{1}{E}\beta_i + \frac{3}{2\pi}\varepsilon^* \xi_s \int\limits_{\Omega} \frac{Q_i}{A} \mathrm{d}\Omega, \qquad i = 1, \ldots, 6 \tag{17.25}$$

Defining $\boldsymbol{N}$ as the Voigt notation representation of N_{ij} tensor, Σ_m as the first stress invariant, and Σ_N as the normal stress to the microplane, respectively, using the following equations:

$$\boldsymbol{N} = \left\{ N_1\ N_2\ N_3\ N_4\ N_5\ N_6 \right\}^T = \left\{ n_1^2\ n_2^2\ n_3^2\ n_1 n_2\ n_1 n_3\ n_2 n_3 \right\}^T \tag{17.26}$$

$$\Sigma_m = \Sigma_1 + \Sigma_2 + \Sigma_3 \tag{17.27}$$

$$\Sigma_N = \boldsymbol{\Sigma} \cdot \hat{\boldsymbol{N}} \tag{17.28}$$

$\boldsymbol{\beta}$, $\boldsymbol{Q}$, and A can be formulated as:

$$\beta_i = \begin{cases} (1+\nu)\Sigma_i - \nu\Sigma_m & i \le 3 \\ 2(1+\nu)\Sigma_i & i > 3 \end{cases} \tag{17.29}$$

$$\boldsymbol{Q} = \boldsymbol{R} - \Sigma_N \hat{\boldsymbol{N}} \tag{17.30}$$

$$A^2 = \boldsymbol{P} \cdot \hat{\boldsymbol{N}} - (\Sigma_N)^2 \tag{17.31}$$

Using some algebraic calculations and simplifications, the utilized vectors $\boldsymbol{P}$, $\hat{\boldsymbol{N}}$, and $\boldsymbol{R}$ are obtained as:

$$\boldsymbol{P} = \begin{Bmatrix} \Sigma_1^2 + \Sigma_4^2 + \Sigma_5^2 \\ \Sigma_2^2 + \Sigma_4^2 + \Sigma_6^2 \\ \Sigma_3^2 + \Sigma_5^2 + \Sigma_6^2 \\ \Sigma_4(\Sigma_1 + \Sigma_2) + \Sigma_5\Sigma_6 \\ \Sigma_5(\Sigma_1 + \Sigma_3) + \Sigma_4\Sigma_6 \\ \Sigma_6(\Sigma_2 + \Sigma_3) + \Sigma_4\Sigma_5 \end{Bmatrix} \tag{17.32}$$

$$\hat{\boldsymbol{N}} = \left\{ N_1 \; N_2 \; N_3 \; 2N_4 \; 2N_5 \; 2N_6 \right\}^T \tag{17.33}$$

$$\boldsymbol{R} = \begin{Bmatrix} (\Sigma_1 N_1 + \Sigma_4 N_4 + \Sigma_5 N_5) \\ (\Sigma_2 N_2 + \Sigma_4 N_4 + \Sigma_6 N_6) \\ (\Sigma_3 N_3 + \Sigma_5 N_5 + \Sigma_6 N_6) \\ \Sigma_4(N_1 + N_2) + N_4(\Sigma_1 + \Sigma_2) + \Sigma_5 N_6 + \Sigma_6 N_5 \\ \Sigma_5(N_1 + N_3) + N_5(\Sigma_1 + \Sigma_3) + \Sigma_4 N_6 + \Sigma_6 N_4 \\ \Sigma_6(N_2 + N_3) + N_6(\Sigma_2 + \Sigma_3) + \Sigma_4 N_5 + \Sigma_5 N_4 \end{Bmatrix} \tag{17.34}$$

In the relations above, $(\cdot)$ denotes the dot product operator. Since $E_i = E_i(\boldsymbol{\Sigma}, T)$, the incremental form of the constitutive equations can be expressed as:

$$\Delta E_i = \sum_{j=1}^{6} \frac{\partial E_i}{\partial \Sigma_j} \Delta \Sigma_j + \frac{\partial E_i}{\partial T} \Delta T \tag{17.35}$$

where (∂) denote the partial differentiation, $\left(\frac{\partial E_i}{\partial \Sigma_j}\right)^{-1}$ the continuum tangent stiffness matrix and $\left(\frac{\partial \mathrm{E}_i}{\partial \Sigma_j}\right)^{-1}\left(\frac{\partial \mathrm{E}_i}{\partial T}\right)$ the tangent thermal moduli vector. Differentiating Eq. (17.25) yields:

$$\frac{\partial \mathrm{E}_i}{\partial \Sigma_j} = C_{ij} = C_{ij}^{e(1)} + C_{ij}^{e(2)} + C_{ij}^{tr(1)} + C_{ij}^{tr(2)} \tag{17.36}$$

$$\frac{\partial \mathrm{E}_i}{\partial T} = \begin{cases} \left(\frac{1}{E_M} - \frac{1}{E_A}\right)\frac{\partial \xi}{\partial T}[(1+\nu)\Sigma_i - \nu\Sigma_m] + \frac{3}{4\pi}\varepsilon^* \frac{\partial \xi_s}{\partial T}\int\limits_{\Omega} \frac{Q_i}{A} \mathrm{d}\Omega & i \le 3 \\ 2\left(\frac{1}{E_M} - \frac{1}{E_A}\right)\frac{\partial \xi}{\partial T}(1+\nu)\Sigma_i + \frac{3}{2\pi}\varepsilon^* \frac{\partial \xi_s}{\partial T}\int\limits_{\Omega} \frac{Q_i}{A} d\Omega & i > 3 \end{cases} \tag{17.37}$$

in which

$$\boldsymbol{C}^{e(1)} = \frac{1}{E}\begin{bmatrix} 1 & -\nu & -\nu & 0 & 0 & 0 \\ -\nu & 1 & -\nu & 0 & 0 & 0 \\ -\nu & -\nu & 1 & 0 & 0 & 0 \\ 0 & 0 & 0 & 2(1+\nu) & 0 & 0 \\ 0 & 0 & 0 & 0 & 2(1+\nu) & 0 \\ 0 & 0 & 0 & 0 & 0 & 2(1+\nu) \end{bmatrix} \tag{17.38}$$

$$C_{ij}^{e(2)} = \begin{cases} \frac{1}{2\bar{\sigma}}\beta_i\left(3\Sigma_j - \Sigma_m\right)\left(\frac{1}{E_M} - \frac{1}{E_A}\right)\frac{\partial\xi}{\partial\bar{\sigma}} & j \le 3 \\ \frac{3}{\bar{\sigma}}\beta_i\Sigma_j\left(\frac{1}{E_M} - \frac{1}{E_A}\right)\frac{\partial\xi}{\partial\bar{\sigma}} & j > 3 \end{cases} \tag{17.39}$$

$$C_{ij}^{\mathrm{tr}(1)} = \frac{3}{2\pi\bar{\sigma}}\varepsilon^*\frac{\partial\xi_s}{\partial\bar{\sigma}}\begin{cases} \int_{\Omega}\frac{Q_i}{2A}\left(3\Sigma_j - \Sigma_m\right) & j \le 3 \\ \int_{\Omega}\frac{3Q_i}{A}\Sigma_j & j < 3 \end{cases} \tag{17.40}$$

$$C_{ij}^{\mathrm{tr}(2)} = \frac{3}{2\pi}\varepsilon^*\xi_s\int_{\Omega}\frac{1}{A}\left(\frac{\partial Q_i}{\partial\Sigma_j} - \frac{Q_i}{A}\frac{\partial A}{\partial\Sigma_j}\right)\mathrm{d}\Omega \tag{17.41}$$

$$\begin{bmatrix} N_1(1-N_1) & -N_1N_2 & -N_1N_3 & N_4(1-2N_1) & N_5(1-2N_1) & -2N_1N_6 \\ -N_1N_2 & N_2(1-N_2) & -N_2N_3 & N_4(1-2N_2) & -2N_2N_5 & N_6(1-2N_2) \\ -N_1N_3 & -N_2N_3 & N_3(1-N_3) & -2N_3N_4 & N_5(1-2N_3) & N_6(1-2N_3) \\ N_4(1-2N_1) & N_4(1-2N_2) & -2N_3N_4 & N_1+N_2-4N_4^2 & N_6-4N_4N_5 & N_5-4N_4N_6 \\ N_5(1-2N_1) & -2N_2N_5 & N_5(1-2N_3) & N_6-4N_4N_5 & N_1+N_3-4N_5^2 & N_4-4N_5N_6 \\ -2N_1N_6 & N_6(1-2N_2) & N_6(1-2N_3) & N_5-4N_4N_6 & N_4-4N_5N_6 & N_2+N_3-4N_6^2 \end{bmatrix} \tag{17.42}$$

$$\frac{\partial A}{\partial\Sigma} = \frac{1}{A}\begin{Bmatrix} (\Sigma_1N_1 + \Sigma_4N_4 + \Sigma_5N_5 - N_1\Sigma_N) \\ (\Sigma_2N_2 + \Sigma_4N_4 + \Sigma_6N_6 - N_2\Sigma_N) \\ (\Sigma_3N_3 + \Sigma_5N_5 + \Sigma_6N_6 - N_3\Sigma_N) \\ \Sigma_4(N_1+N_2) + N_4(\Sigma_1+\Sigma_2) + \Sigma_5N_6 + \Sigma_6N_5 - 2N_4\Sigma_N \\ \Sigma_5(N_1+N_3) + N_5(\Sigma_1+\Sigma_3) + \Sigma_4N_6 + \Sigma_6N_4 - 2N_5\Sigma_N \\ \Sigma_6(N_2+N_3) + N_6(\Sigma_2+\Sigma_3) + \Sigma_4N_5 + \Sigma_5N_4 - 2N_6\Sigma_N \end{Bmatrix} \tag{17.43}$$

Using forward Euler implementation scheme, the stress vector at the end of an increment can be obtained as:

$$\boldsymbol{\Sigma}^{(n+1)} = \boldsymbol{\Sigma}^{(n)} + \Delta\boldsymbol{\Sigma}^{(n)} = \boldsymbol{\Sigma}^{(n)} + \left(\boldsymbol{C}^{(n)}\right)^{-1}\left(\Delta\mathrm{E}^{(n)} - \frac{\partial\mathrm{E}^{(n)}}{\partial T}\Delta T\right) \tag{17.44}$$

As it can be seen, numerical integration over the surface of unit hemisphere is needed for the implementation. Various techniques have been so far proposed for numerical calculation of such integrals, and efficiency as well as error percentage of various schemes have been evaluated in several works (Jalalpour et al. 2019; Huang et al. 2017; Verron 2015; Badel and Leblond 2003; Němeček et al. 2002). Considering

$f(x_1, x_2, x_3)$ as the function to be integrated, the integral may be approximated using (Bažant and Oh1986):

$$\int_{\Omega} f(x_1, x_2, x_3) d\Omega = \sum_{k=1}^{M} w_k f\left(\zeta_1^k, \zeta_2^k, \zeta_3^k\right) \tag{17.45}$$

in which w_k are weights, and $\zeta_1^k, \zeta_2^k, \zeta_3^k$ the direction cosines. The values for these parameters are presented in Karamooz-Ravari and Shahriari (2017).

17.2.1 *Introduction of Tension–Compression Asymmetry*

In order to introduce tension–compression asymmetry into the previously developed constitutive model, it is supposed that the shear strain on each microplane is composed of elastic strain, ε_e, compressive transformation strain, $\varepsilon_{\text{tr}}^-$, and tensile transformation strain, $\varepsilon_{\text{tr}}^+$ (Karamooz Ravari et al. 2015):

$$\varepsilon_T = \varepsilon_e + \varepsilon_{\text{tr}}^+ + \varepsilon_{\text{tr}}^- \tag{17.46}$$

Referring to the previously developed relationships (Poorasadion et al. 2013), the shear strain can be formulated as:

$$\varepsilon_T = \frac{1+\nu}{E}\sigma_T + \varepsilon_+^* \xi_s^+ + \varepsilon_-^* \xi_s^- \tag{17.47}$$

Experimental observations show that the elastic modulus of austenite phase might differ from martensite phase. In addition, the elastic modulus of martensite can be different in tension and compression. The value of SMAs' modulus differs gradually when the phase transformation or reorientation occurs. In order to take this variation into account, the elastic modulus of SMAs is defined as a function of elastic modulus of full austenite, E_A, full martensite in tension, E_M^+, full martensite in compression, E_M^-, full temperature-induce martensite, E_T, and stress- and temperature-induced, ξ_T, martensite volume fractions (Karamooz Ravari et al. 2015):

$$\frac{1}{E} = \frac{1}{E_A} + \xi_s^+\left(\frac{1}{E_M^+} - \frac{1}{E_A}\right) + \xi_s^-\left(\frac{1}{E_M^-} - \frac{1}{E_A}\right) + \xi_T\left(\frac{1}{E_M^T} - \frac{1}{E_A}\right) \tag{17.48}$$

In the equations above, it is necessary to specify the evolution of $\xi_s^{\pm}$ and ξ_T as a function of stress and temperature state. Referring to the phase diagram shown in Fig. 17.3, the phenomenological relationships of Eq. (17.49) might be used:

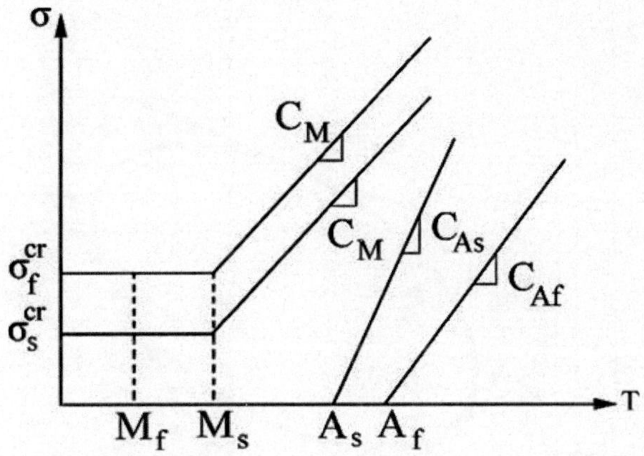

Fig. 17.3 A stress–temperature diagram for martensite transformation of shape memory alloys (Karamooz Ravari et al. 2015)

$$
\begin{cases}
\text{if } T \le M_s \text{ and } \sigma_s^{\mathrm{cr}} < \hat{\sigma} < \sigma_f^{\mathrm{cr}}: \\
Y = \cos\left(\frac{\pi}{\sigma_f^{\mathrm{cr}}-\sigma_s^{\mathrm{cr}}}\left(\hat{\sigma} - \sigma_s^{\mathrm{cr}}\right)\right) \\
\xi_s^r = \frac{1-Y}{2} + \xi_{s0}^r\frac{1+Y}{2},\ \xi_s^p = \xi_{s0}^p\frac{1-\xi_s^r}{1-\xi_{s0}^r},\ \xi_T = (\xi_{T0}+\Delta)\frac{(1-\xi_s^r)}{1-\xi_{s0}^r} \\
\text{if } T < M_f : \Delta = 0 \\
\text{else} : \Delta = \frac{1-\xi_0}{2}(1 - Y_{\mathrm{MT}}) \\
\text{if } T > M_s \text{ and } \sigma_s^{cr} + C_M(T - M_s) < \hat{\sigma} < \sigma_f^{\mathrm{cr}} + C_M(T - M_s): \\
Y = \cos\left(\frac{\pi}{\sigma_f^{cr}-\sigma_s^{cr}}\left(\hat{\sigma} - \sigma_s^{cr} - C_M(T - M_s)\right)\right) \\
\xi_s^r = \frac{1-Y}{2} + \xi_{s0}^r\frac{1+Y}{2},\ \xi_s^p = \xi_{s0}^p\frac{1-\xi_s^r}{1-\xi_{s0}^r},\ \xi_T = \xi_{T0}\frac{1-\xi_s^r}{1-\xi_{s0}^r} \\
\text{if } T > A_s \text{ and } C_{Af}\left(T - A_f\right) < \hat{\sigma} < C_{As}(T - A_s): \\
Y = \cos\left(\frac{\pi}{C_{As}(T-A_s)-C_{Af}\left(T-A_f\right)}\left(C_{As}(T - A_s) - \hat{\sigma}\right)\right) \\
\xi_s^r = \frac{\xi_{s0}^r}{2}(1+Y),\ \xi_s^p = \frac{\xi_{s0}^p}{2}(1+Y),\ \xi_T = \frac{\xi_{T0}}{2}(1+Y) \\
\text{if } M_f < T < M_s \text{ and } \hat{\sigma} < \sigma_s^{cr} \\
\xi_s^r = \xi_{s0}^r,\ \xi_s^p = \xi_{s0}^p,\ \xi_T = \frac{1-\xi_0}{2}(1 - Y_{\mathrm{MT}}) + \xi_{T0}
\end{cases}
\tag{17.49}
$$

in which superscripts "r" and "p" are as (r =+, p = −) in ension and (r = −, p = +) in compression, and $Y_{\mathrm{MT}} = \cos\left(\pi(T - M_s)/\left(M_f - M_s\right)\right)$. To take the material asymmetry into account, the equivalent stress must be modified. Considering J_2 and J_3 as the second and third invariants of deviatoric stress tensor, the following equivalent stress might be used:

$$
\hat{\sigma} = \frac{1}{1+\alpha}\left\{\sqrt{3J_2} + \frac{9}{2}\alpha\frac{J_3}{J_2}\right\}
\tag{17.50}
$$

where α is a real number between 0 and 1, which determines the level of asymmetry. As demonstrated in Fig. 17.4, for $\alpha = 0$, the transformation surface introduces no

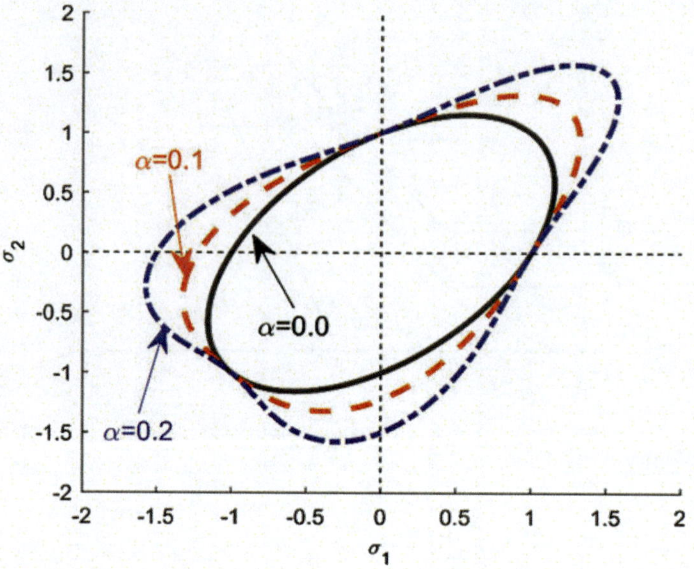

Fig. 17.4 Transformation surface in plane stress state for different typical values of α

asymmetry into the model. However, as its value increases, the level of asymmetry increases too. The level of asymmetry might be different in loading and unloading. Without loss of generality, the value of this parameter can be distinguished for these two loading strategies:

$$\alpha = \begin{cases} \alpha_1 & \text{if} \quad \dot{\sigma} > 0 \\ \alpha_2 & \text{if} \quad \dot{\sigma} < 0 \end{cases} \tag{17.51}$$

By applying the principle of complementary virtual work, the following relation is obtained for the strain tensor:

$$\begin{aligned} \varepsilon_{ij} = & -\frac{\nu}{E}\sigma_{mm}\delta_{ij} + \frac{1+\nu}{E}\sigma_{mn} \cdot \frac{3}{2\pi}\int_{\Omega}\left(N_{mn}N_{ij} + T_{mn}T_{ij}\right)\mathrm{d}\Omega \\ & + \left(\varepsilon_{+}^{*}\xi_{s}^{+} + \varepsilon_{-}^{*}\xi_{s}^{-}\right) \cdot \frac{3}{2\pi}\int_{\Omega} T_{ij}\mathrm{d}\Omega \end{aligned} \tag{17.52}$$

To find how efficient the model is for simulating the tension–compression asymmetry, tension–compression uniaxial test, and four-point bending are simulated and the obtained results are compared against experimental measurements. Here, all the simulations are considered at constant temperature so that there is no temperature-induced martensite and the value of E_M^T is optional. For the sake of brevity, only one experimental case, performed by Reedlunn et al. (2014), is reported here and readers are referred to (Karamooz Ravari et al. 2015) for more information and case studies.

The experimental measurements, presented in Reedlunn et al. (2014), were performed isothermally using cold-drawn, slightly Ni-rich NiTi tube from Memry Corporation. Considering a tube with the length of L, the gage length of L_e, outer diameter of D, and thickness of t, the four-point bending test can be done using the method shown in Figure 17.5. In this method, application of load, F, applies a moment to the tube and the gage attached on the middle of the tube will measure the rotation angle, θ. The mean curvature of the tube, κ, might be calculated as:

$$\kappa = \frac{\theta}{L_s} \tag{17.53}$$

For the present case, $L_e = 9.58$ mm, $D = 3.176$ mm, and $t = 0.318$ mm. The test temperature is higher than austenite finish temperature so that the material is in the austenite phase, and the initial value for ξ_{s0}^{+}, ξ_{s0}^{-}, and ξ_{T0} is zero. The material parameters of the sample are calibrated using the stress–strain response at the temperature of 298 K and are reported in Table 17.1. The predicted stress–strain response of the material is verified against experimental one in Fig. 17.6 showing a good agreement.

Considering M as the applied moment, I the corresponding area moment of inertia of the cross section, C the outer radius of the tube, and Y_0 the position of the neutral axis measured from the centerline of the tube, the normalized moment, MC/I, and Y_0 are plotted versus the dimensionless curvature, $C\kappa$, in Fig. 17.7a and b, respectively. As can be seen, the conformity decreases as the curvature increases and the model

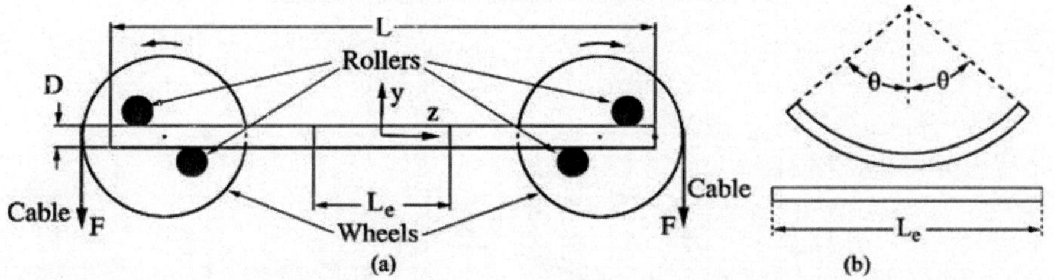

Fig. 17.5 **a** Schematic configuration of the four-point bending test **b** illustration of curvature calculation method (Karamooz Ravari et al. 2015)

Table 17.1 Material parameters obtained by calibration of stress–strain response in tension and compression (Karamooz Ravari et al.2015)

E_A(MPa)	E_M^{+}(MPa)	E_M^{-}(MPa)	ν	M_f(K)	M_s(K)	A_s(K)	A_f(K)	σ_s^{cr}(MPa)
65,300	28,000	87,000	0.45	126	210	248	292	90

σ_f^{cr}(MPa)	C_M(MPa/K)	C_{As}(MPa/K)	C_{Af}(MPa/K)	ε_{+}^{*}	ε_{-}^{*}	α_1	α_2	T(K)
170	3.65	4.5	32	0.054	−0.035	0.23	0.23	298

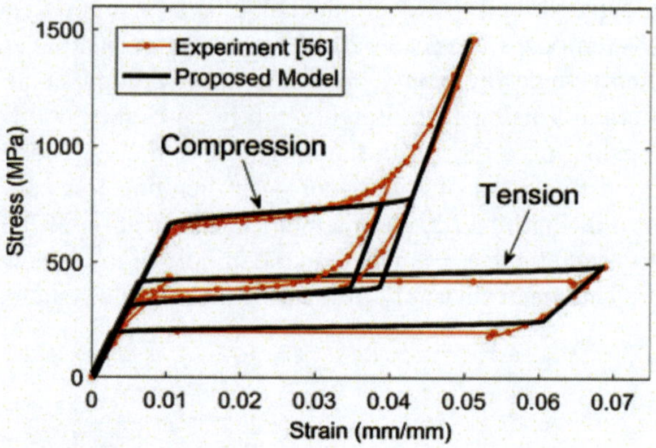

Fig. 17.6 Experimental and numerical stress–strain curve of NiTi tube in tension and compression (Karamooz Ravari et al. 2015)

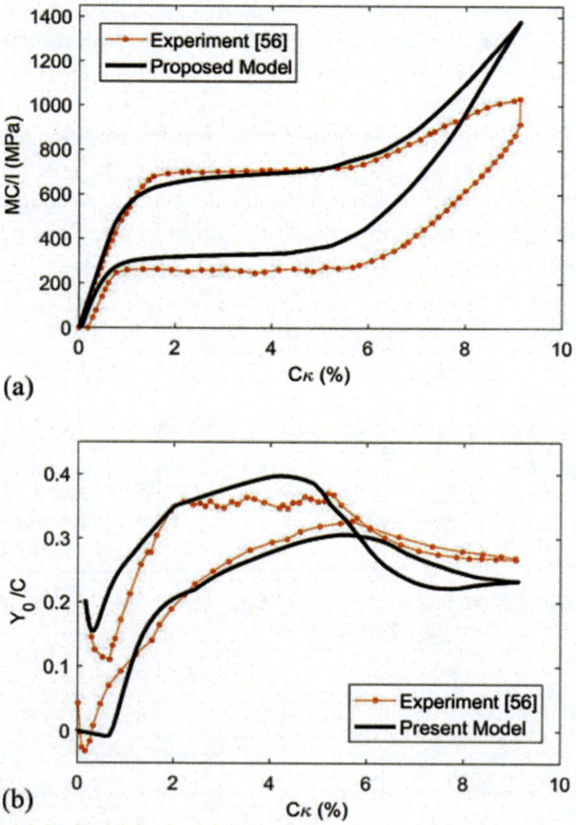

Fig. 17.7 **a** Normalized moment versus dimensionless curvature **b** normalized position of the natural axis versus dimensionless curvature (Karamooz Ravari et al. 2015)

overpredicts the level of applied moment, especially for high values of curvature. This difference might be due to the effects of strain localization in experimental measurements (Reedlunn et al. 2014), over-constraint nature of the finite element model which causes the cross-sectional planes of the tube to remain planar, and the errors in finding the material parameters, especially maximum recoverable strain, due to the transformation-induced plasticity (Qidwai and Lagoudas 2000).

17.2.2 Modeling Plasticity and Cyclic Responses

Contemplating the framework of small strains, the total strain tensor increment, $d\varepsilon_{ij}$ can be decomposed into elastic, $d\varepsilon_{ij}^{\mathrm{e}}$, transformation, $d\varepsilon_{ij}^{\mathrm{tr}}$, and plastic, $d\varepsilon_{ij}^{\mathrm{p}}$, strain tensor increments (Karamooz-Ravari et al. 2018):

$$d\varepsilon_{ij} = d\varepsilon_{ij}^{\mathrm{e}} + d\varepsilon_{ij}^{\mathrm{tr}} + d\varepsilon_{ij}^{\mathrm{p}} \tag{17.54}$$

The elastic and transformation parts are previously formulated using microplane theory. To take the plastic strains into account, it is just necessary to develop a formulation for $d\varepsilon_{ij}^{\mathrm{p}}$. Defining $C_{ijkl} = \frac{\partial\sigma_{ij}}{\partial\varepsilon_{ij}^{\mathrm{e}}}$, the stress increment might be obtained using the following relation:

$$d\sigma_{ij} = C_{ijkl}d\varepsilon_{kl} = C_{ijkl}\left(d\varepsilon_{kl} - d\varepsilon_{kl}^{\mathrm{tr}} - d\varepsilon_{kl}^{\mathrm{p}}\right) = C_{ijkl}\left(d\varepsilon_{kl}^{\mathrm{inp}} - d\varepsilon_{kl}^{\mathrm{p}}\right) \tag{17.55}$$

Utilizing the associate flow rule and considering λ as the plastic multiplier:

$$d\varepsilon_{ij}^{\mathrm{p}} = M_{ij}d\lambda \tag{17.56}$$

If $\bar{\sigma}$ is the von Mises equivalent stress and $\sigma_y(\lambda)$ the yield stress, the yield surface is defined as $f(\sigma_{ij}, \lambda) = \bar{\sigma} - \sigma_y$ and $M_{ij} = \frac{\partial f}{\partial\sigma_{ij}}$. The consistency condition yields:

$$\frac{\partial f}{\partial\sigma_{ij}}d\sigma_{ij} + \frac{\partial f}{\partial\lambda}d\lambda = 0 \tag{17.57}$$

Substitution of Eqs. (17.55) and (17.56) into Eq. (17.57) yields:

$$d\varepsilon_{ij}^{\mathrm{p}} = \frac{1}{A}B_{ijkl}^{-1}M_{kl}M_{mn}d\sigma_{mn} = P_{ijmn}d\sigma_{mn} \tag{17.58}$$

in which $A = M_{ij}C_{ijkl}M_{kl} - \frac{\partial f}{\partial\lambda}$, and $B_{ijkl} = \delta_{ik}\delta_{jl} - \frac{1}{A}M_{pq}C_{pqkl}M_{ij}$. Now, the strain tensor increment can be related to the stress tensor increment as:

$$d\varepsilon_{ij} = \left(C_{ijkl}^{-1} + P_{ijkl} + \frac{\partial\varepsilon_{ij}^{\mathrm{tr}}}{\partial\sigma_{kl}}\right)d\sigma_{kl} \tag{17.59}$$

Based on experiments, the peak strain, residual strain, dissipation energy, transformation stresses, and austenite and martensite elastic moduli accumulate to a specific value during the cyclic loading. Referring to Eq. (17.58), the plastic strain is a function of transformation strain. To take the cyclic response into account, it is supposed that the plastic strain and the transformation response are related to the plastic strain of the previous cycle, and the material parameters change as a function of transformation state. Considering $\varepsilon_{p0}^{(n)}$ as the plastic strain at the beginning of each cycle, some evolution relations are assumed for material parameters during the cyclic loading as follows:

$$\sigma_s^{\mathrm{cr}} = \sigma_{s0}^{\mathrm{cr}} + \left(\sigma_{\mathrm{se}}^{\mathrm{cr}} - \sigma_{s0}^{\mathrm{cr}}\right)\Gamma \tag{17.60}$$

$$\sigma_f^{\mathrm{cr}} = \sigma_{f0}^{\mathrm{cr}} + \left(\sigma_{\mathrm{fe}}^{\mathrm{cr}} - \sigma_{f0}^{\mathrm{cr}}\right)\Gamma \tag{17.61}$$

$$C_M = C_{M0} + (C_{\mathrm{Me}} - C_{M0})\Gamma \tag{17.62}$$

$$C_{\mathrm{As}} = C_{\mathrm{As0}} + (C_{\mathrm{Ase}} - C_{\mathrm{As0}})\Gamma \tag{17.63}$$

$$C_{\mathrm{Af}} = C_{\mathrm{Af0}} + (C_{\mathrm{Afe}} - C_{\mathrm{Af0}})\Gamma \tag{17.64}$$

$$\varepsilon^* = \varepsilon_0^* + \left(\varepsilon_e^* - \varepsilon_0^*\right)\Gamma \tag{17.65}$$

$$E_A = E_{A0} + (E_{\mathrm{Ae}} - E_{A0})\Gamma \tag{17.66}$$

$$E_{\mathrm{M}} = E_{\mathrm{M0}} + (E_{Me} - E_{\mathrm{M0}})\Gamma \tag{17.67}$$

In the relations above, the subscripts "0" and "*e*" denote the first cycle and the accumulated one, respectively, and Γ is the evolution function defined as:

$$\Gamma = \left(1 - \mathrm{e}^{-\mu\left(\frac{\varepsilon_{p0}^{(n)}}{\varepsilon_p^{(\max)}}\right)^m}\right) \tag{17.68}$$

where $\varepsilon_p^{(\max)}$ is the maximum residual strain, and m and μ are material parameters defining the rate of accumulation. Here, for the sake of simplicity and without loss of generality, linear isotropic hardening is considered for the yield surface and the yield stress is defined as:

$$\sigma_y = \sigma_y^0 + h\lambda \tag{17.69}$$

in which h is a constant, and σ_y^0 the initial yield stress which is a function of initial plastic strain and temperature, T, as follows:

$$\sigma_y^0(T) = \sigma_{y0}^0(T) + \left(\sigma_{ye}^0(T) - \sigma_{y0}^0(T)\right)\Gamma \tag{17.70}$$

where the superscript "0" indicates the initial yield stress before linear isotropic hardening happens. Referring to the previous studies, the yield stress of martensite, σ_y^{M0}, is higher than that of austenite, σ_y^{A0}. In addition, the yield stress is constant for $T \le M_f$ and $T \ge A_f$ and varies linearly between these two:

$$\sigma_{yi}^0 = \begin{cases} \sigma_{yi}^{M0} & \text{if} \quad T \le M_f \\ \frac{(T-M_f)\sigma_y^{A0}+(A_f-T)\sigma_{yi}^{M0}}{A_f-M_f} & M_f < T < A_f \\ \sigma_{yi}^{A0} & \text{if} \quad T \ge A_f \end{cases} \tag{17.71}$$

Here, the superscripts M and A denote martensite and austenite, respectively, and subscript i stands for "0" and "e".

Hereafter, the constitutive model is validated against experimental observations of the compressive cyclic responses of NiTi in superelastic and shape memory regimes. Electrode induction melting inert gas atomization process by TLS Technique GmbH (Bitterfeld Germany) was used to produce NiTi powder from $Ni_{50.8}Ti_{49.2}$(at. %) ingots obtained from Nitinol Devices & Components, Inc. (Fremont, CA). A Phenix PXM (3D Systems) SLM machine equipped with a 300 W Ytterbium fiber laser was used to fabricate the fatigue specimens from NiTi powder. The oxygen level inside the chamber was set to 1000 ppm during the fabrication to reduce the impurity level in the final parts. Table 17.2 presents the SLM process parameters used for fabrication of the samples.

In each case, first, the material parameters must be calibrated. The calibration of material parameters is presented in Karamooz-Ravari et al. (2018) in detail and is ignored here for brevity.

The material parameters associated with the superelastic case are presented in Table 17.3. Because the material temperature is above austenite finish temperature, the value of σ_{y0}^{M0} and σ_{ye}^{M0} is optional and $\xi_{T0} = \xi_{s0} = 0$. For the sake of clarity, only the first, ninth, 42nd, and 43rd cycles are depicted in Fig. 17.8. As can be seen, the first and ninth cycles are well reproduced by the model when comparing to the experiment. The 42nd and 43rd cycles are almost coincided with each other, and

Table 17.2 Process parameters used for fabrication of fatigue specimens (Karamooz-Ravari et al. 2018)

Laser power (W)	Scanning speed (m/s)	Hatch spacing (μm)	Layer thickness (μm)	Energy density (J/mm^3)
250	1.25	80	30	83.34

Table 17.3 Utilized material parameters for simulation of cyclic loading in superelastic regime (Karamooz-Ravari et al. 2018)

M_f(°C)	σ_{s0}^{cr}(MPa)	C_{M0}(MPa/°C)	C_{Af0}(MPa/°C)	E_{M0}(MPa)	$\sigma_{\text{ye}}^{\text{M0}}$(MPa)	ε_0^*
36.17	220	11.03	11.0	74,000	NA	0.035
M_s(°C)	$\sigma_{\text{se}}^{\text{cr}}$(MPa)	C_{Me}(MPa/°C)	C_{Afe}(MPa/°C)	E_{Me}(MPa)	$\sigma_{\text{ye}}^{\text{A0}}$(MPa)	ε_e^*
47.87	50	8.0	1.0	34,000	1002	0.013
A_s(°C)	σ_{f0}^{cr}(MPa)	C_{As0} (MPa/°C)	E_{A0}(MPa)	σ_{y0}^{M0}(MPa)	h(MPa)	μ
21.52	650	12.0	60,000	NA	75,000	0.5
A_f(°C)	$\sigma_{\text{fe}}^{\text{cr}}$(MPa)	C_{Ase}(MPa/°C)	E_{Ae}(MPa)	σ_{y0}^{A0}(MPa)	$\varepsilon_p^{(\max)}$	m
51.0	950	14.0	30,000	750	0.019	3.5

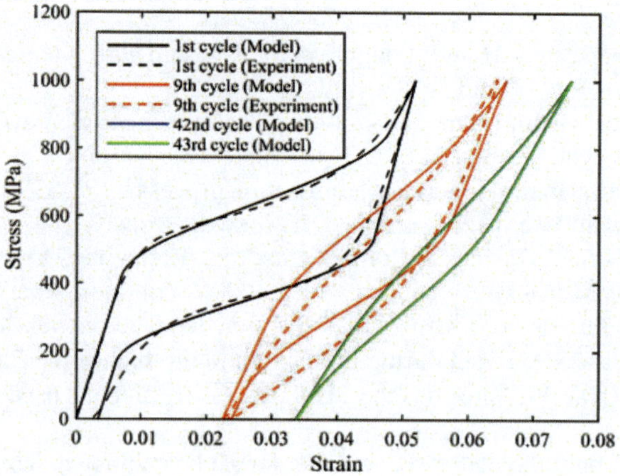

Fig. 17.8 Comparison of the obtained stress–strain responses with experimental ones (Karamooz-Ravari et al. 2018)

by further increasing, no significant changes are observed, demonstrating that the stress–strain response accumulates.

Evolution of the residual strain, peak strain, and dissipation energy during cycling loading is depicted in Fig. 17.9a–c both experimentally and numerically. The prediction of the model is well validated by the first nine experimental cycles. In addition, all the quantities converge to specific values as the number of cycles increase.

Table 17.4 shows the material parameters used for the shape memory regime cyclic loading. The test temperature is about 23 °C which is lower than M_f, so that $\xi_{s0} = 0$ and $\xi_{T0} = 1$. The prediction of the model is compared against the experimental findings for the first and eighth cycles as presented in Fig. 17.10 which shows a good agreement.

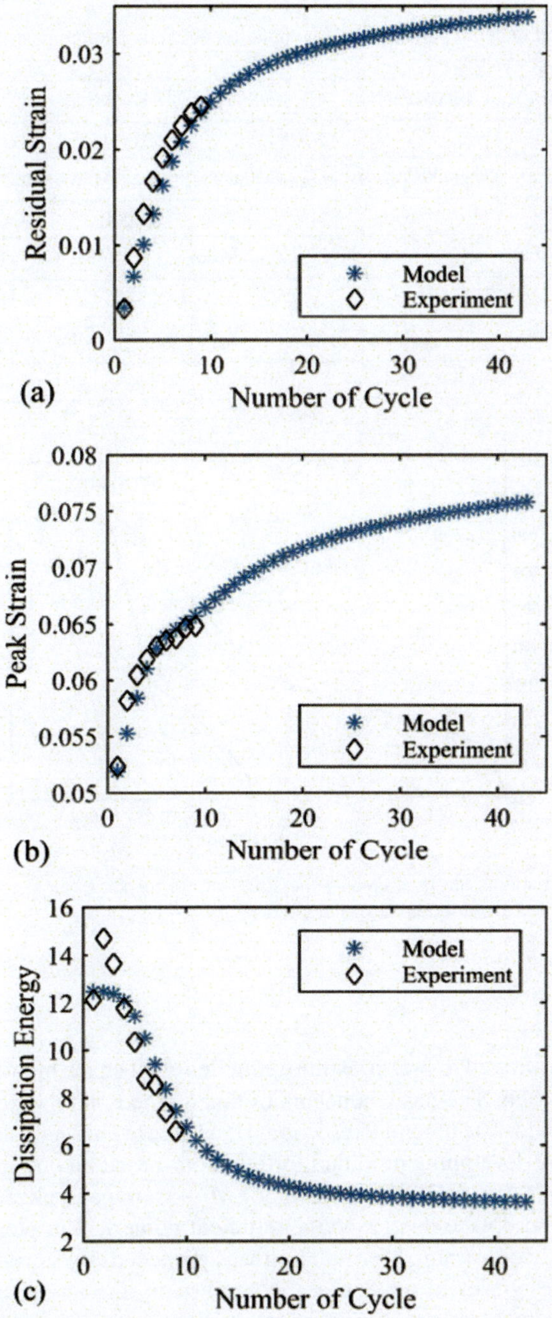

Fig. 17.9 Evolution curves for **a** residual strain **b** peak strain **c** dissipation energy versus the number of cycles (Karamooz-Ravari et al. 2018)

Table 17.4 Utilized material parameters for simulation of cyclic loading in shape memory regime (Karamooz-Ravari et al. 2018)

M_f (°C)	σ_{s0}^{cr} (MPa)	C_{M0} (MPa/°C)	C_{Af0} (MPa/°C)	E_{M0} (MPa)	σ_{ye}^{M0} (MPa)	ε_0^*
26	100	NA	NA	80,000	600	0.0315
M_s (°C)	σ_{se}^{cr} (MPa)	C_{Me} (MPa/°C)	C_{Afe} (MPa/°C)	E_{Me} (MPa)	σ_{ye}^{A0} (MPa)	ε_e^*
54.1	5	NA	NA	62,000	NA	0.015
A_s (°C)	σ_{f0}^{cr} (MPa)	C_{As0} (MPa/°C)	E_{A0} (MPa)	σ_{y0}^{M0} (MPa)	h (MPa)	μ
59	380	NA	NA	350	30,000	6.5
A_f (°C)	σ_{fe}^{cr} (MPa)	C_{Ase} (MPa/°C)	E_{Ae} (MPa)	σ_{y0}^{A0} (MPa)	$\varepsilon_p^{(max)}$	m
84.5	600	NA	NA	NA	0.054	2.8

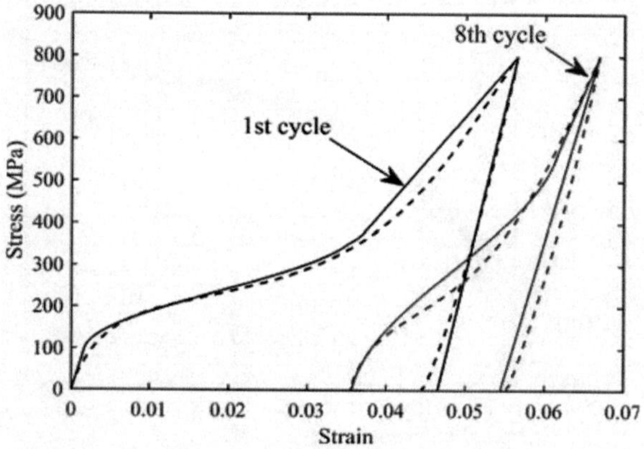

Fig. 17.10 Comparison of the predicted cyclic stress–strain responses in shape memory regime with experimental ones (Karamooz-Ravari et al. 2018)

17.3 Conclusions

This chapter was allotted to 3-D constitutive modeling of shape memory alloys using microplane approach. The basic concepts of this method and its different formulations were presented. The potential directional bias in conventional microplane theory was eliminated by developing modified formulations based on one shear component on each plane. 1-D constitutive equations of SMAs were generalized for multiaxial loadings using microplane theory, and an efficient numerical implementation technique for statically constrained formulation was proposed. By introducing different transformation responses in tension and compression, the model was then modified in such a way that tension–compression asymmetry was taken into account. The obtained results were compared against experimental findings in uniaxial tension and compression tests as well as four-point bending, and good agreements were observed. Finally, considering plastic strains as the source of residual strains upon unloading, a model for prediction of plastic and cyclic response of SMAs was developed. Some austenitic and martensitic samples were fabricated by additive manufacturing, and the specimens were examined under cyclic compressive loadings. The model was validated in cyclic loadings as well, and the developed microplane formulation was proved to be a reliable method in constitutive modeling of shape memory alloys.

The main benefit in application of microplane modeling is that 1-D responses are just required to develop 3-D constitutive equations through the presented homogenization scheme. This not only reduces complexities of constitutive modeling but also effectively assists in determining the required material parameters since only 1-D loadings need to be conducted. This technique has been utilized in many areas and can be extended for several new applications in the future. For instance, the presented approach is suitable for modeling particular classes of artificial materials, such as metamaterials, which can possess extraordinary performance and properties (Barchiesi et al. 2019; De Angelo et al. 2019; Yang et al. 2018; Alibert et al. 2003; dell'Isola et al. 2016). Shape memory alloys can be employed in the structure of such materials, and microplane theory can be applied to study various behaviors of the products.

References

Alibert, J. J., Seppecher, P., & dell'Isola, F. (2003). Truss modular beams with deformation energy depending on higher displacement gradients. *Mathematics and Mechanics of Solids, 8*(1), 51–73.

Badel, P.-B., & Leblond, J.-B. (2003). A note on integration schemes for the microplane model of the mechanical behaviour of concrete. *Communications in Numerical Methods in Engineering, 20*(1), 75–81.

Badnava, H., Mashayekhi, M., & Kadkhodaei, M. (2016). An anisotropic gradient damage model based on microplane theory. *International Journal of Damage Mechanics, 25*(3), 336–357.

Barchiesi, E., Spagnuolo, M., & Placidi, L. (2019). Mechanical metamaterials: a state of the art. *Mathematics and Mechanics of Solids, 24*(1), 212–234.

Bažant, Z. P. (1984). Microplane model for strain controlled inelastic behavior. In C. S. Desai & R. H. Gallagher (Eds.), *Mechanics of Engineering Materials* (pp. 45–59) Wiley.

Bažant, Z. P., & Di Luzio, G. (2004). Nonlocal microplane model with strain-softening yield limits. *International Journal of Solids and Structures, 41*(24–25), 7209–7240.

Bažant, P., & Oh, B. (1986). Efficient numerical integration on the surface of a sphere. *ZAMM-Journal of Applied Mathematics and Mechanics/Zeitschrift für Angewandte Mathematik und Mechanik, 66*(1), 37–49.

Bažant, Z. P., & Prat, P. C. (1987). Creep of anisotropic clay: New microplane model. *Journal of Engineering Mechanics, 113*(7), 1050–1064.

Bažant, Z. P., & Prat, C. P. (1988a). Microplane model for brittle-plastic material: I. Theory. *Journal of Engineering Mechanics, 114*(10), 1672–1688.

Bažant, Z., & Prat, P. (1988b). Microplane model for brittle plastic material: II. Verification. *Journal of Engineering Mechanics, 114*(10), 1689–1699.

Bažant, Z. P., Xiang, Y., & Prat, P. C. (1996). Microplane model for concrete. I: Stress-strain boundaries and finite strain. *Journal of Engineering Mechanics, 122*(3), 245–254.

Bažant, Z. P., & Zi, G. (2003). Microplane constitutive model for porous isotropic rocks. *International Journal for Numerical and Analytical Methods in Geomechanics, 27*(1), 25–47.

Bažant, Z. P., et al. (2000a). Fracturing rate effect and creep in microplane model for dynamics. *Journal of Engineering Mechanics, 126*(9), 962–970.

Bažant, Z. P., et al. (2000b). Large-strain generalization of microplane model for concrete and application. *Journal of engineering mechanics, 126*(9), 971–980.

Brinson, L. C. (1993). One-dimensional constitutive behavior of shape memory alloys: Thermomechanical derivation with non-constant material functions and redefined martensite internal variable. *Journal of Intelligent Material Systems and Structures, 4*(2), 229–242.

Brocca, M., & Bažant, Z. P. (2000). Microplane model and metal plasticity. *ASME Applied Mechanics Reviews, 53*(10), 265–281.

Brocca, M., Brinson, L., & Bažant, Z. (2002). Three-dimensional constitutive model for shape memory alloys based on microplane model. *Journal of the Mechanics and Physics of Solids, 50*(5), 1051–1077.

Caner, F. C., & Carol, I. (2006). Microplane constitutive model and computational framework for blood vessel tissue. *Journal of Biomechanical Engineering, 128*(3), 419–427.

Caner, F. C., de Carlos Blasco, V., & Egido, M. G. (2019). Microplane models for elasticity and inelasticity of engineering materials. In *Handbook of nonlocal continuum mechanics for materials and structures* (pp. 1065–1097). Springer.

Caner, F. C., et al. (2007). Hyperelastic anisotropic microplane constitutive model for annulus fibrosus. *Journal of Biomechanical Engineering, 129*(5), 632–641.

Carol, I., & Bazant, Z. P. (1997). Damage and plasticity in microplane theory. *International Journal of Solids and Structures, 34*(29), 3807–3835.

Carol, I., Bažant, Z. P., & Prat, P. C. (1991). Geometric damage tensor based on microplane model. *Journal of engineering mechanics, 117*(10), 2429–2448.

Carol, I., Jirásek, M., & Bažant, Z. P. (2004). A framework for microplane models at large strain, with application to hyperelasticity. *International Journal of Solids and Structures, 41*(2), 511–557.

Chang, K.-T., & Sture, S. (2006). Microplane modeling of sand behavior under non-proportional loading. *Computers and Geotechnics, 33*(3), 177–187.

Chatti, M., et al. (2019). Modelling of the viscoelastic behaviour with damage induced anisotropy of a plastic-bonded explosive based on the microplane approach. *International Journal of Solids and Structures, 168,* 13–25.

De Angelo, M., Spagnuolo, M., D'Annibale, F., Pfaff, A., Hoschke, K., Misra, A., et al. (2019). The macroscopic behavior of pantographic sheets depends mainly on their microstructure: experimental evidence and qualitative analysis of damage in metallic specimens. *Continuum Mechanics and Thermodynamics, 31*(4), 1181–1203.

Dean, A., et al. (2016). A thermodynamically consistent framework to couple damage and plasticity microplane-based formulations for fracture modeling: development and algorithmic treatment. *International Journal of Fracture, 203*(1–2), 115–134.

dell'Isola, F., Giorgio, I., Pawlikowski, M., & Rizzi, N. L. (2016). Large deformations of planar extensible beams and pantographic lattices: heuristic homogenization, experimental and numerical examples of equilibrium. *Proceedings of the Royal Society A: Mathematical, Physical and Engineering Sciences, 472*(2185), 20150790.

Etse, G., Nieto, M., & Steinmann, P. (2003). A micropolar microplane theory. *International Journal of Engineering Science, 41*(13–14), 1631–1648.

Huang, L. C., et al. (2017). Numerical aspects of microplane constitutive models for concrete. *Applied Mathematical Modelling, 41,* 530–548.

Indriyantho, B. R., Zreid, I., & Kaliske, M. (2019). Finite strain extension of a gradient enhanced microplane damage model for concrete at static and dynamic loading. *Engineering Fracture Mechanics, 216.*

Jalalpour, A., et al. (2019). *On the effects of numerical integration on accuracy of microplane modeling of shape memory alloys, 24ème Congrès Français de Mécanique.* France: Brest.

Jin, C., et al. (2016). Elastic microplane formulation for transversely isotropic materials. *Journal of Applied Mechanics, 84*(1).

Kadkhodaei, M., et al. (2007a). Modeling of shape memory alloys based on microplane theory. *Journal of Intelligent Material Systems and Structures, 19*(5), 541–550.

Kadkhodaei, M., et al. (2007b). Microplane modelling of shape memory alloys. *Physica Scripta, T129*, 329–334.

Karamooz Ravari, M. R., Kadkhodaei, M., & Ghaei, A. (2015). A microplane constitutive model for shape memory alloys considering tension–compression asymmetry. *Smart Materials and Structures, 24*(7).

Karamooz-Ravari, M. R., & Shahriari, B. (2017). Numerical implementation of the microplane constitutive model for shape memory alloys. In *Proceedings of the Institution of Mechanical Engineers, Part L: Journal of Materials: Design and Applications.*

Karamooz-Ravari, M. R., et al. (2018). Modeling the cyclic shape memory and superelasticity of selective laser melting fabricated NiTi. *International Journal of Mechanical Sciences, 138–139*, 54–61.

Kirane, K., Su, Y., & Bažant, Z. P. (2015). *Strain-rate-dependent microplane model for high-rate comminution of concrete under impact based on kinetic energy release theory. Proceedings of the Royal Society A: Mathematical, Physical and Engineering Sciences, 471*(2182).

Kuhl, E., et al. (2001). Microplane modelling and particle modelling of cohesive-frictional materials. In P. A. Vermeer, et al. (Eds.) *Continuous and Discontinuous Modelling of Cohesive-Frictional Materials* (pp. 31–46). Berlin, Heidelberg: Springer Berlin Heidelberg.

Kuhl, E., & Ramm, E. (1998). On the linearization of the microplane model. *Mechanics of Cohesive-frictional Materials, 3*, 343–364.

Leukart, M., & Ramm, E. (2003). A comparison of damage models formulated on different material scales. *Computational Materials Science, 28*(3–4), 749–762.

Leukart, M., & Ramm, E. (2006). Identification and Interpretation of Microplane Material Laws. *Journal of Engineering Mechanics, 132*(3), 295–305.

Li, C., et al. (2017). Spherocylindrical microplane constitutive model for shale and other anisotropic rocks. *Journal of the Mechanics and Physics of Solids, 103*, 155–178.

Mehrabi, R., & Kadkhodaei, M. (2013). 3D phenomenological constitutive modeling of shape memory alloys based on microplane theory. *Smart Materials and Structures, 22*(2).

Mehrabi, R., Kadkhodaei, M., & Chirani, S. A. (2012). Microplane modeling of martensite reorientation in shape memory alloys. In *European Congress on Computational Methods in Applied Sciences and Engineering*. Vienna, Austria: ECCOMAS.

Mehrabi, R., Kadkhodaei, M., & Elahinia, M. (2014a). A thermodynamically-consistent microplane model for shape memory alloys. *International Journal of Solids and Structures, 51*(14), 2666–2675.

Mehrabi, R., Kadkhodaei, M., & Ghaei, A. (2012b). Numerical Implementation of a Thermomechanical Constitutive Model for Shape Memory Alloys Using Return Mapping Algorithm and Microplane Theory. *Advanced Materials Research, 516–517*, 351–354.

Mehrabi, R., et al. (2014b). Microplane modeling of shape memory alloy tubes under tension, torsion, and proportional tension–torsion loading. *Journal of Intelligent Material Systems and Structures, 26*(2), 144–155.

Němeček, J., et al. (2002). Microplane models: computational aspects and proposed parallel algorithm. *Computers & Structures, 80*(27–30), 2099–2108.

Ožbolt, J., Li, Y., & Kožar, I. (2001). Microplane model for concrete with relaxed kinematic constraint. *International Journal of Solids and Structures, 38*(16), 2683–2711.

Poorasadion, S., et al. (2013). An improvement on the Brinson model for shape memory alloys with application to two-dimensional beam element. *Journal of Intelligent Material Systems and Structures, 25*(15), 1905–1920.

Prat, P. C., & Bažant, Z. P. (1991). Microplane model for triaxial deformation of saturated cohesive soils. *Journal of Geotechnical Engineering, 117*(6), 891–912.

Qidwai, M., & Lagoudas, D. (2000). On thermomechanics and transformation surfaces of polycrystalline NiTi shape memory alloy material. *International Journal of Plasticity, 16*(10), 1309–1343.

Reedlunn, B., et al. (2014). Tension, compression, and bending of superelastic shape memory alloy tubes. *Journal of the Mechanics and Physics of Solids, 63*, 506–537.

Salviato, M., Ashari, S. E., & Cusatis, G. (2016). Spectral stiffness microplane model for damage and fracture of textile composites. *Composite Structures, 137*, 170–184.

Steinke, C., Zreid, I., & Kaliske, M. (2019) Modelling of nonlinear concrete behavior-microplane and phase-field approaches. In *7th International Conference on Structural Engineering, Mechanics and Computation*. Cape Town, South Africa.

Taylor, G. I. (1938). Plastic strain in metals. *Journal of Institute of Metals, 62*, 307–324.

Verron, E. (2015). Questioning numerical integration methods for microsphere (and microplane) constitutive equations. *Mechanics of Materials, 89*, 216–228.

Vrech, S., Etse, G., & Caggiano, A. (2016). Thermodynamically consistent elasto-plastic microplane formulation for fiber reinforced concrete. *International Journal of Solids and Structures, 81*, 337–349.

Yang, H., Ganzosch, G., Giorgio, I., & Abali, B. E. (2018). Material characterization and computations of a polymeric metamaterial with a pantographic substructure. *Zeitschrift für Angewandte Mathematik und Physik, 69*(4), 105.

Zhou, T., et al. (2019). A new microplane model for non-proportionally multiaxial deformation of shape memory alloys addressing both the martensite transformation and reorientation. *International Journal of Mechanical Sciences, 152*, 63–80.

Zreid, I., & Kaliske, M. (2016). An implicit gradient formulation for microplane Drucker-Prager plasticity. *International Journal of Plasticity, 83*, 252–272.

Chapter 18
A Plausible Description of Continuum Material Behavior Derived by Swarm Robot Flocking Rules

Ramiro dell'Erba

Abstract In this chapter, we are considering a material continuum, discretized as two-dimensional lattice of particles, undergone a prefixed strain of some its parts, and we calculate its time evolution without using Newton's laws but using position-based dynamics rules. This means that the new position of a particle is determined by the spatial position of its neighbors without defining forces. The aim of the model is to reproduce the behavior of deformable bodies with standard or generalized (Cauchy or second gradient) deformation energy density. The tool that we have realized gives a plausible simulation of continuum deformation also in fracture case. It can be useful to describe final and sometime intermediate configuration of a continuum material under assigned strain of some of its points; the advantages are in saving computational time, with respect to solving classical differential equation. It is very flexible to be adapted for complex geometry samples. The numerical results suggest that the system can effectively reproduce the behavior of first and second gradient continua. We checked coherence with the principle of Saint Venant, and it is able to manage complex effects like lateral contraction, anisotropy or elastoplasticity. Its origin lies in our experience in evolution and control of robotic swarm; for a swarm robotics, just as for an animal swarm in Nature, one of the aims is to reach and maintain a desired geometric configuration. One of the possibilities to achieve this result is to see what its neighbors are doing. This approach generates a rules system governing the movement of the single robot just by reference to neighbor's motion that we have used to describe the continuum deformation. Many aspects have to be still investigated, like the relationships describing the interaction rules between particles and constitutive equations and some results, like beam under shear stress, do not sound very good.

Keywords Continuum mechanics · Swarm robotics · Flocking rules

---

R. dell'Erba (✉)
ENEA Technical Unit Technologies for Energy and Industry—Robotics Laboratory, Rome, Italy
e-mail: ramiro.dellerba@enea.it

F. dell'Isola and L. Igumnov (eds.), *Dynamics, Strength of Materials and Durability in Multiscale Mechanics*, Advanced Structured Materials 137,
https://doi.org/10.1007/978-3-030-53755-5_18

18.1 Introduction

It is well known that time evolution of a material particles system is determined by Newton's dynamics laws; however, in recent years, especially with the evolution of computer graphics driven by videogames applications, there has been great interest in studying the evolution of a particle system whose motion is simply determined by their relative position in a frame, without solving the differential equations of dynamics. The name of this method is position-based dynamics (PBD) (Bender et al. 2015; Bender et al. 2014). This method does not determine forces and solve differential equations but use a position-based approach, where the new position of a particle is determined by its neighbor's positions and can be easily be used to describe complex object behavior. The physically based simulation of deformable solids has been an active research topic of computer graphics since many years: The aim is to simulate the behavior of real materials to achieve graphically realistic results. The credibility requirements of user interfaces in videogames have generated a technology, user interface physic, where physical principles are partially enforced through ad hoc heuristics assumptions and are often implemented without the usual calculus. The PBD methods result in a physically plausible behavior of the continuum but suffer from limitations, modeling complex material properties and describing interactions between heterogeneous bodies (Macklin et al. 2016). But this approach has many practical applications. For example, a touch screen phone contact list can be scrolled, by fingers, with a motion based on velocity and list length. Reaching the end of the list, the motion will bounce as if a collision occurred. The user feels such behavior very realistic even if the effects are heuristically reproduced and are not a solution of Newton's law. Aim of the PBD is not to compute physical process but to generate visually plausible simulation results with low computational cost (Bender et al. 2014), sacrificing some accuracy, with respect of solution of heavy equations by finite element methods (FEM). Some other disadvantages of PBD include low fidelity, poor adaptability and low interactivity. Therefore, sometimes, a physics engine, working through integration techniques that are based on Newton's laws of motion, is added. The advantage of using PBD is in computational simplicity and time machine leading to results similar to those obtainable with FEM but in a much shorter time; it also provides a useful point of view that could be able to help in understanding what features are important in the deformation without solving differential equations. They are fast, robust, simple, efficient and easily configurable. In the beginning, simulations for videogame applications widely used continuum mechanical methods, solving FEM. Classical methods are based on discretization (Lagrangian or Eulerian) of Newton's second law and formulate forces for each mechanical effect. To obtain robust simulations, very small time steps are required by these methods; therefore, they cannot be used in interactive situations owing to the large machine time used. In spite of this, still now, the first approach to simulate deformable objects by continuum mechanics is to discretize equations and to solve them using numerical integration. This can be done in several ways, but many of them are affected by stability problems, arising from stiff differential equations and can be managed using very small time steps, resulting in high computational costs.

In the meantime, graphic processing unit (GPU)-based solvers and adaptive meshes were growing, and PBD methods became popular. Knowledge of traditional forces is avoided in favor of position displacements; the problem is, therefore, transformed in a geometric constraint between configurations. The final positions of the particles are determined by constrained matrix transformation. Many solutions (Rivers and James 2007; Diziol et al. 2011) have been proposed to enhance the efficiency of the method that, owing to the matrix nature, can be parallelized between the cores of GPU to reduce calculation time.

Basically, in PBD, the displacement of a particle (discretized continuum) is determined by the position of its neighbors. Therefore, the compute of new position for a particle set can be considered as a constrained geometrical problem leading to a transformation operator between the matrices representing the particles configuration, C_t, for a discrete set of time steps t_1, t_2, … t_n. It should be noted as most of PBD methods hide the dynamics inside their relationship; moreover, they ask for the knowledge of the velocity of the particles that we try to avoid. One of the differences, between our methods and PBD, lies in the use of the velocity of the particles often used in PBD; this, in our opinion, hides the dynamic inside so, up to now we avoid using them.

In our approach, we try to combine the advantages of both the continuum mechanical and the position-based approaches to describe complex physical phenomena, trying to keep the simulation easy to implement and customizable. We have written a complete customizable and modular algorithm easily expandable to every new feature we would like to introduce. One of the main advantages of the proposed algorithm is the fact that it automatically takes into account large deformation elasticity, which is a topic having an increasing role in today's research (Ladevèze 2012; Steigmann 2009, 2010, 2013; Dell'Isola et al. 2016; Della Corte et al. 2017; Gabriele et al. 2014). The aim of the proposed model was to develop a suitable more general numerical tool capable of modeling the behavior of deformable bodies and to take into account higher gradient constituent relations.

The approach tries to combine continuum mechanical material models with a position-based method using an explicit time integration scheme to manage complex physical effects like isotropic and anisotropic elastic behavior as well as the effects of lateral contraction. To reply behaviors, described by constitutive equations of the materials (e.g., Poisson's effect), we introduce geometric constraint on the lattice and rules ad hoc on the displacements of the points.

Since finite element method (FEM) is a reliable and well-known numerical approach for both classical and generalized continua (see Greco and Cuomo 2013; Ern and Guermond 2013; Greco and Cuomo 2014; Contrafatto et al. 2012; dell'Isola et al. 2015; Cuomo and Greco 2012; Cazzani et al. 2016; Bilotta and Turco 2009 for applications), we compared the results with the corresponding classic mechanical continuum case, whose equations had been solved by FEM simulations with good agreement.

Finite element analysis is a well-established method, but these numerical techniques are usually computationally expensive. The proposed algorithm offers the advantage of limited computational costs. Since the algorithm is based on a linear operation, that is, the computation of the barycenter, its computational cost increases only linearly with the number of particles in the system. In (dell'Erba 2018a, b), it is shown how the model may exhibit a rich range of behaviors, such as asymptotic convergence to the equilibrium configuration, instabilities of various kinds and, in well-determined circumstances, spontaneous growth of the crack length after an almost steady state. So far we have tried to describe the deformation of a continuum medium by this tool useful for complex microstructures not easily analyzed by Cauchy continuum theory generating and generating big quantity of experimental data. Classical Cauchy continua are not able to give accuracy prediction in highly non-homogeneous microstructure, though generalizations have to be introduced either considering additional degrees of freedom, to account for the kinematics at the level of the microstructure (Seddik et al. 2008; Pietraszkiewicz and Eremeyev 2009; Altenbach et al. 2010, 2013; Eremeyev et al. 2012; Altenbach et al. 2010), or including in the deformation energy density higher gradients of the displacement than the first one (Abali et al. 2017; Cuomo et al. 2017; Turco et al. 2016; Dell'Isola et al. 2016, 2015; Javili et al. 2013; Seppecher et al. 2011; Forest et al. 2011; Placidi 2015; Rosi et al. 2013). The latter is a particularly relevant topic if you consider the technological interest in developing exotic mechanical metamaterials able to perform targeted tasks (Dell'Isola et al. 2015; Bückmann 2012; Dell'Isola 2019; Barchiesi et al. 2018; Carcaterra et al. 2015; Turco et al. 2017; Dell'Isola et al. 2016; Milton and Seppecher 2012); therefore, the investigation of new and efficient algorithms is of great interest at the moment.

This approach seemed particularly promising considering the emerging role of microstructured continua, manufactured with computer-aided methods, as a technological resource (see Altenbach et al. 2010; Eremeyev et al. 2012; Altenbach et al. 2013; Dell'Isola et al. 2015; Placidi et al. 2017; Placidi et al. 2010; Altenbach and Eremeyev 2009; Altenbach and Eremeyev 2013; Eremeyev and Pietraszkiewicz 2012; Forest 2009) because the presence of a complex microstructure often leads to macroscopic behaviors that require generalized continua for their accurate modeling (see Abali et al. 2017; Cuomo et al. 2017; Turco et al. 2016; Dell'Isola et al. 2015, 2016; Javili et al. 2013; Forest et al. 2011; Placidi 2015; Rosi et al. 2013; Alibert et al. 2003) for more details and (Andreaus and Placidi 2013) for a historical survey on the subject). For interesting results in nth gradient theory, the reader can see (dell'Isola et al. 2015; Javili et al. 2013; Rosi et al. 2013; Carcaterra et al. 2015; Alibert et al. 2003; Andreaus and Placidi 2013; Madeo et al. 2013; Madeo et al. 2008; Dell'Isola et al. 2011; Auffray et al. 2015; Dell'Isola et al. 2009; Dell'Isola and Seppecher 1995; Dell'isola and Seppecher 1997; dell'Isola et al. 2012; Alibert and Della Corte 2015; Placidi et al. 2013); on the importance of this topic, see (Placidi et al. 2017), where a general overview of recent results is provided. Note that nth gradient theories can be contextualized in a more general framework of micromorphic/microstructured continua, which is a very active research field (see, for instance, Eringen 2012; Germain 1973; Mindlin 1964) for classical references (Seddik et al.

2008; Placidi et al. 2010; Andreaus et al. 2015; Chang and Misra 1990; Masiani et al. 1995; Cecchi and Rizzi 2001; Goda et al. 2014; Placidi et al. 2004; Scerrato et al. 2015; Dos Reis and Ganghoffer 2011; Placidi and Hutter 2006; Giorgio et al. 2015; Enakoutsa et al. 2016) for interesting applications (Altenbach et al. 2010; Eremeyev et al. 2006; Goda et al. 2012; Eremeyev et al. 2007; Pietraszkiewicz and Eremeyev 2009; Berezovski et al. 2016) for recent theoretical results and (Dell'Isola et al. 2015) for a recent review. The main reason of the interest in these theoretical models can be explained because they have been useful in mathematical description of objects whose richness at the microscale cannot be captured by classical continuum models, that is, metamaterials (see, e.g., (Dell'Isola et al. 2015; Del Vescovo and Giorgio 2014) for reviews of recent results and (Seppecher et al. 2011; Dell'Isola et al. 2016; d'Agostino et al. 2015; Eremeyev et al. 2010; Madeo et al. 2014; Steigmann and Pipkin 1991; Steigmann 2008 for interesting examples). The development of new techniques, such as three-dimensional (3D) printing or electro-spinning, gives the possibility to obtain increasingly complex and exotic microstructures, which can provide a reasonably experimental basis. On the other hand, the amount of new experimental data opens several deep and complex theoretical problems. It is clear that, in this context, numerical tools are essential in order to have a suitable mediation between theoretical and experimental results. In particular, in our opinion, a numerical investigation should be a good compromise between computational cost and accuracy of the results, as is required in rapid prototyping processes typical of modern technological research.

18.2 The Origin of the Problem

Systems control of a robotic swarm is often derived from Nature teaching, like fish school behavior; in our laboratory, we are working on underwater robotic swarm, and on topics close to this, since many years (see Fig. 18.1). In (dell'Erba 2015; Moriconi and dell'Erba 2012), the author was investigating the calculation of the geometric configuration of submarine swarm robots by the single elements; this is very important because the swarm, like school fish, adapt its configuration depending on the mission assigned. The concept of robot swarms has been a study theme, for the scientific community, for several years. Swarm research has been inspired by biological behaviors, like those of bees (Karaboga 2005; Passino et al. 2007; Janson et al. 2005) for a long time to take advantage by social activities concepts (Khatib et al. 2008) labor division, task cooperation and information sharing. A single-robot approach is affected by failures that may prevent the success of the whole task. On the contrary, a multi-robot approach can benefit from the parallelism of the operation and by the redundancy given by the usage of multiple agents. Moreover, the operator has the possibility to have multiple views simultaneously and to follow pattern by gradient techniques. In a swarm, the members operate with a common objective, sharing the job workload; the lack of one member can be easily managed by redistributing the job among the others. This feature is especially useful if we

Fig. 18.1 VENUS, element of the swarm realized in our laboratory

consider application as discovery and surveillance of a submarine area. A swarm can be considered as a single body, offering the advantage of a simple way of interfacing with the human end-users and overcoming the problem of the control of a large number of individuals. In the swarm, there is no central brain, mainly because of the excess needs in band pass requested by such a brain. Instead, each individual must possess an intelligent local control system capable of managing its choices according to that of the neighbors on the basis of the available data. Data coherence along the swarm, being affected by the position of the member and by the data propagation speed, is also a research topic. What makes swarms interesting is their capability to fill and control large volume of water by means of a network of cooperating sensors and their capability to move in the most interesting zones, increasing density where a major need is required. The member's geometrical distribution is flexible and adaptable to the task and environmental characteristics. As an example, if priority is to maximize exploration volume, the swarm has to maintain a great spatial dispersion and communication band pass could be slowed down; conversely, if the priority is around the risk that the amount of exchanged information becomes inadequate to ensure the correct behavior of the multi-body, the system itself can physically react by changing geometry despite of the drop in performances for the assigned task. For this reason, it is of primary importance that a single element of the swarm knows, at least locally, its configuration and can move to reach the desired one. Like birds in nature, the element of the swarm can decide its movements according to what its neighbors are doing. To this end, a positioning and control algorithm has been developed so that it reaches the desired configuration. It was then noted that a quite similar algorithm could adapt to describing PBD problems, because the movement was quite similar to deformation of a viscoelastic body. Therefore, starting from the flocking rules governing the behavior of single elements in underwater robotic swarm to reach an assigned geometric configuration, we have adapted the control algorithm in PBD problems (dell'Erba 2018a, c). Like the robot swarm behavior (dell'Erba 2015; Dell'Erba and Moriconi 2014), it does not determine forces and solve differential equations but use a position-based approach, where the new position of a particle is determined by its neighbor's positions and can be easily be used to describe complex objects.

18.3 The Algorithm

The two-dimensional continuum bodies are discretized into a finite number of particles occupying, in their initial configuration, the nodes of a lattice. The kind of lattice is chosen between the five-plane Bravais lattices as you can see in Fig. 18.2 but sometime could be useful to use also honeycomb lattice. This is the first choice we have

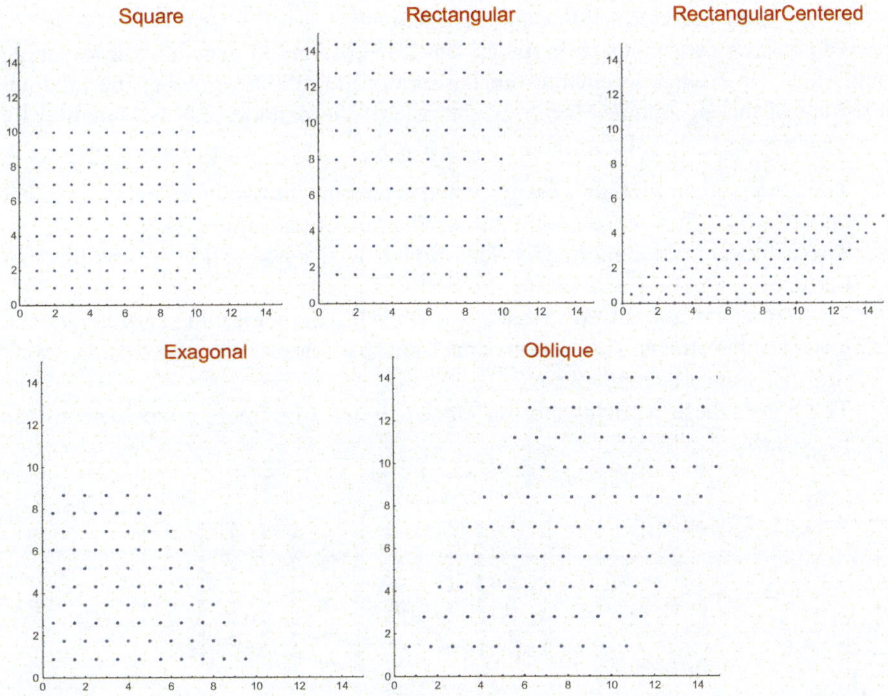

Fig. 18.2 Bravais plane lattice

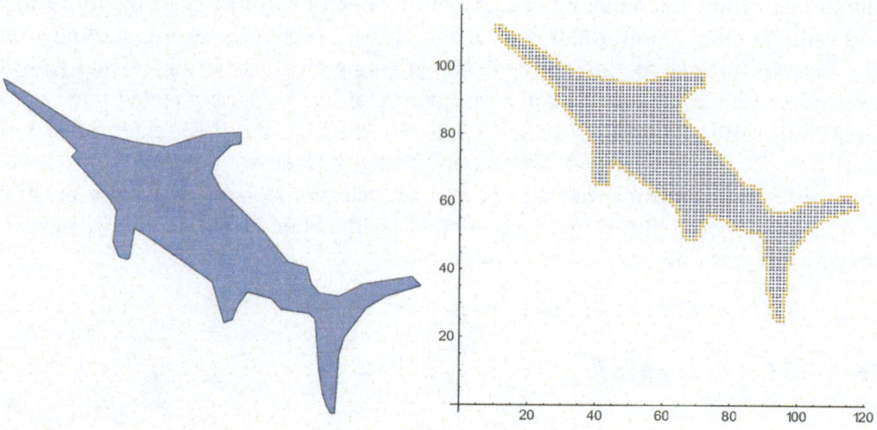

Fig. 18.3 Example of discretized object in the chosen lattice with frame (yellow points). Square lattice has been used

to do (Choice 1); changing lattice, and other parameters we shall see later, as we can obtain different results with the same initial conditions.

In Fig. 18.3, a discretization example is shown.

We consider four kinds of particles, but it is possible to generalize introducing new kind of particles, to enlarge our description possibilities, owing the modular structure of the algorithm. Moreover, the role of the particles can be dynamically changed during body's deformation. They are:

1. The leaders; their motion is assigned and determine the displacement of the other particles (i.e., they represent the imposed strain of part of the body).
2. The followers; their motion is determined by the interaction rule with other particles.
3. The frame; it is introduced so that any particle has the same number of neighbors, to avoid edge effects. The motion of the frame particles is determined by the frame rule (see Figs. 18.4 and 18.5).
4. The ghost; these particles are introduced to describe fracture mechanism (see Fig. 18.6).

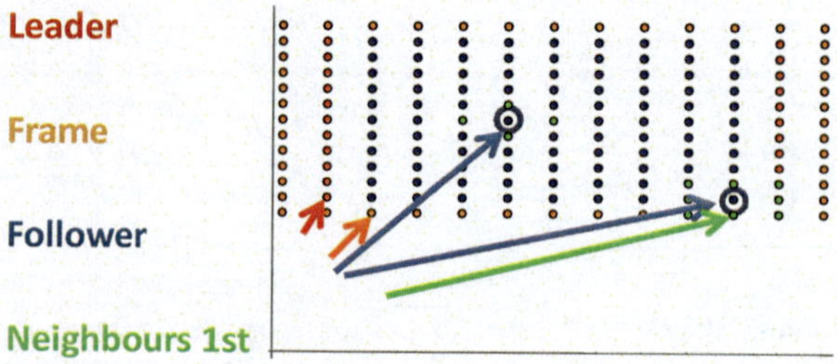

Fig. 18.4 Kind of particles

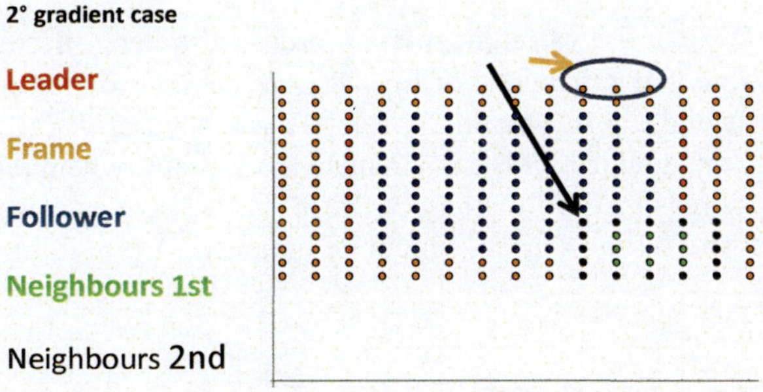

Fig. 18.5 Kind of particles (2° gradient case)

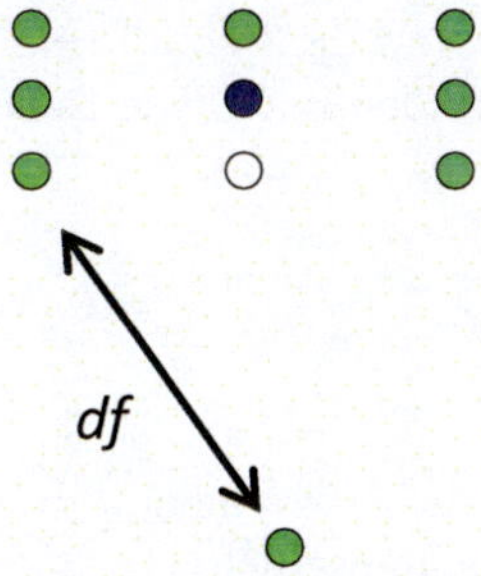

Fig. 18.6 Fracture mechanism: ghost

The displacements of the leaders are assigned so do not need any explanation.

How we determine the displacement of the followers? First we have to choice the neighbors of any particles (Choice 2). Typically, we used the first *nc* particles, where *nc* is the coordination number of the lattice. This is the case of first gradient theory; but we can choose to use a larger set of neighbors, like the neighbors of the neighbors, and this is the second gradient theory case. So far we enlarge the set of points with a supplementary shell, and this can be generalized to nth-order interaction (see Figs. 18.4 and 18.5).

Later, we have to choice the interacting rule (Choice 3) between the particles. The rule describes the position of a particle as function of the neighbor's positions. As example, we can decide to use the center of gravity rule where the new x coordinate of the particle j is

$$x_j(t) = \frac{\sum_{k=1}^{\text{all neighbours of } j} x_k(t)}{N} \tag{18.1}$$

where N is the total number of neighbors; similar equation can be used for the y coordinate. By this way, displacement of a follower point is the average value of the displacements of its neighbors; the number of shells determines the order of interaction. We can use different rules in order to imitate different constitutive equations. Possible generalizations of Eq. 18.1 are geometric, power and weighted mean. Possible weight is the particles' Euclidean distances *dis*(k, j) between the particles k and its neighbor j. This can simulate Hook's law, where recalling force is increased with increasing deformation. By Eq. 18.1, we note as x and y coordinates are independent so Poisson's effect cannot be obtained. A possibility to obtain it is to use

$$y_j(t) = K * \left(x_j(t) - x_j(t_0)\right) * da + \frac{\sum_{k=1}^{allneighboursofj} y_k(t)}{N} \tag{18.2}$$

where *da* is a function of the distance from the central axis, K a parameter determining the response force and $x(t_0)$ the x coordinate at time t_0. So far expansion of x coordinate has effect on the y coordinate. The Euclidean distance, *dis*(k,j), can be used as weight.

$$x_j(t) = \frac{\sum_{k=1}^{allneighboursofj} dis(k, j) x_k(t)}{\sum_{k=1}^{allneighboursofj} dis(k, j)} \tag{18.3}$$

We can also force the follower's movement to overcome the barycenter equilibrium position, leading the lattice to oscillate.

$$\begin{aligned} x_j(t) = {} & \frac{\sum_{k=1}^{\text{all neighbours of } j} w(k, j) x_k(t)}{\sum_{k=1}^{\text{all neighbours of } j} w(k, j)} \\ & + fd\left(\frac{\sum_{k=1}^{\text{all neighbours of } j} w(k, j) x_k(t)}{\sum_{k=1}^{\text{all neighbours of } j} w(k, j)} - MT(j, t_0)\right) \end{aligned} \tag{18.4}$$

where $w(k,j)$ is the weight, *fd* is a feedback factor and $MT(i,t_0)$ is the x coordinate of j point at t_0 to have memory of the initial configuration.

The compute of new position for a particle set can be considered as a constrained geometrical problem using a transformation operator between the matrices describing particles configuration, C_t, for a discrete set of time steps $t_1, t_2, \ldots t_n \ldots$.

In our algorithm, the neighbors can dynamically change at every time step. Actually, we choice to fix the neighbors of every particle at the initial time t_0, and not to change them during time evolution of the configurations; this has the mean to consider a crystalline lattice and therefore to deal with solid phase materials. The concept of neighbors is Lagrangian, and neighborhood is preserved during the time evolution of the system—the only exceptions arising with the fracture algorithm, as shown later. Also the definition of neighbors is customizable by changing metric; for example, we can consider points whose Euclidean distance (weighted or not is another possibility to take into account anisotropies) is less than a threshold, instead of the coordination number of the lattice.

Starting from the leader's motion, each time step the displacement propagates of one shell, determined by the neighbors up to involve all the particles.

To avoid edge effects, we build a frame surrounding the body by an external shell of point, so that any follower interacts with the same number of elements. Our objective is homogeneity of the boundary conditions for all the followers. Without the frame, a corner point has fewer neighbors, with respect to an internal point; so far if its coordinates are determined, as example, by barycenter of its neighbors, this point will be attracted toward the inside and the lattice and will collapse on the other points.

The motion of the frame is simple: It only has to follow the motion of an assigned follower of its competence; in case the assigned followers are more (i.e., in a corner), then an average displacement, or a more generic complex rule (Choice 4), is considered as can be seen in Fig. 18.5. In this case, you have more than a possibility, and the frame can be something more complex than a single shell; as an example, if we are considering second gradient interaction, we need a double shell to reach. Later we shall see as in hexagonal lattice you can choose more than one kind of frame, and the obtained results are completely different.

The process stops when all the elements of the system have moved, and then restarts at every following time step (for a more detailed description the reader is referred to dell'Erba 2018a, c).

The model exhibits pronounced nonlinear behaviors, as shown in (Dell'isola and Seppecher 1997), since composing motions for the leader does not lead to a simple superposition of effects in the configuration of the system.

To manage fracture phenomena, we assume the interactions are decreasing with increasing distance between particles. Therefore, when Euclidean distance between points is "great," they lose their interaction. To address the problem, we start simply considering a threshold effect between neighbor elements, so that when the distance overcomes the threshold, these elements stop to influence each other so they are no longer taken into account in the calculation of the follower position. To preserve symmetry of the Lagrangian neighbors, we introduce ghost points with the purpose of balancing the calculations of the point's displacements, just to balance the equations. They have the purpose of balancing the calculations of the point's displacements. Where are these ghost elements posed? Typical position, where we put ghost points (Choice 6), is that is able to recover the original shape of the lattice (see Fig. 18.6). Anyway other choices lead to different results. All the properties of these ghost elements are the same of the followers, but their motion is not considered, because they are not in the list of the followers. They are just in the right position to balance the cell. As we have seen (Battista et al. 2016), a change in their position produces effects such as the contraction or loosening of the lattice in the deformed configuration. In fact, varying the distances of the ghost elements after fracture from the true elements, plastic-like and elastic-like behaviors can be obtained. As elastic behavior in fracture, we mean the property of the fracture edges or of the disconnected pieces originated after fracture has occurred, to recover its original shape. The algorithm can be easily generalized to second gradient by introducing two different thresholds for the two shells of neighbors.

Practically, you have to decide constraints of the lattice and the interaction rules between the followers, in order to describe the correct behavior of the constitutive equations of the materials.

Reassuming the process is the following (seethe flow chart in Fig. 18.7). We choose a two-dimensional body and one of the Bravais lattice and discretize it to obtain a discrete matrix to represent it. We now decide the constraints of the lattice and the interaction rules between the followers, in order to describe the correct behavior of the constitutive equations of the materials. As an example, we can decide that the lattice has no constraints and displacement of a follower point is the average value of the displacements of its first neighbors (first gradient). We build an adequate frame

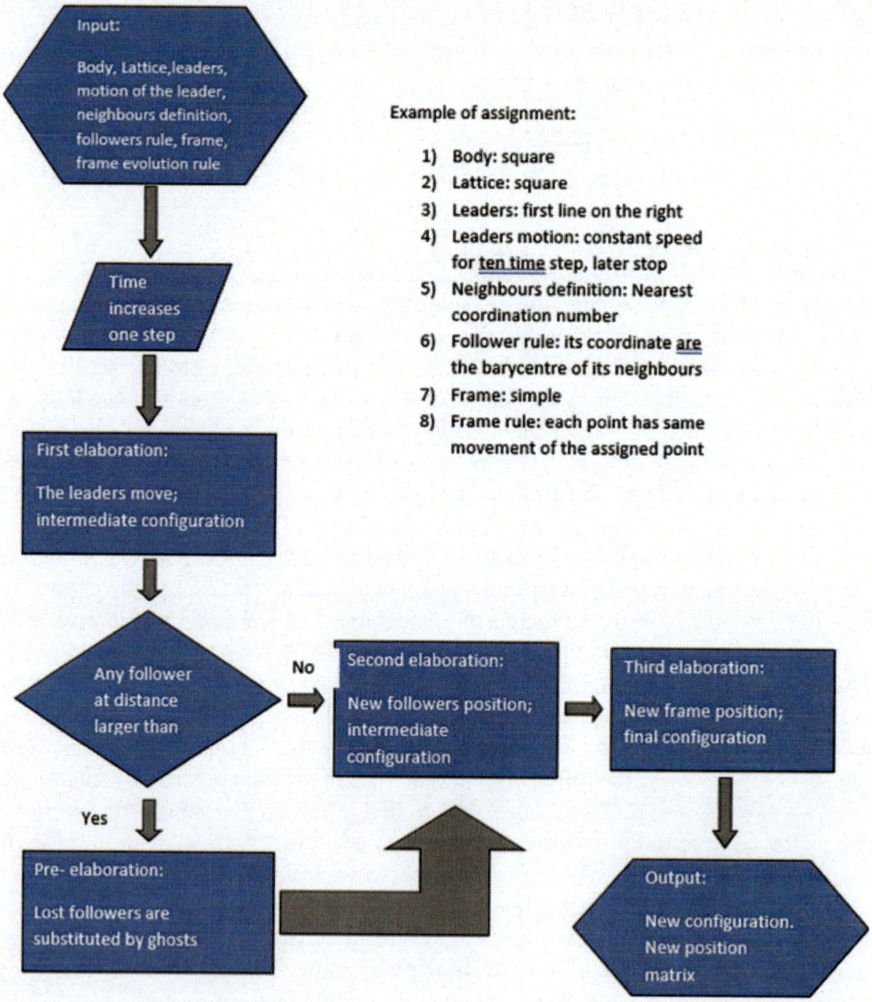

Fig. 18.7 Flowchart of the process

to avoid board effects. We decide the motions of some points, called leaders, for all the time windows we are investigating; we can also decide that they will be leader only for a certain time and late become followers (category change).

Now we can calculate, for each time step, the new configuration of the lattice in three separate operations. When time increases from t_0 to t_1, the leaders change their position from initial configuration according to the prescribed equation. So far we build a new intermediate lattice where only the leaders have been moved. Now we take care that the followers are no longer in equilibrium position owing to the leader's displacement. How we can calculate it? As an example, if the interactions rule establishes that a follower has to be in the barycenter of all its neighbors, we calculate the new position of each follower, taking into account the leader displacement. So far note as at this stage, only the leader's neighbors are involved. Finally, we take into account the rules governing the frame displacement. This is our new configuration at time t_1. It is important to note as reached the configuration is not an equilibrium one, because the three operations must be repeated for many time steps, after the leaders stop. To be more clear if at time step one the leaders have moved, we calculate the follower's displacement. This operation involves only the neighbors of the leaders and not the other far followers. Later, we calculate the frame displacement to close the loop. Now there are some followers (the neighbors of the leader's neighbors) that there are no longer in equilibrium because there has been the displacement of the leader's neighbors. So we need another time step to adjust the configuration and so on. At a certain time, all the followers are involved in the calculation. The followers will suffer the leader's motion after $(k - 1)$ time steps where k is the distance from the leaders, measured in layers. In this meaning, the leader motion "propagates" through the lattice to influence the position of all the followers in a time depending on the lattice dimensions and how many shells of points are being considered in the neighbor's definition. In the same way when leaders stop the followers continue to adjust their position in many time steps.

At this point, we can open a long discussion on the concept of time which is present here only as "step." So our question is concerning if we have to consider some virtual configuration between time t_i, and $t_{i+1,}$ until the equilibrium is reached or not. In the second case, we have used in this work, the second movement of the leaders happens when the second shell of neighbors is just interested from the first displacements of the leaders. There are some conceptual differences in the two methods we are still investigating.

In the next future, we are considering the possibility to discuss the proposed model in a fully variational setting, which is by no means trivial but would provide clear methodological advantages (see (Lanczos 2012) for an introduction and (Placidi et al. 2008; Dell'Isola and Placidi 2011; Dell'Isola et al. 2016; Dell'Isola and Gavrilyuk 2012; dell'Isola et al. 2014) for illustrative cases concerning continua with non-classical properties); therefore, we like to introduce pseudoenergetic considerations by two formulations *PE1* and *PE2* to give a contour plot of the strain distribution. The first is the sum, extended to the neighbors, of squares of the differences between the distances of the point from its neighbors minus the distance in the initial configuration, i.e.,

$$PE1(t, j) = \sum_{k=1}^{\text{all neighbours of } j} (dis(t, k, j) - dis(t_0 k, j))^2$$

Where *dis*(t, k, j) is the Euclidean distance between points k and j at time t. The reason for this choice lies in the attempt to emulate potential energy of material point subject to Hook's law. It is a sort of square distance between the actual configuration C_t and the reference configuration C_0.

To compare time contiguous configuration C_t and C_{t-1}, we define for each point j and each time t

$$PE2(t, j) = ||C_t - C_{t-1}||$$

where || is the norm of the vector defined by the point j at time t and $t - 1$.

It must be underlined that this artifice has no direct connection with the usual energy definition (this is the reason we use the term pseudoenergy) but could be useful to understand deformation.

Moreover, an algorithm based on the geometric barycenter of the neighbors of a given particle is consistent with the idea of locally minimizing an elastic potential, as the centroid has the well-known properties of minimizing the sum of the squared distances from a set of given points in an Euclidean space. Therefore, the proposed algorithm seems a natural discrete approach from the variational point of view. Another possibility, close to these concepts, we are considering is to substitute Eqs. 18.1–18.4 by potential field able to determine particle displacements and avoid collisions.

18.4 Some Examples

Aim of this section is to show the coherence of the model and its adaptability in showing different physical phenomena by changing some parameters. Therefore, we approach some bidimensional problems relative to simple shape object subject to imposed strain of some leaders which are significant; for some of the tests, we shall show and discuss the movement of the particles, the *XY* movement of a significant particle (if present) and some pseudoenergetic considerations by PE1 or PE2. See preceding works (dell'Erba 2018a, c; Battista et al. 2016). The behavior of some more complex ASTM samples and the respect of Saint Venant principle have also been described; moreover, looking for the limits of the tool, we shall discuss some not satisfactory results.

Case (a1) Simple stress test

This was the first simple test we have investigated. Consider a square-shaped specimen subjected to pull and release in tensile test. We are considering a sample undergoing strain from one side (the other side is clamped) at constant velocity in x-direction (speed 0.6 unit length/step time), with a square lattice 10 × 10 particles unit. At a certain time, the pull is released and the leaders return to original configuration (this means that leaders have changed category and now they are followers) attracted by the other points. The simple rule, governing follower's motion is that every point must be placed in the barycenter of its neighbors (Eq. 18.1); the neighbors are determined by the coordination number of the square lattice; therefore, the leader's motion implies a displacement of the first layer that propagates in successive time steps to the other particles. Therefore, the displacements, at each time step, involve a larger shell of points until to regards all the lattice points. In second gradient (Diziol et al. 2011; Ladevèze 2012; Steigmann 2013), we consider also the neighbors of the first neighbors. In Fig. 18.8, we can see the configuration of the lattice over different time together with the PE1 contour plot. Red points are the leaders; blue the followers and orange the frame. From the figure, we can outline that the x displacement of the points seems not depending on the y coordinate; however, looking at the PE1 picture, we can note a light convexity that does mean this is not true.

A deeper examination of the point's displacements confirms as close to the frame the displacements, along x coordinate, are lower with respect to central points. This can be explained as an edge effect. In fact if we consider points on the same vertical lines those that are close to the frame follow the neighbors with a little delay owing to the different rule determining the displacement of the frame and of the followers. So they see a different situation with respect to a central point. Moreover, we can note as the maximum value of PE1 (red area) is not on the leader line but just one

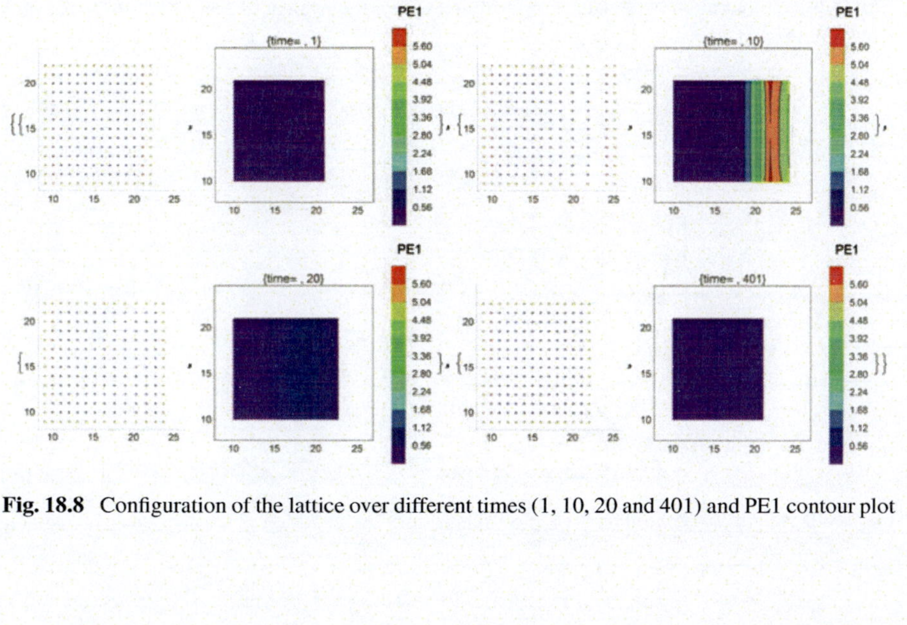

Fig. 18.8 Configuration of the lattice over different times (1, 10, 20 and 401) and PE1 contour plot

line on its left; this because, in this case, the leaders have in their neighbors, some points of the frame that always are close to them. We can avoid this convexity effect using a different frame or mirroring the followers to obtain an infinite sample. This is also evident in Fig. 18.9 where contiguous configurations are compared using the PE2 formula.

Now we put attention on a single point of the lattice. Consider a central point $j = 67$ (sixth column, seventh row, points are numbered from left to right and from bottom to up). The value of the PE1 increases notably when points are pulled, after a delay owing to the propagation time as can be seen in Figs. 18.10 and 18.11; it decreases when the leaders become followers subjected only to the rules leading to equilibrium barycenter position. If we change point, the shape of the curve remains the same but can be less or more flared as can be seen in Figs. 18.12 and 18.13 where we consider a point closer to the leaders ($j = 115$). Also in this picture we can recognize the coordinate x increases linearly (velocity is constant), after a delay (less for $j = 115$), owing to the propagation time and later decrease to the original position.

A light modification can generate instabilities and oscillations; as example, we can add to Eq. 18.1 a feedback term proportional to the difference between actual

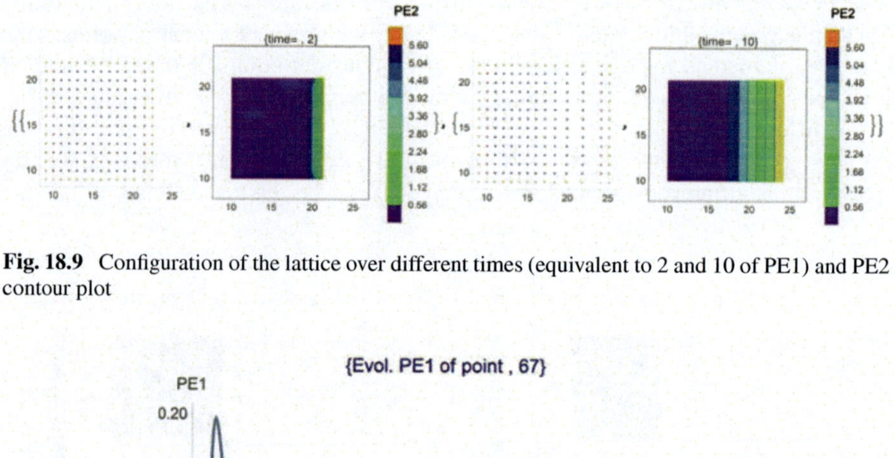

Fig. 18.9 Configuration of the lattice over different times (equivalent to 2 and 10 of PE1) and PE2 contour plot

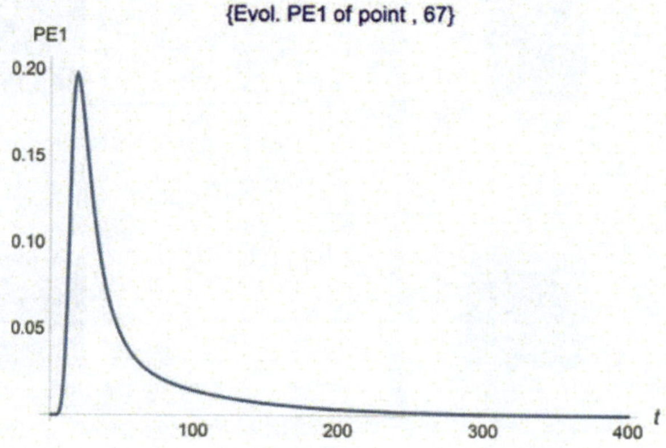

Fig. 18.10 PE1 of the central points $j = 67$ versus time

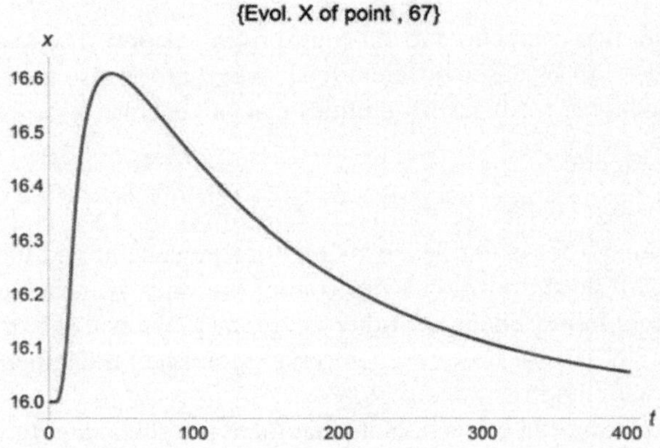

Fig. 18.11 X Evolution of the point $j = 67$ versus time

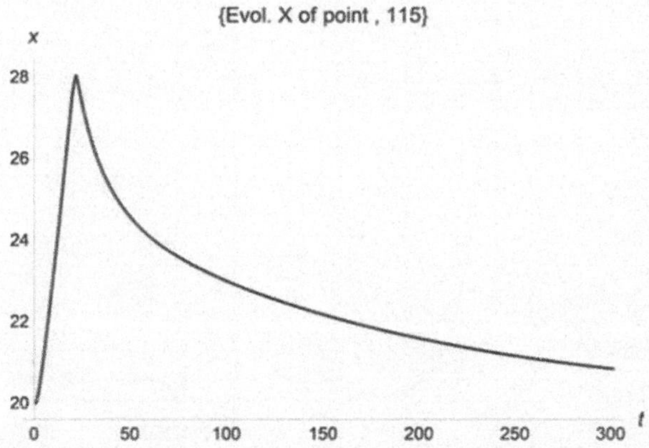

Fig. 18.12 X Evolution of the point $j = 115$ versus time

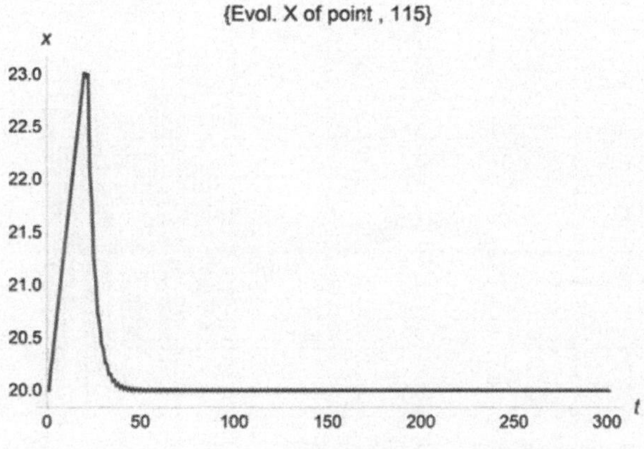

Fig. 18.13 X Evolution $j = 115$ versus time (modified rule with feedback)

and initial position to overshoot the old equilibrium position. The result is shown in Figs. 18.14 and 18.15 (we used different feedback); note as also the reaction time is changed. Using higher feedback instabilities can be generated (see Fig. 18.15).

Case (a2) Poisson effect

In the second case, we change the interaction rules; the case is the same as the preceding (without release, the leaders stop after movement and the motion steps are 100), but the follower position is determined not only by a barycentric equation but the rule takes into account the other coordinate. We call this method "mixed coordinate." In Fig. 18.16, we can see the configuration of the lattice over different times together with the PE1 contour plot.

Lateral contraction can be seen. Note that there is a relaxation time, because the followers need time to adapt themselves. This is due to the rules expression and can

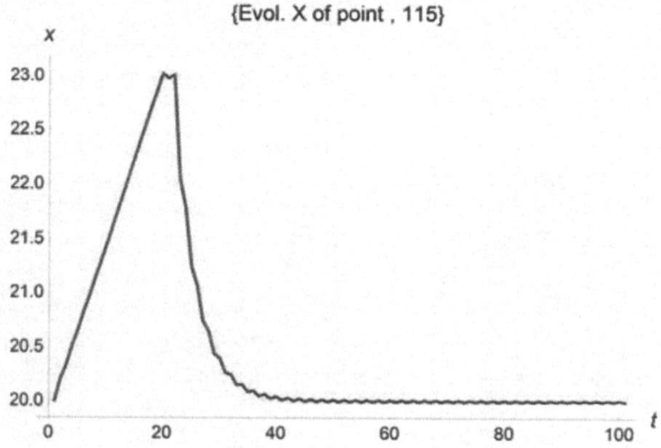

Fig. 18.14 X Evolution $j = 115$ versus time (modified rule with feedback)

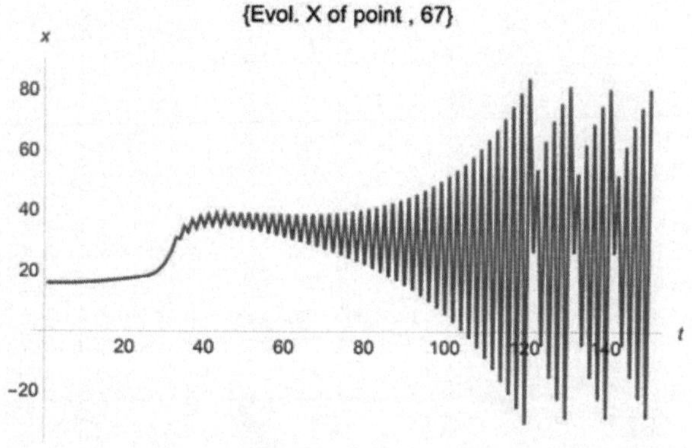

Fig. 18.15 X Evolution of the point $j = 67$ versus time (modified rule with feedback)

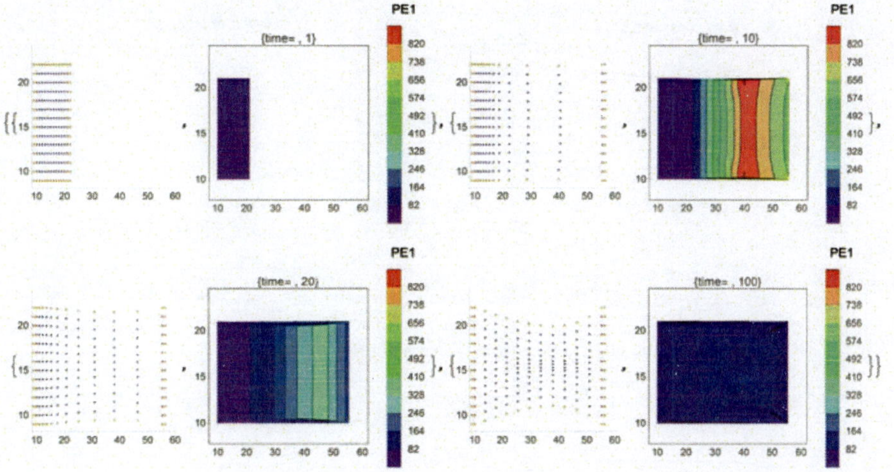

Fig. 18.16 Configuration of the lattice over different times (1, 10, 20 and 100) and PE1 contour plot

be tuned as you desire. The PE1 value seems to follow the configuration; here is more evident the edge effect leading to concavity effect toward left side.

In Fig. 18.17, we can see the configuration of the lattice over different times together with the PE1 contour plot in second gradient case. In second gradient case, lateral contraction can be seen in a less pronounced way; this is because if we take into account a larger number of neighbors, the effect is dumped. The PE1 plot enhances

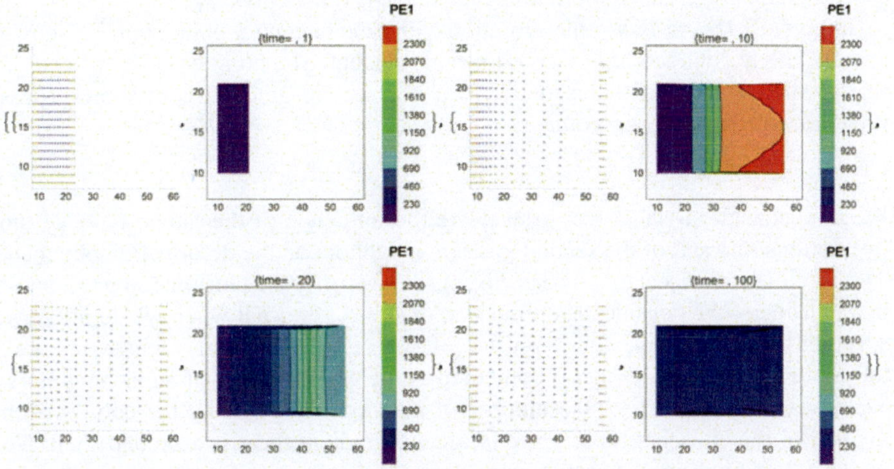

Fig. 18.17 Configuration of the lattice over different times (1, 10, 20 and 100) and PE1 contour plot. Second gradient

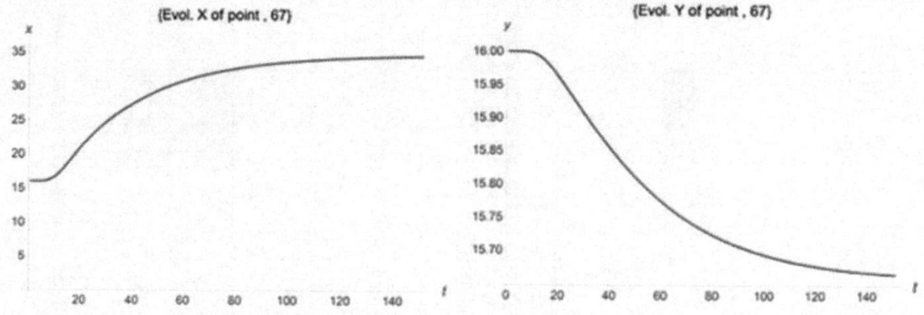

Fig. 18.18 Evolution of the central point $j = 67$ versus time

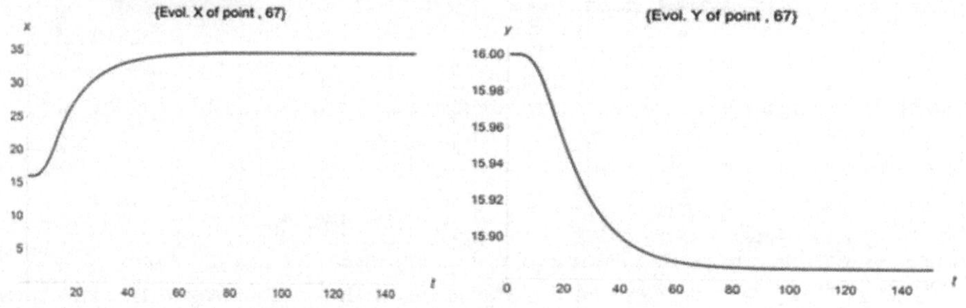

Fig. 18.19 Evolution of the central point $j = 67$ versus time; second gradient case

the convexity of the first line of follower. We shall see better as, in the fracture case, second gradient has much more influence in some other cases.

In Figs. 18.18 and 18.19, the evolution with time of central point $j = 67$ is shown and the lateral contraction in y coordinate, is evident. The point is above the central line so their y coordinates decrease. Note the second gradient case shows a less pronounced effect on y coordinate.

Case (b1) Shear test

We now consider a shear test to evaluate the influence of the lattice type, of the interactions rule and of the second gradient neighbors on the deformation obtained. The specimen is subject to a shear with constant velocity 0.1 unit/time step in x axes, by the leaders. We consider 100 time step of strain. As usual when we do not have specified interaction rules between the followers, we use the barycenter rule, i.e., the coordinate of the follower is computed as the gravity center of all its neighbors. Moreover, if not specified, the number of neighbors is given by the coordination number of the chosen lattice, while in second gradient there is a second shell. No fracture is still considered in this case.

In Fig. 18.20, we can see the configuration of the lattice together with the PE2 contour plot, to compare contiguous configurations. As usual the leaders are red, the followers blue and the frame is in orange color. Lateral deformation are nonlinear and

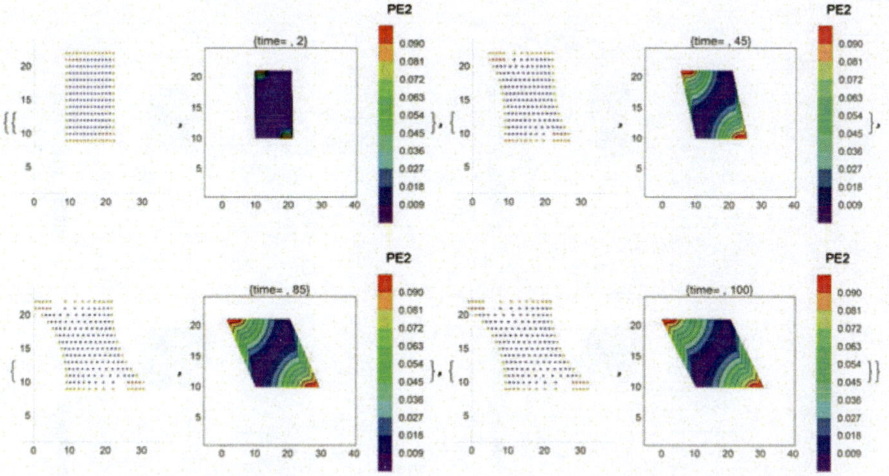

Fig. 18.20 Configuration of the lattice over different times (2, 45, 85 and 100) in shear test square lattice together with PE2 contour plot, indicating differences between contiguous configurations

comparing contiguous configuration, by PE2 function, differences are larger close to the leaders. No deformation of the top and bottom lines can be outlined.

A more marked lateral deformation curve can be seen if we use a honeycomb lattice (see Fig. 18.21). This is an example of how different deformed configurations can be obtained by changing lattice type holding all the other conditions. We remark that this is not a Bravais lattice, but we have used it owing to its large practical applications.

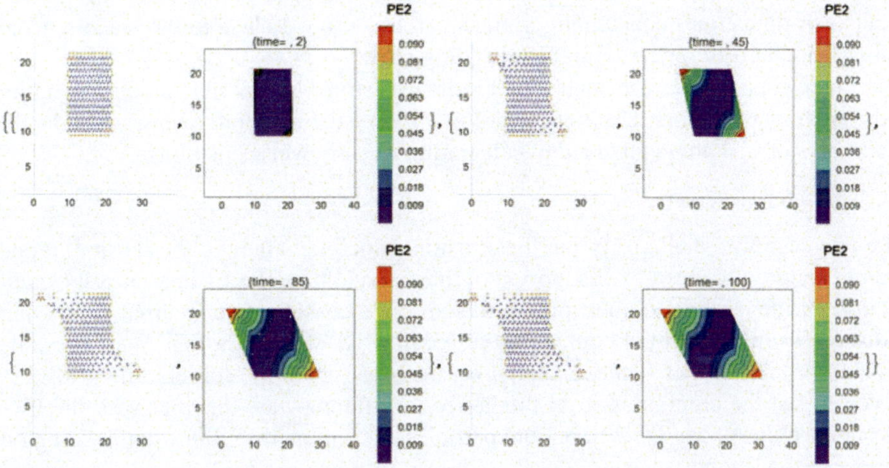

Fig. 18.21 Configuration of the honeycomb lattice over different times (2, 45, 85 and 100) in shear test square lattice together with PE2 contour plot, indicating differences between contiguous configurations

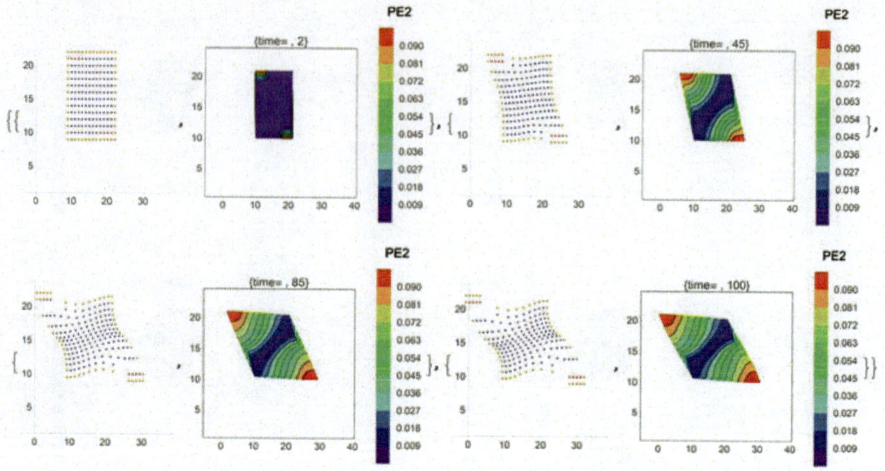

Fig. 18.22 Configuration of the lattice with Poisson's effect over different times (2, 45, 85 and 100) in shear test square lattice together with PE2 contour plot, indicating differences between contiguous configurations

As in the previous case, PE2 contour plot shows that differences between contiguous configurations are larger close to the leaders

A more interesting case, using a square lattice, is shown in Fig. 18.22. Here we have used a rule for the follower making use of "mixed coordinate," which means the y coordinate is dependent on the evolution of the x coordinate. This allows us to obtain lateral contraction, i.e., Poisson's effect. The result of the shear test is a strange "window" flag. Once again PE2 contour plot shows that large differences between time contiguous configurations can be outlined close to the leaders, from no particular differences with the preceding plots.

Quite similar behavior can be observed if we use second gradient model, changing the shell of neighbors. Differences are in a more stiff reaction, owing to the larger numbers of neighbors involved in calculating the follower's positions.

Case (c) Saint Venant

In this case, we shall verify the local action principle, whereby the tension at one point is not influenced by the motion of the external particles to an arbitrarily small circle of the particle in question. To this aim, we consider four internal points that diverge from their initial configuration; practically, we choose four internal points as leaders with opposite movement along the bisectors of the corners. In Fig. 18.23, we can see the configuration of the lattice in different time together with the PE1 contour plot. As can be seen remote particles are not interested in what is going on close to the leaders.

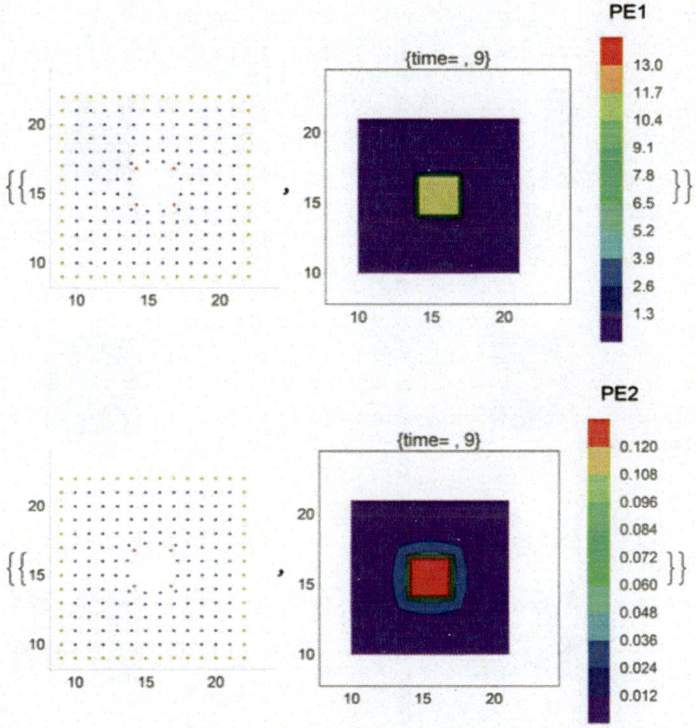

Fig. 18.23 Configuration of the lattice at time (9) with PE1 and PE2 contour plot

Case (d) Fracture test

For the next example, we shall consider a square sample undergone a tensile test with fracture. Fracture distances are 10 units, and speed is 0.5 units/step, for 150 steps long. When a distance between the points is larger than fracture distances, the sample is broken and the followers go to equilibrium position; if no followers remain attached to the leaders (it depends on the distances, we shall see later in other cases), they return to their initial position. As explained in a preceding work (dell'Erba 2018a, c; Battista et al. 2016), the convexity, in the fracture mechanism, is related to the presence of the frame. Analysis of PE1 plot shows that, before fracture, there are areas of stress concentration. Higher stress areas are close to the leaders. The trend of the follower points is quite linear during traction, but it becomes nonlinear when the followers remain alone and return back. This can be explained because the traction is imposed with constant speed, while the reassembly of the points is driven by the follower's rules. Once again involving a larger number of neighbors leads to a more stiff behavior as can be seen in Fig. 18.25 (second gradient case). We can see as the vertical fracture line is different in the case of first (see Fig. 18.24) or second gradient (see Fig. 18.25). Points close to the frame are detached before the others from the leaders, and this effect is more marked in second gradient case. This can be

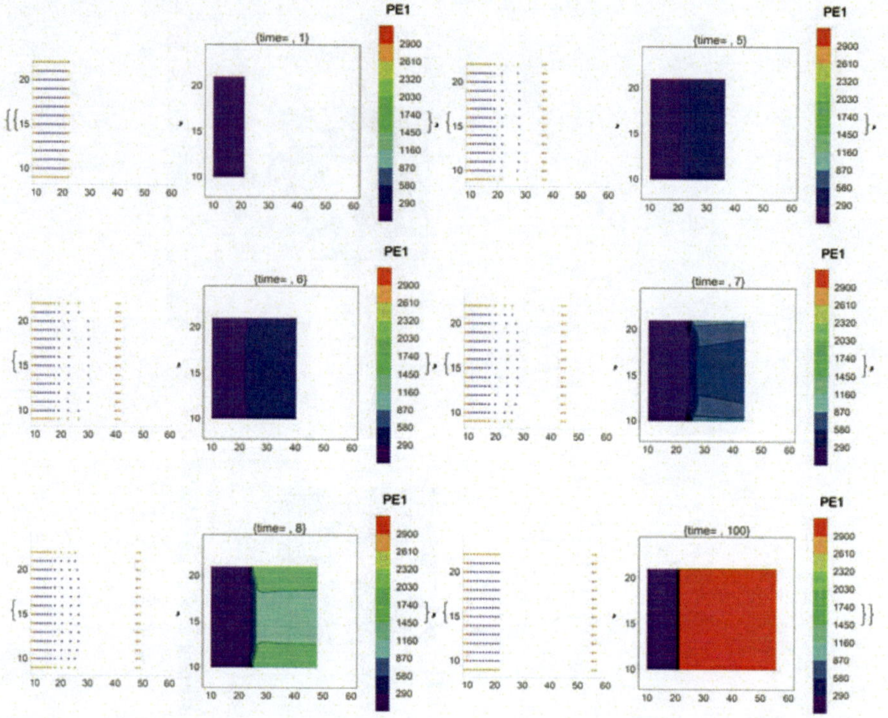

Fig. 18.24 Configuration of the lattice over different times (1,5, 6,7 and 100) and PE1 contour plot, fracture in tensile test. First gradient

explained with the different neighbor's number and also with the larger influence of the frame with respect to the first gradient case.

Remember that in the fracture case, the pseudoenergy plot is less indicative, because we are calculating it using distance between points greater than the fracture threshold. In a future work, we will consider a better definition of this parameter.

The importance of second gradient must be outlined in Fig. 18.26 where x coordinate evolutions versus time of x coordinate of point 103 is showed. The point is situated in middle value as Y coordinate and two lines on the left of the leader's line; differently from first gradient mode, a complex behavior can be observed because after the fracture the x coordinate has a sort of rebound. This can be explained as follows. After the fracture, the point tries to return its initial position (on the left), but later some fast point on its right tries to deviate it to the right. When the group is compacted, they go back all together to initial configuration. So far change in the parameters can lead to complex evolution behavior of the lattice.

The fracture mechanism is strongly dependent on lattice characteristics; in the next case we use a hexagonal lattice, instead of square. In this case, we can choice two different rules for the frame, as can be seen from Fig. 18.27. In the first case, a frame point displacement is just the same of the corresponding follower. But in some

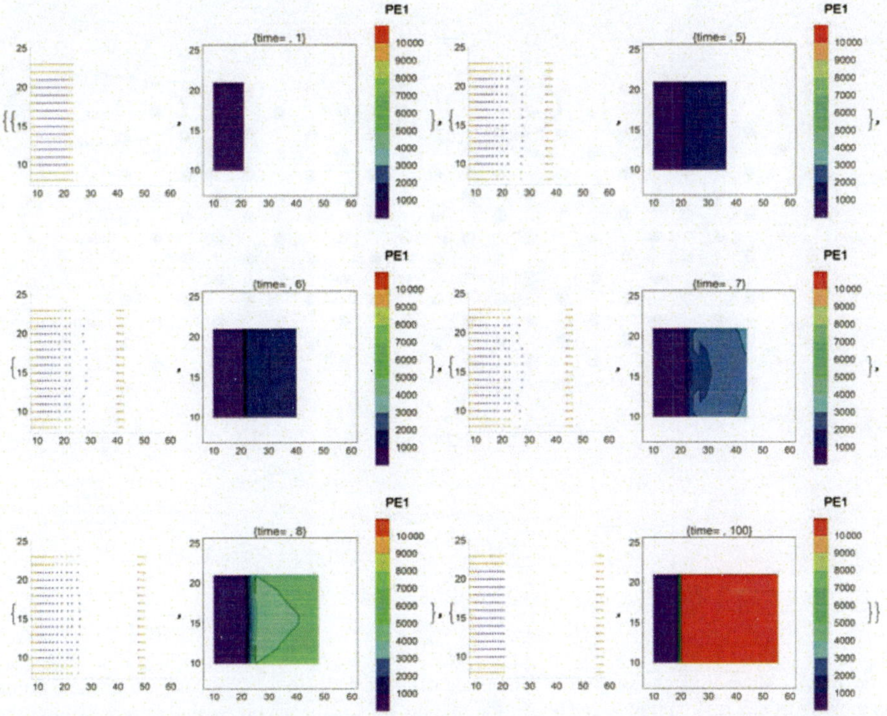

Fig. 18.25 Configuration of the lattice over different times (1, 5, 6,7,8 and 100) and PE1 contour plot, fracture in tensile test. Second gradient

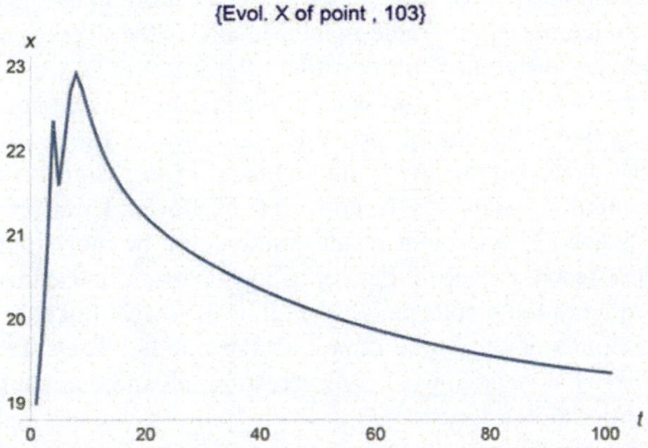

Fig. 18.26 Evolution of the second follower line point $j = 103$ versus time

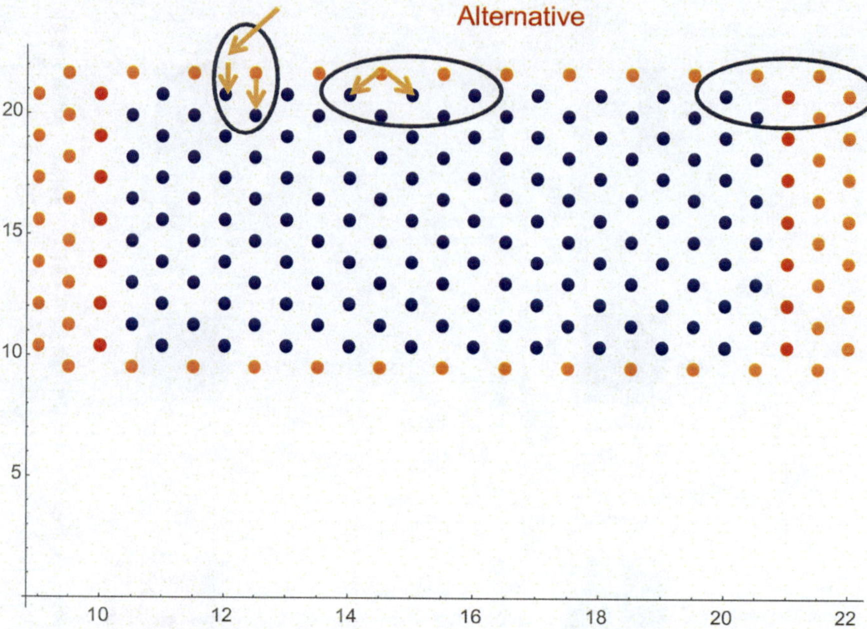

Fig. 18.27 Alternative choice for the frame rule displacement. In the first case, each frame point is moving as a corresponding follower. In the second case as the average value of more followers assigned

case, like hexagonal lattice, you can decide there are more than one corresponding followers and choice the point frame displacement as the average value of them. The behavior is very different from the square lattice case as we expected. It can be noted that a point of the frame remains in the middle of the displacement, if we use the average value. This can be explained as follows. The rules regarding the frame are simple; each point of the frame is linked to an assigned follower and its displacement from time t to $t + 1$ is the copy of the follower. However, in some cases the followers assigned to one point of the frame could be more than one. In such cases, we can choose to take one of them or to consider the displacement of the point as the average value of the displacements of all its followers linked to it. This is the reason that two points of the frame remain in the middle: They are stressed from two opposite sides. The behavior is very interesting; it can be noted that a different equilibrium configuration is reached because the frame is changed and the fictitious, introduced to manage fracture, are not in the list of the followers. In the case of Fig. 18.28, this results in a concave final surface, owing to the modified frame.

If we use the other frame rule, the fracture is similar to square lattice. This leads to different final deformed configuration as can be outlined in Fig. 18.28. The presence of a frame point in the middle leads the final configuration to a concavity. The absence of followers on the right side, once again, is depending on the leader's speed.

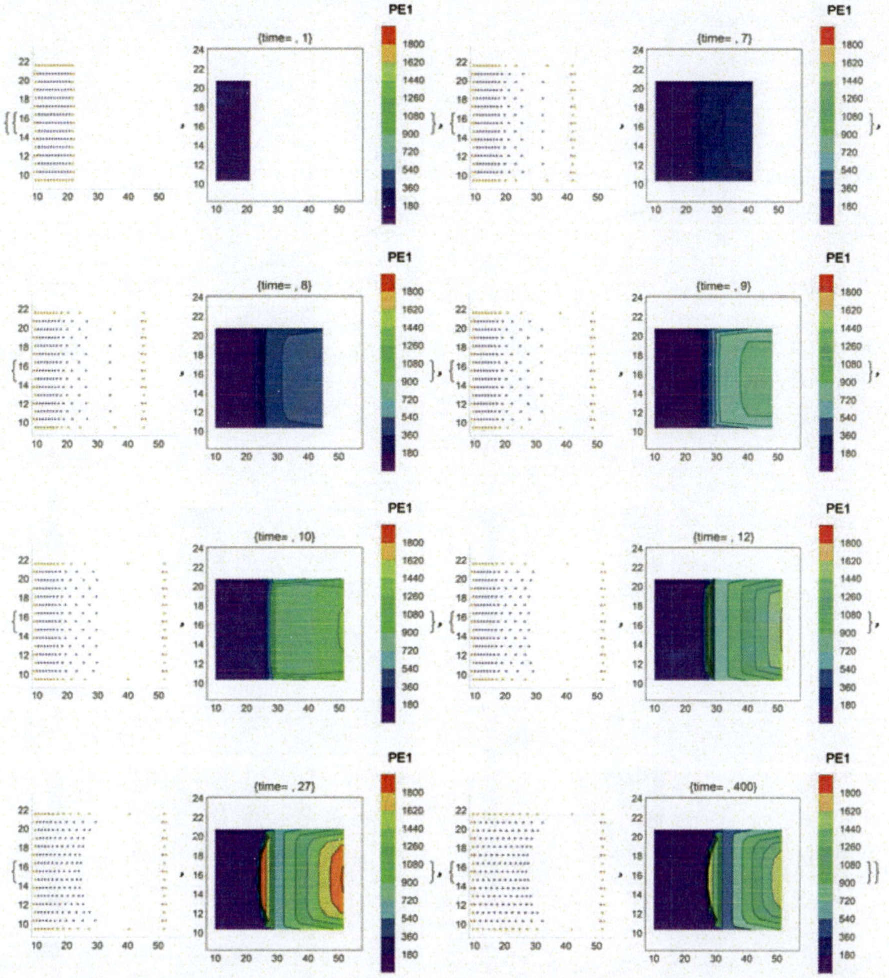

Fig. 18.28 Configuration of the lattice over different times (1, 7, 8, 9, 10, 12, 27 and 400) and PE1 contour plot

A rebound mechanism can be outlined from Figs. 18.29 and 18.30. This is typical of a hexagonal lattice; you can note that in the square lattice this behavior does not exist.

Also in this case, second gradient mode attenuates the effects; see Fig. 18.30. No rebound can be outlined and the equilibrium configuration, after the fracture is similar to the original. Cuspidal point in X coordinate, after about 100 step, can be attributed to a second fracture happening.

Finally in Fig. 18.31, fracture mechanism of simple square lattice under shear stress is shown. It can be noted that the leaders bring with them some of the followers;

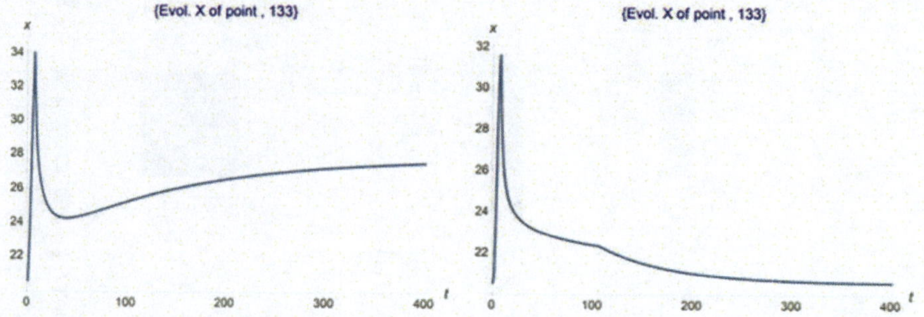

Fig. 18.29 Evolution of the central point $j = 133$ versus time; first and second gradient, see Figs. 18.28 and 18.30

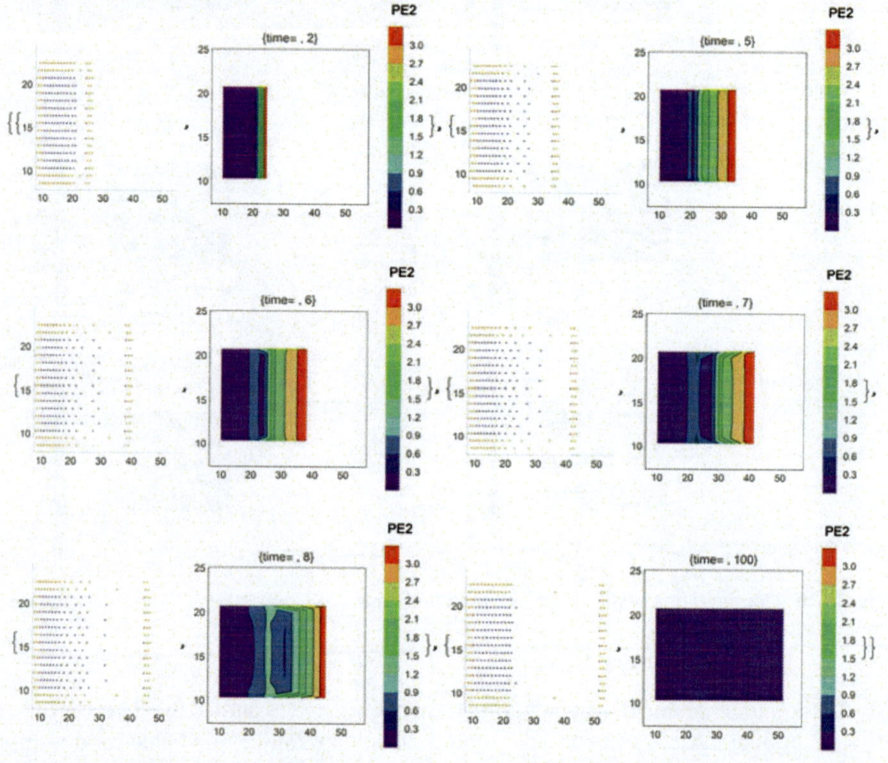

Fig. 18.30 Configuration of the lattice over different times (1, 7, 8, 9, 10, 12, 27 and 401) and PE1 contour plot

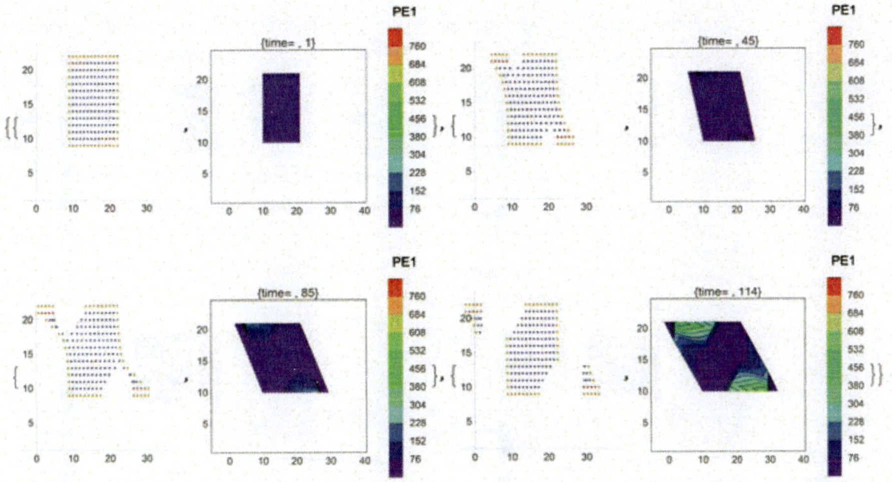

Fig. 18.31 Shear configuration of the lattice over different times (2, 45, 85 and 114) in shear test square lattice with fracture together with PE1 contour plot, of pseudoenergy

this depends on a complex balance between the leader's attraction and the resistance offered by the followers. Changing condition results in changing the number of the "attached" followers. After the fracture, the particles return back to their equilibrium position. Note that if we would position the fictitious in another location, we would obtain a different result. Pseudoenergy has symmetric behavior, as expected. Remember that the pseudoenergy concept was considered on not fractured sample, and it is not calculated on the fictitious points but on the followers so it is not significant.

Another example can be obtained if we consider a rectangular centered lattice but we the neighbor's number, n_c, to five and consider a first gradient interaction; we obtain a completely different result. In Fig. 18.32, we have a tensile test as the preceding; the lower number of particles involved in the calculation of the relative position makes the sample much more fluid, allowing detachment of a larger number of particles, as we can see on the right side of the pictures. The fracture mechanism also is different with respect to the preceding case.

This gives more mobility to the model leading to a more plastic behavior and increasing the number of the detached points; the fracture mechanism is quite different together as well as the final configuration (see Fig. 18.32). This example shows, once again, that change in model parameters leads to different behaviors.

An interesting phenomenon can be seen if we consider an oblique lattice (Fig. 18.33). Owing to the asymmetry (see look at the five red leaders on the right) of the leaders with respect to the frame, a particular breakage fracture can be observed. In fact if we consider a symmetry axes in x-direction, we can note two leaders close to the frame in the upper level and only one close to the bottom. This leads to a fracture starting from the bottom where the attraction of the leaders is lower. It seems to rip

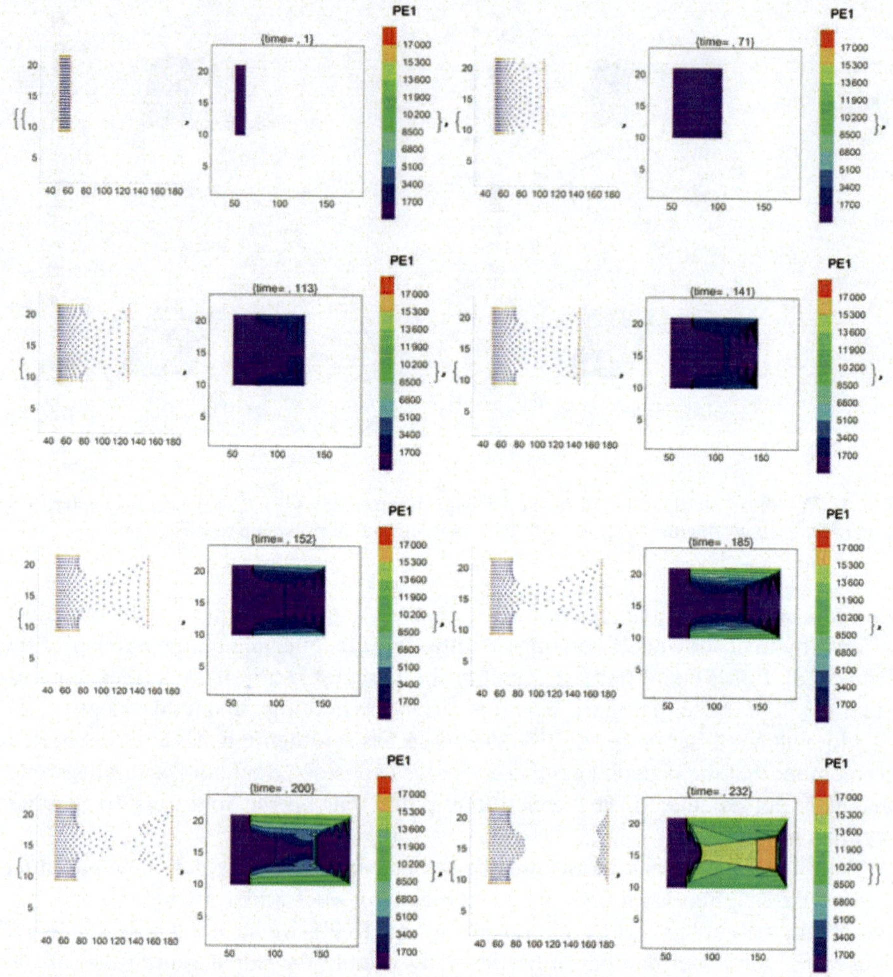

Fig. 18.32 Tensile test with fracture rectangular centered lattice, coordination number 5. Configuration over different times (1, 23, 33, 38, 40 and 63) with PE1 contour plot, of pseudoenergy

a piece of paper. The fracture distance is 10 units, and the speed is 0.4 unit/time step (Fig. 18.34).

Case (c) ASTM test

In order to investigate sample behavior of a well-known shape, we have carried out some calculations considering a specimen quite similar to the ASTM D638 standards for tensile tests. The specimen is clamped at both ends on a surface and pulled on one side, so, in this case, the leaders are many; the test speed is of 2.5 unit/step, in x positive direction, for 150 traction step e 2500 relaxation step. We have considered simple tensile tests with different lattice, lateral contraction effect and fracture cases.

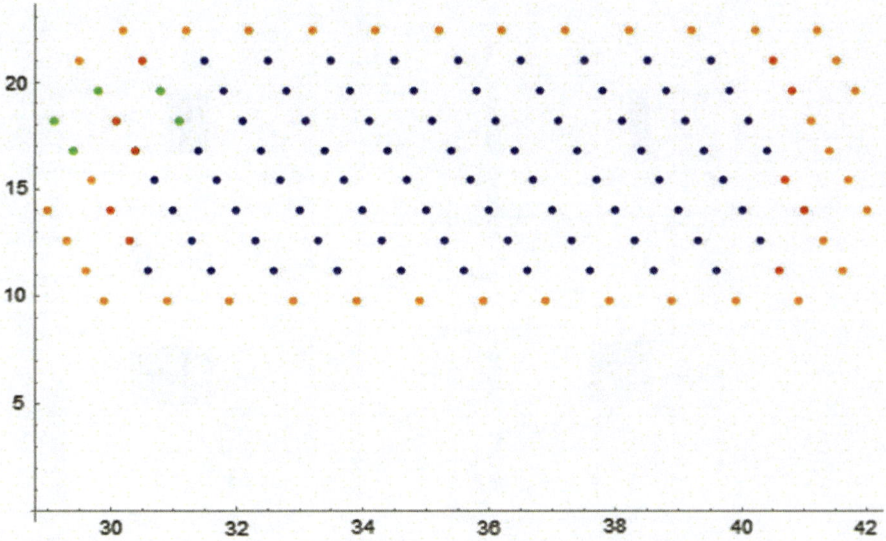

Fig. 18.33 Oblique lattice tensile test. Red points are leaders, blue point followers and yellow the frame. First gradient case. Note the asymmetry of the leaders with respect of the frame

Generally, a larger number of time steps are required for relaxation, owing to the larger number of points employed to describe the specimen; this does not mean a longer relaxation time, because unit time is arbitrary, only that the influence of the displacement propagates at one shell (first gradient case) each time step so we need many steps to involve the whole sample. Once again the second gradient case seems to be stiffer with respect of the first gradient.

It should be noted that, in our pictures, the sample does not reach a symmetric final configuration as we can expect because of the long time required.

If we consider the simple tensile tests (see Figs. 18.35 and 18.36), little quantitative differences in point distribution can be observed during the classical elongation of the specimen in different cases. We can observe differences in the internal distribution on the points and in the convexity of the propagation front of the deformation (see Fig. 18.41), i.e., see the convexity of the points between the two figures. We are working on this and on higher gradient computations.

On the contrary if we use a rectangular shape sample, it does (the points are equally spaced) as can be seen in Fig. 18.37. There is no physical reason for this, our opinion is that this effect is linked to the particular equilibrium condition generate by the geometry. It can be outlined as final configuration is more similar to a symmetric one, in second gradient case, owing to the larger number of points involved in the computation.

In the two cases, we are considering Poisson's effect (Figs. 18.38 and 18.39); it is possible to see lateral contraction. It seems the points cluster to create islands, but this effect must be investigated better. In case of second gradient interaction, this

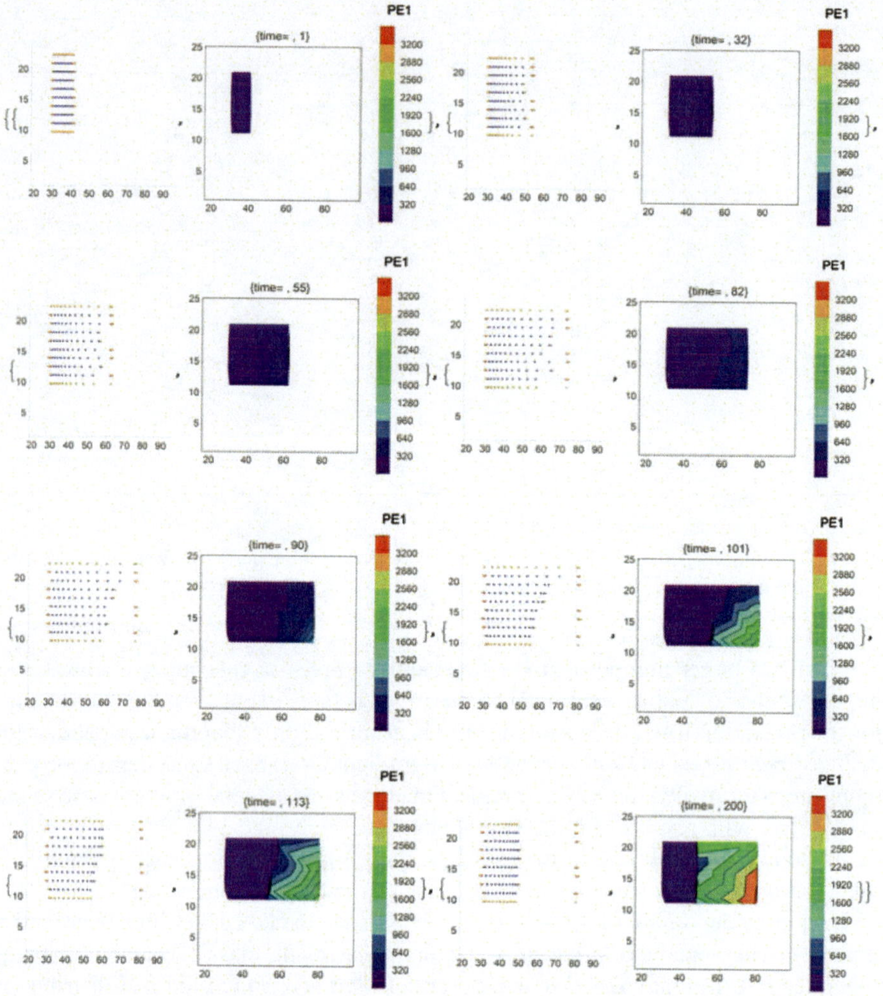

Fig. 18.34 Tensile test with oblique lattice: breakage fracture. Configuration over different times (1, 32, 55, 82, 90,101,113 and 200) with PE1 contour plot of pseudoenergy

does not occur, as can be seen in Fig. 18.39. The fracture test (see Figs. 18.40, 18.41 and 18.42) is considered for the rectangular-centered lattice. Distance fracture is 11 units, and the speed was 0.6 step/unit time. As can be seen, the fracture occurs close to the top of the profile and not in the central area. Studies, in progress, show as the fracture zone can be moved by varying working conditions. We can render the fracture more or less brittle changing the model parameters like neighbors' number, type of lattice, speed, etc. As example in second gradient, the same sample has a more brittle behavior; or if we use a speed of 2.5 step/unit time in the same condition we will get no followers on the right side of the fractured sample. We can observe differences in

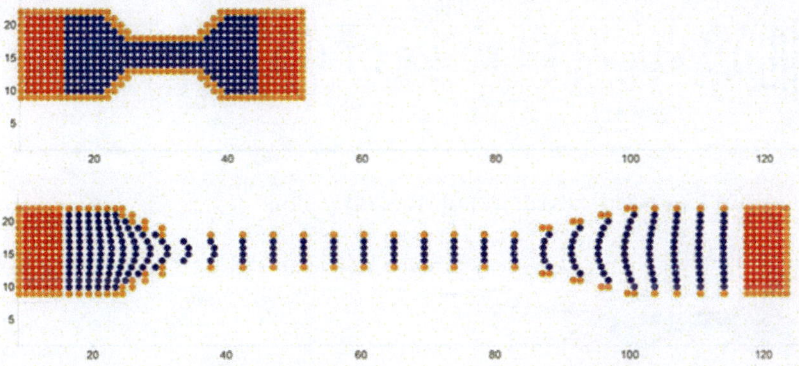

Fig. 18.35 ASTM tensile test square lattice. Configuration over different times (1, 40, 160 and 2500) with PE1 contour plot of pseudoenergy

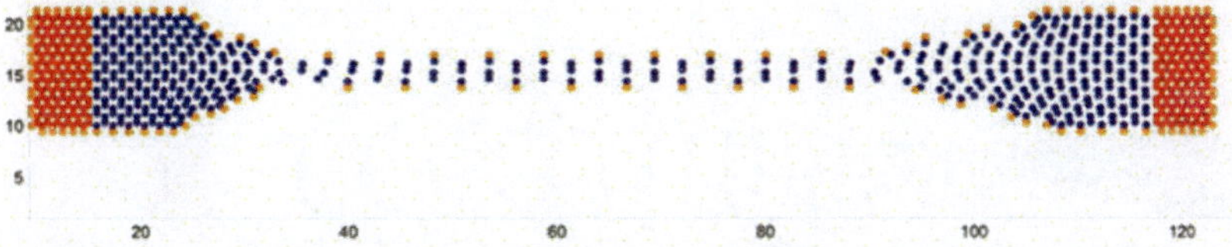

Fig. 18.36 ASTM tensile test rectangular honeycomb lattice. Configuration over different times (1, 40, 160 and 2500) with PE1 contour plot of pseudoenergy

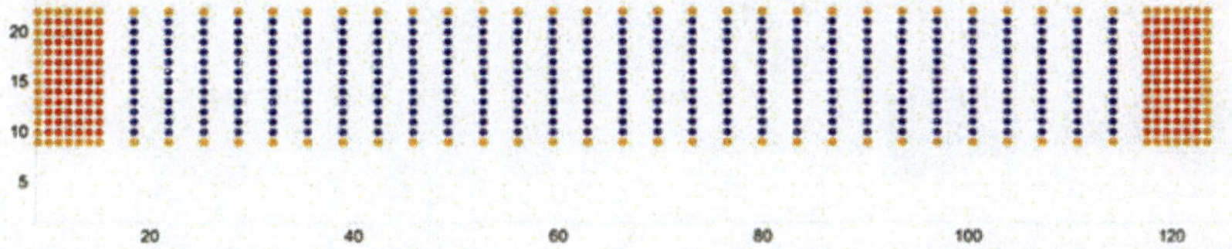

Fig. 18.37 Rectangular sample square lattice

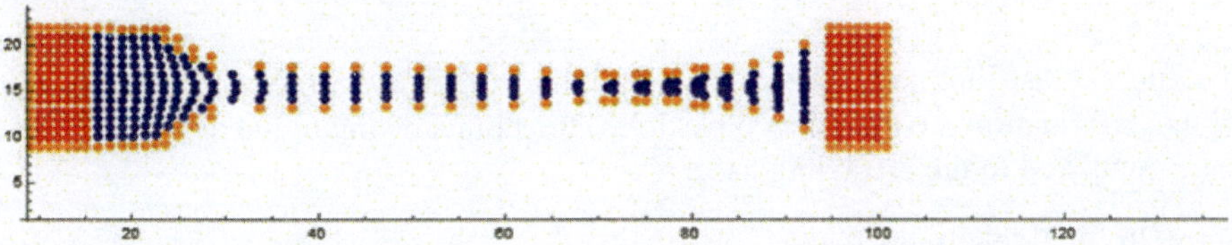

Fig. 18.38 ASTM sample Poisson's effect square lattice

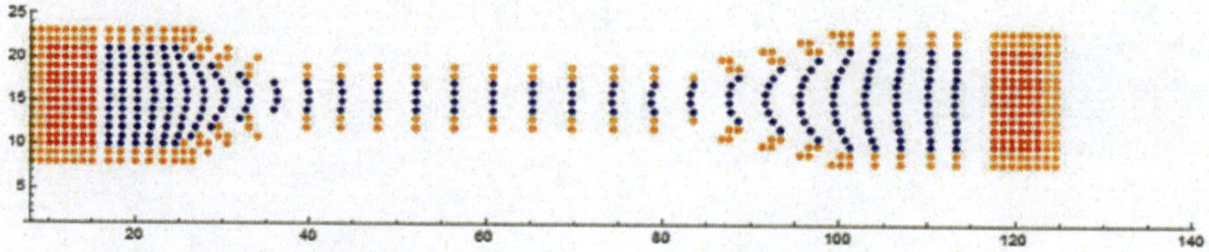

Fig. 18.39 ASTM sample Poisson's effect square lattice 2° gradient

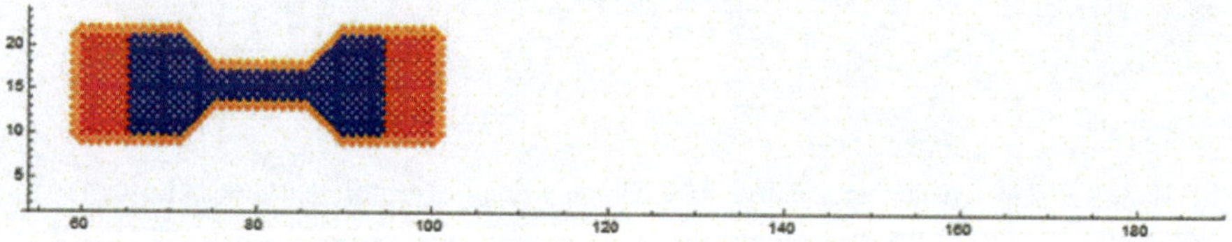

Fig. 18.40 ASTM rectangular-centered lattice before tensile test

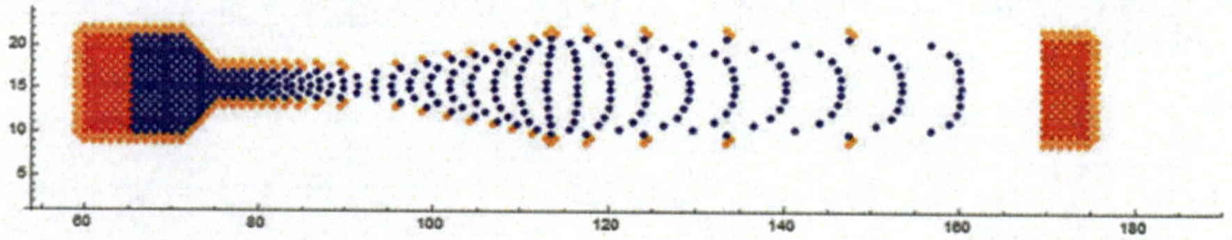

Fig. 18.41 ASTM rectangular-centered lattice after tensile test

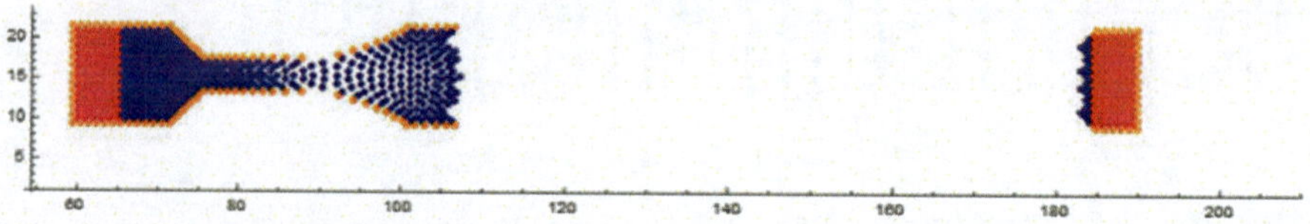

Fig. 18.42 ASTM square lattice fracture

the internal distribution on the points and in the convexity of the propagation front of the deformation, i.e., see the convexity of the points owing to the different lattice used compared to the preceding case.

Case (h) Short beam, the plate

Now we want show some differences between the solution of a bending plate undergoing a shear load and what we have obtained by our tool. In the preceding works, we got some success but we need to go deeply to a better understanding so we are looking for some cases that do not fit on what we are expecting. We consider a bidimensional square plate (X and Y coordinates from 10 to 21) with materials parameters $Y = 1000$ and $\nu = 0.33$, where Y is Young's modulus, ν Poisson's coefficient; boundary load is on the right (Neumann's condition) and no displacements on the left (Dirichlet conditions).

The equations to be solved are:

$$\frac{Y}{2(1+\nu)}\nabla^2\mathbf{u} + \frac{Y}{2(1-\nu)}\left(\frac{\partial}{\partial x}\mathbf{u} + \frac{\partial}{\partial y}\mathbf{v}\right) = 0$$

$$\frac{Y}{2(1+\nu)}\nabla^2\mathbf{v} + \frac{Y}{2(1-\nu)}\left(\frac{\partial}{\partial x}\mathbf{u} + \frac{\partial}{\partial y}\mathbf{v}\right) = 0$$

$\mathbf{u}(x, y)$ and $\mathbf{v}(x, y)$ are the displacements function. We pose as boundary conditions 50 Pa as shear stress on the plate (Neumann's condition for $x = 21$) and $\mathbf{u}(10, y) = \mathbf{v}(10, y) = 0$ as Dirichlet condition. Note that we are using Bernoulli's equation while Timoshenko model should be more appropriated. Anyway this is just a first attempt so we reserve the right to use it in a next paper. These equations can be solved numerically if we discretize our plate by a 10×10 square lattice; the solution is shown in Fig. 18.43 and the von Mises plot in Fig. 18.44; deformed mesh is plotted in red color.

Our intention is to compare the strain of the plate, obtained by FEM solutions, with that we can compute by our tool. Therefore, we have to assign the displacements of some points, the leaders make some choice about the algorithm (lattice, interaction

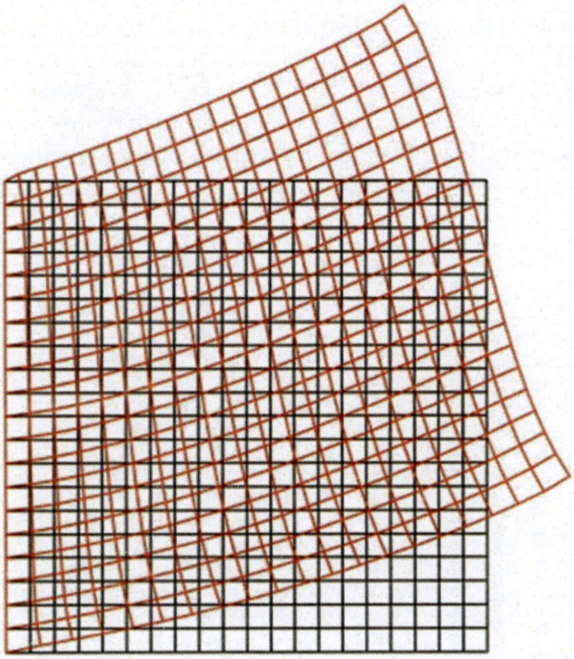

Fig. 18.43 FEM solutions of bidimensional square

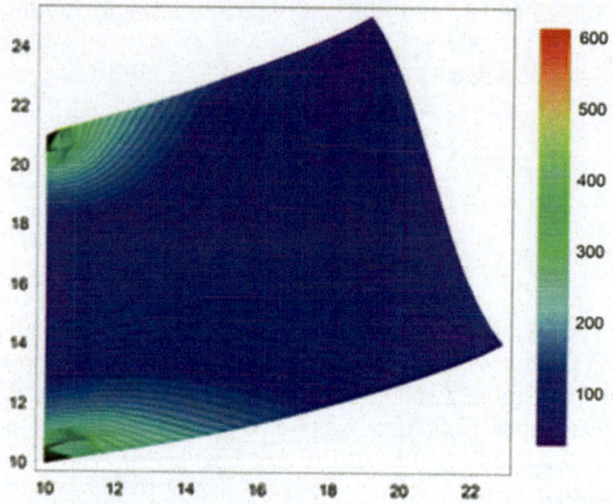

Fig. 18.44 Von Mises plot of FEM solutions

rules between the followers, etc.) and compute the strain when the followers readjust themselves, after a while. As leaders we choose the right and left side of the plate, so we impose the displacements of these points as computed from the FEM equations and investigate the arrangement of the other points. It is important to remark that we have no criteria about the choice of what lattice, interaction law between followers, etc.... So as first attempt we use a square lattice and no weight in the computational of the followers coordinate. In Fig. 18.45, the obtained configuration, together with the FEM solution (red points), is shown; in Fig. 18.46 the corresponding von Mises plot. The points on the left and on the right of the plate are overlapped because of the leaders and we have imposed their displacement as the FEM solution of the plate

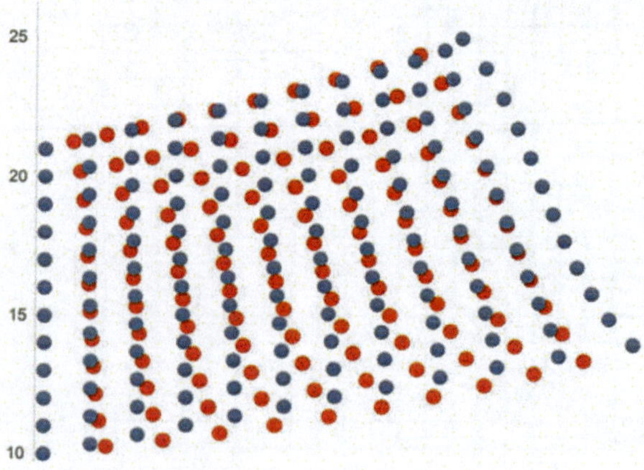

Fig. 18.45 Beam deformation

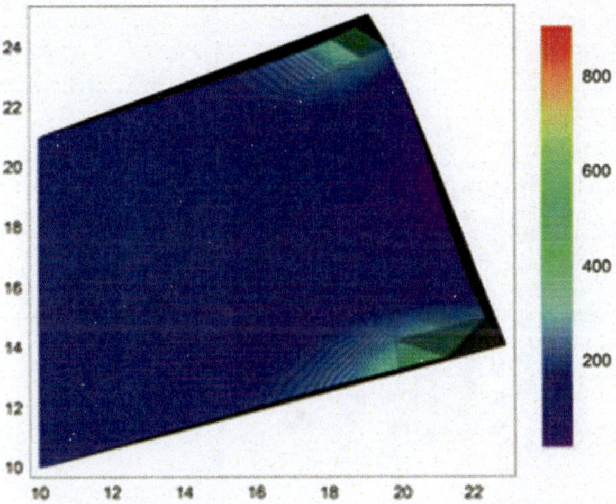

Fig. 18.46 Corresponding von Mises plot

deformation. It can be outlined as external configuration of the plate is quite the same but the internal displacement of the points, i.e., the strain, is different; this can be highlighted if we plot the von Mises stress. Change in tool's parameters lead to different configuration, corresponding to different strain of the plate, almost none of them are satisfactory as shown in the following pictures. Second gradient case has showed no appreciable differences; we have tried to give a quantitative measurement of the discrepancies with the FEM solutions, using the average value of the sum of the square differences between coordinates, in many cases. Nobody emerges as the best match so we can conclude that the plate deformation can be sometimes very similar to the FEM solution, but the von Mises stress plot is quite always unsatisfactory. This point needs to be studied again to a better understanding of the physic behind the tool and what should be drive our choices in the tool to describe material continuum. We have to remember that the materials parameters Y and ν do not appear explicitly in our algorithm, but they are hidden into the interaction relationship between the followers, the neighbors and the choice of the lattice. So far we have no idea on how to select our choices to match the well-known problem. Changing the parameters of our tool, we obtain different results nobody of them perfectly coincident with classical solution, but owing to the flexibility of the tool, we believe it exist a parameter combination to fit the deformed beam; but this is meaningless until we do not understand how to choose the parameters. The reason of this lies in the fact that, up to now, we do not start from the constitutive equations of the materials leading to the rules governing points displacement. We have to work on how to connect the rules of our model with classical physical proprieties of the material.

In Figs. 18.47 and 18.48, we are considering the case with square lattice, first gradient and we used power mean in Eq. 18.1 with power factor -1 (harmonic mean). Here a strong difference with the FEM solution (red points) can be noted

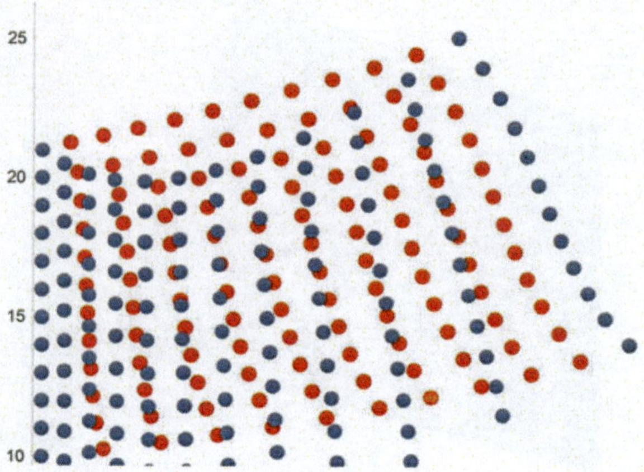

Fig. 18.47 Plate deformation

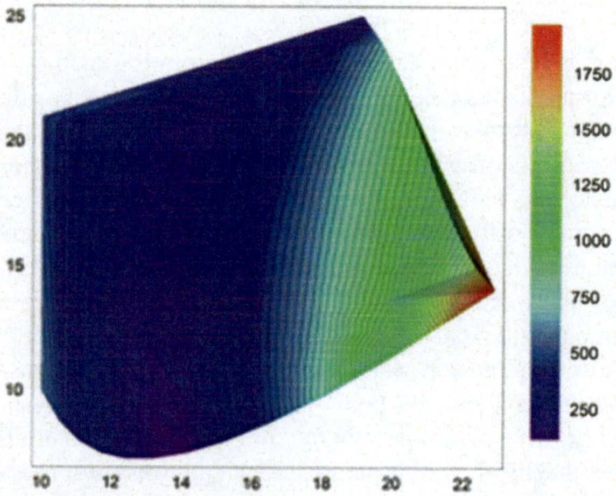

Fig. 18.48 Corresponding von Mises plot

but similar to the Timoshenko Beam deformation shape where the cross sections, perpendicular to the neutral axis before deformation, stay plane after deformation but are not necessarily perpendicular to the neutral axis after deformation. This was expected because the -1 power parameter changes deeply the Eq. 18.1 into the following equation:

$$x_j(t) = \left(\frac{\sum_{k=1}^{\text{all neighbours of } j} (w(k)x_k(t))^p}{\sum_{k=1}^{\text{all neighbours of } j} w(k)^p}\right)^{1/p}$$

where *w(k)* is the weight, in this case 1, of the k-element and *p* the mean parameter, −1.

In Figs. 18.49 and 18.50, we are considering honeycomb lattice that sounds as one of the best results.

Case (i) Long beam

Also in the case of longer beam, we can try many combinations of the parameters tool to fit the deflection of our beam, but it is useless: We need to connect the constitutive parameters with the tool. If we increase the ratio length/width of our beam from

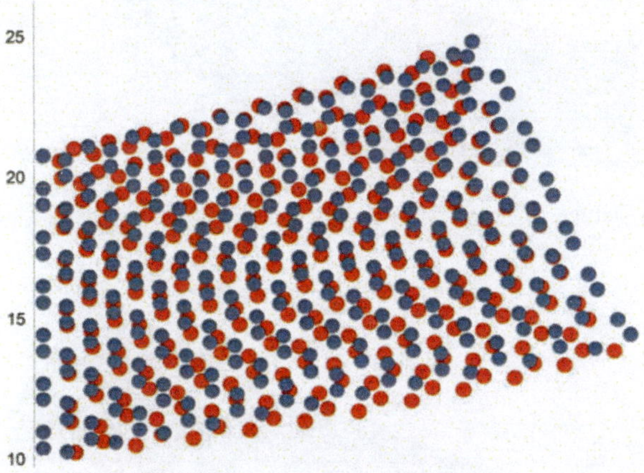

Fig. 18.49 Plate deformation

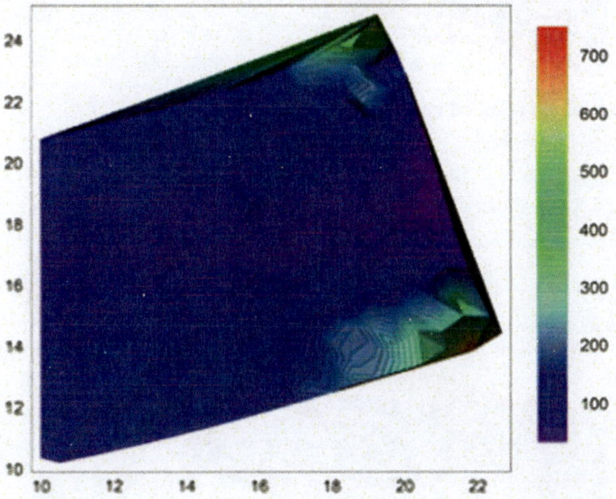

Fig. 18.50 Corresponding von Mises plot

1 to 5, the results are still different (see Figs. 18.51, 18.52, 18.53 and 18.54). As can be seen the results do not behave as we expected and show a wide range of possibilities in which we could choose. Similar results were obtained if we increase

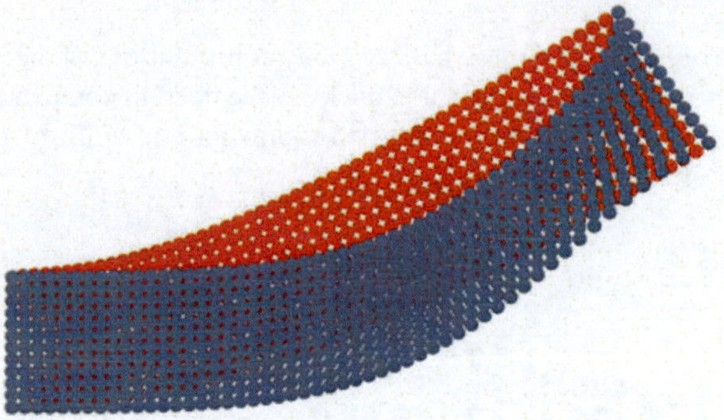

Fig. 18.51 Square lattice

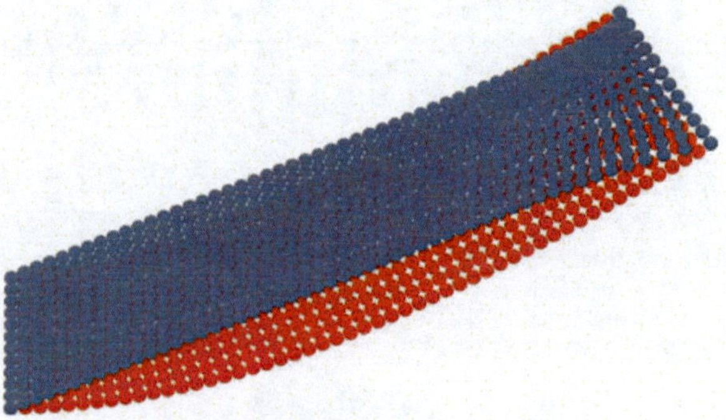

Fig. 18.52 Square lattice reduced neighbor numbers

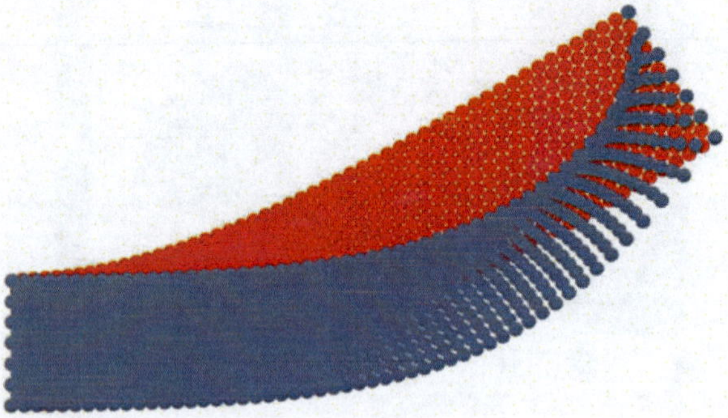

Fig. 18.53 Hexagonal lattice

Fig. 18.54 Hexagonal lattice reduced neighbor numbers

the ratio length/width up to 10 or more. Although the external shape of the beam can sometimes be acceptable, once again we can see as the displacement of the internal particles is quite different from the FEM solutions; it is clear that the differences are sometimes noticeable, as outlined by the von Mises' graphs. So far we stop trial and errors to think about a better understanding of the physic behind the tool and what should drive our choices in the tool to describe material continuum. This will be the object of the next paper.

18.5 Future Work

The presented results are interesting, but they still are at a preliminary stage. We have collected some success showing plausible deformation in different conditions but when we consider a beam under loading, the need to connect constitutive equations with the parameters of our tool emerges powerfully. We have stressed what still need to investigate as a beam under shear stress. Changing the parameters of our tool, we obtain different results nobody of them perfectly coincident with classical solution. The reason of this lies in the fact that, up to now, we do not start from the constitutive equations of the materials to obtain the rules governing points displacement. So far the most important topic to be investigated is concerning how to relate materials parameters with the choices we are doing in our tool. This must be done to achieve a connection with the usual methods of continuum mechanics. As matter of fact, we, actually, have not criteria about how direct our tool's choices to describe a particular material continuum. In the beam deformation, we have assigned Young's modulus and Poisson's coefficient, but there is no relationship between them and the parameters of our model, i.e., the follower's rules. This is the reason of the discrepancy in the resulting deformation compared to FEM solution.

Generalization in 3D of the tool is quite easy but still needs some optimization in the code to keep the computation time in the order of seconds, by using a normal PC desktop.

We are relaxing the hypothesis that the neighbors always are the same to describe liquid and gas; this needs an intermediate calculation step because you have to compute who the neighbors are, defined in this case as the particles inside a specified volume, at each time step.

We are introducing constrained on the particle's motion to describe structured object like pantograph (Giorgio et al. 2016; Boutin et al. 2017; Giorgio et al. 2018; Spagnuolo et al. 2017; De Angelo 2019; Andreaus et al. 2018; Turco et al. 2016; Turco and Rizzi 2016). It can be described as a set of beams with constrained point in the pivot or, in the Hencky vision, can be conceived as a set of points interconnected by springs.

Further developments are concerning different fracture mechanism, different frame to avoid edge effects, other interactions rules and adaptive lattice.

A generalization of the interaction algorithm, to encompass the richness of behavior of different materials like metal or plastic is auspicable, including potentially complex biological tissues (Andreaus et al. 2013; Andreaus et al. 2012; Andreaus et al. 2008; Lekszycki and Dell'Isola 2012; Giorgio et al. 2016). An appropriate potential interaction could take into account different deformation regimes, such as elastic and plastic ones together with fracture. To this aim, pseudoenergetic considerations are introduced also to achieve a better understanding of the process. This is preliminary to introducing potential descriptive interactions depending on the relative distance between the particles, which are able to reproduce the well-known physical behavior.

Cellular automata seems to be a good candidate to enhance our work; a cellular automata is a simple computational mechanism that, for example, changes the color of each cell on a lattice based on the color of neighbors' cells according to a transformation rule. Some attempts, to use them in mechanics, have been done (Dong et al. 2013; Psakhie 2010). Principal limit of cellular automata is regarding as it does not evolve sufficiently, so they quickly reach a limited asymptote in their order of complexity and this will be the object of a future paper. An evolutionary process involving conflict and competition is needed, like in biology systems to overcome this difficult. Moreover, there is no way to predict the outcome of a cellular process without actually running the process. So even though our decisions are determined, there is no way to predetermine what these decisions will be. But the system has succeeded, especially in fluid dynamics to describe complex behavior. The question posed here is concerning if we can work on patterns of information, rather than matter and energy; this question is important and still open. We would like to make a connection with our tool.

Moreover, how stable and robust is the model? What is possible to describe with this model and what are the physical reasons of its success? Is there a hidden dynamic inside? Does a connection exist between pseudoenergy and a real potential? Finally, the mathematical study of the homogenization of lattice systems like the one here considered seems to pose interesting problems, and will probably require non-trivial ideas in the field of functional convergence (Alibert and Della Corte 2015; Dos Reis and Ganghoffer 2011; Dos Reis and Ganghoffer 2012; Rahali et al. 2015; Goda et al. 2013; Alibert et al. 2017). These, and many others, are the object of a next job.

18.6 Conclusion

In this work, we have discussed about a tool, presented in previous works, able to describe strain deformation of a continuum medium taking in account complex physical effects in a plausible way. Flocking rule, used in robot swarm, can be used to describe deformation of bidimensional continuum medium by a simple algorithm highly customizable and able to adapt to take in account complex physical effects in a plausible way. The tool is based on position-based dynamics. Differently from the PBD methods used in computer graphics, we still do not ask for the knowledge of the velocity and do not introduce some kind of forces to take in account mechanical effects. The strain is imposed on some particles (leaders) whose motion is assigned and the other particles (followers) move according to some rules governing particle position. The motion of the followers is determined, like in a bird swarm, by the position of their neighbors. So far the deformed configuration is calculated not by Newton's law but only by the relative positions between the particles of the system, the characteristics of the lattice and by rules describing the how a particle would like to place with respect to its neighbors. Changes of some parameters like, lattice, interaction rules, fracture distance, numbers of neighbors (i.e., to introduce first and second gradient theory) lead to different behavior. One of the principal advantages is in the saving machine time for computing. Computational costs are low because we do not solve differential equations but only algebraically equation systems.

Being based on a linear operation, the computational cost of the algorithm increases linearly with the number of the elements; on the contrary, usually, cost usually associated with FEMs has over-exponential growth.

Essentially, we have to compute the action of a transformation operator between matrices and the job can be parallelized between the GPU cores of the powerful video card, saving computational cost need to solve FEMs.

The tool we propose could be considered as just a graphic representation of a plausible behavior because, actually, we imitate a known behavior adjusting the algorithm parameters.

Anyway, the proposed algorithm is intrinsically accounting for geometrically nonlinear deformations, which is a crucial theme in modern structural mechanics.

This is not a new kind of physics, just a graphic representation of a plausible behavior; keep in mind that, up to now, you do not start from the constitutive equations of the materials leading to the rules governing points displacement. Actually, we just imitate a known behavior adjusting the algorithm parameters. Anyway the results are also interesting even if still in a preliminary form.

The presented results are interesting, but they still are at a preliminary stage. Edge effects are taken in account by a frame, and fracture mechanism is described by a threshold effect. We have showed as changing some parameters like, lattice, interaction rules, fracture distance, numbers of neighbors much different behavior can be described.

In a preceding work, we verified the results of this tool that are in accordance with results obtained by FEM, also in the fracture case; the results showed good similarity between the prediction of the discrete model and the predictions of standard FEM simulations. In particular, some aspects in the geometry of the deformed configurations in the continuous case are found for the discrete system as well. Moreover, a fracture algorithm was introduced, and some results, including periodic crack formation, were provided. Also in this case, a comparison with the continuous case shows a good agreement between the fracture front geometry (discrete system) and the most stressed area (continuous system).

We have also used an ASTM shape whose results will be compared in a real tensile test experiment, but many questions are still open. We have collected some success showing plausible deformation in different conditions, but when we consider a beam under loading, we fail and the need to connect constitutive equations with the parameters of our tool emerges powerfully. However, the tool has demonstrated enough flexibility to give chances that, once connected with the constitutive parameters, it can describe the richness of behavior of different materials, including potentially complex biological tissues.

References

Abali, B. E., Müller, W. H., & Dell'Isola, F. (2017). Theory and computation of higher gradient elasticity theories based on action principles. *Arch. Appl. Mech. 87*(9), 1495–1510.

Alibert, J.-J., & Della Corte, A. (2015). Second-gradient continua as homogenized limit of pantographic microstructured plates: a rigorous proof. *Z. Für Angew. Math. Phys., 66*(5), 2855–2870.

Alibert, J.-J., Della Corte, A., Giorgio, I., & Battista, A. (2017). Extensional Elastica in large deformation as $$\Gamma $$-limit of a discrete 1D mechanical system. *Z. Für Angew. Math. Phys., 68*(2), 42.

Alibert, J.-J., Seppecher, P., & Dell'isola, F. (2003). Truss modular beams with deformation energy depending on higher displacement gradients. *Math. Mech. Solids, 8*(1), 51–73.

Altenbach, J., Altenbach, H., & Eremeyev, V. A. (2010). On generalized Cosserat-type theories of plates and shells: a short review and bibliography. *Arch. Appl. Mech., 80*(1), 73–92.

Altenbach, H., Bîrsan, M., & Eremeyev, V. A. (2013). Cosserat-type rods. In *Generalized Continua from the Theory to Engineering Applications* (pp. 179–248). Heidelberg: Springer.

Altenbach, H., & Eremeyev, V. A. (2009). On the linear theory of micropolar plates. *ZAMM-Journal Appl. Math. Mech. Für Angew. Math. Mech., 89*(4), 242–256.

Altenbach, H., & Eremeyev, V. A. (2013). *Generalized Continua From the Theory to Engineering Applications, CISM Courses and Lectures* (Vol. 541). Udine: Springer.

Altenbach, H., Eremeyev, V. A., Lebedev, L. P., & Rendón, L. A. (2010). Acceleration waves and ellipticity in thermoelastic micropolar media. *Arch. Appl. Mech., 80*(3), 217–227.

Andreaus, U., Colloca, M., & Iacoviello, D. (2012). An optimal control procedure for bone adaptation under mechanical stimulus. *Control Eng. Pract., 20*(6), 575–583.

Andreaus, U., Colloca, M., & Toscano, A. (2008). Mechanical behaviour of physiological and prosthesized human femurs during stair climbing: A comparative analysis via 3D numerical simulation. *Minerva Ortop. E Traumatol., 59*(4), 213–220.

Andreaus, U., Giorgio, I., & Lekszycki, T. (2013). *A 2-D Continuum Model of a Mixture of Bone Tissue and Bio-resorbable Material for Simulating Mass Density Redistribution Under Load Slowly Variable in Time* [Online]. Available at: http://www.scopus.com/inward/record.url?eid=2-s2.0-84882523299&partnerID=40&md5=e34b49c6b02b11f8e584e3cb57b5db43.

Andreaus, U., Giorgio, I., & Madeo, A. (2015). Modeling of the interaction between bone tissue and resorbable biomaterial as linear elastic materials with voids. *Z. Für Angew. Math. Phys., 66*(1), 209–237.

Andreaus, U., & Placidi, L. (2013). At the origins and in the vanguard of peri-dynamics, non-local and higher gradient continuum mechanics. An underestimated and still topical contribution of Gabrio Piola. ArXiv Prepr. ArXiv13105599 [Online]. Available at: http://arxiv.org/abs/1310.5599. Consultato: 17-gen-2014.

Andreaus, U., Spagnuolo, M., Lekszycki, T., & Eugster, S. R. (2018). A Ritz approach for the static analysis of planar pantographic structures modeled with nonlinear Euler–Bernoulli beams. *Contin. Mech. Thermodyn., 30*(5), 1103–1123.

Auffray, N., dell'Isola, F., Eremeyev, V. A., Madeo, A., & Rosi, G. (2015). Analytical continuum mechanics à la Hamilton–Piola least action principle for second gradient continua and capillary fluids. *Math. Mech. Solids, 20*(4), 375–417.

Barchiesi, E., Spagnuolo, M., & Placidi, L. (2018). Mechanical metamaterials: A state of the art. *Math. Mech. Solids, 24*(1), 212–234.

Battista, A., Rosa, L., dell'Erba, R., & Greco, L. (2016). Numerical investigation of a particle system compared with first and second gradient continua: Deformation and fracture phenomena. *Math. Mech. Solids*, p. 1081286516657889.

Bender, J., Koschier, D., Charrier, P., & Weber, D. (2014). Position-based simulation of continuous materials. *Comput. Graph, 44*, 1–10.

Bender, J., Müller, M., & Macklin, M. (2015). Position-based simulation methods in computer graphics. In *Eurographics (Tutorials)* [Online]. Available at: https://www.researchgate.net/profile/Jan_Bender/publication/274940214_Position-Based_Simulation_Methods_in_Computer_Graphics/links/552cc4a40cf29b22c9c466df/Position-Based-Simulation-Methods-in-Computer-Graphics.pdf. Consultato: 06-set-2017.

Berezovski, A., Giorgio, I., & Corte, A. D. (2016). Interfaces in micromorphic materials: Wave transmission and reflection with numerical simulations. *Math. Mech. Solids, 21*(1), 37–51.

Bilotta, A., & Turco, E. (2009). A numerical study on the solution of the Cauchy problem in elasticity. *Int. J. Solids Struct., 46*(25–26), 4451–4477.

Boutin, C., Giorgio, I., & Placidi, L. (2017). Linear pantographic sheets: Asymptotic micro-macro models identification. *Math. Mech. Complex Syst., 5*(2), 127–162.

Bückmann, T., et al. (2012). Tailored 3D mechanical metamaterials made by dip-in direct-laser-writing optical lithography. *Adv. Mater., 24*(20), 2710–2714.

Carcaterra, A., Dell'Isola, F., Esposito, R., & Pulvirenti, M. (2015). Macroscopic description of microscopically strongly inhomogenous systems: A mathematical basis for the synthesis of higher gradients metamaterials. *Arch. Ration. Mech. Anal., 218*(3), 1239–1262.

Cazzani, A., Stochino, F., & Turco, E. (2016). An analytical assessment of finite element and isogeometric analyses of the whole spectrum of Timoshenko beams. *ZAMM-Journal Appl. Math. Mech. Für Angew. Math. Mech., 96*(10), 1220–1244.

Cecchi, A., & Rizzi, N. L. (2001). Heterogeneous elastic solids: A mixed homogenization-rigidification technique. *Int. J. Solids Struct., 38*(1), 29–36.

Chang, C. S., & Misra, A. (1990). Application of uniform strain theory to heterogeneous granular solids. *J. Eng. Mech., 116*(10), 2310–2328.

Contrafatto, L., Cuomo, M., & Fazio, F. (2012). An enriched finite element for crack opening and rebar slip in reinforced concrete members. *Int. J. Fract., 178*(1–2), 33–50.

Cuomo, M., Dell'Isola, F., Greco, L., & Rizzi, N. L. (2017). First versus second gradient energies for planar sheets with two families of inextensible fibres: investigation on deformation boundary layers, discontinuities and geometrical instabilities. *Compos. Part B Eng., 115*, 423–448.

Cuomo, M., & Greco, L. (2012). *Isogeometric Analysis of Space Rods: Considerations on Stress Locking* (pp. 5094–5112) [Online]. Available at: http://www.scopus.com/inward/record.url?eid=2-s2.0-84871627441&partnerID=40&md5=48d09dd7e5493bafe0ef2bb10904d094.

d'Agostino, M. V., Giorgio, I., Greco, L., Madeo, A., & Boisse, P. (2015). Continuum and discrete models for structures including (quasi-) inextensible elasticae with a view to the design and modeling of composite reinforcements. *Int. J. Solids Struct., 59*, 1–17.

De Angelo, M., et al. (2019). The macroscopic behavior of pantographic sheets depends mainly on their microstructure: Experimental evidence and qualitative analysis of damage in metallic specimens. *Contin. Mech. Thermodyn.*, 1–23.

Del Vescovo, D., & Giorgio, I. (2014). Dynamic problems for metamaterials: review of existing models and ideas for further research. *Int. J. Eng. Sci., 80*, 153–172.

dell'Erba, R. (2015). Determination of spatial configuration of an underwater swarm with minimum data. *Int. J. Adv. Robot. Syst., 12*(7), 97.

dell'Erba, R. (2018a). Swarm robotics and complex behaviour of continuum material. *Contin. Mech. Thermodyn.*, Mag. 2018 (Online). Available at: https://doi.org/10.1007/s00161-018-0675-1.

dell'Erba, R. (2018b). Position-based dynamic of a particle system: a configurable algorithm to describe complex behaviour of continuum material starting from swarm robotics. *Contin. Mech. Thermodyn.*, 1–22.

dell'Erba, R. (2018c). Position-based dynamic of a particle system: A configurable algorithm to describe complex behaviour of continuum material starting from swarm robotics., *Contin. Mech. Thermodyn., 30*(5), 1069–1090.

Dell'Erba, R., & Moriconi, C. Bio-inspired Robotics — it. [Online]. Available at: http://www.enea.it/it/produzione-scientifica/edizioni-enea/2014/bio-inspirede-robotics-proceedings. Consultato: 15-dic-2014.

Dell'Isola, F., et al. (2019). Pantographic metamaterials: An example of mathematically driven design and of its technological challenges. *Contin. Mech. Thermodyn., 31*(4), 851–884.

dell'Isola, F., Auffray, N., Eremeyev, V. A., Madeo, A., Placidi, L., & Rosi, G. (2014). Least action principle for second gradient continua and capillary fluids: A Lagrangian approach following Piola's point of view. In *The Complete Works of Gabrio Piola: Volume I* (pp. 606–694). Heidelberg: Springer.

Dell'Isola, F., Bucci, S., & Battista, A. (2016). Against the fragmentation of knowledge: The power of multidisciplinary research for the design of metamaterials. In *Advanced Methods of Continuum Mechanics for Materials and Structures* (pp. 523–545). Heidelberg: Springer.

Dell'Isola, F., d'Agostino, M. V., Madeo, A., Boisse, P., & Steigmann, D. (2016). Minimization of shear energy in two dimensional continua with two orthogonal families of inextensible fibers: The case of standard bias extension test. *J. Elast., 122*(2), 131–155.

Dell'Isola, F., Della Corte, A., Greco, L., & Luongo, A. (2016). Plane bias extension test for a continuum with two inextensible families of fibers: A variational treatment with Lagrange multipliers and a perturbation solution. *Int. J. Solids Struct., 81*, 1–12.

Dell'Isola, F., & Gavrilyuk, S. (2012). *Variational Models and Methods in Solid and Fluid Mechanics* (Vol. 535). Springer Science & Business Media.

dell'Isola, F., Giorgio, I., & Andreaus, U. (2015). Elastic pantographic 2D lattices: A numerical analysis on the static response and wave propagation. In *Proc. Est. Acad. Sci.* (Vol. 64, N. 3, p. 219).

Dell'Isola, F., Giorgio, I., Pawlikowski, M., & Rizzi, N. L. (2016). Large deformations of planar extensible beams and pantographic lattices: Heuristic homogenization, experimental and numerical examples of equilibrium. In *Proc. R. Soc. Math. Phys. Eng. Sci.* (Vol. 472, No 2185, p. 20150790).

Dell'Isola, F., Madeo, A., & Placidi, L. (2011). Linear plane wave propagation and normal transmission and reflection at discontinuity surfaces in second gradient 3D continua. *ZAMM Z. Angew. Math. Mech., 92*(1), 52–71.

Dell'Isola, F., Madeo, A., & Seppecher, P. (2016). Cauchy tetrahedron argument applied to higher contact interactions. *Arch. Ration. Mech. Anal., 219*(3), 1305–1341.

Dell'Isola, F., & Placidi, L. (2011). Variational principles are a powerful tool also for formulating field theories. In *Variational models and methods in solid and fluid mechanics* (pp. 1–15). Heidelberg: Springer.

Dell'Isola, F., Sciarra, G., & Vidoli, S. (2009). Generalized hooke's law for isotropic second gradient materials. *Proc. R. Soc. Math. Phys. Eng. Sci., 465*(2107), 2177–2196.

Dell'Isola, F., & Seppecher, P. (1995). *The Relationship Between Edge Contact Forces, Double Forces and Interstitial Working Allowed by the Principle of Virtual Power*.

Dell'isola, F., & Seppecher, P. (1997). Edge contact forces and quasi-balanced power. *Meccanica, 32*(1), 33–52.

Dell'Isola, F., Seppecher, P., & Corte, A. D. (2015). The postulations á la D'Alembert and á la Cauchy for higher gradient continuum theories are equivalent: A review of existing results. In *Proc. R. Soc. Math. Phys. Eng. Sci., 471*(2183), 20150415.

dell'Isola, F., Seppecher, P., & Madeo, A. (2012). How contact interactions may depend on the shape of Cauchy cuts in Nth gradient continua: Approach "à la D'Alembert". *Z. Angew. Math. Phys., 63*(6), 1119–1141.

Dell'Isola, F., Steigmann, D., & Della Corte, A. (2015). Synthesis of fibrous complex structures: Designing microstructure to deliver targeted macroscale response. *Appl. Mech. Rev., 67*(6), 060804.

Della Corte, A., dell'Isola, F., Esposito, R., & Pulvirenti, M. (2017). Equilibria of a clamped Euler beam (Elastica) with distributed load: Large deformations. *Math. Models Methods Appl. Sci., 27*(08), 1391–1421.

Diziol, R., Bender, J., & Bayer, D. (2011). Robust real-time deformation of incompressible surface meshes. In *Proceedings of the 2011 ACM SIGGRAPH/Eurographics Symposium on Computer Animation*, New York, NY, USA (pp. 237–246) [Online]. Available at: http://doi.acm.org/10.1145/2019406.2019438.

Dong, Y., Zhang, G., Xu, A., & Gan, Y. (2013). Cellular automata model for elastic solid material. *Commun. Theor. Phys., 59*(1), 59–67.

Dos Reis, F., & Ganghoffer, J.-F. (2011). Construction of micropolar continua from the homogenization of repetitive planar lattices. In *Mechanics of generalized continua* (pp. 193–217). Heidelberg: Springer.

Dos Reis, F., & Ganghoffer, J. F. (2012). Equivalent mechanical properties of auxetic lattices from discrete homogenization. *Comput. Mater. Sci., 51*(1), 314–321.

Enakoutsa, K., Corte, A. D., & Giorgio, I. (2016). A model for elastic flexoelectric materials including strain gradient effects. *Math. Mech. Solids, 21*(2), 242–254.

Eremeyev, V. A., Ivanova, E. A., & Indeitsev, D. A. (2010). Wave processes in nanostructures formed by nanotube arrays or nanosize crystals. *J. Appl. Mech. Tech. Phys., 51*(4), 569–578.

Eremeyev, V. A., Ivanova, E. A., Morozov, N. F., & Solov'ev, A. N. (2006). On the determination of eigenfrequencies for nanometer-size objects. *Doklady Physics, 51*, 93–97.

Eremeyev, V. A., Ivanova, E. A., Morozov, N. F., & Strochkov, S. E. (2007). The spectrum of natural oscillations of an array of micro-or nanospheres on an elastic substrate. *Doklady Physics, 52*, 699–702.

Eremeyev, V. A., Lebedev, L. P., & Altenbach, H. (2012). *Foundations of Micropolar Mechanics*. Springer Science & Business Media.

Eremeyev, V. A., & Pietraszkiewicz, W. (2012). Material symmetry group of the non-linear polar-elastic continuum. *Int. J. Solids Struct., 49*(14), 1993–2005.

Eringen, A. C. (2012). *Microcontinuum Field Theories: I. Foundations and Solids*. Springer Science & Business Media.

Ern, A., & Guermond, J.-L. (2013). *Theory and Practice Of Finite Elements* (vol. 159). Springer Science & Business Media.

Forest, S. (2009). Micromorphic approach for gradient elasticity, viscoplasticity, and damage. *J. Eng. Mech., 135*(3), 117–131.

Forest, S., Cordero, N. M., & Busso, E. P. (2011). First vs. second gradient of strain theory for capillarity effects in an elastic fluid at small length scales. *Comput. Mater. Sci., 50*(4), 1299–1304.

Gabriele, S., Rizzi, N. L., & Varano, V. (2014). A one-dimensional nonlinear thin walled beam model derived from Koiter shell theory. *Civ.-Comp Proc. 106*.

Germain, P. (1973). The method of virtual power in continuum mechanics. Part 2: Microstructure. *SIAM J. Appl. Math., 25*(3), 556–575.

Giorgio, I., Andreaus, U., Scerrato, D., & Dell'Isola, F. (2016). A visco-poroelastic model of functional adaptation in bones reconstructed with bio-resorbable materials. *Biomech. Model. Mechanobiol., 15*(5), 1325–1343.

Giorgio, I., Della Corte, A., dell'Isola, F., & Steigmann, D. J. (2016). Buckling modes in pantographic lattices. *Comptes Rendus Mécanique, 344*(7), 487–501.

Giorgio, I., Galantucci, L., Della Corte, A., & Del Vescovo, D. (2015). Piezo-electromechanical smart materials with distributed arrays of piezoelectric transducers: Current and upcoming applications. *Int. J. Appl. Electromagn. Mech., 47*(4), 1051–1084.

Giorgio, I., Harrison, P., Dell'Isola, F., Alsayednoor, J., & Turco, E. (2018). Wrinkling in engineering fabrics: A comparison between two different comprehensive modelling approaches. *Proc. R. Soc. Math. Phys. Eng. Sci., 474*(2216), 20180063.

Goda, I., Assidi, M., Belouettar, S., & Ganghoffer, J. F. (2012). A micropolar anisotropic constitutive model of cancellous bone from discrete homogenization. *J. Mech. Behav. Biomed. Mater., 16*, 87–108.

Goda, I., Assidi, M., & Ganghoffer, J.-F. (2013). Equivalent mechanical properties of textile monolayers from discrete asymptotic homogenization. *J. Mech. Phys. Solids, 61*(12), 2537–2565.

Goda, I., Assidi, M., & Ganghoffer, J.-F. (2014). A 3D elastic micropolar model of vertebral trabecular bone from lattice homogenization of the bone microstructure. *Biomech. Model. Mechanobiol., 13*(1), 53–83.

Greco, L., & Cuomo, M. (2013). B-Spline interpolation of Kirchhoff-Love space rods. *Comput. Methods Appl. Mech. Eng., 256*, 251–269.

Greco, L., & Cuomo, M. (2014). An implicit G1 multi patch B-spline interpolation for Kirchhoff-Love space rod. *Comput. Methods Appl. Mech. Eng., 269*, 173–197.

Janson, S., Middendorf, M., & Beekman, M. (2005). Honeybee swarms: how do scouts guide a swarm of uninformed bees? *Anim. Behav., 70*(2), 349–358.

Javili, A., Dell'Isola, F., & Steinmann, P. (2013). Geometrically nonlinear higher-gradient elasticity with energetic boundaries. *J. Mech. Phys. Solids, 61*(12), 2381–2401.

Karaboga, D. (2005). *An Idea Based on Honey Bee Swarm for Numerical Optimization.* Technical Report-tr06, Erciyes University, Engineering Faculty, Computer Engineering Department, [Online]. Available at: http://www-lia.deis.unibo.it/Courses/SistInt/articoli/bee-colony1.pdf. Consultato: 07-nov-2014.

Khatib, O, Kumar, V., & Rus, D. (2008). *Experimental Robotics: The 10th International Symposium on Experimental Robotics.* Heidelberg: Springer.

Ladevèze, P. (2012). *Nonlinear Computational Structural Mechanics: New Approaches and Non-incremental Methods of Calculation.* Springer Science & Business Media.

Lanczos, C. (2012). *The Variational Principles of Mechanics.* Courier Corporation.

Lekszycki, T., & Dell'Isola, F. (2012). A mixture model with evolving mass densities for describing synthesis and resorption phenomena in bones reconstructed with bio-resorbable materials. *ZAMM Z. Angew. Math. Mech., 92*(6), 426–444.

Macklin, M., Müller, M., & Chentanez, N. (2016). *XPBD: Position-Based Simulation of Compliant Constrained Dynamics* (pp. 49–54) [Online]. Available at: http://dl.acm.org/citation.cfm?doid=2994258.2994272. Consultato: 06-ott-2017.

Madeo, A., Dell'Isola, F., & Darve, F. (2013). A continuum model for deformable, second gradient porous media partially saturated with compressible fluids. *J. Mech. Phys. Solids, 61*(11), 2196–2211.

Madeo, A., Dell'Isola, F., Ianiro, N., & Sciarra, G. (2008). A variational deduction of second gradient poroelasticity II: An application to the consolidation problem. *J. Mech. Mater. Struct., 3*(4), 607–625.

Madeo, A., Placidi, L., & Rosi, G. (2014). Towards the design of metamaterials with enhanced damage sensitivity: second gradient porous materials. *Res. Nondestruct. Eval., 25*(2), 99–124.

Masiani, R., Rizzi, N. & Trovalusci, P. (1995). Masonry as structured continuum. *Meccanica, 30*(6), 673–683.

Milton, G., & Seppecher, P. (2012). A metamaterial having a frequency dependent elasticity tensor and a zero effective mass density. *Phys. Status Solidi B, 249*(7), 1412–1414.

Mindlin, R. D. (1964). Micro-structure in linear elasticity. *Arch. Ration. Mech. Anal., 16*(1), 51–78.

Moriconi,C., & dell'Erba, R. (2012). The localization problem for harness: A multipurpose robotic swarm. In *SENSORCOMM 2012, The Sixth International Conference on Sensor Technologies and Applications* (pp. 327–333) [Online]. Available at: http://www.thinkmind.org/index.php?view=article&articleid=sensorcomm_2012_14_20_10138. Consultato: 04-apr-2014.

Passino, K. M., Seeley, T. D., & Visscher, P. K. (2007). Swarm cognition in honey bees. *Behav. Ecol. Sociobiol., 62*(3), 401–414.

Pietraszkiewicz, W. & Eremeyev, V. A. (2009). On natural strain measures of the non-linear micropolar continuum. *Int. J. Solids Struct., 46*(3), 774–787.

Pietraszkiewicz, W., & Eremeyev, V. A. (2009). On vectorially parameterized natural strain measures of the non-linear Cosserat continuum. *Int. J. Solids Struct., 46*(11–12), 2477–2480.

Placidi, L. (2015). A variational approach for a nonlinear 1-dimensional second gradient continuum damage model. *Contin. Mech. Thermodyn., 27*(4–5), 623–638.

Placidi, L., Dell'Isola, F., Ianiro, N., & Sciarra, G. (2008). Variational formulation of pre-stressed solid-fluid mixture theory, with an application to wave phenomena. *Eur. J. Mech. ASolids, 27*(4), 582–606.

Placidi, L., Faria, S. H., & Hutter, K. (2004). On the role of grain growth, recrystallization and polygonization in a continuum theory for anisotropic ice sheets. *Ann. Glaciol., 39*, 49–52.

Placidi, L., Giorgio, I., Della Corte, A., & Scerrato, D. (2017). Euromech 563 Cisterna di Latina 17–21 March 2014 Generalized continua and their applications to the design of composites and metamaterials: A review of presentations and discussions. *Math. Mech. Solids, 22*(2), 144–157.

Placidi, L., Greve, R., Seddik, H., & Faria, S. H. (2010). Continuum-mechanical, Anisotropic Flow model for polar ice masses, based on an anisotropic flow enhancement factor. *Contin. Mech. Thermodyn., 22*(3), 221–237.

Placidi, L., & Hutter, K. (2006). Thermodynamics of polycrystalline materials treated by the theory of mixtures with continuous diversity. *Contin. Mech. Thermodyn., 17*(6), 409–451.

Placidi, L., Rosi, G., Giorgio, I., & Madeo, A. (2013). Reflection and transmission of plane waves at surfaces carrying material properties and embedded in second-gradient materials. *Math. Mech. Solids, 19*(5), 555–578.

Konovalenko, Ig. S., Smolin, A. Yu., & Psakhie, S. G. (2010). Multilevel simulation of deformation and fracture of brittle porous materials in the method of movable cellular automata. *Phys. Mesomech., 13*(1–2), 47–53.

Rahali, Y., Giorgio, I., Ganghoffer, J. F., & Dell'Isola, F. (2015). Homogenization à la Piola produces second gradient continuum models for linear pantographic lattices. *Int. J. Eng. Sci., 97*, 148–172.

Rivers, A. R., & James, D. (2007). *FastLSM: Fast Lattice Shape Matching for Robust Real-Time Deformation* (Vol. 26).

Rosi, G., Giorgio, I., & Eremeyev, V. A. (2013). Propagation of linear compression waves through plane interfacial layers and mass adsorption in second gradient fluids. *ZAMM-Journal Appl. Math. Mech. Für Angew. Math. Mech., 93*(12), 914–927.

Scerrato, D., Giorgio, I., Della Corte, A., Madeo, A., & Limam, A. (2015). A micro-structural model for dissipation phenomena in the concrete. *Int. J. Numer. Anal. Methods Geomech., 39*(18), 2037–2052.

Seddik, H., Greve, R., Placidi, L., Hamann, I., & Gagliardini, O. (2008). Application of a continuum-mechanical model for the flow of anisotropic polar ice to the EDML core, Antarctica. *J. Glaciol., 54*(187), 631–642.

Seppecher, P., Alibert, J.-J., & Isola, F. D. (2011). Linear elastic trusses leading to continua with exotic mechanical interactions. *Journal of Physics: Conference Series, 319*, 012018.

Spagnuolo, M., Barcz, K., Pfaff, A., Dell'Isola, F., & Franciosi, P. (2017). Qualitative pivot damage analysis in aluminum printed pantographic sheets: Numerics and experiments. *Mech. Res. Commun., 83*, 47–52.

Steigmann, D. J. (2008). Two-dimensional models for the combined bending and stretching of plates and shells based on three-dimensional linear elasticity. *Int. J. Eng. Sci., 46*(7), 654–676.

Steigmann, D. J. (2009). A concise derivation of membrane theory from three-dimensional nonlinear elasticity. *J. Elast., 97*(1), 97–101.

Steigmann, D. J. (2010). Applications of polyconvexity and strong ellipticity to nonlinear elasticity and elastic plate theory. In *Poly-, Quasi-and rank-one convexity in applied mechanics* (pp. 265–299). Heidelberg: Springer.

Steigmann, D. J. (2013). Koiter's shell theory from the perspective of three-dimensional nonlinear elasticity. *J. Elast., 111*(1), 91–107.

Steigmann, D. J., & Pipkin, A. C. (1991). Equilibrium of elastic nets. *Philos. Trans. R. Soc. Lond. Ser. Phys. Eng. Sci., 335*(1639), 419–454.

Turco, E., Barcz, K., Pawlikowski, M., & Rizzi, N. L. (2016). Non-standard coupled extensional and bending bias tests for planar pantographic lattices. Part I: numerical simulations. *Z. Für Angew. Math. Phys., 67*(5), 122.

Turco, E., Dell'Isola, F., Cazzani, A., & Rizzi, N. L. (2016). Hencky-type discrete model for pantographic structures: Numerical comparison with second gradient continuum models. *Z. Für Angew. Math. Phys., 67*(4), 85.

Turco, E., Giorgio, I., Misra, A., & Dell'Isola, F. (2017). King post truss as a motif for internal structure of (meta) material with controlled elastic properties. *R. Soc. Open Sci., 4*(10), 171153.

Turco, E., & Rizzi, N. L. (2016). Pantographic structures presenting statistically distributed defects: Numerical investigations of the effects on deformation fields. *Mech. Res. Commun., 77*, 65–69.

Chapter 19
Mud Shrinkage and Cracking Phenomenon Experimental Identification Using Digital Image Correlation

Mahdia Hattab, Said Taibi, and Jean-Marie Fleureau

Abstract The main goal of this research was to try, experimentally, to identify the cause of initiation and propagation of cracks network related to desiccation. The work may provide a better understanding of the drying behavior in clayey soils. Kaolin, which is a little swelling "nearly pure" kaolinite, was studied in this context. The experimental method is based on the determination of the local two-dimensional strains and displacements fields using the softwares Vic-2D and Vic-3D on thin layers of the initially saturated clay. Using digital image correlation technique at the macroscopic level, the analyses of strain fields during drying allow to clearly identify and characterize different phenomena such as: shrinkage, stress concentration, initiation, and propagation of cracks.

19.1 Introduction

Soils cracking phenomenon due to desiccation is among the more complex issues that arise in the geotechnical engineering field. The phenomenon, related to the clayey soil behavior, can have a severe impact on the bulk properties of soils, reducing strength, affecting compression ability, increasing permeability, etc. Consequently, the mechanical performances, as well as the hydric properties of the material, weaken, causing in many circumstances, sometime dramatically, damages in geotechnical structures, earthen structures, and soil-supported structures. One can

M. Hattab (✉)
Laboratoire d'Etude des Microstructures et de Mécanique des Matériaux, Université de Lorraine, CNRS UMR 7239, Arts et Métiers ParisTech, 57000 Metz, France
e-mail: mahdia.hattab@univ-lorraine.fr

S. Taibi
Laboratoire Ondes et Milieux Complexes, CNRS UMR 6294, Université Le Havre Normandie, 53 Rue de Prony, BP540, 76058 Le Havre CEDEX, France

J.-M. Fleureau
Laboratoire de Mécanique des Sols, Structures et Matériaux, CNRS UMR 8579, Université Paris-Saclay, CentraleSupélec, 3 Rue Joliot Curie, 91190 Gif-sur-Yvette, France

F. dell'Isola and L. Igumnov (eds.), *Dynamics, Strength of Materials and Durability in Multiscale Mechanics*, Advanced Structured Materials 137,
https://doi.org/10.1007/978-3-030-53755-5_19

quote for instance structures such as engineering barriers, consisting of clayey barrier system that seals wastes; the latter might be with different dangerousness level. The barrier is used to ensure the isolation and the confinement of wastes (Albrecht and Benson 2001). It thus appears obvious that any variation in the initial expected performances of this kind of structure, especially of the compressibility and of the permeability, risks a loss of sealing and leaks of dangerous substances into the environment. Hence, the clayey barriers behavior has to be precisely followed and mastered. In common geotechnical engineering practice, desiccation cracks are widely reported in clayey soils, see for instance Longwell (1928), Willden and Mabey (1961), Mitchell (1986), Morris et al. (1992), Kong (1994).

Because of ecological and environmental concerns, eco-geo-materials such as raw earth building materials, having a minimum embodied energy, are currently receiving special attention (Gallipoli et al. 2017). Treated with a low percentage of binders, the raw earth becomes a suitable eco-material in the raw earth eco-structures. One of the used techniques concerns, for instance, the casted raw earth for structural and non-structural walls. The technique in fact is not new, it has been developed all over the centuries in the past to build small structures. Today the material finds a real keen interest and is more and more used in the eco-construction sector. Casted raw earth requires the material to be prepared at water contents reaching the liquid limit of the used clayey earth. Consequently, many problems related to the preparation method can appear such as shrinkage desiccation resulting in an increased risk of cracking. To correct the imperfections of this type of eco-materials, aiming to generalize their use, it is necessary to understand the mechanisms of shrinkage and of the initiation and propagation of desiccation cracking. All these issues involve the shrinkage-swelling phenomenon, defined as the volume changes of the soils under the action of capillary stress. However, the essential mechanisms of desiccation cracking are not well understood in clayey materials, and consequently, predictive tools are inadequately developed. It should be pointed out here that works on other materials such as concrete, for instance, can provide large lighting on how to model such complex material during drying. Modeling based on the high ability of the material to dissipate excess energy could be an interesting approach to consider and to adapt to clayey soils under high suctions (Giorgio and Scerrato 2017; Giorgio et al. 2019).

One of the hypotheses put forward in the literature to explain the initiation of cracks in soft mud clays (the material being initially close to saturation and under drying) is related to a blockage of the shrinkage process. This rather local mechanism, related to different parameters of structural origin and/or geometrical boundary conditions, gives rise to a concentration of stresses which, when they reach the failure limit of the clay, lead to crack initiation.

Experimental investigations showed that the locally developed stresses are often of tensile type (Corte and Higashi 1960; Lachenbruch 1961; Lloret et al. 1998; Péron 2008; Péron et al. 2009; Wei 2014; Eid et al. 2015), and that cracking propagates in mode I. Assuming anisotropic shrinkage in a specimen of small thickness and rectangular surface as suggested in Wei et al. (2016), the works of Wei (2014), Ighil Ameur (2016) and more recently Cheng et al. (2020) highlighted the effect of stress concentration development. Experimentally, the process can be indirectly observed through the evolution of a local strain field of mechanical type, obtained by difference between the measured total local strain field (using digital image correlation) and the estimated anisotropic shrinkage strain field. This approach permits to evaluate the risk of crack initiation during drying.

19.2 Basic Notions—Hypotheses and Soil–Water Characteristic Curve

During drying with no external mechanical stresses, the volume of a soil sample changes (shrinkage) because the water in the soil becomes subjected to a tension state (the air being at the atmospheric pressure). More generally, the difference between the air (u_a) and the water (u_w) pore pressures, or capillary pressure (u_c), is the main macroparameter causing the movements of the fluids in the porous medium:

$$u_c = u_a - u_w \tag{19.1}$$

Thus, the shrinkage in clayey soils is strongly related to local mechanisms that induce water movements between pores, quite similar to that defined in granular materials (Scholtès et al. 2009; Yuan and Chareyre 2017). At the scale of the pore, the difference in pressure (Eq. 19.1) results in meniscus forming between air and water, characterized by the surface tension of the water and the interface curvature (Laplace's law). As mentioned for instance by Biarez et al. (1987), the concept of capillary pressure may be defined as a global notion, at larger scale than the molecular scale. The concept includes both capillary properties of the medium through the pore dimensions, and the adsorption properties (in clays cases) through the shape of the solid–liquid–air membrane. Considering the electrical forces (of van der Waals type) in the vicinity of the clay particles, the connection angle itself may be considered as a macroscopic parameter involving the wetting concept of the soil material (Fleureau et al. 1988; Santamarina 2001; Lu et al. 2010).

Let us consider in this study macroscopic behavior, which we know is strongly dependent on the microscopic mechanisms. The shrinkage phenomenon may be placed in the formal framework of continuous media mechanics. Without external mechanical stresses, the capillary pressure becomes similar to suction. The shrinkage in clayey mud during drying can be addressed in different steps as illustrated on Fig. 19.1.

(i) Drying of the initially saturated clay provokes a progressive loss of mass caused by water evaporation. Hence, the water content changes, leading to (ii) suction development in the material, which induces (iii) volumetric strains and the decrease of the volume of the specimen. During the process, one can assume that the saturation of the material is maintained as long as the point of air entry is not reached yet. Meanwhile,

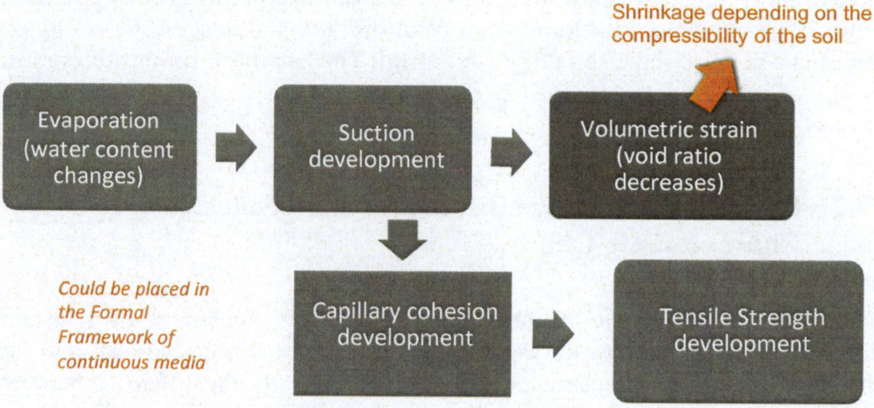

Fig. 19.1 Shrinkage phenomenon during drying—macroscopic considerations

from the suction increasing results (iv) the capillary cohesion development, which in turn increases the (v) strength of the clay.

Steps (i), (ii), and (iii)

The relationships between the water content change and the suction development in the clayey mud can be deduced experimentally using several techniques, depending on the range in which suction has to be applied (Fleureau et al. 1993). The tests are generally performed on small samples of about 1 cm^3 in volume. The determination of the soil–water characteristic curve (SWCC) can thus be defined in terms of void ratio as a function of suction (Fleureau et al. 1993; Wei 2014), or in terms of volumetric water content versus suction as shown in Fig. 19.2. The tests here were performed by different authors, in kaolin clay, and the results were summarized by Yang (2020).

Van Genuchten (1980) proposed an equation linking the volumetric water content θ with suction s (Eq. 19.2 and Fig. 19.2) where θ_s, θ_r, are the volumetric water contents in the saturated state and in the residual state (very dry state of the clay), respectively, n and α are material parameters.

$$\theta = \theta_r + \frac{\theta_s - \theta_r}{[1 + (s/\alpha)^n]^{1-1/n}} \tag{19.2}$$

Fredlund (2002) equation links the void ratio variation (which can also be written in terms of volumetric strain) to the gravimetric water content variation (Eq. 19.3):

$$\begin{cases} e = a_0\left[\left(\frac{w}{b_0}\right) + 1\right]^{1/c_0} \\ b_0 = \frac{S_{r0}a_0}{G_s} \end{cases} \tag{19.3}$$

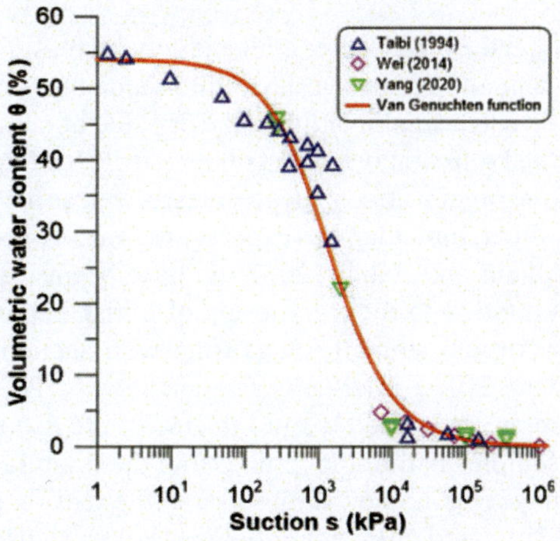

Fig. 19.2 (s–θ) change during drying relative to *steps i–ii*—water content at $w_0 = 2\ w_L$ (Yang 2020)

where a_0 is the minimum void ratio upon complete drying (corresponding generally to shrinkage limit), c_0 and b_0 are material parameters. S_{r_0} is the initial degree of saturation, and G_s is the specific gravity.

By combining Eq. 19.3 with Eq. 19.2, one can express the void ratio (or volumetric strains knowing that: $\varepsilon_v = \Delta e/(1 + e_0)$) as a function of suction. The SWCC curve can thus be expressed in terms of deformation as a function of suction.

Steps (ii), (iv), and (v)

It is well known that shrinkage is related to the development of effective stresses in the material. In the absence of an external mechanical stress tensor ($\sigma_{ij} = 0$), Bishop and Blight (1963) relation can be expressed in Eq. 19.4, linking the effective stress to the capillary stress via $\chi\ S_r$, where χ is Bishop's parameter and S_r the degree of saturation.

$$\sigma'_{ij} = -u_a\delta_{ij} + \chi S_r(u_a - u_w)\delta_{ij} \tag{19.4}$$

As reported by Biarez et al. (1987), the main difficulty in this relationship is the experimental determination of the χ parameter. Numerous works (Blight 1967; Barden et al. 1969) demonstrated that χ parameter rather depends on the loading path applied to approach the hydro-mechanical behavior. The cohesion developed by the capillary stress (Haines 1923) changes during drying and leads to the increase of the material strength obeying the Mohr–Coulomb criterion. Lu et al. (2010) indicated that a macroscopic continuum representation of suction stress is the tensile stress that can be determined from the direct tensile test. Through different flexure and tensile tests, strong relations linking the strength with water content (and then suction) and the nature of the clay were highlighted (Avila 2004; Wei 2014; Tang et al. 2014; Ighil Ameur and Hattab 2017). From tensile tests, results on compacted clayey soil, Tang et al. (2014) showed that the tensile strength generally increases with the decreasing water content and eventually reaches a maximum value at a critical water content named w_c. Then, it decreases significantly on the dry side of optimum water content. These results highlighted different behaviors under tensile loading, which the authors related to microstructure organization of aggregates as well as the different regimes of unsaturated soils (pendular, funicular, and capillary regimes). Flexure tests (Fig. 19.3) performed by Ighil Ameur and Hattab (2017) on little-beams of one-dimensionally consolidated clays submitted to different levels of initial imposed suction showed a maximum peak (maximum strength) increasing with suction. This corroborates quite well the Tang et al. (2014) observations in tensile tests. The marked drops after the peak, showing rupture of brittle type in Fig. 19.3a, are due to the high suctions imposed to the clay samples before the tests. Using the digital image correlation at the peak of maximum strength, one can observe in the tension zone, at the bottom of the beam (Fig. 19.3b), a red color zone showing large extension with an average of horizontal local strains ε_{xx} equal to $+0.2\%$. This zone defines the crack initiation by indirect tensile strength.

19.3 Characterization of Free Shrinkage Development

19.3.1 Material and Methods

The used material is kaolin K13, a commercial synthetic clay from Sibelco, France. Table 19.1 summarizes the geotechnical and mechanical parameters. The grain size distributions of kaolin K13, obtained by laser analysis, show that the clay fraction (<2 μm) is about 83%. The mineralogical analysis, carried out by X-ray diffraction, reveals the presence of various secondary minerals, in addition to kaolinite, such as illite and quartz. SEM image in Fig. 19.4a (Hammad et al. 2013) shows kaolinite particles in the form of hexagonal flat and rigid platelets.

Desiccation tests have been carried out in the laboratory under controlled temperature ($T = 20°$) and relative humidity (RH) conditions, using the salt solution technique. The tests were performed on a thin square layer of clay slurry (prepared at a water content $w_0 = 1.2\ w_L$) of 5 mm thickness, and 200 mm of length and width (Fig. 19.4b). The sample is spread on a smooth support and placed in the RH-controlled chamber. Notice that the little samples of slurries for the determination of the SWCC were prepared in the same way as this larger square slurry sample. The experimental device presented in Fig. 19.5a ensures the control of relative humidity in the chamber (for instance at RH = 5.8% when the equilibrium is reached as shown in Fig. 19.5b). This setup allows to control the kinetic of the drying process.

To precisely follow the displacements and strains field during the deformation of the clay, digital image correlation (DIC) technique was used. Using this method requires a great rigor through each step of the test, from the preparation of the sample surface to observe, to the interpretation of the results. The calibration (i.e., choice

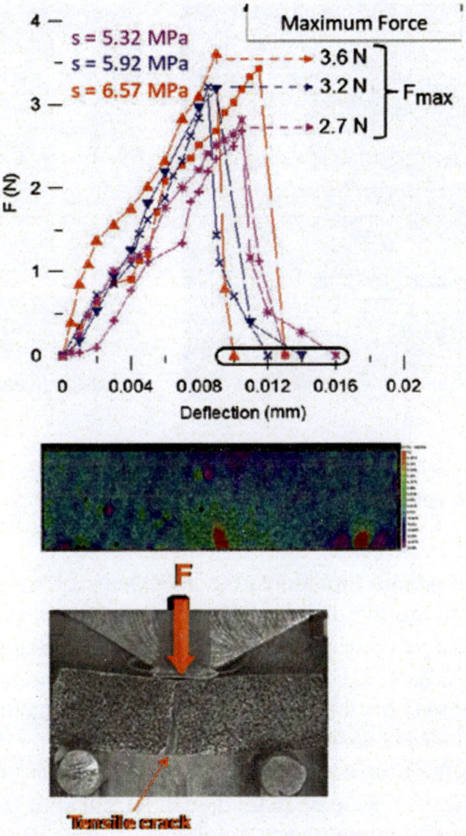

Fig. 19.3 Flexure tests on little-beams of clays by Ighil Ameur and Hattab (2017). **a** Flexure test results under different imposed initial suctions. **b** Digital image correlation performed at the peak of maximum strength (with the initial suction of $s = 5.32$ MPa, 35 mm, 10 mm, and 10 mm of dimensions)

Table 19.1 Physical and mechanical properties of kaolin K13

Material	Liquid limit w_L (%)	Plastic limit w_P (%)	Shrinkage limit w_s (%)	Solid grain density	<% 80 μm	<% 2 μm	Compression index C_c	Swelling index C_s
Kaolin K13	42	21	17–24	2.63	100	83	0.28	0.09

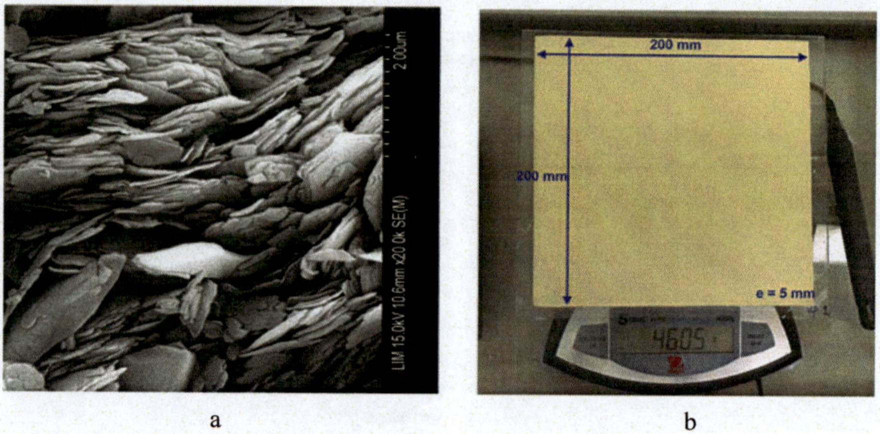

Fig. 19.4 Clay properties and sample preparation. **a** SEM photographs of kaolinite (Hammad et al. 2013). **b** Dimensions of clay sample for drying tests (Yang 2020)

of the lighting type and directions, the speckled pattern, etc.) must be adapted to the material all along the drying. DIC technique consists of capturing a series of pictures of a specimen during the strain process with a digital camera (Fig. 19.5a). The specimen needs to be prepared with an adequate random speckled dot pattern, so that the program may be able to calculate the displacements of the dots. During the tests, the camera takes one initial picture, then others after deformation, the program computes the differences between the pixels of the different images, and correlates them to derive the displacement field, then the strain field. The equipment consists of computer software, lighting device previously calibrated, and a digital camera with appropriate lens and resolution (Fig. 19.5a).

Hence, using computer program, the images are analyzed to create contour maps of displacements and strains fields all along the drying test. Two programs, Vic-2D and Vic-3D, were used here. Vic-2D provides two-dimensional strain maps of a planar specimen (similar to Fig. 19.6a) for each drying state of the specimen. Using two cameras, with Vic-3D, the system is able to provide enough information to obtain the three-dimensional strain and displacement fields (similar to Fig. 19.6b).

19.3.2 DIC Experimental Method and Principle of Measurement

With Vic-2D, the outputs are:

1. u_x (mm)—displacement along the longitudinal x-axis, with respect to the reference image.
2. u_y (mm)—displacement along the transversal y-axis.

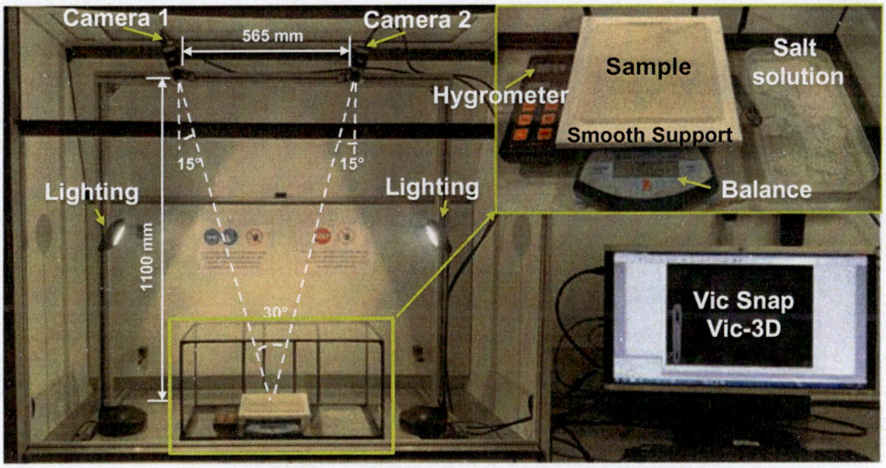

a

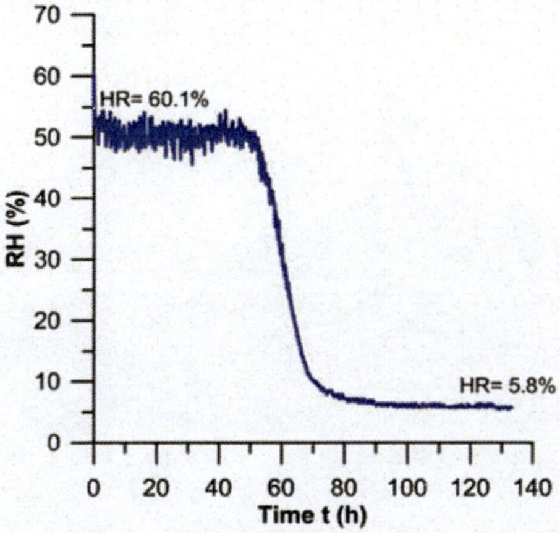

b

Fig. 19.5 Drying test setup by Yang (2020). **a** Sample setup and controlled chamber. **b** Relative humidity control and equilibrium in the test chamber

Strain variables are also generated in post-processing, with positive values corresponding to extensions and negative values corresponding to compressions

1. ε_{xx} (%)—normal component of strain in the x direction
2. ε_{yy} (%)—normal component of strain in the y direction
3. ε_{xy} (%)—distortion.

With Vic-3D, the outputs are:

The strain variables ε_{xx}, ε_{yy}, and ε_{xy} are generated as in Vic-2D, and u_z (mm), which is the displacement along the z-axis (Fig. 19.6b).

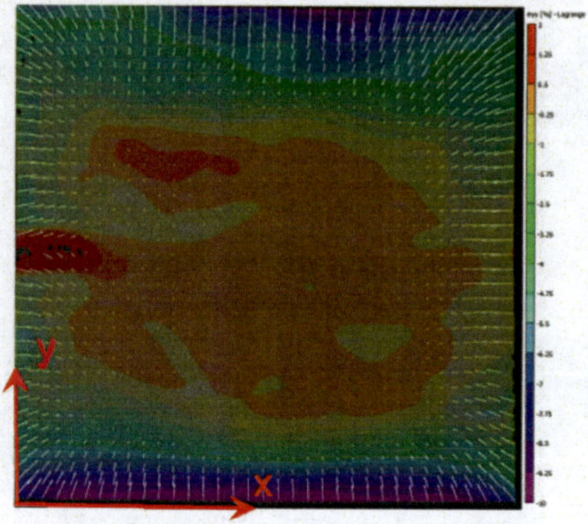

a

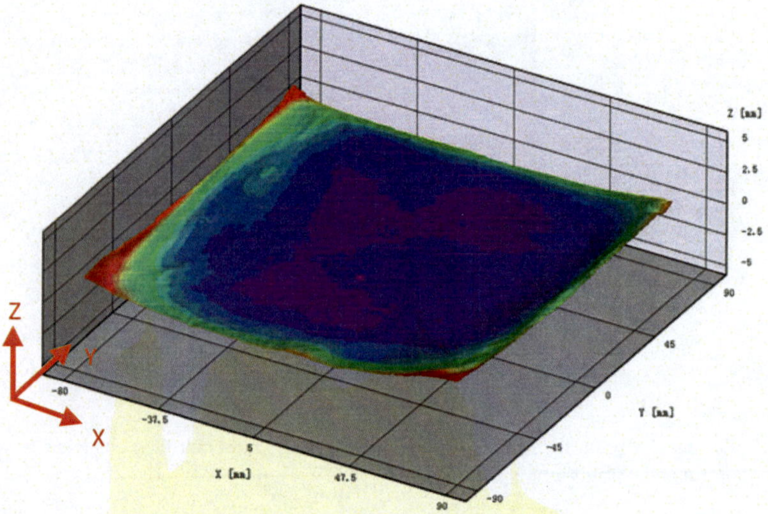

b

Fig. 19.6 DIC maps Yang (2020). **a** Vic-2D strain field map. **b** Vic-3D vertical displacement field—topography shape of the sample surface

Images were taken at intervals of 2 min during the test. The dimensions of the images for analysis are 2452-2056 pixels and the analysis with Vic-2D is done with a subset of 21 pixels and a step of 7 pixels. Calibration tests gave an accuracy of $\Delta\varepsilon/\varepsilon$ of about 5% in both x-axis and y-axis, and an accuracy of 7% in the vertical direction (z-axis) with Vic-3D.

19.3.3 Shrinkage Development in a Square Form Sample of Kaolin K13

As mentioned previously, the shrinkage is represented by the volumetric strains related to the drying process and occurs independently of external imposed stress tensor. Micro-macro-analyses of clay shrinkage during desiccation performed by Wei et al. (2013) showed that, in most cases, the strains in the three directions can be considered as isotropic when the soil microstructure organization is isotropic. However, the anisotropy of strains in the three directions during shrinkage remains an open question in the community and is still under discussion (Bronswijk 1990; Cornelis et al. 2006; Tang et al. 2011; Péron et al. 2009; Auvray et al. 2014). The geometric boundary conditions basically seem to be an important factor that has a large impact on the anisotropy strains development (Bronswijk 1990; Rijniersce 1983; Wei et al. 2016).

Let us consider a square sample of kaolin K13 (Fig. 19.4b), set up initially as slurry fully saturated clay at 1.2 w_{L} of water content, as explained in the Materials and Methods section. The plane smooth support in contact with the clay slurry was lubricated by means of a fine grease layer so that the shrinkage can freely develop. The RH-controlled chamber imposes RH = 5.8% and the temperature is maintained at 20 °C.

Hence, the volumetric strain due to the shrinkage is:

$$\varepsilon_{\mathrm{v}} = \varepsilon_{xx} + \varepsilon_{yy} + \varepsilon_{zz} \tag{19.5}$$

If one considers isotropic shrinkage, the shrinkage tensor can be expressed as:

$$\varepsilon_{ij}^{\mathrm{sh}} = \frac{\varepsilon_{ii}}{3}\delta_{ij} = \varepsilon^{\mathrm{sh}}\delta_{ij} \tag{19.6}$$

And, in terms of void ratio:

$$\varepsilon^{\mathrm{sh}} = \frac{1}{3}\frac{\Delta e}{1 + e_0} \tag{19.7}$$

The 2D map presented Fig. 19.7 shows the principal major strain field deduced from Vic-3D program at the end of drying. Compression is obtained in every part of the sample with negative value of ε_1 ranging from −6.3 to −3.7%. The white arrows

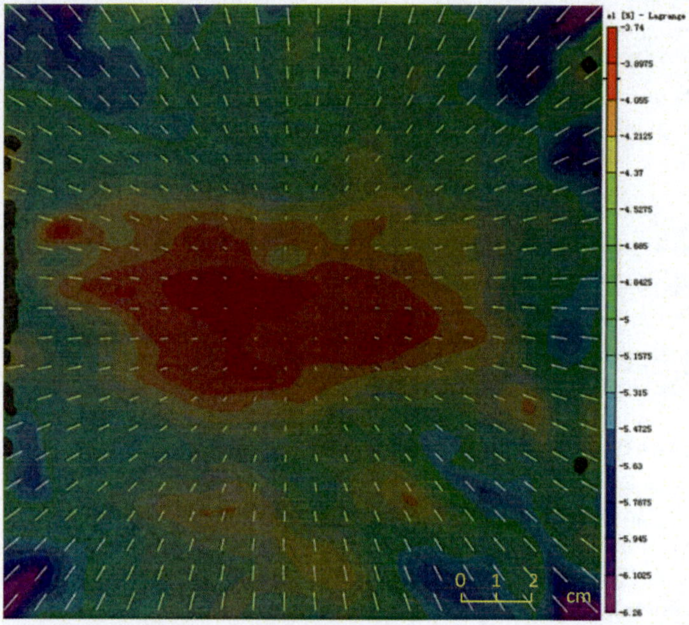

Fig. 19.7 2D (x y) map of ε_x principal major strain field—Vic-3D program results (Yang 2020)

represent the displacement vectors related to the shrinkage process; they are radial, all oriented toward the center of the sample, larger on the edges and negligible in the center. Global strains of the sample can be deduced from the DIC with Vic-3D results as follows, considering the displacement in the edges and in the surface of the sample:

$$\varepsilon_{xx} = \frac{(u_x)_1 + (u_x)_2}{h_{0h}}; \quad \varepsilon_{yy} = \frac{(u_y)_1 + (u_y)_2}{h_{0v}}; \quad \varepsilon_{zz} = \frac{\overline{u}_z}{e_{0z}} \tag{19.8}$$

where $(u_x)_1$ and $(u_x)_2$ are the average displacements measured on the longitudinal sides of the sample, $(u_y)_1$ and $(u_y)_2$ are the average displacements on the transversal sides, and u_{zi} is the average displacements of the surface at the vertical direction. The initial sample dimensions are: $h_{0h} = 200$ mm length, $h_{0v} = 200$ mm width, and $e_{0z} = 5$ mm thick.

Figure 19.8 shows global axial strains ε_{xx}, ε_{yy} and ε_{zz} during the desiccation test; a similarity of ε_{xx} and ε_{yy} curves is highlighted (orthotropic behavior). However, the ε_{zz} curve is well above the two others showing that the vertical shrinkage strain is much larger. This appears quite consistent with the fact that the evaporation occurs more intensively through the larger surface of the sample in contact with air. Figure 19.8 shows a volumetric strain ε_v curve evolution deduced from the DIC measurements very close to that calculated from Eq. 19.3, giving the theoretical shrinkage curve. This means that the shrinkage process of the clay slurry sample happened without

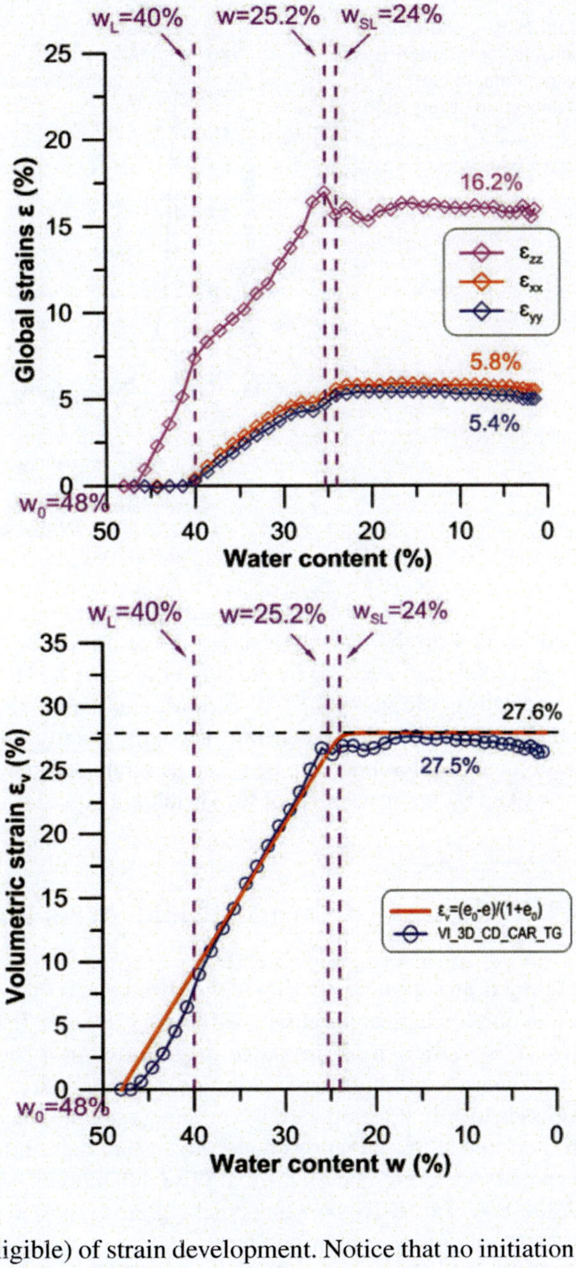

Fig. 19.8 Global strains and volumetric variation versus water content change during desiccation. Vic-3D calculation (Yang 2020)

any blockage (or they are negligible) of strain development. Notice that no initiation of cracks was observed.

In this case of geometry chosen for the sample, if we assume $\varepsilon_{xx} = \varepsilon_{yy}$ as shown in Fig. 19.8, the coefficient of anisotropy defined by $\eta = \varepsilon_{zz}/\varepsilon_{xx}$ depends on the drying state of the material as we can observe in Fig. 19.9. The curve seems to

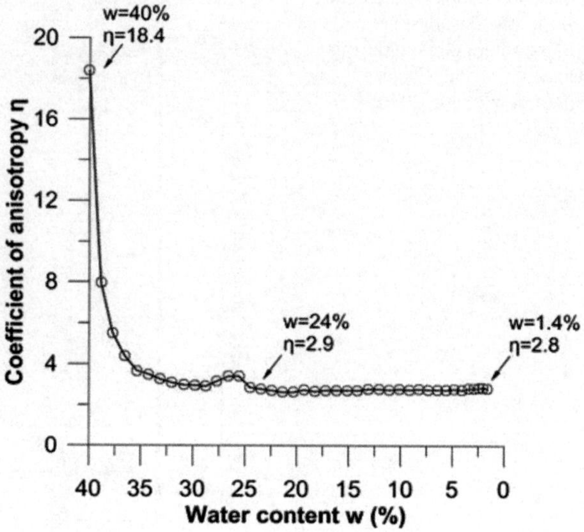

Fig. 19.9 Coefficient of anisotropy variation versus water content during desiccation (Yang 2020)

follow a power law reaching $\eta = 2.8$ at the end of drying. This value is quite well consistent with the values given in the literature: see for instance $\eta = 2.5$ by Péron et al. (2009) and $\eta = 2.0$ by Auvray et al. (2014). However, to define precisely the anisotropy coefficients in all directions, and their evolution, the geometry boundary conditions have to be taken into account. Finally, the precise characterization of shrinkage development can be taken as reference to deduce any strain development provoked by the restriction of the shrinkage process.

19.4 Restrained Shrinkage and Stress Concentration

During desiccation, restriction to the shrinkage process can appear in a zone leading to a complex deformation with different velocities from that of shrinkage. In practice, this restriction to shrinkage may be due to a friction or adhesion on support, heterogeneity of boundary conditions, local porosity, mineralogy, etc. Accordingly, stresses locally develop and, when they reach the maximum strength, the soil fails. Experimental investigations have shown that these internal stresses are often of the tensile type (Corte and Higashi 1960; Lachenbruch 1961; Lloret et al. 1998; Péron 2008; Wei et al. 2016), and that cracking propagates in mode I. It is this phenomenon that we are trying to highlight here.

Figure 19.10 represents DIC results with Vic-2D program, on kaolin K13 clay during drying under controlled temperature (24 °C) and relative humidity (RH = 76%) (corresponding to an imposed suction at the boundary conditions of the sample of about 38 MPa). The clay slurry layer was prepared at a water content $w_0 = 1.2\ w_L$, with dimensions of 300 mm × 200 mm and 3 mm thickness, which allows to

a

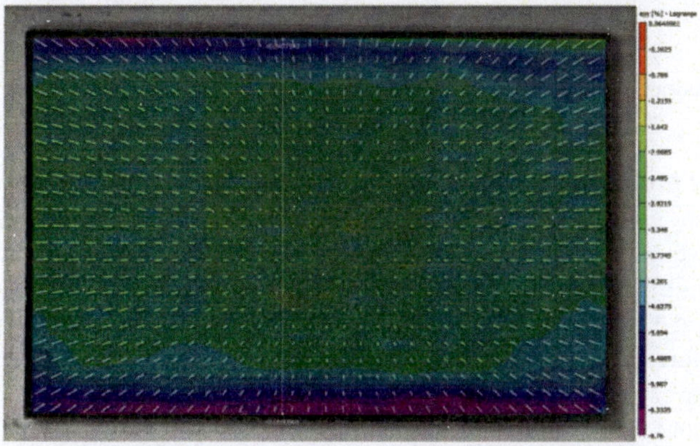

b

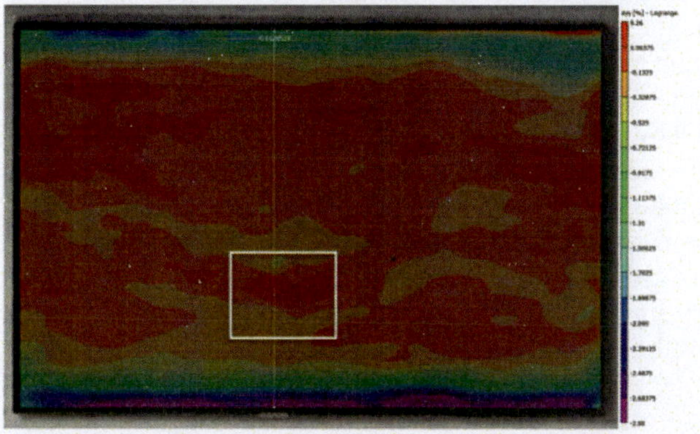

c

Fig. 19.10 DIC results using Vic-2D program, $w_0 = 48\%$, $T = 24$ °C, RH = 76%, sample dimensions of 3 × 300 × 200 mm^3 (Ighil Ameur 2016). **a** ε_{yy} DIC results at $t = 3$ h of drying. **b** ε_{yy} DIC results at the end of drying $t = 381$ h. **c** ε_{yy} DIC results at $t = 40$ h of drying. **d** Maps of principal strain directions at $t = 40$ h of drying (extensions in red color and compression in blue color)

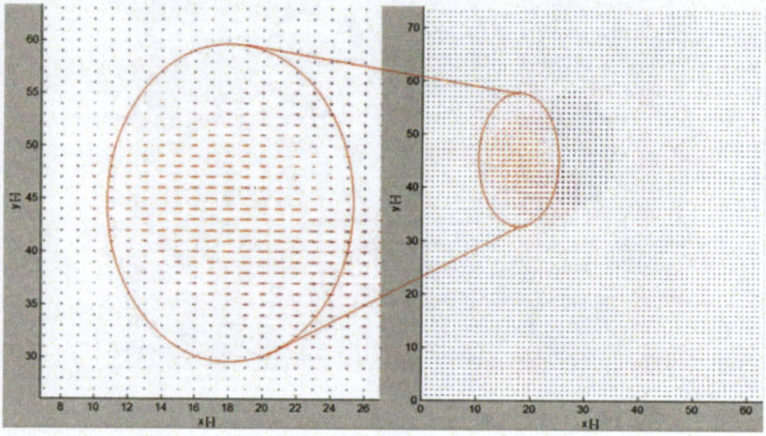

d

Fig. 19.10 (continued)

consider a 2D problem. The sample is placed on a rectangular smooth support; the surface was not greased so that a slight default might exist on the support surface as shown in Wei et al. (2016).

At the beginning of drying (Fig. 19.10a, $t = 3$ h), except in very few and limited zones that appear in extension (positive values), the results show the map of total ε_{yy} strains in compression (negative values); the shrinkage takes place everywhere in the sample. On the other hand, at $t = 381$ h, at the end of drying, the entire sample is in compression and, as we can see in Fig. 19.10b, the zone of extension has completely disappeared.

Let us see what happened in the meantime at $t = 42$ h of drying: one can observe in Fig. 19.10c the zone in extension already formed at the beginning of drying, the water content and the suction are about 38% and $s = 0.17$ kPa, respectively, the shrinkage strain, calculated by Eq. 19.7, is about −3.6%. The map of principal strains in Fig. 19.10d shows the direction of principal major strains all toward the x direction, resulting from a tensile stress concentration in the area. Theoretically, as mentioned, a crack can initiate when concentrated tensile stresses exceed the capillary cohesion strength of the material.

Figure 19.11 gives more insights by presenting the curves of ε_{yy} strain distribution along a straight line in y direction, during drying (characterized by time variation). The line passes through the extension zone (circle in Fig. 19.10b) located at about $y = 38$ mm.

One can observe very clearly here that in the zone around $y = 40$ mm, an extension has developed in the earlier stages of the drying process. In this area, at $t = 3$ h and $t = 9$ h, the curves show an evolution of a peak of extension, and at $t = 40$ h of drying, the peak becomes more marked, pointing the risk of soil failure by tensile. Shrinkage afterward overrides this local singularity and the curves entirely (including the peak)

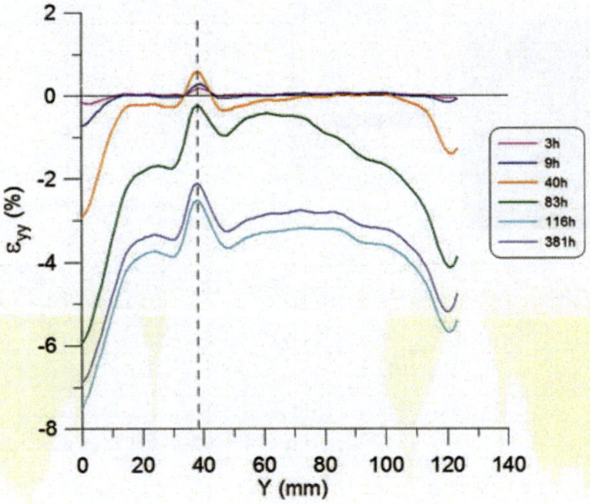

Fig. 19.11 ε_{yy} strain distribution along straight line in y direction passing through the extension zone (Ighil Ameur 2016)

move to the compression zone. Finally no cracking occurred in this case, and the tensile stress remaining is always below the capillary cohesive strength. Figure 19.12 shows another test performed by Ighil Ameur (2016) showing the effect of the strain field in extension developed in the earlier stages of drying, which disappears at the end of the drying. We can observe in Fig. 19.13 the process in terms of principal strain directions, with extensions in red and compressions in blue. Globally, the process is similar to that described in Fig. 19.11, the contraction due to shrinkage taking place in the entire sample. However, one can observe, particularly in the last stages of drying (t = 144 h and t = 381 h), the impact of the earlier tensile development in the zone, with the appearance of four blocks highlighting different principal strain directions limited by a progressive rotation of principal strains.

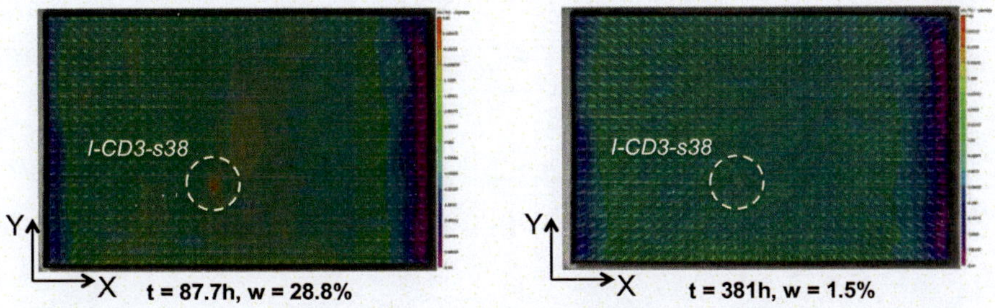

Fig. 19.12 Stress concentration captured by ε_{xx} DIC results using Vic-2D program. $w_0 = 48\%$, T = 24 °C, RH = 76%, sample dimensions of 3 × 300 × 200 mm^3 (Ighil Ameur 2016)

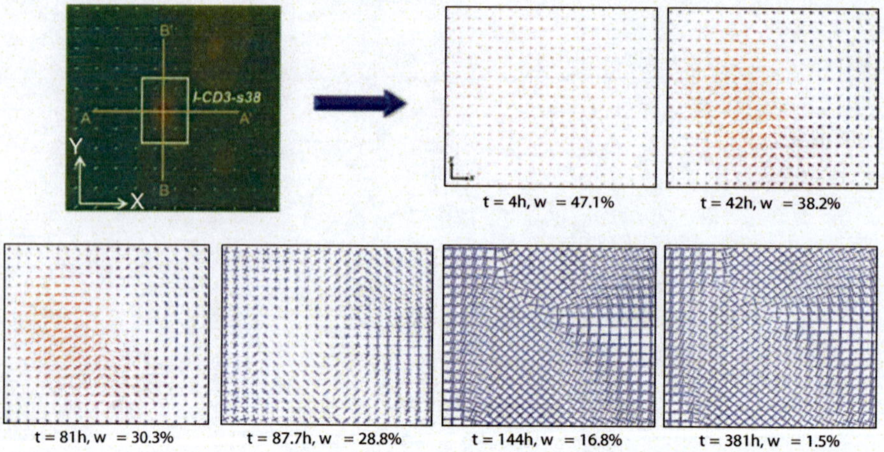

Fig. 19.13 Total principal strain field—stress concentration during shrinkage (Ighil Ameur 2016)

Assuming that the free shrinkage is anisotropic, the effect of the stress concentration can be regarded as the result of the development of local "mechanical" strain field $\varepsilon_{ij}{}^{*}$ (Bazant and Wittmann 1982; Amarasiri and Kodikara 2013), obtained by difference between the local total strain field ε_{ij}, measured by digital image correlation, and the estimated anisotropic free shrinkage strain field $\varepsilon_{ij}{}^{sh}$ discussed above and calculated with (Eq. 19.9).

$$\varepsilon_{xx}^{\text{sh}} = \varepsilon_{yy}^{\text{sh}} = \varepsilon^{\text{sh}} \text{ and } \frac{\varepsilon_{zz}^{\text{sh}}}{\varepsilon_{xx}^{\text{sh}}} = \eta = 2.8,$$

$$\varepsilon_{\text{v}}^{\text{sh}} = 2\varepsilon_{xx}^{\text{sh}} + \eta\varepsilon_{xx}^{\text{sh}} \text{ hence } \varepsilon^{\text{sh}} = \frac{1}{2+\eta}\varepsilon_{\text{v}} \tag{19.9}$$

This calculation technique allows experimentally to approach the risk of cracking development by drying. In term of increment, the local total strain increment is:

$$\mathrm{d}\varepsilon_{ij} = \mathrm{d}\varepsilon_{ij}^{*} + \mathrm{d}\varepsilon_{ij}^{\text{sh}} \tag{19.10}$$

19.5 Cracks Initiation and Propagation in Opening Mode (Mode I)

Figure 19.14a shows the DIC of ε_{xx} results, at $t = 47$ h, which represents the end of drying, the water content being equal to $w = 2.2\%$. The sample of kaolin K13 was placed on unlubricated smooth support, in the chamber controlled at low relative humidity (RH = 15%). At this state of drying, global ε^{sh} (Eqs. 19.7 and 19.8) is

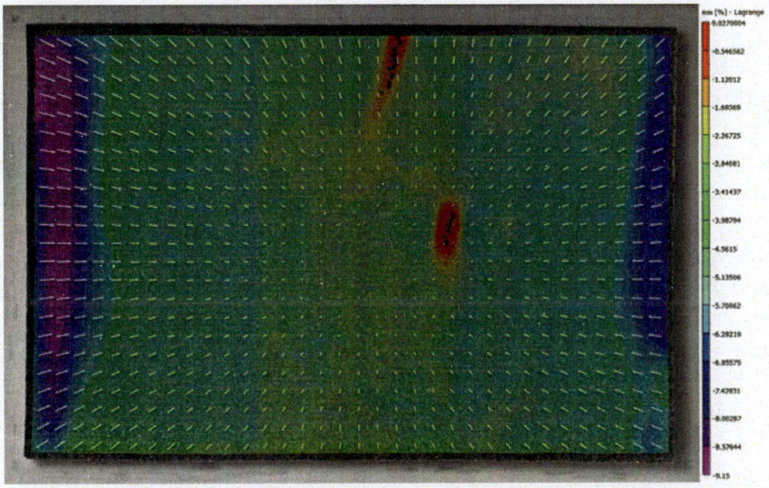

a Crack initiation in opening mode - DIC in the final state of drying, w = 2.2%, t = 47 h of drying

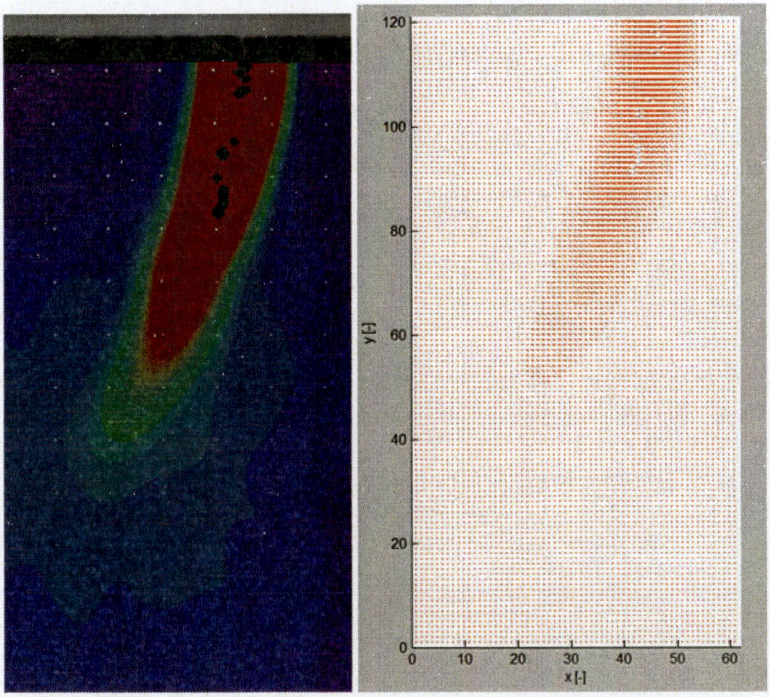

b « Mechanical strains » ε^*_{xx} field and principal strain map

Fig. 19.14 ε_{xx} DIC results using Vic-2D program; $w_0 = 48\%$, $T = 21$ °C, controlled RH = 15%, smooth support sample dimensions of $3 \times 300 \times 200$ mm^3 (Ighil Ameur 2016). **a** Crack initiation in opening mode—DIC in the final state of drying, $w = 2.2\%$, $t = 47$ h of drying. **b** «Mechanical strains» ε^*_{xx} field and principal strain map equal to −12%. In this map, two cracks were developed, one from the longitudinal (x-axis) edge of the sample, the other at the center.

Figure 19.14b shows the crack initiated from the top edge of the sample, the map being represented in terms of mechanical strains. One can see here a zoom of ε^*_{xx} field map as well as the map of the principal strain directions. The zone appears in extension toward the x direction, especially around the crack, where larger strains are captured, which appear perpendicular to the development of the crack. This result draws the signature of cracking in Mode I. The crack is developed in opening mode and was initiated when the concentrated tensile stresses exceeded the material capillary cohesive strength.

Figure 19.15 exhibits another example of drying test performed by Wei et al. (2016), carried out on the same kaolin placed on a rough support. A complex network cracking can be observed in this figure, representing the final state of drying obtained at $t = 7$ h. The zone of interest (Zone I) was chosen from the longitudinal strain ε_{xx} map to analyze the propagation of cracks through DIC of images captured at different drying stages (Fig. 19.16). Minimum and maximum ε_{xx}, as well as shrinkage ε^{sh}, were calculated at different times. The suction values in Table 19.2 were deduced from the (s–w) curve of Fig. 19.2b.

Figure 19.16 Analysis

Stage a, $t = 2$ h: beginning of desiccation, the strains ε_{xx} in this zone range from −0.83 to 1.32%.

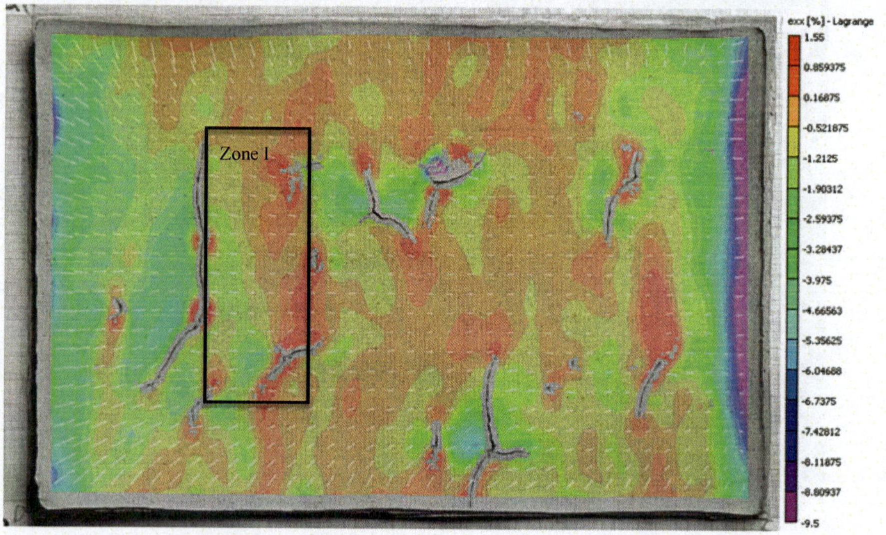

Fig. 19.15 Cracking network at the end of drying $t = 7$ h. Sample of kaolinite P300, 3 × 300 × 200 mm^3 of dimensions. Rough support, $w_0 = 40\%$, $T = 20$ °C drying in the laboratory atmosphere with about 50% of RH (Wei et al. 2016)

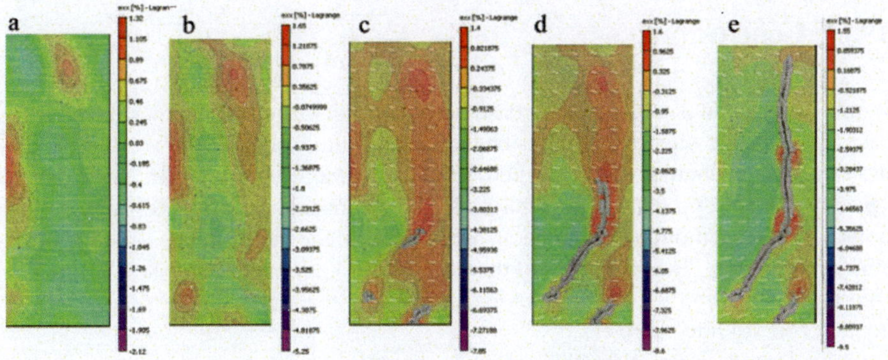

Fig. 19.16 Drying process (Table 19.2) in zone I of interest (Wei 2014; Wei et al. 2016)

Table 19.2 Parameters variation during the drying process

Time	$t = 2$ h	$t = 5$ h	$t = 6.33$ h	$t = 6.67$ h	$t = 7$ h
w: %	37	31.5	28.5	27	26.5
s: kPa	1	60	1000	2000	2300
ε^{sh}: %	−0.96	−2.75	−3.74	−4.2	−4.39
Max. ε_{xx}: %	1.32	1.65	1.4	1.6	1.55
Min. ε_{xx}: %	−2.12	−5.25	−7.85	−8.6	−9.5

Stage b, $t = 5$ h: the ε_{xx} strains increase, the maximum and minimum ε_{xx} in the chosen zone are 1.65 and −1.8%, respectively. The red zones indicate extensions, whereas the green–blue zones indicate compressions.

Stage c, $t = 6.33$ h: the cracks begin to appear. Local displacements become important in the zone (displacement vectors are longer). In the vicinity of the cracks, there are mainly extensions.

Stage d, $t = 6.67$ h: this is 20 min later than stage c, the small cracks connect together and more cracks are formed. Maximum and minimum ε_{xx} are 1.6 and −4.77%, respectively, which are larger than those at stage c.

Stage e, $t = 7$ h: small cracks have gone through the sample, forming a network of cracks. Displacements remain large in the chosen zone and minimum ε_{xx} evidently decreases, to a value of −5.35%. At this stage, after the cracks are formed in the zone, ε_{xx} decreases. This is because energy is released, stresses are relaxed, and strains decrease toward the shrinkage contraction value.

Two mechanisms are highlighted in these observations; first it is the cause behind the initiation of cracks that seems to be often related to the soil failure in mode I. The second is the rapid connection between different small cracks just after their formation, which adds complexity to the crack network development.

19.6 Conclusions

In this experimental work, mainly based on the observations of a clay slurry during drying, shrinkage phenomenon has been placed in the formal framework of continuous media mechanics and assumed to evolve independently from the external imposed stresses. Cracking propagation has been considered from the macroscopic point of view. Without external mechanical stresses, the capillary pressure becomes similar to suction. The properties of the unsaturated kaolin were precisely characterized by establishing the drying curves, taking into account both the suction, water content and volume changes.

Digital image correlation (DIC) technique has been used on a saturated slurry clay sample during drying, to precisely follow the changes in displacements and strains field.

In the shrinkage phenomenon, for a given clayey material, the geometric boundary conditions seem to be an important factor having a large impact on the anisotropic strains development. In the current study (square and rectangular samples), shrinkage that appears anisotropic with ${\varepsilon_{xx}}^{sh} \approx {\varepsilon_{yy}}^{sh}$ and ${\varepsilon_{zz}}^{sh}/{\varepsilon_{xx}}^{sh} = \eta$. η is the coefficient of anisotropy. During desiccation, restriction to the shrinkage process can appear, leading to the development of local stress concentration. Accordingly, when the stresses locally developed reach the maximum strength, the soil fails.

The stress concentration and its development can be regarded as the result of the evolution of a local “mechanical” strain field ${\varepsilon_{ij}}^{*}$, obtained by difference between the local total strain field ε_{ij}, measured by digital image correlation, and the estimated anisotropic shrinkage strain field ${\varepsilon_{ij}}^{sh}$. The current observations show that stress concentration often concerns the tensile stress development. Consequently, crack occurs when a concentrated tensile stress exceeds the material capillary cohesive strength, the latter increasing with the capillary stress development. The direction of the crack is perpendicular to that of the principal extension strains in opening mode. The extension strains remain very important in the positions where the opening of the crack is not large enough to relax them. One of the key issues, that deserves to be thoroughly analyzed in the perspectives of this study, relates to the local stress development inducing the local strains that can be captured by the DIC method (see for instance works of Hild and Roux 2006, and dell’Isola et al. 2019). In our problem, the point is of going back to the capillary stresses, which are rather difficult to measure directly in the material during the drying process. A possible method is to perform complementary relevant tests allowing to couple the measurements, that is to say a simultaneous measurement of stresses and strains during the test such as Ighil Ameur and Hattab (2017), and El Hajjar et al. (2020).

References

Albrecht, B. A., & Benson, C. H. (2001). Effect of desiccation on compacted natural clays. *Journal of Geotechnical & Geoenvironmental Engineering, 127,* 67–76. https://doi.org/10.1061/(ASCE)1090-0241(2001)127:1(67).

Amarasiri, A. L., & Kodikara, J. K. (2013). Numerical modelling of a field desiccation test. *Géotechnique, 63*(11), 983–986. https://doi.org/10.1680/geot.12.P.010.

Auvray, R., Rosin-Paumier, S., Abdallah, A., & Masrouri, F. (2014). Quantification of soft soil cracking during suction cycles by image processing. *European Journal of Environmental and Civil Engineering, 18*(1), 11–32.

Avila, G. (2004). *Study of shrinkage and cracking of clays—Application to clay in Bogota* (Doctoral dissertation). Polytechnic University of Catalunya, Barcelona, Spain.

Barden, R. J., Madedor, A. O., & Sides, G. R. (1969). Volume change characteristics of unsaturated clays. *Journal of Soil Mechanics & Foundations Div, 95*(1), 33–51.

Bazant, Z. P., & Wittmann, F. H. (1982). *Creep and shrinkage in concrete structures.* New York: Wiley.

Biarez, J., Fleureau, J.-M., Zerhouni, M.-I., & Soepandji, B. S. (1987). Variations de volume des sols argileux lors de cycles de drainage-humidification.*Revue Française de Géotechnique, 41,* 63–71.

Bishop, A. W., & Blight, G. E. (1963). Some aspects of effective stress in saturated and partly saturated soils. *Geotechnique, 13,* 177–197.

Blight, G. E. (1967). Effective stress evaluation for unsaturated soils. *Journal of the Soil Mechanics and Foundation Division, Proceedings of the A.S.C.E., 93*(S.M.2, Mars), 125–148.

Bronswijk, J. J. B. (1990). Shrinkage geometry of a heavy clay soil at various stresses. *Soil Science Society of America Journal, 54*(5), 1500–1502.

Cheng, W.-Q., Yang, Z-T., Bouchemella, S., Hattab, M., & Bian, H.-B. (2020). Desiccation shrinkage process in clayey soils—Experimental and numerical analysis. Paper in progress.

Cornelis, W. M., Corluy, J., Medina, H., Diaz, J., Hartmann, R., Meirvenne, M. V., & Ruiz, M. E. (2006). Measuring and modeling the soil shrinkage characteristic curve. *Geoderma, 137*(1–2), 179–191.

Corte, A., & Higashi, A. (1960). *Experimental research on desiccation cracks in soils.* Research Report No. 66. Wilmette, IL, USA: U.S. Army Snow Ice and Permafrost Research Establishment, Corps of Engineers.

dell'Isola, F., Seppecher, P., Spagnuolo, M., Barchiesi, E., Hild, F., Lekszycki, T., et al. (2019). Advances in pantographic structures: Design, manufacturing, models, experiments and image analyses. *Continuum Mechanics and Thermodynamics, 31*(4), 1231–1282.

Eid, J., Taibi, S., Fleureau, J. M., & Hattab, M. (2015). Drying, cracks and shrinkage evolution of a natural silt intended for a new earth building material Impact of reinforcement. *Construction and Building Materials, 86*(13), 120–132.

El Hajjar, A., Ouahbi, T., Eid, J., Hattab, M., & Taibi, S. (2020). Shrinkage cracking of unsaturated fine soils: New experimental device and measurement techniques. *Strain*, e12352. https://doi.org/10.1111/str.12352.

Fleureau, J. M., Kheirbek-Saoud, S., Taibi, S., & Soemitro, R. (1993). Behaviour of clayey soils on drying—Wetting paths. *Canadian Geotechnical Journal, 30*(2), 287–296.

Fleureau, J. M., Wignyodarsono, L., & Zerhouni, M. I. (1988). Effects of surfactants on the mechanical properties of a kaolinite in relation to the solid-liquid contact angles. *Canadian Geotechnical Journal, 25,* 675–683.

Fredlund, D. G. (2002). Use of the soil-water characteristic curve in the implementation of unsaturated soil mechanics. In *UNSAT 2002, Proceedings, Third International Conference on Unsaturated Soils*, Recife, Brazil, March 10–13, 3.

Gallipoli, D., Bruno, A. W., Perlot, C., & Mendes, J. (2017). A geotechnical perspective of raw earth building. *Acta Geotechnica, 12,* 463–476. https://doi.org/10.1007/s11440-016-0521-1.

Giorgio, I., De Angelo, M., Turco, E., & Misra, A. (2019). A Biot–Cosserat two-dimensional elastic nonlinear model for a micromorphic medium. *Continuum Mechanics and Thermodynamics*, 1–13.

Giorgio, I., & Scerrato, D. (2017). Multi-scale concrete model with rate-dependent internal friction. *European Journal of Environmental and Civil Engineering, 21*(7–8), 821–839.

Haines, W. B. (1923). The volume changes associated with variations of water content in soil. *Journal of Agricultural Science, 13*, 296–311.

Hammad, T., Fleureau, J. M., & Hattab, M. (2013). Kaolin/montmorillonite mixtures behaviour on oedometric path and microstructural variations. *European Journal of Environmental and Civil Engineering, 17*(9), 826–840.

Hild, F., & Roux, S. (2006). Digital image correlation: From displacement measurement to identification of elastic properties—A review. *Strain, 42*(2), 69–80.

Ighil Ameur, L. (2016). *Étude expérimentale de l'endommagement et de la fissuration d'une matrice poreuse* (Doctoral dissertation). Université de Lorraine, Metz, France (in French).

Ighil Ameur, L., & Hattab, M. (2017). Crack initiation and propagation of clays under indirect tensile strength test by bending related to the initial suction. In *Advances in Laboratory Testing and Modelling of Soils and Shales* (pp. 173–180). Cham: Springer.

Kong, D. F. (1994). *Properties of fissured clay* (1st ed.). Beijing, China: Geological Publishing Company.

Lachenbruch, A. H. (1961). Depth and spacing of tension cracks. *Journal of Geophysical Research, 66*, 4273–4292.

Lloret, A., Ledesma, A., Rodríguez, R. L., Sánchez, M. J., Olivella, S., & Suriol, J. (1998). Crack initiation in drying soils. In *Proceedings of the Second International Conference on Unsaturated Soils*, Beijing, China, August 27–30, 1998 (Vol. 1, pp. 497–502). Beijing, China: International Academic Publishers.

Longwell, C. R. (1928). Three common types of desert mud cracks. *American Journal of Science, 5th Series, XV*(86), 136–145.

Lu, N., Godt, J. W., & Wu, D. T. (2010). A closed form equation for effective stress in unsaturated soil. *Water Resources Research, 46*, W05515.

Mitchell, J. K. (1986). Practical problems from surprising soil behavior. *Journal of Geotechnical Engineering, 112*(3), 255–289.

Morris, P. H., Graham, J., & Williams, D. J. (1992). Cracking in drying soils. *Canadian Geotechnical Journal, 29*(2), 262–277.

Péron, H. (2008). *Desiccation cracking of soils* (Doctoral dissertation). Ecole Polytechnique Fédérale de Lausanne, Lausanne, Switzerland.

Péron, H., Hueckel, T., Laloui, L., & Hu, L. B. (2009). Fundamentals of desiccation cracking of fine grained soils: Experimental characterization and mechanisms identification. *Canadian Geotechnical Journal, 46*(10), 1177–1201.

Rijniersce, K. (1983). A simulation model for physical soil ripening in the Ijsselmeerpolder. Lelystad, The Netherlands: Rijksdienst voor de Ijsselmeerpolders.

Santamarina, J. C. (2001). Soils behavior at microscale: Particle forces. In *Proceedings of a Symposium on Soil Behavior and Soft Ground Construction, in Honor of Charles C. Ladd*, October 2001, MIT.

Scholtès, L., Hicher, P., Nicot, F., Chareyre, B., & Darve, F. (2009). On the capillary stress tensor in wet granular materials. *International Journal for Numerical and Analytical Methods in Geomechanics, 33*(10), 1289–1313.

Tang, C.-S., Pei, X.-J., Wang, D.-Y., Shi, B., & Li, J. (2014). Tensile strength of compacted clayey soil. *Journal of Geotechnical & Geoenvironmental Engineering, 141*(4). https://dx.doi.org/10.1061/(ASCE)GT.1943-5606.0001267.

Tang, C.-S., Shi, B., Liu, C., Suo, W. B., & Gao, L. (2011). Experimental characterization of shrinkage and desiccation cracking in thin clay layer. *Applied Clay Science, 52*(1–2), 69–77.

van Genuchten, M. T. (1980). A closed-form equation for predicting the hydraulic conductivity of unsaturated soils. *Soil Science Society of America Journal, 44*(89), 2–898.

Wei, X. (2014). *Etude micro-macro de la fissuration des argiles soumises à la dessiccation* (Doctoral dissertation). Ecole Centrale Paris, Châtenay-Malabry, France (in French).

Wei, X., Hattab, M., Bompard, P., & Fleureau, J.-M. (2016). Highlighting some mechanisms of crack formation and propagation in clays on drying path. *Geotechnique, 66*(4), 287–300.

Wei, X., Hattab, M., Fleureau, J. M., & Hu, R. L. (2013). Micro-macro-experimental study of two clayey materials on drying paths. *Bulletin of Engineering Geology and the Environment, 72*(3), 495–508.

Willden, R., & Mabey, D. R. (1961). Giant desiccation fissures on the black rock and smoke creek deserts, Nevada. *Science, 133*(3461), 1359–1360.

Yang, Z.-T. (2020). *Étude expérimentale et modélisation de la fissuration hydrique d'un sol argileux 2020* (Doctoral dissertation in progress). Université de Lorraine, Metz, France (in French).

Yuan, C., & Chareyre, B. (2017). A pore-scale method for hydromechanical coupling in deformable granular media. *Computer Methods in Applied Mechanics and Engineering, 318,* 1066–1079. https://doi.org/10.1016/j.cma.2017.02.024.